International Conference on Advances in Materials, Mechanical and Manufacturing (AMMM 2019)

IOP Conference Series: Materials Science and Engineering
Volume 576

Beijing, China
22 – 24 March 2019

ISBN: 978-1-5108-9451-8
ISSN: 1757-8981

Printed from e-media with permission by:

Curran Associates, Inc.
57 Morehouse Lane
Red Hook, NY 12571

Some format issues inherent in the e-media version may also appear in this print version.

Institute of Physics
Dirac House, Temple Back
Bristol BS1 6BE UK

Phone: 44 1 17 929 7481
Fax: 44 1 17 920 0979

techtracking@iop.org

Additional copies of this publication are available from:

Curran Associates, Inc.
57 Morehouse Lane
Red Hook, NY 12571 USA
Phone: 845-758-0400
Fax: 845-758-2633
Email: curran@proceedings.com
Web: www.proceedings.com

Table of contents

Volume 576

2019 International Conference on Advances in Materials, Mechanical and Manufacturing (AMMM 2019)

22–24 March 2019, Beijing, China

Accepted papers received: 26 June 2019
Published online: 7 August 2019

Preface

Papers

Mechanical Design, Manufacturing and Automation

Mechanical System Control and Simulation

Structural Design and Precision Engineering

Materials Science and Engineering

IOP Conf. Series: Materials Science and Engineering **576** (2019) 011001 doi:10.1088/1757-899X/576/1/011001

Preface

We are very pleased to introduce the proceedings of the 2019 International Conference on Advances in Materials, Mechanical, and Manufacturing (AMMM 2019), it took place in Beijing, China from March 22-24, 2019. The Conference provided a setting for discussing recent developments in a wide variety of topics including Material Structure Design, Liquid Mechanics and Thermal Engineering, Mechanical Engineering and Advanced Manufacturing etc.

The Conference has been an excellent opportunity for participants coming from many countries to present and discuss their topics in their respective research areas. A major conference theme was related to Material Structure Design and Performance Analysis; this session also included papers on, material properties measurement and mathematical predictions of inspection performance through modeling. A second session was devoted to the theme of Liquid Mechanics and Thermal Engineering; the third session was concerned with Intelligent Robot and Artificial Intelligence Technology, the Fourth session was involved with Mechatronics, after that the fifth session in Mechanical Engineering and Advanced Manufacturing, finally the Sixth session Industrial Machinery Design and Key Technologies. In the event, the conference was highly successful. The 50 presented papers maintained the high promise suggested by the written abstracts and the programme was chaired in a professional and efficient way by the session chairmen who were selected for their international standing in the subject. The delegates, was also highly gratifying, showing the high level of international interest in the matter. This is also indicated by a large number of countries, represented by the delegates.

We would like to thank all participants for their contributions to the Conference program and for their contributions to these Proceedings. Many thanks go as well to the School of Mechanical Engineering and Automation, Beihang University, for their support, which allowed all foreign participants to feel more at home, and especially, the opening remark from the Dean gave a warm welcome to all the participants. Our special thanks go to our Conference Committees for their devoted assistance in the overall organization of the conference. It is our pleasant duty to acknowledge the support from the Beihang University, Beijing, China; technically support from National Sun Yat-sen University, Taiwan and the University of Alberta, Canada.

We are looking forward to the next International Conference on Advances in Materials, Mechanical and Manufacturing. We hope that it will be interesting and enjoying at least as all of its predecessors.

Conference Chair: Prof. Lang Lihui
April 3, 2019

IOP Conf. Series: Materials Science and Engineering **576** (2019) 011002 doi:10.1088/1757-899X/576/1/011002

Conference Committees

International Advisory Committee
Yeong-Maw Hwang, National Sun Yat Sen University, Taiwan

Conference Chairs
Sergei Alexandrov, Beihang University, China

Lang Lihui, Beihang University, China

Local Organizing Chair
Blala Hamza,The Algerian Space Agency, Algeria&Beihang University, China

Program Chairs
Ken-ichi Manabe, Tokyo Metropolitan University, Japan

Zengtao Chen, ASME Fellow, University of Alberta, Canada

Local Technical Committees
Li Dongsheng, Beihang University,China

Li Xiaoxing, Beihang University,China

Li Jianjun, Huazhong University of Science and Technology, China

Zhan Mei, Northwestern Polytechnical University,China

Zhan Lihua, Central South University, China

He Zhubin, Harbin Institute of Technology, China

Yu Haiping, Harbin Institute of Technology, China

Wang Baoyu, University of Science and Technology of Beijing, China

Wenzhao Li, Beijing Institute of Astronauctical Systems Engineering, China

International Technical Committees
Wen-Xue Wang, Kyushu University, Japan

Andrew Rae, University of the Highlands and Islands, UK

Joao Manuel R. S. Tavares, Universidade do Porto (FEUP), Portugal

Rajan Ambat, Technical University of Denmark (DTU), Denmark

Rosa Cervera Sarda, Alcala University, Spain

Joerg Niemann, University of Applied Sciences Dusseldorf, Germany

Dimitris Drikakis, University of Nicosia, Cyprus

Hamid Bahai, Brunel University London, UK

IOP Conf. Series: Materials Science and Engineering **576** (2019) 011002 doi:10.1088/1757-899X/576/1/011002

IOP Conf. Series: Materials Science and Engineering **576** (2019) 011003 doi:10.1088/1757-899X/576/1/011003

Peer review statement

All papers published in this volume of *IOP Conference Series: Materials Science and Engineering* have been peer reviewed through processes administered by the proceedings Editors. Reviews were conducted by expert referees to the professional and scientific standards expected of a proceedings journal published by IOP Publishing.

IOP Conf. Series: Materials Science and Engineering **576** (2019) 012001 doi:10.1088/1757-899X/576/1/012001

Early Faults and Characteristics of Large Capacity Non-salient Pole Synchronous Electric Machines

Andong Wang[1], Jihao Wang[1], Junqing Li[2], Yangshuo Ma[2] and Fuchun Sun[1]

[1]State Grid Shandong Electric Power Research Institute, 250002 Jinan, China
[2]School of Electric and Electronics Engineering, North China Electric Power University, 071003 Baoding, China

Abstract. Large-capacity synchronous machines are prone to failure due to their complicated structure. The timely detection and processing in the early stage of the failure can avoid further expansion of the fault and avoid huge economic losses. By studying the structure of water-hydrogen-hydrogen turbine generators, this paper analyzes the types and characteristics of early faults of every component, and provides reference for the operation and state maintenance of synchronous machines.

1. Introduction

In recent years, the requirements for operational safety and reliability of the synchronous generators are increasing with the capacity. Synchronous generators faults are inevitable in the course of operation due to the large capacity and the complicated structure, which affect the safe operation of the unit and bring huge economic losses to the equipment manufacturing enterprises and users of synchronous machine, especially in the event of a serious accident.

At present, scholars have analyzed the causes and treatment of accidents of turbine generators and provide some typical failure cases which are repaired with the huge cost [1-17]. Therefore, it is very important to master the common faults and characteristics of the generators, which is helpful for detecting and treating them in the early stage of fault, and prevents the slight fault evolving to serious accident.

On the basis of consulting a large number of references related to generators fault cases, the early fault types, fault causes and fault characteristics of large non-salient pole synchronous machines are analyzed. The work can provide reference for the operation and condition maintenance of large capacity synchronous generators.

2. Structure of large capacity Non-salient synchronous electric Machines

The large non-salient pole synchronous machine is generally used as a turbo-generator. At present, the capacity of the turbo generator has reached 1000MW [1]. The increase in synchronous generators capacity is mainly determined by its cooling system. The common cooling methods for large-capacity synchronous generators are water-hydrogen-hydrogen and fully air-cooled. Since the cooling effect of water and hydrogen is better than that of air, the weight and volume of the water-hydrogen-hydrogen turbine generator are less than that of the fully air-cooled synchronous generator under the same capacity, while its operation and maintenance are more complicated than that of the fully air-cooled generator [2].

The major components of the synchronous electric machines, such as the stator, rotor, cooling system and bearing, are all composed of different parts. Water-hydrogen-hydrogen turbine generator is a widely used type of unit, in which the stator windings are cooled by water flowing in hollow-strands and the stator cores and rotors are cooled by hydrogen. In this paper, the water-hydrogen-hydrogen non-salient pole synchronous generator is taken as an example to analyze its early fault types, causes and characteristics.

3. Fault and characteristic Analysis of synchronous generator

3.1. Stator section

The stator of the synchronous generator is mainly composed of stator core, stator winding, cooling water system, stator end, etc., and they are analyzed one by one in the following parts.

3.1.1. Stator core. Common faults in the stator core are interlamination short-circuit, partial blockage of the ventilating duct, loosening of the core, wear, cracks, and so on [3-5,11]. Manufacturing and assembly process reasons, which include uneven paint film of silicon steel sheet in the manufacturing process, the quality problem of the insulating paint, and the damage of the silicon steel sheet insulating paint when the core is stacked etc., may cause short circuit between the sheets, which may increase the eddy current loss in the core and cause local overheating under the action of alternating magnetic field. During the manufacturing process, the ventilating duct between the core segments flattened and the debris entering the ventilating duct can cause partial blockage of the ventilating duct. The consequence is that the cooling capacity of the gas is weakened, resulting in local overheating and temperature increasing of the core. Therefore, monitoring the core tooth temperature, yoke temperature, gas temperature of inlet and outlet of ventilating duct, and comparing the temperature difference between different monitoring points can help to determine the short circuit and poor ventilation faults online.

During the manufacturing process of the machine, the iron core is not tightly packed, and the metal foreign matter is mixed, which may cause the stator core to resonate under the action of alternating electromagnetic force during the operation of the machine, resulting in looseness, wear and crack of the stator core. The vibration caused by core loosening, wear and crack is double frequency one. When the excitation current is removed, the vibration disappears [6]. The fault types and characteristics of stator cores are shown in Table 1.

In off-line condition, the core failure can also be found by experiment of the loss or measuring the surface temperature of the inner core of the stator by using the infrared thermal imager.

Table 1. Stator core faults and their characteristics

Component	Fault type	Fault characteristics
Stator core	Interlamination short-circuit	The local temperature rises, and the temperature difference between different temperature points of the stator core increases
	Local blockage of ventilator duct	Local temperature rises, temperature difference between different gas outlets increases
	Loosening of the core, wear, cracks	Double frequency vibration occurs and is related to the excitation current. When there is no excitation, the amplitude is zero

3.1.2. Stator windings. The early failure of the stator winding mainly includes insulation defects, wear and insulation performance degrading. Insulation faults in the stator windings occur mostly at the end-windings, such as the hand-making insulation worn at the hydroelectric connector due to vibration and the stator bar worn by poor fixing at the nasal tip. Vibration during operation and long-term erosion of oil and moisture can easily result in fatigue fracture and insulation damage of the connecting copper wire. More seriously, if the early insulation degradation can not be detected, it maybe develop to insulation breakdown which will cause accidents such as inter-turn short circuit, grounding, and even phase-to-phase short circuit in the stator winding [3-5，8-14].

Therefore, during the operation of the electric machines, it is very important to strengthen the insulation monitoring of the stator windings, and to detect the deterioration of the stator winding insulation by partial discharge analysis and insulation overheat monitoring. For the older units without partial discharge and insulation over-temperature on-line monitoring equipment, the insulation performance of the windings should be evaluated by testing the insulation resistance, AC leakage current, dielectric loss, maximum partial discharge, etc. [7]. The stator winding fault and its characteristics are shown in Table 2.

Table 2. Stator winding faults and their characteristics

Component	Fault type	Fault characteristics
Stator winding	Insulation deterioration	Insulation resistance drop
		Increase in dielectric loss
		Leakage current increase
		Partial discharge increase

3.1.3. Stator internal cooling water system. Early failures in the stator cooling water system are leaking and clogging. The leakage of internal cold water is usually caused by material defects, substandard internal cooling water leading to corrosion and cracks of hollow copper wire, abnormal vibration of unit, etc. The blockage of waterway isusually caused by the corrosion and scaling of hollow copper wire which is produced by the entry of sundries into the internal cooling water system or the low PH value of the inner cooling water [8].

When the stator water system leaks, the flow rate of internal cooling water increases, and hydrogen with the higher pressure seeps into the water and accumulates at the top of the stator inner chiller. When the stator waterway is blocked, the inlet pressure increases and the temperature rise. The temperature varies according to the location of the blockage. The temperature rise of the stator windings and the water outlets generally increases when the blockage is in the busbar [9]. When the blockage occurs inside the hollow conductor, the temperature of the outlet of the bar and temperature of the layers' insulation of that slots rise, which causes the temperature difference between slots and the temperature difference between the outlet of the bars to increase [10]. The faults and their characteristics of the stator inner cooling water system are shown in Table 3.

3.2. Rotor section

The rotor of the synchronous generator is mainly composed of rotor winding, rotor shaft, collector ring and brush, rotor retaining ring, bearing etc. [11]. The rotor rotates at the power frequency. When a vibration abnormality occurs, the variation amplitude is positively correlated with the rotor speed.

3.2.1. Rotor winding. Early failures of the rotor winding mainly include insulation degradation of the rotor winding, thermal deformation, and slight inter-turn short circuit [12]. The characteristics of the rotor winding insulation degradation are the same as those of the stator winding. The thermal deformation of the rotor winding is usually caused by local overheating of the rotor winding due to the inter-turn short circuit or the clogging of the ventilating duct. The thermal deformation of the rotor winding causes the rotor power frequency vibration to increase. The turn-to-turn short circuit of the rotor windings causes the excitation current to increase, the output reactive power to drop, the shaft voltage to rise, and the power frequency vibration to increase [13-14].

The thermal deformation of the rotor winding causes the thermal stress distribution on the rotor to be uneven, and the degree of deformation is related to the magnitude of the excitation current.

Therefore, the magnitude of the power frequency vibration caused by the thermal deformation of the rotor is related to the rotor speed and the magnitude of the excitation current.

Table 3. The faults and their characteristics of stator inner cooling water system

Component	Fault type	Fault characteristics
Stator winding	Leak water	Internal cooling water flow increased
		Hydrogen at the top of the cooling water tank in the stator increase
		Leakage parameter changes
	Blockage	Increased Inlet pressure
		Increased temperature difference between the total inlet and outlet pipes of the stator
		Increased interlayer temperature difference between upper and lower layer bars in different slots
		Increased interlayer temperature of the plugged bar
		The water quality of the cooling water in the stator deteriorates (conductivity, oxygen content, pH value and copper ion content are out of range)

The turn-to-turn short circuit of the rotor winding reduces the effective ampere-turn of the rotor. When the excitation voltage is constant, the excitation current increases and the reactive power output decreases. In addition, the short circuit between turns can cause the rotor to be heated unevenly and the thermal stress distribution changes, resulting in vibration to increase, and the vibration amplitude increases with the increase of the excitation current.

The early failure of the rotor winding and its characteristics are shown in Table 4.

Table 4. Faults and characteristics of rotor windings

Component	Fault type	Fault characteristics
Rotor winding	Thermal deformation	The power frequency vibration increases
		The vibration amplitude changes with the rotation speed and the excitation current
		The vibration has a gradual change feature
	Inter-turn short circuit	Excitation current increases
		Output reactive power drop
		Axis voltage rise
		The power frequency vibration increases
		The vibration amplitude changes significantly with the change of the rotational speed and the excitation current
		The vibration has a fast change feature

3.2.2. Large shaft of the rotor. The rotor shaft of the turbo generator is forged from integral steel, and its early failures include mass imbalance caused by manufacturing process, thermal imbalance caused by inter-turn short circuit or partial blockage of ventilating duct [15]. The amplitude of the power frequency vibration caused by the imbalance of the rotor mass is only related to the magnitude of the speed. When the speed is constant, the vibration amplitude remains unchanged. Partial blockage of the rotor ventilating duct causes local overheating of the rotor, resulting in thermal imbalance of the rotor shaft. The degree of imbalance is related to the degree of clogging and cooling parameters, and also to the magnitude of the excitation current. The large rotor shaft failure and its characteristics are shown in Table 5.

Table 5. The faults and their characteristics of The large rotor shaft

Component	Fault type	Fault characteristics
Large rotor shaft	The local blockage of rotor ventilatting duct	The power frequency vibration increases,
		The vibration amplitude changes with the change of the rotation speed, the excitation current, and the cooling medium;
		The vibration has a gradual change feature
	Rotor mass imbalance	The power frequency vibration increases,
		The vibration amplitute changes significantly with the rotation speed
		The vibration has stability

3.2.3. Collector-ring and brush. The reliable contact between the collector ring and the brush is the key to ensure the normal operation of the synchronous electric machine [16]. The surface of the collector ring and the brush is rough and dirty, carbon powder is accumulated, the ventilation is poor, the gap between the brush holder and the collector ring or the brush holder and the brush is too large, and the brush is stuck, etc., all of which may cause poor contact between the collector ring and the brush.

The bad contact between the collector ring and the brush can produce spark. In severe cases, the ring fire is formed, resulting in accidents such as destroying the collector ring, the brush holder and the brush carrier [11-14].

The operation and maintenance of the collector ring and the brush are generally carried out by regular inspections, and the problematical one are promptly cleaned and replaced with new carbon brushes. The collector ring and brush faults and their characteristics are shown in Table 6.

Table 6. Faults and characteristics of collector-ring and brush

Component	Fault type	Fault characteristics
Collector-ring and brush	Poor contact	rough surface
		The gap between the brush holder and the collector ring or the brush is too large
	Wear	Toner accumulation
	Poor ventilation	Increase in temperature

3.2.4. Rotor retaining ring. The retaining ring failure mainly manifests as stress corrosion, crack and deformation. The larger humidity of the hydrogen in the generator is very likely to cause its stress corrosion and cracking. In addition, faults such as the inter-turn short circuit of the rotor winding are also likely to cause damage to the retaining ring [14-16]. The rotor retaining ring is mainly used to protect the rotor end coil, so the failure of the retaining ring may cause damage to the rotor winding and thermal imbalance of the rotor, causing power frequency vibration.

The rotor retaining ring fault can be detected by ultrasonic inspection and lamination metallographic examination during regular inspection.

The rotor retaining ring failure and its characteristics are shown in Table 7.

Table 7. Faults and characteristics of rotor retaining ring

Component	Fault type	Fault characteristics
Rotor retaining ring	Stress corrosion, cracks and deformation	The power frequency vibration increases
		The vibration amplitude changes with the change of the rotational speed and the excitation current
		The vibration has a gradual change feature

3.2.5. Bearing. The problem of the bearing is mainly caused by the lubricating oil and the misalignment of the bearing. The pressure of the top shaft oil film drops, the oil quality of the lubricating oil is substandard, the oil film whirls, the oil film oscillates, etc., which may cause the bearing oil temperature to rise, the bearing bush to wear, and the vibration to increase. The vibration amplitude varies with the speed and bearing oil temperature and is abrupt [6]. The misalignment of the bearing causes the double frequency vibration, and the amplitude changes significantly with the rotation speed, which varies with the change of the bearing oil temperature too. Bearing failure and its characteristics are shown in Table 8.

Table 8. Faults and their characteristics of bearing

Component	Fault type	Fault characteristics
Bearing	Oil film whirl	Vibration with (0.40~0.48) times the power frequency occurs
		The vibration amplitude changes obviously with the rotation speed, and the bearing oil temperature
		Vibration has sudden change feature
		Bearing oil temperature rise
	Oil film oscillation	The vibration of the first critical speed of the rotor increases
		The vibration amplitude changes obviously with the rotation speed, and the bearing oil temperature
		Vibration has sudden change feature
		Bearing oil temperature rise
	Bearing misalignment	Double frequency vibration occurs
		The vibration amplitude changes obviously with the rotation speed, and the bearing oil temperature

It can be seen from the above analysis that the early failures and their characteristics of large synchronous electric machines show diversity and crossover, as shown in Table 9. Therefore, comprehensive analysis using multi-information fusion method for fault diagnosis can improve the accuracy of fault diagnosis.

Table 9. Early failures and its characteristics of water-hydrogen-hydrogen Non-salient Pole Synchronous Machines

Unit	Main section	Component	Fault type	Fault characteristics
Water-hydrogen-hydrogen non-salient pole synchronous electric machines	Stator	Stator internal cooling water system	Leak water	Internal cold water flow Increased
				Hydrogen at the top of the cold water tank in the stator
				Leakage parameter changes
			Blockage	The water quality of the cold water in the stator deteriorates (conductivity, oxygen content, pH value and copper ion content are out of range)
		Stator winding	Insulation deterioration	Insulation resistance drop
				Increase in dielectric loss
				Leakage current increase
				Partial discharge increase
		Stator core	interlamination short-circuit	The local temperature rises, and the temperature difference between different temperature points of the stator core increases.
			Local blockage of ventilator	Local temperature rises, Temperature difference between different gas outlets increases.
			Loosening of the core, wear, cracks	Double frequency vibration occurs and is related to the excitation current. When there is no excitation, the amplitude is zero.
	Rotor	Rotor winding	Inter-turn short circuit	Excitation current increases
				Output reactive power drop
				Axis voltage rise
				The power frequency vibration increases, and the amplitude changes significantly with the change of the rotational speed and the excitation current; the vibration has a fast degeneration.
			Thermal deformation	The power frequency vibration increases, and the amplitude changes with the rotation speed and the excitation current; the vibration has a gradual change.
		Large rotor shaft	Rotor ventilation duct blockage	The power frequency vibration increases, and the amplitude changes with the change of the rotation speed, the excitation current, and the cooling medium; the vibration has a gradual change.
			Rotor mass imbalance	The power frequency vibration increases, the vibration amplitude changes

				significantly with the rotation speed; the vibration has stability.
		Rotor retaining ring	Stress corrosion, cracks and deformation	The power frequency vibration increases, and the amplitude changes with the change of the rotational speed and the excitation current; the vibration has a gradual change.
		Bearing	Oil film whirl	(0.40~0.48) times the power frequency vibration occurs; the amplitude changes obviously with the rotation speed, and changes with the change of the bearing oil temperature and is sudden
				Bearing oil temperature rise
			Oil film oscillation	The vibration of the first critical speed of the rotor increases; the amplitude changes significantly with the rotational speed, and changes with the change of the bearing oil temperature and is sudden
				Bearing oil temperature rise
			Bearing misalignment	Double frequency vibration, the amplitude changes obviously with the rotation speed, and changes with the change of bearing oil temperature
		Collector-ring and brush	Poor contact	Rough surface
				The gap between the brush holder and the collector ring or the brush is too large
			Wear	Toner accumulation
			Poor ventilation	Increase in temperature

4. Conclusion

Due to the complicated structure and large coupling of electric, magnetic, thermal, fluid and stress during operation, the large capacity non-salient pole synchronous electric machines has many kinds of faults, and the causes of faults are complicated. There is a many-to-many relationship between fault types and fault characteristics. Especially for early faults, the fault characteristics are not obvious, and it needs to be diagnosed from multiple aspects. Therefore, mastering the mapping relationship between the fault type and the fault feature can be detected and dealt with them in the early stage of the fault in time, so as to prevent the fault from further expanding and serious accidents and huge losses.

5. References

[1] He Guoan, Chen Qiang, gan Wenliang, Jiang Bo, Zhang Xueyan, Zhang Weijun, *ELECTRIC POWER* **46**(10), 18-22(2013)

[2] Amrita Sinha, D. N. Vishwakarma, R. K. Srivastava, *Electric Power Components and Systems* **38**(1), 100-114(2010)

[3] Shen Liangwei, *Large Machine and Hydraulic Turbine* (6), 1-9(1998)

[4] LI Wei-qing, LIU Shuang-bao, *Large Machine and Hydraulic Turbine* (3), 11-15(2000)

[5] MAO Guo-guang, *Power System Technology* **24**(11), 1-7(2000)

[6] ZHANG Guo-qiang, YANG Yi, *Large Machine and Hydraulic Turbine* (5), 29-34(2011)

[7] Yan Ying, Yan Guilin, Hao Jianbo, Ren Zhangao, *ELECTRIC POWER* **38**(02), 39-42+52(2018)

[8] Adam Gozdowiak, Piotr Kisielewski, Ludwik Antal, *Power Electronics and Drives* **2**(1), 151-162(2017)

[9] SHAO Yongbin, *Large Machine and Hydraulic Turbine* (6), 27-31(2014)

[10] WANG Shaoyu, SUN Shuwen, HAN Changli, LIU Wenzhong, *Electric Power*, **42**(5):35-40(2009)
[11] JIANG Jian-ming, PU Jun-song, MA Hui, LIU Xi, CAO Jun, Large Machine and Hydraulic Turbine (2), 6-9+24(2009)
[12] Jiang Qian, *Dongfang Electric Machines* (1), 43-50(2014)
[13] SHI Jin-yuan , YANG Yu, SUN Qing , CHEN Dao-lai, *Proceedings of the CSEE* **20**(7), 44-47+51(2000)
[14] HANG Ming, ZHAN Yuezhang, YU Jianfei, *Hubei Electric Power*, **40**(5), 13-15+19(2016)
[15] Jiang Qian, *Dongfang Electric Machines* (1), 43-50(2014)
[16] Amrita Sinha, D. N. Vishwakarma, R. K. Srivastava, *Electric Power Components and Systems*, **37**(9), 957-969(2009)
[17] TSYPKIN M, *Proceedings of 8th International Electric Machines and Drives Conference*, pp 406-411(2011)

Improvements in miniaturized Hall Thrusters by use of high-temperature SmCo magnets and additive manufacturing techniques

A Olano[1], J Ren[1], G Zhang[1], H Tang[2], T Zhang[3] and J Li[4]

[1]Key Laboratory of Spacecraft Design Optimization & Dynamic Simulation Technologies of Ministry of Education, School of Astronautics, Beihang University, Beijing 100191, People's Republic of China
[2]Key Laboratory of Spacecraft Design Optimization & Dynamic Simulation Technologies of Ministry of Education, School of Space and Environment, Beihang University, Beijing 100191, People's Republic of China
[3]School of Materials Science and Engineering, Beihang University, Beijing 100191, People's Republic of China
[4]Science and Technology on Vacuum Technology and Physics Laboratory, Lanzhou Institute of Physics, Lanzhou 730000, People's Republic of China

Correspondence author' e-mail: thb@buaa.edu.cn

Abstract. This work presents the difficulties associated with the miniaturization of Hall Thruster engines and how state of the art materials and new manufacture techniques can potentially solve these problems. Hall Thrusters are electric propulsion systems that require specific magnetic field topography and uniform propellant distribution for optimal operation which is difficult to achieve with typical materials and by conventional manufacturing methods in miniaturized engines. To keep the optimal magnetic field distribution at small thruster sizes, it is shown that new alloys of SmCo permanent magnets can generate the desired magnetic field distribution and that their high-temperature resistance properties assure its operation temperature will be under the maximum operational temperature. In addition, whereas for the small dimensions required for the anode traditional manufacture methods only allow for simple designs, it is explained how the implementation of 3D-printing techniques can improve the uniformity in the azimuthal distribution of the propellant by allowing complex geometries in the design of the anode that are unattainable with traditional manufacturing.

1. Introduction

There is a recent trend in using micro and minisatellites (10-500 kg) for low Earth orbit (LEO) missions such as Earth Observation platforms, large constellations of satellites to provide worldwide internet coverage, drag-compensation, debris removal and de-orbiting at the end of the mission [1]. Hall Thrusters have been a very successful propulsion system for spacecraft [2] and are considered as potential candidates for the increasing market in smallsat missions [3]. The application limits derive from the scarce available power (<500 W) for the system as for smallsats the size of the solar arrays is limited and for LEO missions they can spend around 30% of the operational time in eclipse. In order to adapt the Hall Thruster to lower nominal powers, a process of miniaturization of the thruster has to be made with the technical difficulties associated with the accurate manufacturing of the small components and the increased temperatures due to the reduced radiating surfaces.

This work will be focused in presenting how the advances in magnetic materials and manufacture techniques allow the Hall thruster to be adapted to low powers, in particular to 100 Watts. The paper is divided as follows: In section 2 Hall thruster systems are introduced and it is explained the importance of the magnetic field and anode flow distribution in the performance of the thruster together with the complexities associated with miniaturization. In section 3 the permanent magnets are presented as option to generate the magnetic field and the singularities of their manufacture to the required dimensions. In section 4 it is shown the benefits of using additive manufacturing techniques for the anode which allows complex geometries to achieve a uniform propellant distribution.

2. Hall thruster dimensional constrains

Hall Thruster engines are propulsion systems where the thrust is achieved by electrostatically accelerating ions out of the thruster [4]. An axial electric field is created between an anode and a cathode, which for Hall Thruster is usually a hollow cathode and is the source of the electrons. In order to avoid the electrons streaming directly to the anode, a radial magnetic field is established perpendicular to the electric field creating an ExB drift motion in the electrons (Hall current) perpendicular to the electric and magnetic fields. The propellant, usually Xenon, is injected through the anode into the discharge chamber where collisions are produced between the neutrals and the confined electrons, ionizing the propellant. On one hand the electrons, through collisions, diffuse towards the anode and this allows a sustained discharge of the thruster. On the other hand, the ions are accelerated by an electrostatic force equal to the Lorentz force acting on the Hall current which its corresponding reaction on the magnetic structure of the thruster transmits the thrusting force to the body. The assembly uses an annular geometry to prevent the Hall current of reaching the walls of the thruster which otherwise would compromise the electron confinement.

The magnetic field is usually created using electromagnets as the magnetic field can be easily modified by tuning the current through the coils. A schematic of a conventional Hall thruster where the magnetic field is generated by magnetic windings is shown in Figure 1.

With the requirement of low power propulsion systems for smallsats the resultant thruster has to be a low-mass compact device with a relatively high efficiency. These characteristics cannot be accomplished by operating a large thruster at low power level and hence scaling down of the dimensions has to be done. The parameters of the 100-W thruster are described in Table 1.

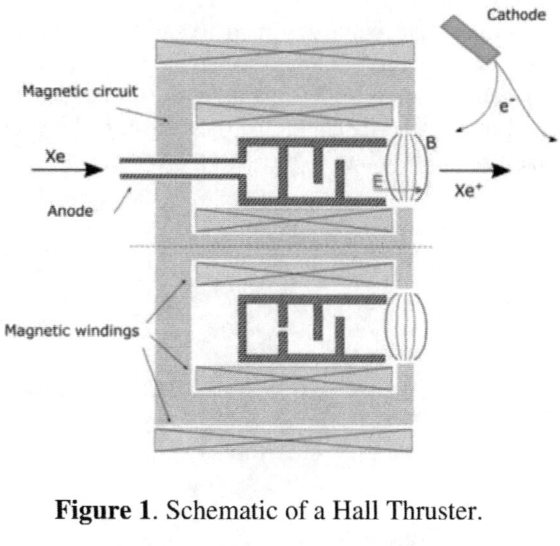

Figure 1. Schematic of a Hall Thruster.

IOP Conf. Series: Materials Science and Engineering **576** (2019) 012002 doi:10.1088/1757-899X/576/1/012002

Table 1. 100-W HT parameters.

100-W HT	
Thruster Diameter (mm)	15.7
Channel Width (mm)	4.9
Channel Length (mm)	4
Discharge Power (W)	100
Discharge Voltage (V)	300
Discharge Current (A)	0.33
Magnetic Field Strength (G)	223
Propellant Flow Rate (mg/s)	0.37
Thrust (mN)	<6.44
Specific Impulse	1786
Efficiency	<0.56

For a correct performance of the thruster, two aspects are of great relevance:
- Topography of the magnetic field
- Uniform distribution of the propellant

The work of Hofer [5] provides a detailed description of the desired settings of the topography of the magnetic lines. The summarised guidelines for a proper magnetic distribution design are:
- Symmetry of the magnetic field along the centre of the discharge channel.
- Notable gradient of the magnetic field from the anode to the location of the maximum
- Concavity of the magnetic field lines.
- Higher value of the magnetic field close to the walls.

While this requirements can be easily satisfied in traditional medium size Hall thrusters (1-2 kW), it becomes increasingly difficult to cope with all of them when scaling down the thruster. In particular the lack of space to place the magnetic coils together with the increased temperature in the windings and the increasing magnetic coil required power to thruster power ratio, make the coils a less attractive option in contrast with the permanent magnets [6].

Permanent magnets, although they allow a compact design and power savings during operation compared to coils, deprive flexibility to the magnetic configuration. To alleviate this situation, several permanent magnet configurations have been designed to take into account different magnetic field intensities and topographies. As explained below, the development of new magnetic materials allow the required sizes of permanent magnets to be manufactured and ensure that maximum operating temperatures are not reached during the discharge of the thruster. An illustration of the 100-W Hall Thruster with permanent magnets is shown in Figure 2.

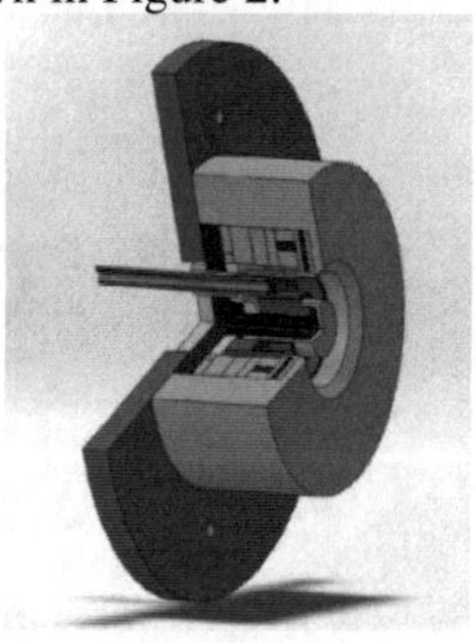

Figure 2. 100-W Hall Thruster.

For Hall Thrusters it has been shown that azimuthal non-uniformity distribution of the flow affects negatively to the thruster performance [7]. Compared to an azimuthally uniform propellant flow distribution, the non-uniformity enhances the electron diffusion incrementing the discharge current. In miniaturized designs, to tackle this problem becomes increasingly tricky as traditional manufacture techniques where welding is required introduce accuracy problems or are unable to reproduce complex features. This is one of the problems considered in the reduced performance of the MHT-9 [8]. To solve this contingency, additive manufacture will be used for the anode.

3. Modified SmCo alloy for the permanent magnets

The requirements shown in section 2 for the magnetic topography can be substantially satisfied by the use of permanent magnets as shown in the nominal design in Figure 3. The design uses one permanent magnet ring of 24.2 mm inner diameter and 2mm width with a thickness of 2mm and another ring of 34.6mm inner diameter and 2 mm width with a thickness of 4 mm.

The material used in the manufacture is a new recently developed high-temperature SmCo series 2:17 (Sm_2Co_{17}) permanent magnet [9]. This permanent magnet was chosen because its superior performance at high temperatures compared to previously reported SmCo magnets. They were manufactured in the traditional powder metallurgy process; Figure 4 depicts all the steps of the process and the time expended in each of them. Alloys with nominal composition Sm(Co,Fe,Cu,Zr)z were prepared by arc-melting under Argon atmosphere at 0.6 bar. The alloy powders were obtained by ball milling with a particle size of 3-5 μm that were aligned and pressed in a magnetic field up to 20 kOe

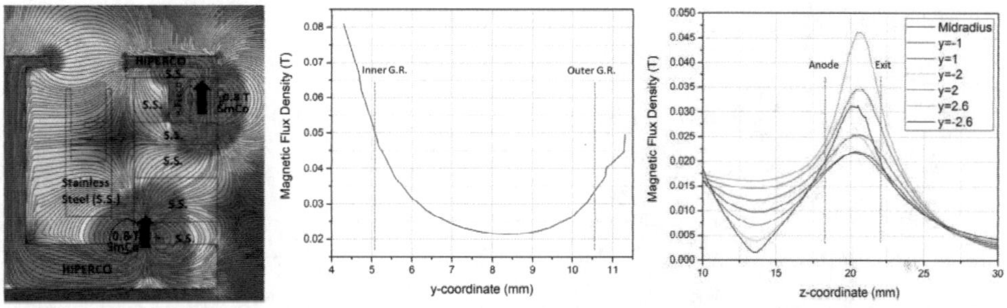

Figure 3. Magnetic field topography (left), magnetic field at halfway of the discharge channel along the radial coordinate (center) and magnetic field for different radial coordinates along the axis coordinate(right).

and further compacted by cold isostatic pressing. The green compacts were sintered in a vacuum tube furnace and solution heat treated at 1200-1220°C for 1-4 hours and after they were quenched to room temperature. The magnets were further aged at 780-850°C for 24 hours and slowly cooled down to 500°C at a rate of 0.7°C/min, then aged at 400°C for 10 hours, followed by water quenching. After machining the final shape, the permanent magnet is magnetized. As the magnet will operate in vacuum, coating was not required. The final manufacture of the magnets can be seen in Figure 5.

The basic properties of the high-temperature resistant SmCo permanent magnets can be seen in Table 2. Thermal simulations estimate a value of around 300°C for the two permanent magnet rings in the nominal design and a value of 500°C for the highest temperature operating point in the off-nominal design of 300W. Both values are inside the maximum operating temperature of the modified SmCo permanent magnet.

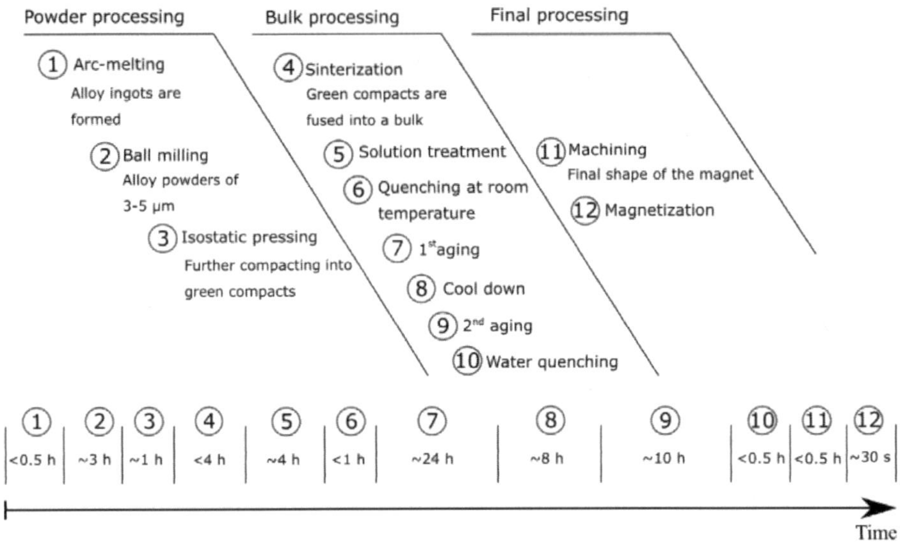

Figure 4. SmCo manufacture process.

Figure 5. High-temperature resistant SmCo permanent magnets. A coin has been used for reference. The two magnets above the coin are used in the nominal design. The bigger magnet in the upper part of the photograph is used for another magnetic configuration with higher magnetic field.

Table 2. Modified SmCo permanent magnet properties.

Material Properties	
Br (room temperature)	0.9~0.95 T
Hcj (room temperature)	30~40 kOe
(BH)max (room temperature)	20~22MGOe
Br (550ºC)	0.65~0.7 T
Hcj (550ºC)	6.5~7.5 kOe
(BH)max (550ºC)	9~10 MGOe
Thermal coefficient of Br	0.04~0.06%/ºC
Thermal coefficient of Hcj	-0.1~-0.2%/ºC
Density	8.3~8.5g/cm^3
Hardness	550-750 HV
Maximum Operating Temperature	550ºC
Curie Point	825º C

4. Additive manufacturing for the anode

Additive manufacture, usually known as 3D printing, has been used during two decades but it was recently when it has been successfully implemented in the aerospace industry. Before, this manufacturing technique was limited to prototypes and for testing purposes but with the refinement of the processes 3D printed materials begin to be used in operational products [10]. Among the different methods of additive manufacturing, when low production quantities are required, selective laser melting (SLM) is the most convenient method to manufacture complex geometries as it has the highest resolution [11,12].

For electric propulsion, additive manufacturing has been recently used for manufacturing the channel and parts of the distributor of the Hall Thruster with PEI and ABS polymers respectively [13]. While this work showed the potential benefits of using 3D printing in Hall Thrusters, the temperature of the discharge was beyond the temperature limits of the materials used, not allowing a steady state operation of the thruster. Thus, our research finds particularly interesting the applications of additive manufacture with metal materials, which could overcome the thermal limitations. The recent research of M. Sangregorio *et al.* [14] shows the advantages of manufacturing the metal grids of ion thrusters with SLM instead of traditional techniques. They show how with accuracies of 20 μm and good mechanical properties they reduce the manufacturing time from 400 hours to less than one day and how the price of the product decreases from several thousands of dollars to a few hundred. Although the geometry complexities of the ion grids and the Hall thruster anode are different, it seems there is no reason the benefits from the SLM techniques could not be achieved in the Hall thruster anode manufacture, where the minimum feature size is limited to 0.5 mm.

By using SLM for the anode, we can just not reduce the cost and time span for the manufacturing but we can design a more complex geometry that provide better azimuthal uniformity in the propellant flow at the exit of the anode. In Hall thrusters, the anode is not only the positive bias for the electrostatic field but it has the additional function of distributing the propellant flow all around the discharge chamber. A manifold with orifices which communicates the flow from the inlet of the anode to the discharge channel is usually used for this purpose. The restrictions in the dimensions enforce the use of a single feed entrance to inject the propellant into the manifold bringing along accentuated azimuthal distribution heterogeneity. With traditional manufacture techniques this problem can be partially solved but the limitations associated to the miniaturized assembly impede the optimal design. The limitation in our design include the minimum diameter of the orifices could not be smaller than 0.5 mm and welding process was restricted to assembly the feeder and anode distributor to the anode cup or body. The cross-sections of the different components of the anode are shown in Figure 6.

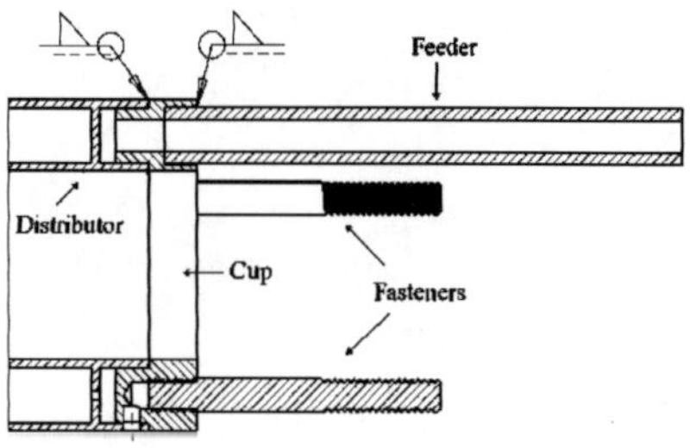

Figure 6. Anode cross section (for conventional manufacturing).

Using SLM for the manufacture of the anode, we can introduce modifications in the geometry with the aim of slowing down the flow and allowing for homogeneous flow diffusion along the channel. Two different geometries have been chosen to illustrate the benefits of using SLM. In design A, it was included just a simple set of baffles that redirect the flow after the propellant is injected through the orifices. In design B, a more complex geometry is used to redirect the flow with similar purposes.

Table 3 shows the three different designs (traditional manufacture, 3D-printing A and 3D-printing B) with the isometric view of the anode, the cross-section of the distributor and the propellant

IOP Conf. Series: Materials Science and Engineering **576** (2019) 012002 doi:10.1088/1757-899X/576/1/012002

distribution at the exit of the anode. The isometric view has been sectioned in order to illustrate the inner geometry in the distributor. It can be seen how the propellant distribution in the traditional manufacture design has a high concentration region corresponding to the location where the feeder is assembled and another concentration peak at the diametrically opposite region. Both 3D-printed designs offer a more uniform azimuthal distribution, specially design B. This can be numerically quantified by taking the densities distribution along the anode´s exit midradius of each design. Making statistical analysis of the distributions, it can be observed in Figure 7 the improvements in the density distribution of each 3D-printed design compared to the traditional manufacture design.

Table 3. Anode designs.

	Traditional manufacturing	3D printing – A	3D printing – B
Isometric view	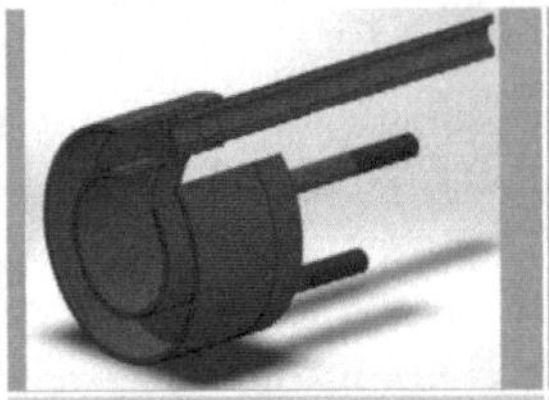		
Cross-section			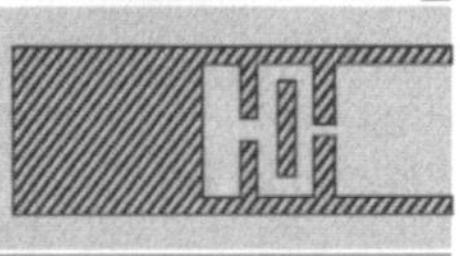
Propellant distribution			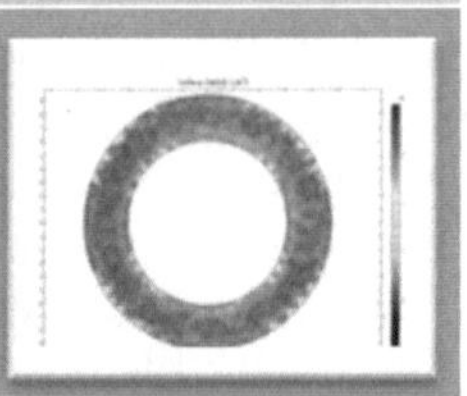

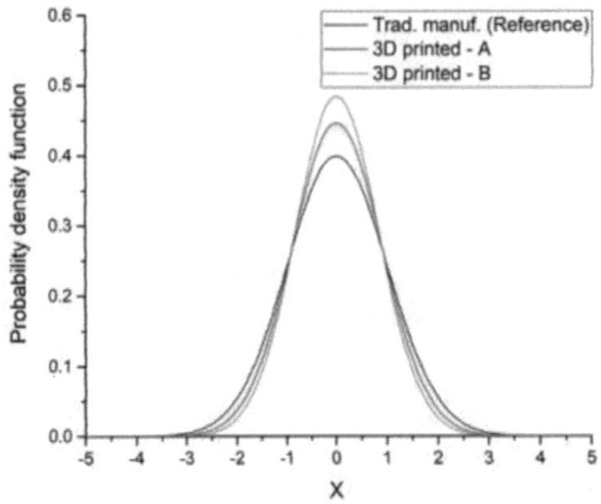

Figure 7. Probability density functions of the flow density at the anode's exit midradius for the three designs. Assuming normal distributions, the probability density function for the anode with traditional manufacturing has been standardized and the 3D-print respective distributions have been accordingly referenced.

After running thermal simulations, stainless steel 316 was chosen for the anode as there is extensive experience in printing this material, its relative permeability is close to one and its melting point (1400ºC) is far above of the operational temperature (680º C). It has been reported that for a deposition layer thickness of 20 μm the properties of SLM stainless steel 316 are similar to the materials obtained by traditional metallurgical methods [15] although significant porosity can be found in the surface. The porosity can be reduced by post-processing methods as finish machining or it may induce fatigue failures of the material [16]. In any case, as the anode is not a structural component of the thruster, it is not a major concern the surface quality of the anode after 3D printing.

5. Conclusions

This work was focused on introducing the associated complexities of miniaturizing a Hall Thruster engine to adapt its nominal power to 100 Watts and how high-temperature resistance permanent magnets and new techniques of manufacturing can potentially solve these problems. As electromagnetic coils have to been ruled out when scaling down the thruster due to increments in the temperature and increments in the magnetic coil required power to thruster power ratio, it was shown how a design with permanent magnets can provide the required magnetic field according to the guidelines for a proper magnetic field distribution. In addition, thermal simulations confirm that the maximum operating temperature of the new alloy of SmCo magnets will be within the nominal operation temperature of the thruster.

Finally, it was explained how additive manufacture, together with the benefits of reducing cost and time of manufacture, allows for geometries that are unattainable with traditional manufacturing techniques resulting in better azimuthal distributions that are expected to improve the performance of the thruster.

6. References

[1] Pergola P, Ruggiero A and Andrenucci M 2011 *32nd Int. Electric Propulsion Conf. 11-15 September (Wiesbaden)*
[2] Saleh J H, Geng F, Ku M and Walker M L R 2017 *Acta Astronaut.* **139** 141-156
[3] Misuri T, Ducci C, Albertoni R, Andrenucci M and Pedrini D 2015 *34th Int. Electric Propulsion Conf. 4-10 July (Hyogo-Kobe)*
[4] Zhurin V V, Kaufman H R and Robinson R S 1999 *Plasma Sources Science and Technology* **8**
[5] Hofer R R 2004 Dissertation Thesis (Ann Arbor: The University of Michigan) pp 95-120
[6] Raitses Y, Merino E and Fish N J 2010 *J. Appl. Phys.* **108**
[7] Kawashima R, Bak J, Komurasaki K, Koizumi H and Hara K 2017 *35th Int. Electric Propulsion Conf. 8-12 October (Atlanta)*
[8] Warner N Z and Martínez-Sánchez M 2007 Dissertation Thesis (Cambridge: Massachusetts Institute of Technology) pp 195-201
[9] Zhang T, Liu H, Liu J and Jiang C 2015 *Apply. Phys. Lett.***106**
[10] Joshi S C and Sheikh A A 2015 *Virtual Phys. Prototyp.* **10** 175-185
[11] Sames W J, List F A, Pannala S, Dehoff R R and Babu S S 2016 *Int. Mater. Rev.* **61**
[12] Kruth J P, Vandenbroucke B, Vaerenbergh J and Mercelis P 2005, *Int. Conf. Polymers and Moulds Innovations 20-23 April (Gent)*
[13] Hopping E P and Xu K G 2017 *35th Int. Electric Propulsion Conf. 8-12 October (Atlanta)*
[14] Sangregorio M, Wang N, Xie K, Zhang Z and Wang X 2018 *Rapid Prototyp. J.* **25**
[15] Zhukov A, Deev A and Kuznetsov P 2017 *Phys. Procedia* **89** 172-178
[16] Kaynak Y and Kitay O 2018 *J. Manuf. Mater. Process.* **2018**

IOP Conf. Series: Materials Science and Engineering **576** (2019) 012003 doi:10.1088/1757-899X/576/1/012003

Key technology and equipment for local heating roll-forming of high strength square tube

Xue-Feng Peng[1], Jing-Tao Han[2] and Zong-Miao Dai[1]

[1]Ninth department, 713th Research Institute of China Shipbuilding Industry Corporation, Zhengzhou 450015, China

[2]School of Materials Science and Engineering, University of Science and Technology Beijing, Beijing 100083, China

Abstract. The application of advanced high strength steel and the development innovation manufacturing technology is an effective way to realize the automobile lightweight and improve the safety performance design of high strength steel products. In this paper, a local heating "square to square" roll-forming process was proposed to solve the difficult problems of the introducing high strength steel into the traditional roll-forming equipment. Two kinds of inductors structure were compared and analyzed. Based on the "round to square" process, the size design of the rollers in forming process was deduced by the mutual relationship of the angle. And the relationship between the frame layout was given. The calculation of example and the key technical equipment were completed. Based on this new technology, the performance test of high strength tubes was carried out. It was found that the mechanical properties of the corner and the dimensional accuracy of the section are significantly improved. The research shows that this key technology and equipment of hot roll-forming can effectively overcome the problems of high strength steel.

1. Introduction

With the development of automobile lightweight, high-strength steel square tubes have been widely used in vehicles. The traditional forming process of square tube mainly has two kinds. One is directly formed by bending the steel plate, after welding, referred to as the "directly to square" process. The other one is first bending the steel plate into a round tube, after welding, then finishing into a square tube, known as the "round to square" process [1]. For the traditional square tube, the two processes have their own characteristics and can be selected according to customer needs. However, for the high-strength steel square tube, because of its high strength and low elongation, the advantage of "directly to square" process in terms of elongation is obvious, but the upper corner defects caused by empty bending and the phenomenon of thinning of bending angle in forming process are inevitable. This paper combines the advantages of "round to square" process and "directly to square" process as well as induction heating technology, and puts forward a local heating "square to square" roll-forming process". Its technological principle is that the steel plate is bent into a preformed tube with a larger fillet radius by using "directly to square" process at first. Only a slight strain occurs at the corner of the square tube. The fillet of the square tube is local heated linearly before finishing, and the forming performance is improved when the fillet reaches a certain temperature. When the heated preformed tube is into finishing pass, the high quality square tube with excellent mechanical properties and small radius or sharp fillet can be obtained by controlling the technological parameters. In this paper, the key

IOP Conf. Series: Materials Science and Engineering **576** (2019) 012003 doi:10.1088/1757-899X/576/1/012003

technology and equipment analysis will be carried out from two aspects of inductor structure design and pass design of "square to square" process.

2. Design of local heating inductor

The structure design of inductor is closely related to the local heating roll-forming process, which determines whether the hot roll-forming technology can be successful. In the process of induction heating, there will be sharp-angle effect. When there is a special sharp angle area in the heated parts, the magnetic induction line is easily gathered at this sharp area, which makes the induction current density larger. The traditional inductor design generally avoids the sharp-angle effect to cause over-burning. This paper makes full use of the sharp-angle effect of induced current to realize local heating at the fillet of high-strength square tube. According to the characteristics of square tube section, the layout principle of the inductor designed is shown in figure 1 (a).

In the design, the "driving" effect of the magnetizer is also utilized. The U-type magnetizer is mounted on the induction coil of inductor, so that more magnetic induction lines can pass through the square tube from the open side of magnetizer, as shown in figure 1 (b), which is called the "driving" effect of the magnetizer. Essentially, the magnetizer changes the direction of the magnetic induction lines, thereby increasing the amount of magnetic flux that through the inner of square tube.

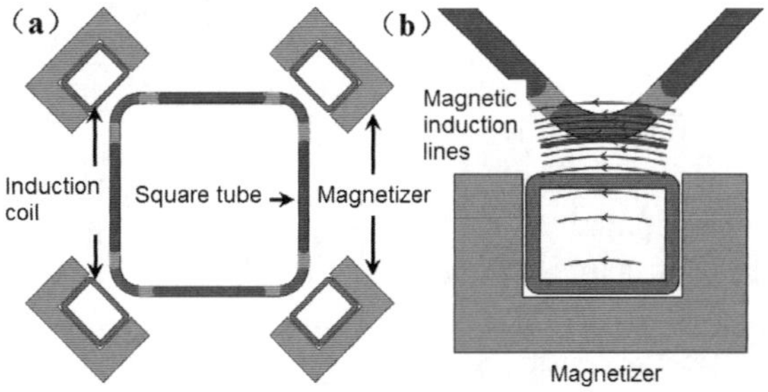

Figure 1. Schematic diagram of inductor structure.

The inductor is made of copper tube by drawing and welding. The copper tube is heated and filled with sand to avoid shrinkage of rectangular section in the forming process. The structure design of inductor is shown in figure 2, which are named as single coil inductor and compound coil inductor, respectively. The single coil inductor is a circular shape, and the silicon lamination is embedded in the heating zone to change the direction of the magnetic induction line to obtain a higher temperature in the corner area. This coil has a higher flux utilization rate, but the heating time is shorter at the same forming speed. The composite coil inductor is also attached with silicon lamination in the heating zone, which can realize a long-term linear heating process, but the magnetic flux efficiency is relatively low.

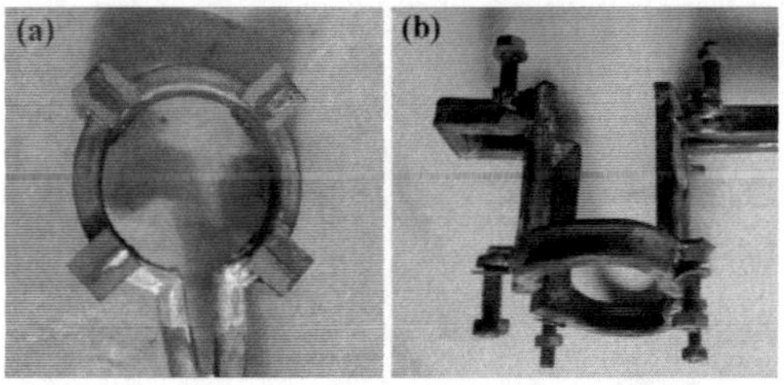

Figure 2. Design of two inductor structures.

Figure 3 shows the temperature distribution of the outer surface of square tube obtained by the two inductor structures in ANSYS simulation analysis software [3]. Figure 4 is a physical diagram of the surface temperature distribution of square tube heated by local induction during the experiment.

Setting target temperature at the corner of square tube is 600 °C. It is found from figure 3 that the maximum temperature is at the corner of square tube, which illustrates that the magnetizer plays the role of "driving" function. Both inductors can achieve the local heating effect of square tube. However, it is found that the heat affected zone of single coil inductor is larger and the temperature is higher than the value of composite coil inductor from figure 3 (a). And the heat affected zone temperatures at the center of the preformed tube plane are 86 °C and 195 °C for both inductors, respectively. For the composite coil inductor, it is clear that there is no blackening in the square tube plane as shown in figure 4 (b). Considering that the high temperature in heat affected zone caused by single coil inductor will affect the microstructure and mechanical properties of the undeformed zone, so the composite coil inductor is selected.

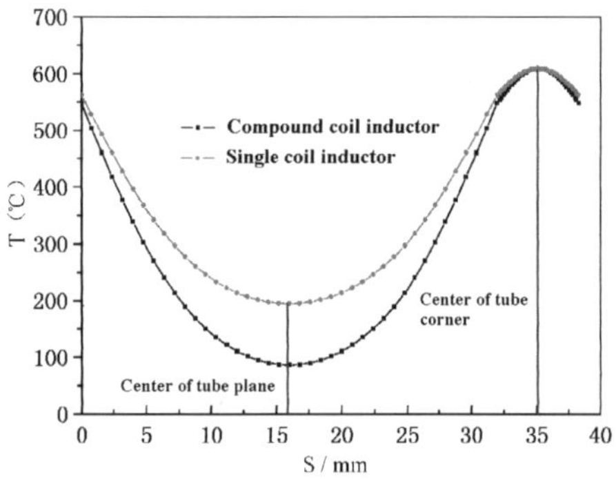

Figure 3. Simulated temperature distribution of two inductor structures.

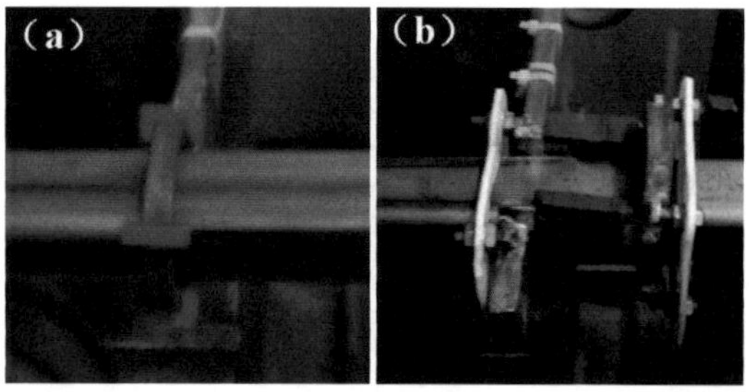

Figure 4. Heating effect diagram of two inductor structures.

3. Design of "square to square" process

The design theory of "square to square" process is based on the pass design of the "round to square" process. Researchers have accumulated a lot of experience about the pass design of "round to square" process. Sun Jia [4] introduced a design method of square tube with deformation angle and deflection coefficient as basic design parameters. This method can accurately reflect the deformation process. Cao Guo-fu [5] proposed the coefficient design method of "round to square" based on the specific coefficient relationship between the sizes of round tube and square tube, and deduced the coefficient μ_i and λ_i of pass design. Liu Rong-pei [6] gave a method of pass design angle mutual method, i.e. the sum of the radians corresponding to the long side, the short side and the fillet of square tube is constant during the roll-forming process. This design method is simple, the process parameters are easy to control and the forming accuracy is high.

3.1. Pass design method of "square to square" process

In this paper, the mutual relationship of angle is used to deduce the design of rollers for every pass in design of "square to square" process. The variation of section dimension of square tube during forming process is shown in figure 5. The design principle is that the linear topological relationship each line section is maintained during the change of preformed tube to square tube. That is to say, it is one-to-one correspondence and unchanged between the long side L_{Ae}, the short side L_{Be} and arc length A_eB_e of the corners of square tub with the distribution lines of preformed tube. The preformed tube has the behavior of reducing diameter and extending after each pass of deformation. Where μ_i is the elongation coefficient of the i pass, then:

$$L_{Ae} = L_{Ai} + \mu_i A_i B_i,\quad L_{Be} = L_{Bi} + \mu_i C_i D_i,\quad A_e D_e = \mu_i B_i C_i \tag{1}$$

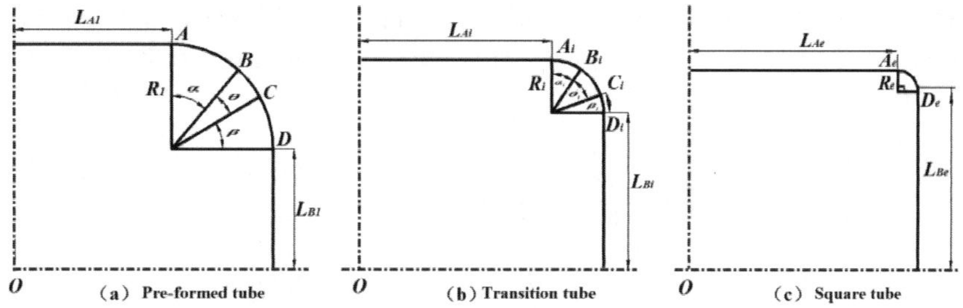

Figure 5. Deformation of square tube in "square to square" process.

When the angle α, θ and β corresponding to the deformed arc AB, BC and CD are determined, there is a unique value corresponding to α_i, θ_i and β_i. It can be proved that the sum of the angle corresponding to the long side half, the short side half and the transition arc of the square tube during deformation is a constant 90°.

3.2. Frame layout for finishing pass

The finishing frame uses four passes, of which only the latter three pass participate in the deformation, and the first one only plays the guiding role. The principle of average distribution of the amount of deformation is adopted. In the method of pass design angle mutual, the amount of deformation is expressed by the change of the angle corresponding to the long side, short side and transition corner of the last square tube.

3.3. Design and calculation of pass

Taking the high-strength square tube with cross-section size of 40×40×2 mm and an outer radius of r=4 mm as an example, the total elongation coefficient is u=1.04 and the number of finishing passes n is three. And the radius of the preformed tube is assumed to be r=10 mm. According to the expansion width of the sheet, the outer diameter of the preformed tube is determined to be D_0=42.88 mm, and the circumference of the preformed tube is:

$$L_0 = 4(D_0 - 2r_0) + 2\pi r_0 = 154.35\,\text{mm} \tag{2}$$

The literature [3] shows that the corner wall of square tube has obvious thickening behavior in the "square to square" process, so there is a certain amount of reduction in the forming process. Then the actual circumference and outer diameter of preformed tube are respectively:

$$L = \mu L_0 = 160.52\,\text{mm},\quad D = 44.42\,\text{mm}$$

Therefore, the total diameter reduction in the forming process is as follows:

$$\Delta D = (D - D_e)/2 = 2.\,\text{mm} \tag{3}$$

During the deformation process, the outer circumference of each pass and the arc of each section are basically unchanged, so as to ensure the forming precision. Although the reduction exists in each forming pass, because of its small size, assuming that the reducing is evenly distributed in every forming pass, the outer circumference of the pass is:

$$L_i = 4\left(D - 2\left(r - \tfrac{i}{n}\Delta D\right)\right) + 2\pi\left(r - \tfrac{i}{n}\Delta D\right) \tag{4}$$

The length of four plane edges of square tube and the arc length corresponding to transition arc are respectively:

$$L_{Ai} = \left(D - 2\left(r - \tfrac{i}{n}\Delta D\right)\right) \text{ and } B_iC_i = \tfrac{\pi}{2}\left(r - \tfrac{\Delta D i}{n}\right) \tag{5}$$

The angle of the transition arc before deformation is:

$$\theta_i = \frac{\pi R_e}{2R_i} \tag{6}$$

According to the angle mutual relationship and the same angle corresponding to the long side and the short side during roll-forming process, the angle corresponding to the length of the tube plane before the deformation is:

$$\alpha_i = \beta_i = \left(90° - \frac{\pi R_e}{2R_i}\right)/2 \tag{7}$$

The calculation results of the roller each pass the "square to square" process are shown in Table 1.

Table 1. Size of the rollers in forming process.

Pass	Circumference L / mm	Side length L_{Ai} / mm	Arc length / mm	Angle/°		Radius / mm
				α_i	θ_i	
0	160.52	12.21	15.71	27°	36°	10.00
1	158.38	13.68	12.24	21.9°	46.2°	7.79
2	156.27	15.16	8.76	12.7°	64.6°	5.57
3	154.19	16.00	6.28	0°	90°	4.00

In the pass design of "square to square" process, the angle corresponding to the straight line section changes from 27° to 0°, and the angle corresponding to the transitional arc section changes from 36° to 90°. Therefore, the values of rollers each pass can be obtained only by setting the varied angle α_i and θ_i during the rolling process. Figure 6 is schematic diagram of pass frame of "square to square" process.

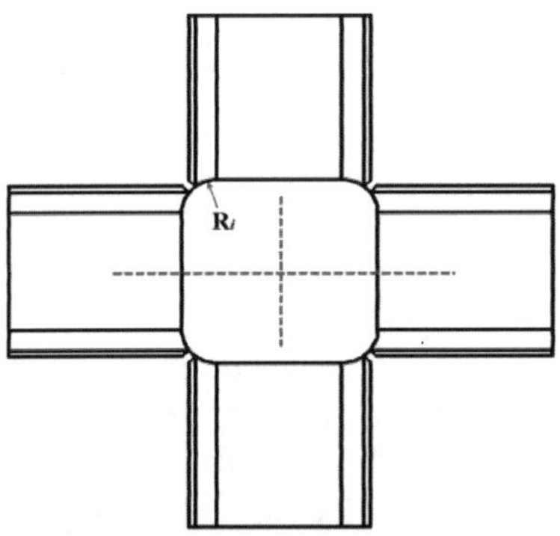

Figure 6. Schematic diagram of pass frame.

4. Equipment of hot roll-forming for high-strength square tube

According to the design of inductor and the analysis of pass of "square to square" process, the equipment of local heating roll-forming process for square tube has been completed. As shown in figure 7, it mainly includes the finishing frame, induction heating equipment, temperature recorder and infrared thermometer. QSTE700TM high strength steel is selected as raw material to prepare preformed tube. The self-designed composite coil inductor is used to realize local heating of four corner parts of square tube, as shown in figure 8. And the target temperature is guaranteed to be constant through the automatic feedback system. The hot roll forming temperature is 650°C. When entering the finishing stage, the square tube is subjected to extrusion force in four directions of the rollers. Due to the influence of temperature, the square tube with excellent mechanical properties and small rounded radius is obtained [7]. Figure 9 gives the mechanical properties of the corner of the hot-formed square tube. It can be seen in figure that the cold work hardening effect of high strength square tubes formed at room temperature has been significantly slowed down. The corner elongation of square tube increased from 6.2% at room temperature to 14.8% after hot roll-forming. Figure 10 shows that the plane section size accuracy of high-strength square tube after hot roll-forming is higher than that at room temperature, which illustrated that there was a significant springback in the corner of the square tube.

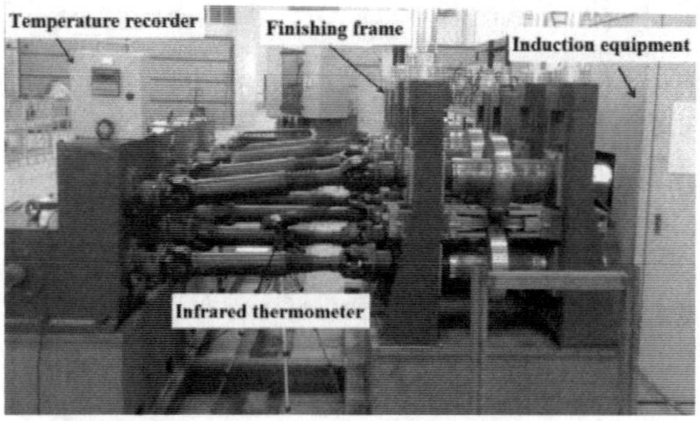

Figure 7. Equipment of local heating roll-forming process for square tube.

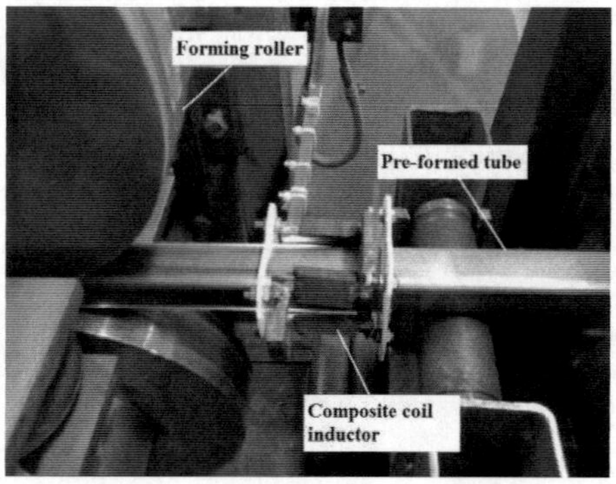

Figure 8. Key forming pass of "square to square" process for square tube.

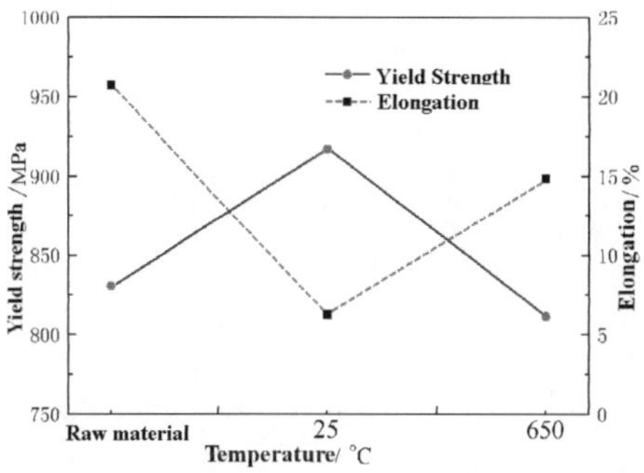

Figure 9. Mechanical properties of hot roll-formed square tube.

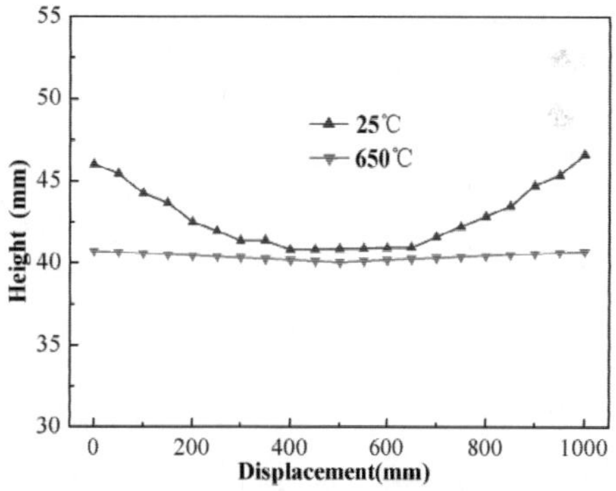

Figure 10. Section sizes of square tube under different forming temperatures.

5. Conclusion

In order to solve the problems of equipment and process caused by traditional roll-forming technology for high strength square tube, a local heating "square to square" forming process for high strength square tube is proposed in this paper. By comparing the two inductor structures, it is found that the heating area of composite coil inductor is more concentrated and the temperature of heat-affected zone is smaller than that single coil inductor. The frame layout of finishing pass and design as well as calculation of rollers of "square to square" process for square tube are given. And the key technical equipment of hot roll-forming process is developed. The mechanical properties and section sizes of high strength square tube manufactured by this new technology are tested. It is found that the mechanical properties of the corner and the dimensional accuracy of the section are significantly improved. The corner elongation of square tube increased from 6.2% at room temperature to 14.8% after hot roll-forming and the dimension accuracy of square tube section was reduced from 6.62 mm to 0.74 mm, which effectively overcomes the production problems of the high-strength square tube.

6. References

[1] Rossi B, Degée H and Boman R 2013 *F. E. Ana. Des.* **72** 13
[2] Xu H 2003 *Weld. Pip. Tub.* **03** 45
[3] Peng X F, Chen Q and Wang Y 2016 *Chin. J. Chem. Eng.* **07** 973

[4] Sun J 1992 *Weld. Pip. Tub.* **05** 40
[5] Cao G F 2014 *Weld. Pip. Tub.* **07** 49
[6] Liu P R and Zhang H L 2000 *Weld. Pip. Tub.* **02** 32
[7] Peng X F, Han J T and Yan P J 2016 *Chin. J. Chem. Eng.* **06** 820

IOP Conf. Series: Materials Science and Engineering **576** (2019) 012004 doi:10.1088/1757-899X/576/1/012004

The Best Preload of Bolted Joint Plate for Anti-resonance

Songjian Wu, Wenguang Liu* and Xiaoting Wang

School of Aeronautical Manufacturing Engineering, Nanchang Hangkong University, 330063, Nanchang, China

Corresponding Author's E-mail: liuwg14@nchu.edu.cn

Abstract. Based on the problem of modal frequencies variation of bolted joint plate caused by the variation of bolt preloads, the purpose of this article is to propose a method of finding the best preload of bolted joint plate. Firstly, a modal analysis experimental system of bolted joint plates was set up. Subsequently, the modal frequency test of some bolted joint plates are completed. A method for designing the best preload of bolted joint plate was proposed in the end. Results indicate that the bolt preload has certain influence on modal frequency of bolted joint plate. The process of applying a preload to the bolt is regarded as the tightening of a bolt from loosen state to tighten state. As two loosen bolts are simultaneously tightened into tighten bolts, the modal frequencies is increased with a big magnitude. However, as one of loosen bolt is tightened to tighten bolt, but the other is unchanged, the modal frequencies is increased with a small magnitude. The sensitivity of first four modal frequencies of the jointed plate are different to the bolt preload. Therefore, working environment and working force of jointed plate must be considered in the dynamic design. Based on the experimental result, a best bolt preload design method of bolted joint plate in the dynamic design can be proposed.

1. Introduction

Due to the advantages of higher standardization, convenient to make, simple structure,lower cost and easier to disassemble and assemble, it is widely used in aerospace, vehicle and agricultural equipment. However, with the development of modern dynamic machinery, the force applied to the machinery is becoming to more and more complex. Vibration, shock, variable force and temperature is the inevitable working environment for many machineries. As a results, bolt looseness often occurs. Aiming at the loosening of bolted joints, researchers in the academic and engineering fields have proposed many factors that affect the looseness of bolted joints. For example, preload, temperature, and external loads. Among them, the bolt preload is particularly important for the bolted joints. If the bolt preload is applied too small, the component cannot bear the working force, and slipping will be introduced. This will affect the performance of the machine. But if bolt preload is too big, the bolt may be deformed plastically. Thereby, the strength of bolted joint structures is reduced. It will reduce the working life of the component. Therefore, a reasonable blot preload not only can play a fastening, anti-slip effect, but also can improve working life of the component. Considering the impact mechanisms of preload on the dynamic characteristics, the study on the vibration, stiffness and strength of bolted joints have been focused by the scholars at home and abroad.

With the development of vibration theory, the measurement method of bolt preload based on vibration characteristics has been paid more and more attention. For the sake of monitoring the preload, Du F et al. proposed a virtual time reversal guided wave method to test the preload of the bolted

L-shaped joint[1]. In order to study the relationship between the bolt preload and the bolt elongation, Jimenez-Pena C et al. gave a proof of the concept of the relationship between bolt elongation and bolt preload based on the DIC method, and proposed a calibration method for the bolt preload-elongation relationship within a given clamping range[2]. Pakdil M et al. studied the effects of preload on the fail response of composite bolted joints with bolt hole clearances[3]. Esmaeili F et al. studied the effects of bolt clamping force on the fatigue life of double-lapped and hybrid-lapped bolt joints. The study revealed the positive effect of clamping force on the fatigue life of simple joints and hybrid joints[4]. Zhao T studied the effect of bolt preload on the bolted joints. It is found that if the fasteners are loosen, the fatigue life may be very short[5]. Liu F carried out random bolt load distribution test based on the single-lapped multi-bolt joints under different parameters. The results shown that the bolt preload and the bolt hole clearance have significant influences on the change of bolt load distribution[6]. Guang Z et al. carried out the influence of bolt preload on the modal frequencies of the bolted joints. It was found that the modal frequency of the bolted joint is increased with the increasing of the bolt preload[7]. Dravid S et al. conducted a research on the tightness of bolted joints by using different washers instead of torsional moments to explore the bolt's resistance to looseness[8]. Zhang M et al. conducted an experiment on the self-loosening behavior of bolted joints caused by thread wear under lateral cyclic load. Results indicated that reasonable preload can reduce the thread wear to improve the stiffness of the joint and enhance the tightening of the bolted joint[9]. Giannopoulos I K et al. studied the effects of preload on the tensile strength of composite plates with bolt hole. Results indicated that the reducing of preload can reduce the stress concentration around hole.and increase the tensile and fatigue strength of the structure, and the increasing preload will cause the slippage on the surface of the mating bolt plate[10-11]. Dhote J X et al. used DIC equipment to extract the strain field of the joint surface of the bolted joint plate under lubrication and drying conditions. Chakherlou T N et al. analyzed the influences of preload on the laws of stress distribution around screw hole by finite element method. When the ratio of external load to bolt preload is too small, increasing the preload will shorten fatigue life[12-13]. Griza S et al. discussed the relationship between the bolt length and fatigue strength of bolt joints. It was shown that the increasing of the length of bolts can increase the fatigue strength of bolt joints, but this conclusion does not apply to short bolt joints[14] . Zhai Y et al. analyzed the influence of clearance between bolt and hole as well as the bolt torque on the strain of contact surface of the carbon fiber epoxy resin countersunk head bolt joint. Results shown that the increasing of the clearance will increase the strain of the contact surface and the increasing of the bolt torque is beneficial to alleviate the strain at the joint contact[15]. Benhamena A et al. studied the impact of the preload on the fretting fatigue characteristics of the bolted joint plate[16].

As above mentioned, a lot of researches on the bolt preload problems have been done. But few works were done on the impacts of different bolt preload on modal frequencies of the bolted joint plate. In the actual working process, once the modal frequency of the structure is close to the working frequency, resonance occurs, which is going to lead to loose bolts and unpredictable disasters. Therefore, in a dynamic design of the bolted joint plate, the modal frequency must be considered to avoid resonance during operation. Moreover, the modal frequency is usually related to the preload. Thus, a proper bolt preload helps the system to avoid resonance effectively and is beneficial to the vibration fatigue design.

In this article, the tests on the influences of different preload on the modal frequency of the bolted joint plate are carried out. Based on the influence mechanisms of preload on the modal frequencies, a preload design method of bolted joint plate is proposed. This method can provide optimization method of the bolt preload.

2. Experiment of bolted joint plate

2.1. Test piece

A single-lapped and double bolts joint plate is designed. The geometry and physical map are shown in Figure 1 and Figure 2. The dimensions are as follows:L_1=242mm, L_2=242mm, W=55mm, t=5 mm and the bolt hole diameter D= 8.8 mm. The major diameter of bolt d=8 mm, The distance from the center

IOP Conf. Series: Materials Science and Engineering **576** (2019) 012004 doi:10.1088/1757-899X/576/1/012004

of the plate hole to the end of the plate e=24 mm, and distance between the center of the two holes c=10mm. The bolted joint plate is made of SUS304 stainless steel, and the bolt and nut are made of LY12CZ aluminum alloy. The mechanical properties of SUS304 in room temperature are: elastic modulus E=193 GPa, Poisson's ratio υ=0.3, material density ρ= 7870 kg/m^3, yield strength σ_s=260 MPa, ultimate strength σ_b=631 MPa. The mechanical properties of LY12CZ aluminum in room temperature are: elastic modulus E=69.5 Gpa, Poisson's ratio υ= 0.33, material density ρ=2770 kg/m^3, yield strength σ_s= 400 MPa, ultimate strength σ_b=443 MPa.

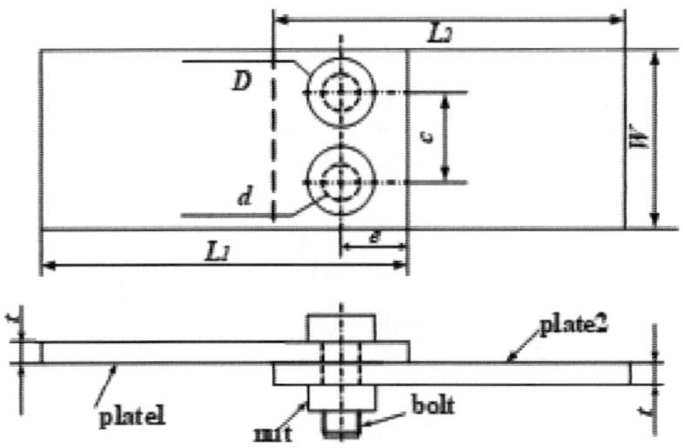

Figure 1. Geometric size of test specimen.

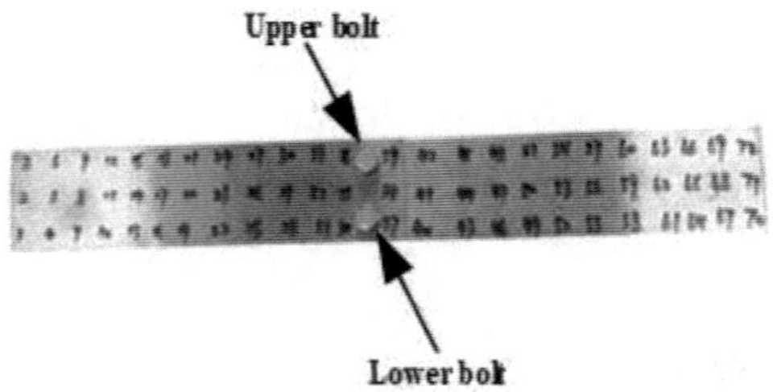

Figure 2. Physical map of test specimen.

2.2. Test procedures

In order to simulate a free-free state of the bolted joint plate, the tested specimen is supported by two balloons during the modal test, as shown in Figure 3. The system mainly includes an incentive hammer, a piezoelectric acceleration sensor, a bolted joint plate, a modal analysis system and a dynamic data acquisition as shown in Figure 4. Place the tested bolted joint plate on two light-weight and soft balloons, install an acceleration sensor at a suitable position on the surface of the test specimen, and use a hammer to strike it. The direction of tap of hammer must be consistent with the vertical direction of sticking position of the acceleration senor. Then, the measured signal is imported into the YMC9800 modal analysis software by using acquisition equipment. The modal frequency response curve is obtained by fast Fourier transform. The vibration mode, damping, and modal frequency of the tested specimen are obtained. The flow chart of the test is shown in Figure 5.

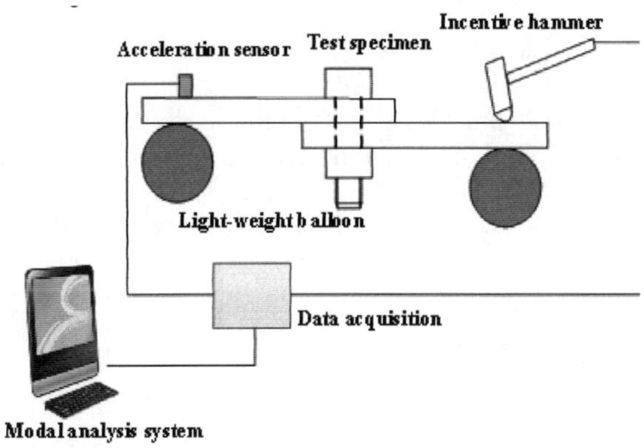

Figure 3. Schematic diagram of the test system.

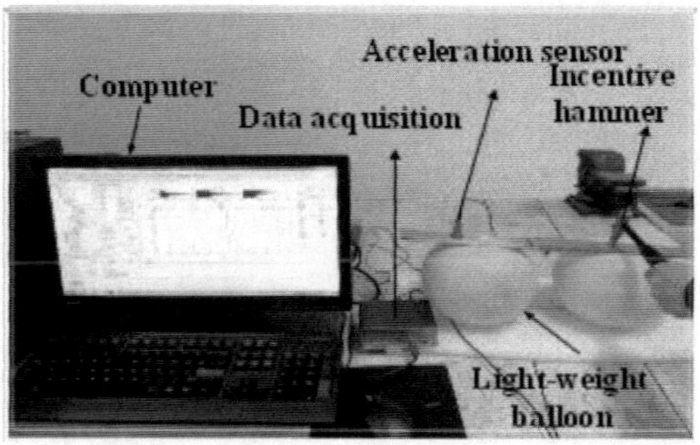

Figure 4. Physical map of modal analysis experimental system of bolted joint plate.

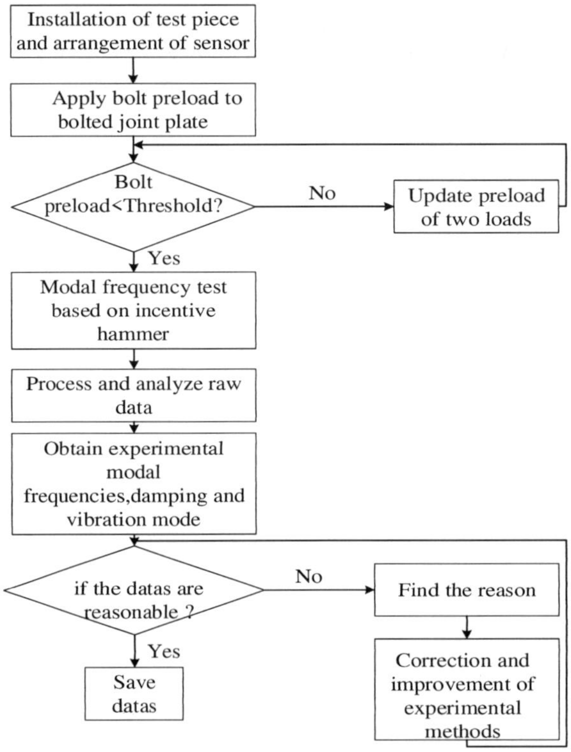

Figure 5. Flow chart of test.

3. Results and discussion

3.1. Preload setting

The bolt preload is limited by the strength of the two bolts and the loading condition of the bolted joint. The bolt preload is usually calculated by

$$F_0 = \eta \times \sigma_s \times A_s \tag{1}$$

Where: η is set as 0.5-0.7; σ_s is the yield strength of bolt; A_s is the cross-sectional area of the bolt in minor diameter.

The torque is usually calculated by

$$M_t = K \times F_0 \times d \tag{2}$$

Where: K is the tightening torque coefficient; F_0 is the bolt preload; d is the middle diameter of the bolts; K is set as 0.1-0.3. The working environment is approximated as a lubricating surface oxidation with K= 0.2 in this article.

Thus, the relationship between the bolt preload F_0 and torque M_t is

$$F_0 = 625 \times M_t \tag{3}$$

The applied torques can be estimated by the equation (3). The torque applied to the bolt are 1.0 N.m, 1.5N.m, 2.0 N.m, 2.5 N.m, 3.0 N.m, 3.5 N.m, 4.0 N.m, 4.5 N.m, 5.0 N.m. And the corresponding preload are 625N, 937.5 N, 1250 N, 1562.5 N, 1875 N, 2187.5 N, 2500 N, 2812.5 N, 3125N.

3.2. Discussion on Factors Affecting Modal Frequency of Bolted joint Plates

In order to better describe the relationship between bolt preload and modal frequency, the definition of frequency reduction rate is introduced, which is defined by the following formula.

$$\tau = \frac{f_i - f_0}{f_0} \times 100\% \tag{4}$$

Where: τ is the rate of decline of the modal frequency; f_0 is the initial modal frequency of the test piece; f_i is the modal frequency of each stage of the test piece.

In order to describe the effect of different bolt preload, the bolt preload is divided into the bolt preload of the upper bolt and lower bolt, and the ratio of the torsional moment of the upper bolt at each stage and the maximum torsional moment of the upper bolt is described by dimensionless m. The definition formula is

$$m = G_i\left(G_{\max}\right)^{-1} \tag{5}$$

Where: G_{max} is maximum torsion moment of upper bolt; G_i is the torsion moment of the upper bolt at each stage of the test piece.

Use the dimensionless n to describe the ratio of the torsion moment of the lower bolt at each stage and the maximum moment of the lower bolt, and define the formula.

$$n = T_i\left(T_{\max}\right)^{-1} \tag{6}$$

Where: T_{max} is the test piece maximum lower end bolt torsion force; T_i is the torsion moment of the lower end bolt at each stage of the test piece.

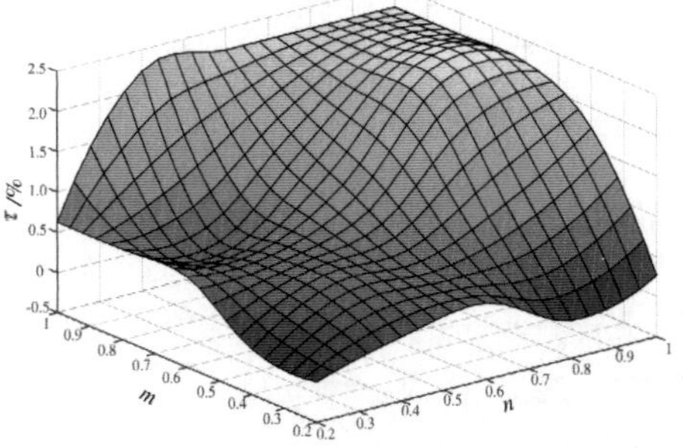

(a) Mode One

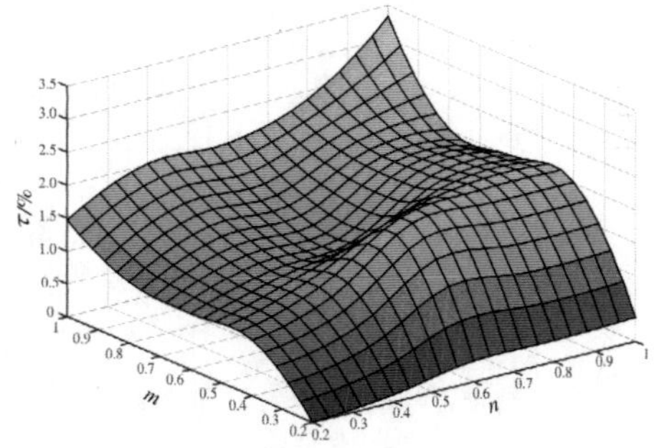

(b) Mode Two

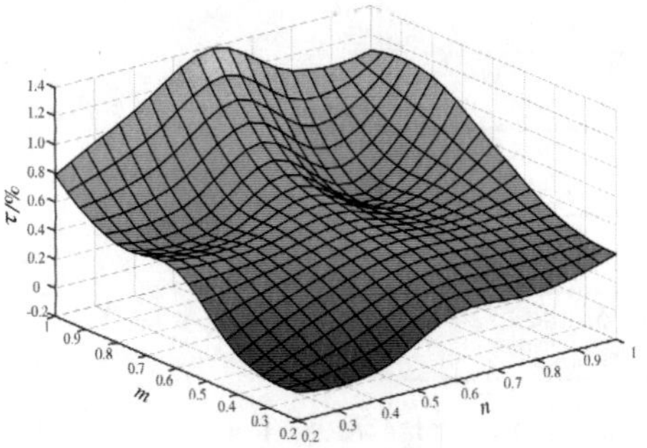

(c) Mode Three

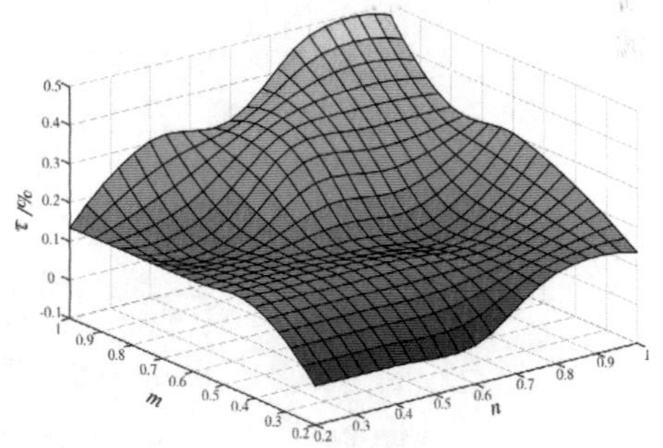

(d) Mode Four

Figure 6. Influence of bolt preload on the first four order modal frequencies.

Figure 6 shows that the modal frequencies varies with the increasing of the preload. The variation law is nonlinear. when the preload of upper and lower bolts are applied simultaneously is compared with the preload of one of two bolts is applied, it is found that the former has a greater influence on the modal frequency of the member. The process of applying the preload to the upper or the lower bolts, it is regarded as the tightening from the "loosen" bolt to the "tighted" bolt. As two "loosen" bolts are tightened to the "tighten" bolt, the modal frequencies is increased and the amplitude is up to 3.3%. As a “loosen” bolt is tightened to “tighten” bolt, but the other is unchanged, the modal frequency is increased but the amplitude is only 1.5%. In dynamic design, as the excitation frequency is equal to the modal frequency of the specimen, resonance occurs. In order to avoid the resonance by adjusting the preload of the bolted joint plate to change its natural modal frequency, the modal frequency is far from the excitation frequency to avoid resonance. These bolt preload can make the modal frequency of the test specimen away from the external excitation frequency, which is called the best bolt preload. The frequency variation of each step of the joint plate is also different. The second-order frequency falling rate is up to about 3%, and the fourth-order frequency falling rate is only about 0.4%. This shows that the sensitivity of each order modal frequency to bolt preload is different. Under different orders of the same test piece, the value of the best bolt preload is also different, which indicates that the working environment of the test piece should be considered in the design, anti-resonant best bolt preload design under the actual working condition provides a basis for anti-resonance design.

3.3. The best preload of bolted joint plate

Under the action of dynamic load, the excitation frequency is close to the modal frequency of the structure to lead resonance has been a concern. In order to solve this problem effectively, this paper designs the best bolt preload of the bolted joint plate from the dynamic point of view based on the change of the modal frequency to achieve the anti-resonance purpose.

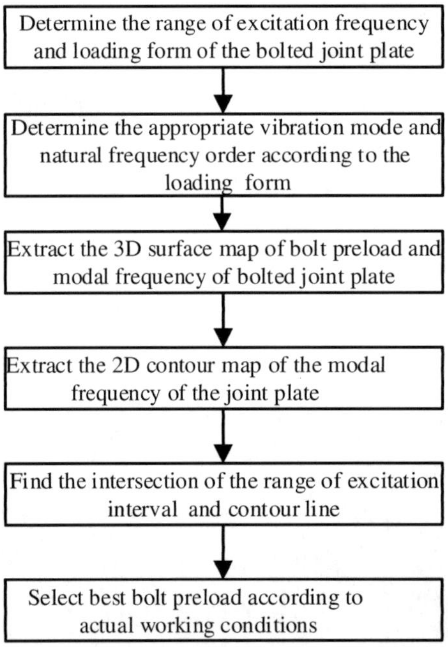

Figure 7. The best bolt preload design based on modal frequency.

In order to explain the application of the scheme shown in Figure 7 in detail, the modal frequencyanalysis of the bolted joint plates are taken as an example. It is assumed that the single-lapped and double bolts joint plate is subjected to a lateral load. The actual working frequency is in the vicinity of the second-order mode. The operation frequency range is set as [285,288] Hz. In the dynamic design, the 3D map of the second-order modal frequency of the component is first extracted, and then the 2D contour map is drawn in the interval corresponding to the operating frequency. As shown in Figure 8, under the action of dynamic load, if the excitation frequency is in the working frequency range of the structural members, resonance is easy to be generated. As the bolt preload of the bolt is too small, the fastening property of the member is not guaranteed. Excessive preload is easy to cause the bolt to be broken, and the bolted joint are crushed, bitten, twisted or broken. In summary, the bolt preload should be reasonably selected based on the working environment in which the component is located. Then the maximum bolt preload of the two bolts of the bolted joint plate should be determined, and the actual bolt preload of the two bolts should be converted according to the relationship shown in Figure 8 to determine the best preload for anti-resonant of bolted joint plate. Through this optimization method, a new idea can be provided for the anti-resonant fatigue dynamics design of the bolted joint.

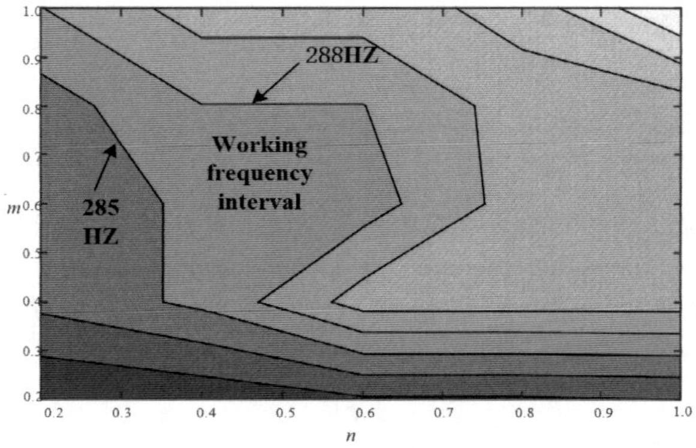

Figure 8. Influence of bolt preload on second-order mode.

4. Conclusions

A single-lapped and double bolted joint plate was choosed as the object, the influence of the preload on the modal frequency of the bolted joint plate is studied. A design method of the best preload for anti-resonant of bolted joint plate is proposed. The main conclusions are as follows:

- The preload has a certain influence on the modal frequency of the joint plate, and the preload applied to two bolts has a greater influence on modal frequencies. This property can be utilized to change the magnitude of the bolt preload so that the modal frequency is kept away from the excitation frequency to achieve anti-resonance purposes.
- The variation of modal frequency at each step of the bolted joint plate is different under the same bolt preload, which shows that the sensitivity of each mode modal frequency to the bolt preload is different. Under different orders of the same test piece, the value of the best preload of the bolt is also different, which indicates that the working environment of the test piece should be considered in the design.
- The design of the best preload of the bolted joint plate should be related to the actual working conditions. According to the actual working condition frequency, find the best preload of the bolt, and reasonably avoid the external excitation frequency to achieve anti-resonance purpose.

5. References

[1] Du F, Xu C and Wu G 2018 *Sensor.* **18**
[2] Jimenez-Pena, Carlos, Lavatelli A, Balcaen R, et al 2018 *Proceedings.* **2**, 514
[3] Pakdil M, Sen F, Sayman O, et al 2007 *Reinf. Plast. Compos.* **26**, 1239-1252
[4] Esmaeili F , Chakherlou T N and Zehsaz M 2014 *ENG FAIL ANAL.* **45**, 406-420
[5] Zhao T, Palardy G, Villegas I F, Rans C and Martinez M 2017 *Compos. Part B-Eng.* **112**, 224-234
[6] Liu F , Zhang J , Zhao L , et al 2015 *Composi. Struct.* **131**, 625-636
[7] Guang Z, Zhiliang X, Xin J, et al 2018 *Tribol. Int.* **127**
[8] Dravid S , Tripathi K and Chouksey M 2014 *Procedia Tech.* **14**, 543-552
[9] Zhang M, Lu L, Wang W, et al 2018 *Wear.*
[10] Hammami C, Balmes E and Guskov M 2016 *MECH SYST SIGNAL PR.* **70**, 714-724
[11] Giannpoulos I K, Doronidawes D, Kourousis K I and Yasaee M 2017 *Compos. Part B-Eng.* **125**, 19-26
[12] Dhote J X, Comer A J, Stanley W F, et al 2013 *Compos. Struct.* **96**, 216-225

[13] Chakherlou T N, Razavi M J and Abazadeh B 2013 *Eng. Fail. Ana.* **29**, 62-74
[14] Griza S, Silva M E G D, Santos S V D, et al 2013 *ENG FAIL ANA.* **34**, 397-406
[15] Zhai Y, Li D, Li X, et al 2015 *Compos. Struct.* **127**, 411-419
[16] Benhamena A, Talha A, Benseddiq N, et al 2010 *MAT SCI ENG R.* **527**, 6413-6421

Acknowledgments
This research work is supported by the National Science Foundation of China (51565039) and the Science and Technology Project of Jiangxi Provincial Education Department (GJJ170584).

IOP Conf. Series: Materials Science and Engineering **576** (2019) 012005 doi:10.1088/1757-899X/576/1/012005

Bolt layout optimization design based on genetic algorithm

B Shen, Z W Pan, H Lin and W Z Li

Beijing Institute of Astronautical Systems Engineering, 100076 Beijing, China

Abstract. When working, electronic equipment is often subjected to mechanical environments such as vibration and shock. The reliability problem of components after experiencing vibration environment is one of the key research contents in the development of electronic equipment. The printed circuit board on which the component is mounted is deformed during vibration, and the vibration is transmitted to the component, so that the component corner has a large acceleration, which is one of the important reasons for the failure of the electronic device. Based on the finite element simulation of the circuit board, the application of genetic algorithm in bolt layout optimization is realized by MATLAB programming. By optimizing the bolt layout, the acceleration value at the component corners is greatly reduced, and the reliability of the electronic equipment is improved.

1. Introduction

When working, electronic equipment is often subjected to mechanical environments such as vibration and shock. The printed circuit board on which the component is mounted is deformed during vibration, and the vibration is transmitted to the component, so that the component pin has a large acceleration, which is one of the important reasons for the failure of the electronic device. Through the finite element simulation method, the acceleration of the board under vibration can be estimated in the design, and the layout of the bolt can be optimized to reduce the acceleration value of the component corner, so that the mechanical environment of the printed circuit board can be improved, and the reliability of the electronic equipment can be improved.

The bolts are typically mounted at the edge of the printed circuit board. Because of the need to determine the layout of each bolt, it is not possible to traverse all conditions. Previous scholars often proposed the bolt layout by calculating a number of typical bolt layout models[1, 2]. However, due to the limited number of typical models selected, the optimality of the optimization results is often not guaranteed. In recent years, intelligent optimization algorithms have been rapidly developed, which can overcome the limitations of traditional optimization algorithms and achieve global optimization[3, 4]. Krishnakumar[5] presents a fixture layout optimization technique that uses the genetic algorithm (GA) to find the fixture layout that minimizes the deformation of the machined surface due to clamping and machining forces over the entire tool path. Huanlin Liu[6] proposes a gene density genetic algorithm (GDGA) to optimize the LED layout. Gene density-based segmented crossover and gene mutation strategy is put forward. A weighted differences function is designed to select the right LED lamps' layout.

This paper improves the genetic algorithm and optimizes the layout of the bolt by finite element simulation. In the process of solving the optimization model, using the idea of parametric modeling, the interaction between MATLAB and NASTRAN is realized, and the purpose of optimizing bolt layout and improving the mechanical environment of components is achieved.

2. Finite element simulation of printed circuit board

In this paper, a simple circuit board is used as the research object, and the finite element software PATRAN is used for modeling. The shape of the board is regular and there is only one component on the board. The finite element model is shown in Figure 1.

Figure 1. Finite element model.

The board is mounted on the platform with 8 bolts. The bolts are located at the four corners of the board and in the middle of four sides. According to the actual working conditions, the load on the board is from the vibration of the platform, so the unit acceleration in the Z direction is applied at the nodes at the bolts position by 1 m/s^2. The linear elastic constitutive equation model is used for the plates and components in the calculation process. The material parameters used in each part are shown in Table 1.

Table 1. Material parameters.

	Material	Length (mm)	Width (mm)	Height or Thickness (mm)	Elastic Modulus(GPa)	Poisson ratio	Density (kg/m3)
PCB	CCL	150	100	2	14	0.28	1800
Component	Ceramics	30	30	5	310	0.22	3980

The modal analysis result of the board model is shown in Table 2 below.

Table 2. Results of modal analysis.

Modal Order	1	2	3	4	5
Frequency(Hz)	557.6	749.7	1012.6	1088.8	1310.2
Modal Order	6	7	8	9	10
Frequency(Hz)	1577.6	1832.6	1987.3	2135.6	2271.9

Apply acceleration sinusoidal frequency excitation to the board, the interval is 20~2000Hz, and the initial modal damping ratio is 0.01. The acceleration or stress frequency response curve of the component corner position can be obtained. In this paper, the optimization is based on the sinusoidal sweep analysis results of the board.

3. Genetic algorithm and improvement

3.1. Genetic algorithm

In the 1960s, Professor John Holland of the University of Michigan in the United States creatively proposed to apply the genetic mechanism of biology to artificial adaptive systems to perform adaptive search in a group manner. In the 1970s, John Holland et al. proposed a more complete theory and method for genetic algorithms[7]. On the basis of this, Dr. De Jong conducted a large number of numerical function optimization experiments, established a framework of genetic algorithms, and the prototype of genetic algorithms was initially formed. To date, genetic algorithm has been widely used in combinatorial optimization, machine learning, adaptive control, planning and design, and artificial life.

IOP Conf. Series: Materials Science and Engineering **576** (2019) 012005 doi:10.1088/1757-899X/576/1/012005

Genetic algorithm is a process that simulates biological evolution during the optimization process. The trait of each individual depends on the genes on the chromosome. The process of biological evolution is also the process of chromosome crossing, genetic variation, and natural selection between individuals. Through the competition mechanism of “survival of the fittest, survival of the fittest”, individuals with poor viability are gradually eliminated, thereby improving the survival level of the whole population and realizing the evolution and development of biology. In the algorithm, the optimization variables represent the genes on the chromosome, and the objective function value represents the viability of the individual.

This paper adopts the roulette strategy to achieve the competition mechanism of "survival of the fittest, survival of the fittest". The probability that each individual will survive is

$$p_i = \frac{F_i}{\sum_{j=1}^{N} F_j} \tag{1}$$

where F is the objective function value, N is the individuals in the population. The purpose of the formula (1) is to allow individuals with higher objective function value to obtain higher survival chances. However, considering that the purpose of optimization in this paper is to reduce the objective function, the above formula is changed to

$$p_i = \frac{\sum_{j=1}^{N} F_j - F_i}{(N-1) \times \sum_{j=1}^{N} F_j} \tag{2}$$

The goal of optimizing the objective function can be achieved by programming the selection, crossover and mutation operations in the genetic algorithm.

3.2. Adaptability improvement
In order to make the genetic algorithm used in this paper more suitable for bolt layout optimization, it has been adaptively improved during programming, mainly in the following two aspects[8].

- Considering that the program calls NASTRAN to perform the calculation for a long time, each calculation result is stored to avoid repeated calculations for the same state, and the iteration time can be shortened.
- Since each current state has a limited neighbourhood, the searched neighbourhoods are recorded during the optimization process, avoiding repeated searches to the same neighbourhood, which can increase the convergence speed.

4. Optimization design of bolt layout

The acceleration value at the component corner is an important indicator to measure the mechanical environment of the component. In the case of material properties meeting the requirements, the reduction of the acceleration value means the improvement of the mechanical environment of the components, which is especially important for the reliability of the board and even the entire electronic equipment.

Changing the mounting position of the bolt can change the position at which the board receives the vibration of the platform, that is, the position of the load input. Each time the bolt position is changed, the board model needs to be redesigned. For rapid modeling, automation of modeling is required. Therefore, the idea of parametric modeling is introduced, and the related data such as nodes and units in the model file are changed by programming to realize the update of the model. On this basis, the program calls the finite element calculation software NASTRAN to calculate the changed model file, and obtains and reads the calculation result of the new model to determine whether the acceleration value is reduced under the new layout. The MATLAB version used in this paper is 2017a, and the CPU is Intel(R) Core(TM) i5-6500 CPU@3.20GHz.

4.1. Objective function
The objective function can be selected in a variety of ways, such as acceleration or stress, and multiple variables also can be coupled as an optimized objective function. In this paper, the average acceleration value at the component corner is selected as the objective function of the optimization. Acceleration can be obtained directly by NASTRAN calculation and MATLAB reading of the calculation result file.

4.2. Restrictions
There are two restrictions as follows.

- The spacing of each bolt is not less than 20mm, which prevents the bolt from being too dense to cause damage to the circuit board.
- The installation of the bolt adopts the principle of center symmetry, that is, if the bolt mounting position of one side is determined, the installation position of the opposite bolt is also determined.

4.3. Decision variables and bounds
The optimization object in this paper is the installation position of the bolts, so the decision variables are the coordinates of the bolts. Under the premise of the restrictions in 4.2, only two bolts' position needs to be determined. So there are four decision variables, which are x1, y1, x2 and y2. Decision variables satisfy the inequalities as follows.

$$22 \le x \le 128 \tag{3}$$

$$22 \le y \le 78 \tag{4}$$

Where x and y are even.

4.4. Optimization results
There are 8 bolts on the board in the model. Except for the bolts on the four corners, there is a bolt on each side that needs to be optimized for layout. Due to the constraint of "central symmetry for the installation of bolts", the number of variables in the optimization process is reduced to two. The selected genetic algorithm optimization parameters are shown in Table 3.

Table 3. Optimization parameters of Genetic algorithm.

Name	Value
Cross rate	0.75
Mutation rate	0.15
Population size	20
Genetic algebra	300

In the current grid density, there are 1566 adjustable positions where the bolts on each side meet the constraints. If the search is traversed, the results of all the positions need be analyzed and calculated. The calculation time for each position is about 8 seconds, and the time of traversed search required is about 3.5 hours. By traversing the search, the optimal solution is found to be 14.39 m/s^2.
The optimization process of the objective function is shown in Figure 2. The abscissa is the number of iterations and the ordinate is the acceleration value. The line below represents the optimal solution found during the optimization process, and the upper line represents the average of the accelerations of

all individuals in the population under the current number of iterations. With the increase of genetic algebra, the average value of the population's acceleration decreases and then fluctuates up and down, and the optimal solution is reduced to a minimum of 14.39m/s^2. Due to the large variation rate and the large population size in the parameters selected in this paper, the upper line always fluctuates above the line below, and the two lines do not intersect. Larger variability and population size can make more inferior solutions in each generation of populations, reducing the possibility of falling into local optimum during the optimization process, and this is also the main reason for the upper line being far from the line below.

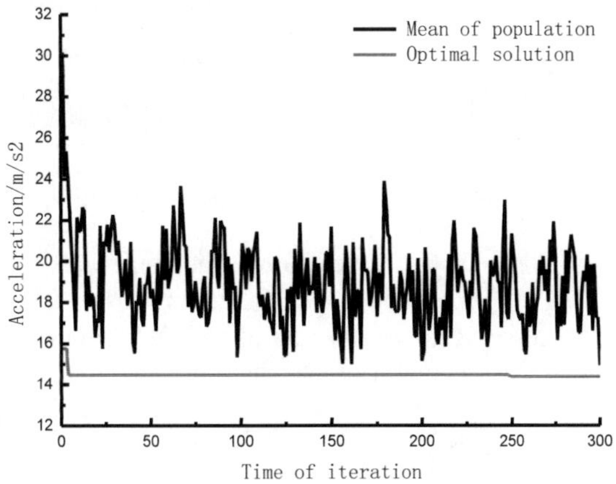

Figure 2. Convergence result of acceleration value optimization.

In order to verify the stability of the program and observe the effect of the initial solution on the optimization of the algorithm, repeat the operation 5 times, record the time and result of each operation, as shown in Table 4.

Table 4. Optimization results.

Num	Time(min)	Results(m/s^2)
1	84.9	14.39
2	68.7	14.39
3	82.4	14.39
4	74.8	14.39
5	81.5	14.39

The average running time of the program is 78.46 minutes, about 1.3 hours. The length of the run time is related to the initial solution, that is, to the initial position of the randomly selected component, and the direction of the search during the optimization process is random.

It can be seen that the average time required for the search optimization using the improved genetic algorithm is 78.46 minutes. Compared with the traversal search, the time is shortened to about one third, and the results of the five searches are global optimal solutions. It shows that using the genetic algorithm to optimize the bolt installation position can jump out of the local optimal solution and can be used to search the global optimal solution.

The schematic diagram of the bolt mounting position before and after optimization is shown in Figure 3 and Figure 4. The distances between the optimized position and other bolts are greater than 20mm, which meets the constraint requirements. Before the optimization, the bolts are equally spaced. After optimization, the bolts on the upper and lower sides are located near the third bisector, and the bolts on the left and right sides are located near the bisector. It can be seen from the figure that the position of the bolt on the two sides closer to the component of the optimized circuit board is closer to the corner position of the component.

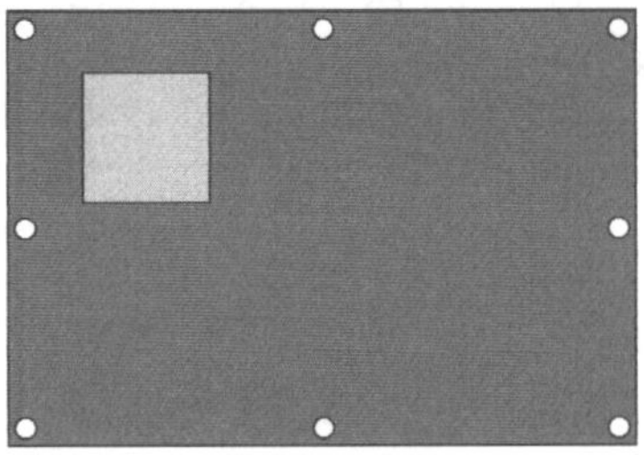

Figure 3. Position before optimization.

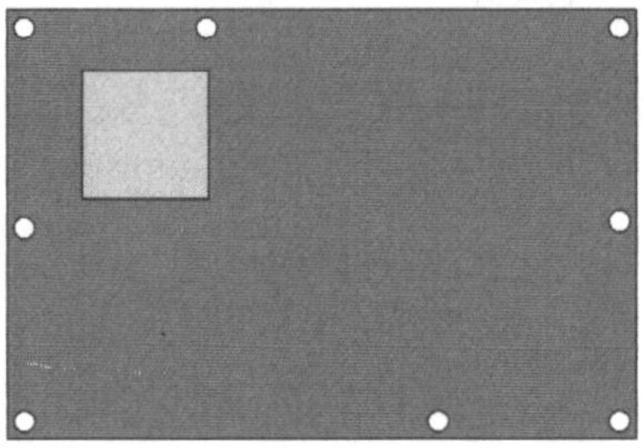

Figure 4. Position after optimization.

The frequency response curve of the average acceleration at the component corners before and after optimization is shown in Figure 5. The peak value of the frequency response curve of the optimized acceleration average is reduced compared to before optimization. Under the bolt layout before optimization, the average value of the acceleration at the component corner is $43.9m/s^2$. After optimization, the average acceleration is only $14.4m/s^2$, and the reduction ratio is 67.2%. After optimization, the peaks in the curve are left-biased, indicating that the modal frequencies of the various stages are reduced after the bolts are not equally spaced.

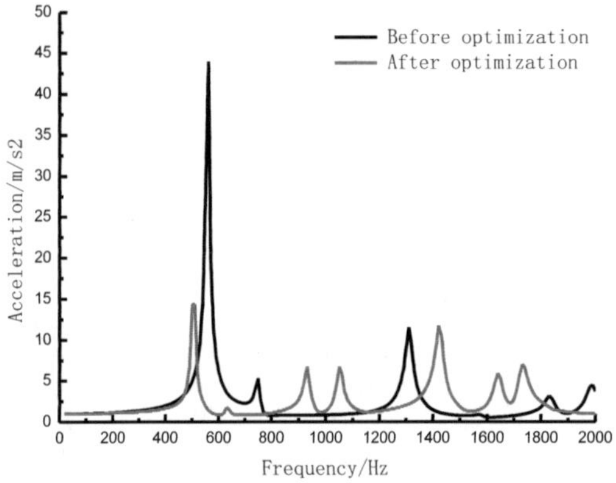

Figure 5. The mean acceleration response curve before and after optimization (Z direction).

The average stress at the component corners of the bolt layouts before and after optimization is calculated, as shown in Figure 6. The peak value of the frequency response curve of the optimized stress average is also reduced compared to before optimization. Under the bolt layout before optimization, the average stress at the component corners is 0.216 MPa. After optimization, the stress average is only 0.172 MPa, and the reduction ratio is 20.4%. The frequency at which each peak is reached is also reduced. The bolt layout position obtained by optimizing the acceleration value at corner also greatly reduces the stress value at corner.

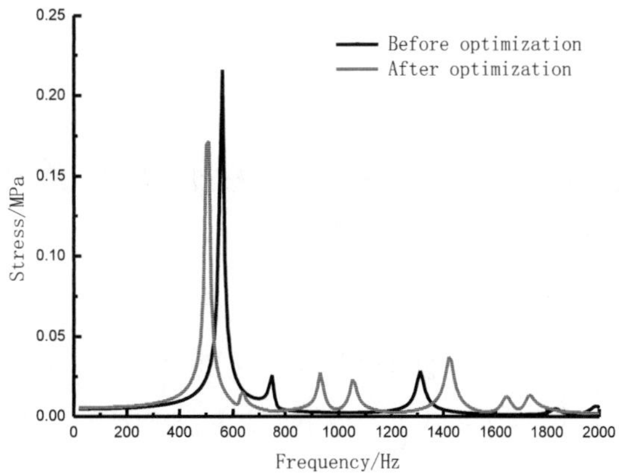

Figure 6. The mean stress response curve before and after optimization (Z direction).

5. Conclusion

Based on the finite element simulation analysis, this paper introduces the idea of parametric modeling, realizes the interaction between MATLAB and NASTRAN, and realizes the application of the improved genetic algorithm in bolt layout through programming. The results of finite element simulation show that optimizing the layout of the bolt can reduce the acceleration value at the component corner and achieve the reduction of the stress value at the component corner, which can improve the overall reliability. The bolt layout optimization method adopted in this paper is reasonable and feasible, and has practical value and promotion significance.

6. References

[1] Tang W C and Yi H 1997 *J. Machine Design* **12** 16-17
[2] Shu T, Zhang K and Ren H G 2010 *J. Projectiles, Rockets, Missiles and Guidance* **30** 178-180
[3] Xu D F 2016 Study on the optimization of clamping of the thin-walled cylinder based on particle swarm optimization and finite element method *Chongqing University*
[4] Luo Q P 2012 Study of genetic algorithm improvement and the application in its structural optimization *North University of China*
[5] Krishnakumar K and Melkote S N 2000 *J. Machine Tools & Manufacture* **40** 579-598
[6] Liu H L, Wang X, Chen Y, Kong D Q and Xia P J 2017 *J. Optics Communications* **390** 76-81
[7] Holland J H 1975 Adaptation in natural and artificial systems *University of Michigan*
[8] Shen B, Pan Z W, Lin H and Li W Z 2019 Component layout optimization design based on genetic algorithm *J. MSAM*(to be published)

Thermal-mechanical coupling analysis of functionally graded cylindrical shells

Xiayao Feng, Wenguang Liu* and Chao Liu

School of Aeronautical Manufacturing Engineering, Nanchang Hangkong University, 330063, Nanchang, China

Corresponding Author's E-mail: liuwg14@nchu.edu.cn

Abstract. Functionally graded materials is a new material for thermal protection in the aerospace field. It is extremely important to study the temperature and stress changes in the thermal environment to ensure the normal operation of the spacecraft and the safety of crews. In this paper, ABAQUS is used to perform sequential thermal coupling and full thermal coupling analysis. The temperature variation and mechanical properties of functionally graded cylindrical shells in thermal environment under temperature and force field and temperature field interacts with force field are studied. Based on the first law of thermoelasticity and thermodynamics, the three-dimensional stress and energy equations of functionally graded cylindrical shells in thermal environment are derived. The example shows that the temperature variation law and mechanical characteristics of the two types of thermo-mechanical coupling analysis are almost the same. The temperature value results of the sequential thermal-mechanical coupling analysis are the same as the results of the full thermal coupling, but the stress value results are lower than the results of the full thermal coupling.

1. Introduction

Functionally graded materials, as a new type of thermal protection material, were first proposed by Japanese scholar Niino and Hirai, etc., which usually consist of two or more materials. The functionally graded materials used in the aerospace field are mainly made of metal and ceramics. The metal has good strength to resist deformation or fracture. The ceramic has good heat resistance, heat insulation and high temperature oxidation resistance. The continuous change in composition between the two eliminates the macroscopic interface in the material. The overall material behaves good thermal stress relaxation characteristics, enabling it can be used under harsh environmental conditions such as large temperature difference, ultra-high temperature, and high-speed heat flow impact. It is expected to be used as the fuselage of the new-generation space shuttle, the inner wall of the combustion chamber, and the ultra-high temperature heat-resistant materials for turbine engines and high-efficiency gas turbines.

In recent years, with the continuous exploration of human beings in the field of extraterrestrial, experts have increasingly explored the application of functionally graded materials in the aerospace field. Domestic and foreign scholars have gradually deepened their research on thermal analysis. Loy and Lam studied the mechanical problems of FGMs and obtained the mechanical properties of metal-based ceramic FGMs[1]. Cho and Oden used the Crank-Nicolson-Galerkin scheme to establish a thermal stress analysis model for FGMs cylindrical shells[2]. Na and others considered the stress and critical temperature, and studied the material fraction by the power-distribution of the functional gradient material volume fraction optimization[3]. Huang and Rao used the classical Donnell shell theory to derive the buckling controlling equations and the analytical expressions of the buckling critical temperature of the FGMs elliptical cylindrical shells, and analyzed the effects of uniform, linear and nonlinear temperature rise environments on the physical properties of FGMs. And the variation of the critical temperature of buckling with the eccentricity of the section and the parameters of the material composition[4]. Xu and Zhang et al. prepared HATi axisymmetric biological FGMs by powder metallurgy method, and determined the mechanical properties and thermal expansion coefficient of HATi composite materials[5]. Lin and Bai developed the axisymmetric WC/Co heavy metal FGMs by microwave sintering method, and observed the microstructure and hardness of the material[6]. Hu and Zhang developed a numerical manifold research method for two-dimensional steady-state heat conduction problems of FGMs. The governing equations and boundary conditions are given, and the quadrature method of correlation matrix is discussed[7]. Lan and Fu proposed the Layered Exponential Precise Method (LPEM) for the steady-state heat conduction equation of FGMs. The effectiveness of the method is illustrated by an example[8]. Amini et al. considered the temperature field, studied the nonlinear free vibration and forced vibration of thick annular FGMs plates based on the first-order shear deformation plate theory and von Karman equation, and studied two kinds of constituent materials in the thermal environment. The effect of functional gradient plate material composition and thermal load on vibration characteristics and stress in a hot environment were investigated[9]. Noda and Jin used a finite inhomogeneous elastic solid containing cracks to study the problem of stable thermal stress, and calculated the thermal stress intensity factor of FGMs[10]. Zhao et al. used the asymptotic solution method to derive the transient temperature field and transient thermal stress field formula of the double-sided functionally graded coating infinite sandwich plate under convective boundary conditions[11]. Zhang et al. used the Laplace transform and the power series method to obtain the dynamic temperature field of the FGMs shell. The differential quadrature method was developed to obtain the transient thermal stress of the FGMs cylindrical shell under thermal shock. And when the volume of the component was changed, it can reduce the results of thermal stress[12]. Ozturk et al. studied the elastoplastic deformation of a functionally graded solid cylinder with fixed and internal heat sources according to Tresca's yield criterion and its associated flow law. The unknown interface radius values were determined by Mathematica 5.2 to analyze the expression of the stress, strain and radial displacement distribution. And the time history of temperature field and thermal stress are obtained by using the residual theorem and the fast Laplace inverse transform method (FLIT) respectively. The transient stress field and the thermoelastic stress wave propagation of the functionally graded thick hollow cylinder under any thermomechanical impact load are studied[13]. Nie and Batra analyzed the axially symmetric deformation of the isotropic FGMs rotating disk and physical properties change in the radial direction. The Airy stress function is used to analyze the hoop stress and radial stress in the constitutive relation of the material[14].

Studies have shown that domestic and foreign scholars have carried out a lot of research on the design optimization and strength of FGMs structure, they mainly focus on the research methods of heat conduction and thermal stress of FGMs, and most of the research objects are plate and shell structures. Rarely involved in finite element software to compare the results of different thermodynamic coupling analysis types. And many structures in the aerospace field, such as the space shuttle's tail nozzle, can be simplified into a cylindrical shell axisymmetric structure. In this paper, through the thermal analysis module of finite element software ABAQUS, sequential thermal coupling and full thermal coupling analysis of functionally graded cylindrical shells are carried out respectively. The law of temperature and stress variation of the functional gradient materials with different thermal couplings analysis type under the thermal environment temperature field and force field are revealed. It will provide a reference for the temperature and stress of the functionally graded material when it is used in actual working conditions.

2. Functional gradient cylindrical shell modeling

Figure 1 is the FGMs cylindrical shell model. The outer surface is pure ceramic ZrO_2, the inner surface is pure metal *Ti-6Al-4V*, and the material distribution is a power exponential function distribution in the thickness direction. In the middle surface, the composition of the two components is the same. Material properties such as modulus of elasticity, density, coefficient of thermal expansion, thermal conductivity, etc., all show continuous change in the thickness direction and are influenced by the external environment and volume fraction index.

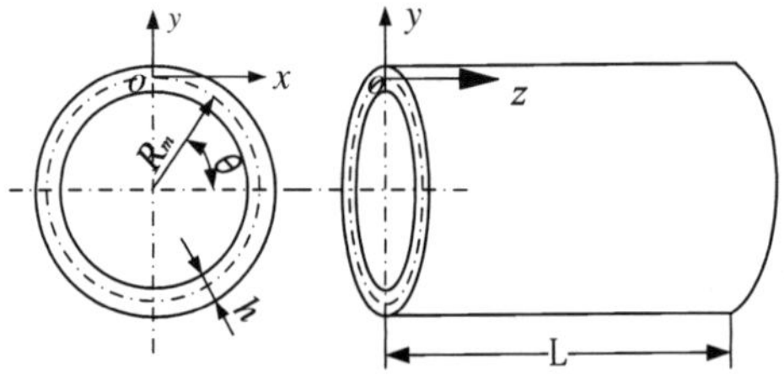

Figure 1. FGMs cylindrical shell model.

The coordinate system shown in figure 1 is established in the middle surface of the FGMs cylindrical shell. The length L = 0.8m, the middle surface radius R_m = 0.405m, and the shell thickness h = 0.01m. The material property parameter expression is:

$$\Gamma = (\Gamma_c - \Gamma_m)(z/h + 0.5)^N + \Gamma_m \tag{1}$$

Where Γ can be expressed as elastic modulus E, poisson's ratio ν, density ρ, thermal expansion coefficient α, thermal conductivity κ, etc., and Γ_c and Γ_m are material properties of ceramics and metals, respectively. N is the volume fraction index and takes a value of $0 < N < \infty$.

Since the material properties are affected by temperature, the specific expression of the material property parameter in the thermal environment is:

$$\Gamma_{(i)} = P_0(P_{-1}T^{-1} + 1 + P_1T + P_2T^2 + P_3T^3) \tag{2}$$

Where i is c or m.P is the temperature sensitivity coefficient of the material, using international units, as shown in table 1. T is the ambient temperature.

Table 1. Temperature sensitivity characteristic parameters.

	Material	P_{-1}	P_0	P_1	P_2	P_3
E (GPa)	ZrO_2	0	132.20	3.805e-4	-6.127e-8	0
	Ti-6Al-4V	0	122.70	-4.605e-4	0	0
ν	ZrO_2	0	0.3330	0	0	0
	Ti-6Al-4V	0	0.2888	1.108e-4	0	0
ρ (kg/m^3)	ZrO_2	0	3657	0	0	0
	Ti-6Al-4V	0	4420	0	0	0
α (1/K)	ZrO_2	0	13.300e-6	-1.421e-3	9.549e-7	0
	Ti-6Al-4V	0	7.4300e-6	7.483e-4	-3.621e-7	0
κ (W/mK)	ZrO_2	0	1.78	0	0	0
	Ti-6Al-4V	0	6.10	0	0	0

3. Basic principles of thermal coupling analysis

3.1. Thermal stress theory

Determining the temperature field, since the FGMs cylindrical shell is an axisymmetric model, the stress, strain and displacement components throughout the interior are symmetrical about the axis and are only related to the parameters R and z, and the shear stress is zero.

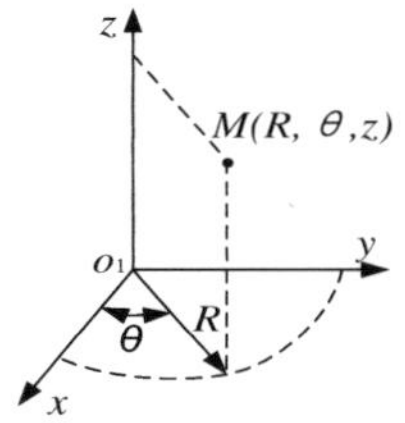

Figure 2. Cylindrical coordinate system.

The shear stress along the z-axis on the cylindrical surface and the shear stress in the *R* direction perpendicular to the z-axis are ignored. Assuming that the temperature *T* is only a function of *R*, a cylindrical coordinate system as shown in figure 2 is established, and *M* is a point of the cylindrical shell space. In this coordinate system, the stress-temperature difference relationship is obtained according to the generalized Hooke's law in thermoelasticity[15].

$$\begin{aligned} \sigma_R &= 2G[I_1\frac{\partial u}{\partial R}+I_2(\frac{u}{R}+\frac{\partial \omega}{\partial z})]-\beta T \\ \sigma_\theta &= 2G[I_1\frac{u}{R}+I_2(\frac{\partial \omega}{\partial z}+\frac{\partial u}{\partial R})]-\beta T \\ \sigma_z &= 2G[I_1\frac{\partial \omega}{\partial z}+I_2(\frac{\partial u}{\partial R}+\frac{u}{R})]-\beta T \end{aligned} \quad (3)$$

Where: *u*, *ω* are the radial displacement in the *R* direction and the axial displacement along the *z* axis, respectively. σ_R, σ_θ, and σ_z are radial normal stress, hoop normal stress, and axial normal stress, respectively. Among them:

$$I_1=\frac{1-v}{1-2v}, I_2=\frac{v}{1-2v}, G=\frac{E}{2(1+v)}, \beta=\frac{\alpha E}{1-2v} \quad (4)$$

3.2. Thermal stress calculation

Due to the temperature field and force field of the FGMs cylindrical shell in the thermal environment, the applied load and temperature are axisymmetric. If the inner diameter is R_i and the outer diameter is R_e, it is assumed that the temperature change *T* is only a function of *R*, and the temperature difference between the inner and outer surfaces is *ΔT*, and is independent of *z*. Then the equilibrium equation of the cylindrical shell is simplified to

$$\frac{\partial \sigma_R}{\partial R}+\frac{\partial R-\sigma\theta}{R}=0 \quad (5)$$

$$\frac{\partial \sigma_z}{\partial z}=0$$

According to the boundary conditions that have been set on the inner and outer surfaces, the equation (3) is substituted into equation (5), then the parameters of the temperature *T* are calculated to obtain the thermal stress expression:

$$\begin{aligned} \sigma R &= \frac{\alpha E \Delta T}{2(1-v)\ln k_1}\left[-\ln k_2 - k_3(1-k_2^2)\ln k_1\right] \\ \sigma\theta &= \frac{\alpha E \Delta T}{2(1-v)\ln k_1}\left[1-\ln k_2 - k_3(1+k_2^2)\ln k_1\right] \\ \sigma z &= \frac{\alpha E \Delta T}{2(1-v)\ln k_1}\left[1-2\ln k_2 - 2k_3\ln k_1\right] \end{aligned} \quad (6)$$

Among them:

$$k_1 = \frac{R_e}{R_i}, k_2 = \frac{R_e}{R}, k_3 = \frac{R_i^2}{R_e^2 - R_i^2} \tag{7}$$

3.3. Energy equation

When the FGMs cylindrical shell acts under external force and thermal load, the external force is converted into deformation energy and stored in the body. In addition, the internal particle motion generates kinetic energy. Due to the input and output of heat accompanying the process, according to the first law of thermodynamics, the deformation energy and heat caused by external load are equal to the sum of the kinetic energy and the internal energy increment of the FGMs cylindrical shell per unit time. Its expression is

$$dK + dU = \delta W + \delta Q \tag{8}$$

Where: dK, dU, δW, δQ are per unit volume, per unit time the kinetic energy increment, the internal energy increment, the work done by the surrounding materials, and the external heating amount.

According to the second theorem of thermodynamics, if the process is reversible, then

$$\delta Q = Tds \tag{9}$$

Where: T and s are the absolute temperature and the entropy per unit volume.

Assume that the volumetric forces of the FGMs cylindrical shell are X, Y, Z, the area forces are $\bar{X}$, $\bar{Y}$, $\bar{Z}$, and the displacements of the mass points are δu, δv, and δw. According to the Green's theorem, the integral operation is used to integrate the area into volume integral, the δW_1 which is external force do work on the cylindrical shell is determined by the boundary conditions:

$$\delta W_1 = \iiint_V \begin{bmatrix} \left(\frac{\partial \sigma_x}{\partial x} + \frac{\partial \tau_{xy}}{\partial y} + \frac{\partial \tau_{zx}}{\partial z} + X\right) du \\ + \left(\frac{\partial \tau_{xy}}{\partial x} + \frac{\partial \sigma y}{\partial y} + \frac{\partial \tau_{yz}}{\partial z} + Y\right) dv \\ + \left(\frac{\partial \tau_{xz}}{\partial x} + \frac{\partial \tau_{yz}}{\partial y} + \frac{\partial \sigma_z}{\partial z} + Z\right) dw \end{bmatrix} dV + \iiint_V \begin{matrix} (\sigma_x d\varepsilon_x + \sigma_y d\varepsilon_y + \sigma_z d\varepsilon_z \\ + \tau_{xy} d\gamma_{xy} + \tau_{xz} d\gamma_{xz} + \tau_{yz} d\gamma_{yz}) \end{matrix} dV \tag{10}$$

$$\begin{aligned} \frac{\partial \sigma_x}{\partial x} + \frac{\partial \tau_{yx}}{\partial y} + \frac{\partial \tau_{zx}}{\partial z} + X = 0 \\ \frac{\partial \sigma_y}{\partial y} + \frac{\partial \tau_{zy}}{\partial z} + \frac{\partial \tau_{xy}}{\partial x} + Y = 0 \\ \frac{\partial \sigma_z}{\partial z} + \frac{\partial \tau_{xz}}{\partial x} + \frac{\partial \tau_{yz}}{\partial y} + Z = 0 \end{aligned} \tag{11}$$

Where: σ_x, σ_y, σ_z, τ_{xy}, τ_{yz}, τ_{zx} are the six stress components of the cylindrical shell, respectively. ε_x, ε_y, ε_z, γ_{xy}, γ_{yz}, γ_{zx} are the six strain components of the cylindrical shell, respectively.

Then, according to the equilibrium differential equation (11), the simplified equation (12) is obtained.

$$\delta W_1 = \iiint_V (\sigma_x d\varepsilon_x + \sigma_y d\varepsilon_y + \sigma_z d\varepsilon_z + \tau_{xy} d\gamma_{xy} + \tau_{xz} d\gamma_{xz} + \tau_{yz} d\gamma_{yz}) dV \tag{12}$$

Then, according to $W_1 = \int_V W dV$, equation (13) is obtained.

$$\delta W = \sigma_x d\varepsilon_x + \sigma_y d\varepsilon_y + \sigma_z d\varepsilon_z + \tau_{xy} d\gamma_{xy} + \tau_{xz} d\gamma_{xz} + \tau_{yz} d\gamma_{yz} \tag{13}$$

For kinetic energy, assuming that the process is reversible, the force and temperature change slowly with time, which can be treated as a quasi-static problem, ignoring kinetic energy, therefore

$$dK = 0 \tag{14}$$

Substituting the equations (9), (13), and (14) into (8), so the energy equation of the FGMs cylindrical shell under thermal coupling is

$$dU = Tds + (\sigma_x d\varepsilon_x + \sigma_y d\varepsilon_y + \sigma_z d\varepsilon_z + \tau_{xy} d\gamma_{xy} + \tau_{xz} d\gamma_{xz} + \tau_{yz} d\gamma_{yz}) \quad (15)$$

4. Thermal coupled finite element analysis

The FGMs cylindrical shell was studied at room temperature of 300 K. The inner surface was subjected to a stable temperature field of 300 K, and the outer surface was subjected to a uniform load of 2000 MPa and a temperature field varying with time $T = 300 + t$, t was 600 s. The density ρ, the thermal conductivity κ, and the temperature-related parameters E, α and specific heat C are set in the ABAQUS material attribute unit, and the temperature gradient when calculating the parameters is 100 K. The Poisson's ratio v is 0.3. At room temperature of 300 K, the specific heat of ZrO_2 is 467 J/kg•K, and the specific heat of *Ti-6Al-4V* is 690 J/kg•K. It is assumed that the specific heat of the material changes linearly and uniformly in the thickness direction with the change of temperature gradient.

In the analysis step module, the set time length is 600 s, the maximum incremental step number is 100, and the incremental step size is 10, and other options are default. In the boundary condition module, constrain the displacement in the z direction and the rotation in the x and y directions. In the load module, set the amplitude table of temperature field. The input time is 0 K at time 0 s and 800 K at 600 s. Figure 3 shows the FGMs cylindrical shell finite element model with a grid number of 1540.

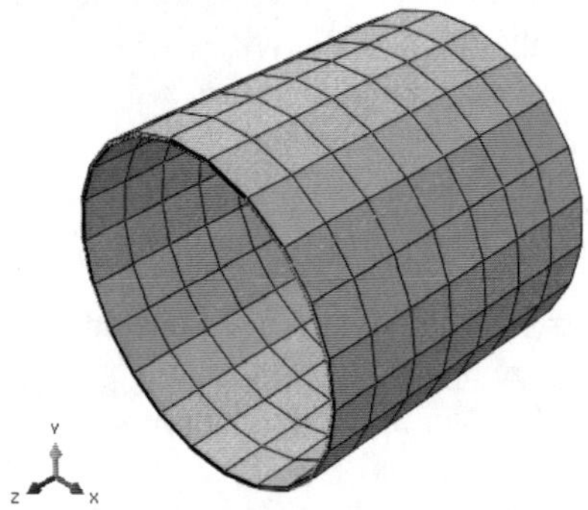

Figure 3. FGMs cylindrical shell finite element model.

4.1. Sequential thermal coupling

The stress-strain field in such an analysis depends on the temperature field, but the temperature field is unaffected by stress strain. In the finite element software, the heat conduction analysis is first performed, and then the obtained temperature field is used as a known condition, and thermal stress analysis is performed to obtain a stress strain field. The grid used to analyze heat transfer problems and the grid of thermal stress analysis can be different. The heat conduction grid unit type is DC3D8, and the thermal stress grid unit type is C3D8R.

4.2. Full thermal coupling

There is a strong interaction between the stress-strain field and the temperature field in this type of analysis, which needs to be solved simultaneously in the finite element software. The grid unit type is C3D8T.

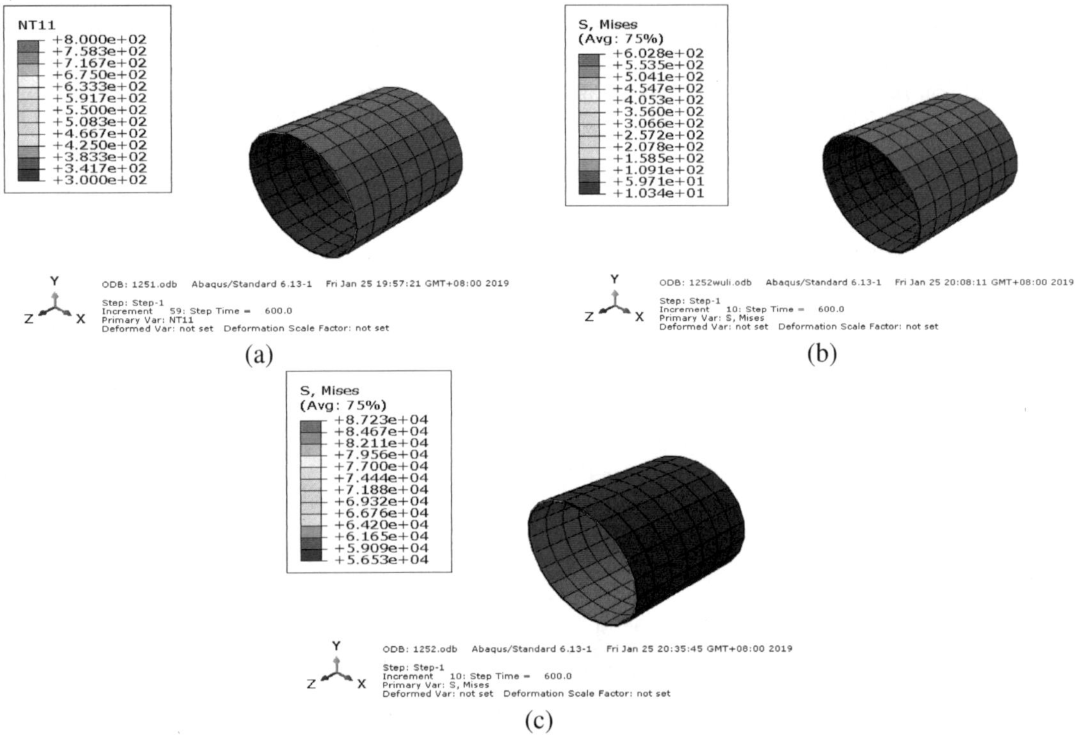

Figure 4. Temperature and stress cloud diagram of sequential thermal coupling (Figure a is the temperature cloud diagram. Figure b is the stress cloud diagram only under the temperature field. Figure c is the stress cloud diagram under the action of temperature field and force field.).

Figure 4 shows that the temperature of the FGMs cylindrical shell gradually increases from the inner surface to the outer surface under the action of the temperature field in the sequential thermal-mechanical coupling analysis. And owing to the temperature field tends to be stable over time, the inner and outer layer temperatures will also tend to stabilize. Eventually it tends to be stable. Only under the action of temperature field, the stress on the outer surface is larger than the stress on the inner surface, which also shows that the outer surface generates thermal stress due to the temperature field. Due to the external surface force loading, the stress after the sequential thermal coupling is higher, and the metal layer is stronger than the ceramic layer, which embodies the advantages of the functionally graded material.

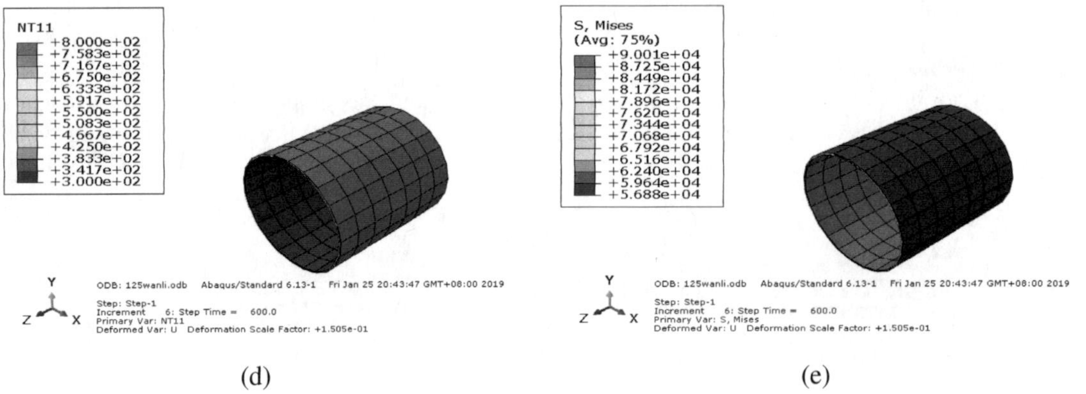

Figure 5. Temperature and stress clouds diagram of full thermal coupling (Figure d is the temperature cloud diagram. Figure e is the stress cloud diagram.).

Figure 5 shows that in the full thermal-mechanical coupling analysis, the temperature of the FGMs cylindrical shell gradually increases from the inner surface to the outer surface under the action of temperature field and force field. It is also shown that the temperature of the inner and outer layers eventually tends to be stable as the temperature field tends to stabilize over time. Under the action of the temperature field and the force field, the stress of the complete thermal coupling is also behaved that the metal layer is larger than the ceramic layer.

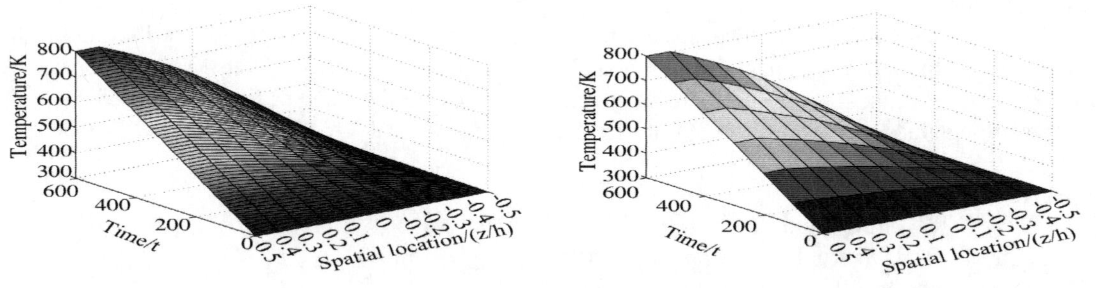

(f) (g)

Figure 6. Temperature map that changes with time and space (Figure f is the sequential thermal coupling. Figure g is the fullthermal coupling.).

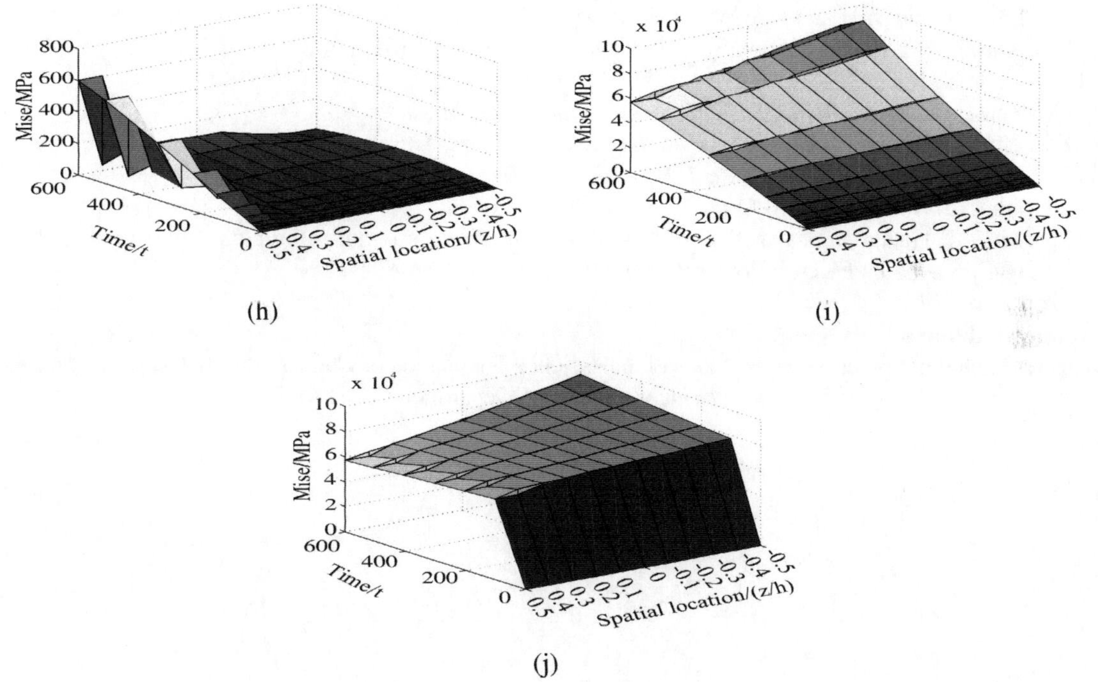

(h) (i)

(j)

Figure 7. Stress map that changes with time and space (Figure h is the stress map only under the temperature field of sequential coupling. Figure i is the stress map under the temperature field and force field of sequential coupling. Figure j is the stress map under the temperature field and force field of full thermal coupling.).

Figure 6 shows that the temperature variation laws of the two types of thermal coupling analysis are basically the same, and the temperature rises gradually from the inner surface to the outer surface with time increasing, and roughly linear increase in the same section. Figure 7 shows that when only under the temperature field of the sequential thermal-mechanical coupling analysis, the stress is the largest at the outer surface ceramic (z/h=0.5), and the internal stress is not much different. When temperature field and force field simultaneously, although the stress values of the two types of analysis are the largest at the inner surface metal, and the stress increases with time, but the rate of increase is different. Moreover, the stress value of full thermal coupling is larger than the sequential thermal coupling, it gradually decreases from the middle to the top at 102 s.

5. Conclusion
For the functionally graded cylindrical shell, the temperature change law and stress distributions of the two types of thermo-mechanical coupling analysis are approximately the same. The temperature values of the sequential thermo-mechanical coupling analysis are the same as the full thermal coupling, but the stress values are lower than the full thermal coupling.

6. References
[1] Loy C T, Lam K Y and Reddy 1999 *INT J. Mech. Sci.* **41**, 3
[2] Cho J R and Oden J T 2000 *Comput. Method. Appl. M.* **188**, 1
[3] Na K S and Kim J H 2009 *Compos. Struct.* **89**, 4
[4] Huang H W and Rao D H 2017 *J. South. China. Univ Techno: Nat. Sci. Ed.* **45**, 5
[5] Xu Q, Zhang X H, Han J C and He X D 2003 *Acta. Mater. Compos. Sin.* **423-425**, 5
[6] Lin W, Bai X D, Ling Y H and Jiang Z H 2003 *Mater. Sci. Forum.* **423-425**
[7] Hu G D, Zhang H H and Tan Y X 2017 *Chin. J. Appl. Mech.* **34**, 2
[8] Lan L H, Fu M H and Gao W L 2011 *Acta. Scientiarum. Nat. U. Sunyatseni:Nat. Sci.* **50** 4
[9] Amini M H, Rastgoo A and Soleimani M 2010 *J. Therm. Stresses.* **33**, 11
[10] Noda N, Jin Z H 1994 *Arch. Appl. Mech.* **64**, 2
[11] Zhao J, Li Y and Ai X 2008 *Thin. Solid. Films.* **516**, 21
[12] Zhang J H, Li G Z, Li S R and Ma Y B 2015 *J. Therm. Sresses.* **38**, 9
[13] Ozturk A, Gulgec M 2011 *Int. J. Eng. Sci.* **49**, 10
[14] Nie G J, Batra R C 2010 *Compos. Struct.* **92**, 3
[15] Li W T, Huang B H and Bi Z B 2004 *Theoretical Analysis and Application of Thermal Stress* (Beijing: China Electric Power Press) p 90

Acknowledgment
This research work is supported by the National Science Foundation of China (51565039) and the Science and Technology Project of Jiangxi Provincial Education Department (GJJ170584).

IOP Conf. Series: Materials Science and Engineering **576** (2019) 012007 doi:10.1088/1757-899X/576/1/012007

Uncertainty Measurement and Analysis for Quality and Reliability in Manufacturing Process

Altaf Ahmed[1,*], Jia Guozhu[2], Arfan Majeed[3] and Abdus Salam[4]

[1,2] Department of Management Science and Engineering, School of Economics and Management, Beihang University (BUAA), Beijing 100191, China
[3] School of Mechanical Engineering, Northwestern Polytechnical University (NPU), Xian 710072, China
[4] School of Aeronautical Science and Engineering, Beihang University (BUAA), Beijing 100191, China

E-mail: altafahmed@buaa.edu.cn

Abstract. The variation and uncertainty existed in every process and considered as an essential element in precise, accurate and high-quality manufacturing, especially in the Industry 4.0 era. Manufacturing quality and reliability is affected by variations and uncertainties in different process parameters and influencing factors. The variations and uncertainties are considered very critical in measurement and monitoring process, i.e., prediction or prognostics. These variations and subsequent uncertainties are needed to be identified, quantified, analyzed and controlled. This paper presents an approach to uncertainty measurement and analysis in manufacturing. The key characteristic parameter is identified which is critical for final product quality and reliability. To determine the uncertainty in the specific parameter, the uncertainty factors of both the manufacturing process and measurement process are determined. The combined uncertainty is determined by considering the influencing factors both in manufacturing and measurement process. The relation of each factor to the key parameter is analyzed and the sensitivity coefficient is determined mathematically or experimentally. The combined and expanded uncertainty is determined considering all related factors. The proposed approach has been applied to an additive manufacturing technique, and useful results are achieved.

1. Introduction

The industrial manufacturing is proliferating leading towards Industry 4.0. Internet of things, cyber-physical systems, intelligent manufacturing, cloud manufacturing, additive manufacturing, big data analytics are important element technologies in Industry 4.0 [1–3]. Industry 4.0 need fast, accurate, precise, reliable, flexible and holistic measurements using advanced measurement technologies and determine measurement uncertainty which describes the accuracy of measurement results quantitatively [4].

The development of new technologies has shifted the quality and reliability from corrective and preventive to predictive and prognostic approach. Therefore, many prognostic techniques are being deployed in the manufacturing sector [5]. The monitoring and measurement constitute a significant part of prognostics and uncertainty in measuring parameters are very critical for effective action [5,6].

The importance and need for uncertainty measurement is also recognized in aerospace measurement processes [7].

Additive manufacturing (AM) is a major key player element in Industry 4.0 which offers rapid, efficient, customization and intelligent solution for building complex products through automation and digitization [8,9]. It is rapidly adopting by aerospace, military, automobile, oil and gas, and medical sectors which demand high accuracy, precision, and reliability, along with desired repeatability and reproducibility [10]. Therefore, research in AM quality and reliability has a particular focus and an active research area [11].

Most uncertainty measurement and analysis research work focused on testing and calibration domain and it has not explored for other manufacturing processes. Therefore, there is a necessity to explore and develop some quantitative approaches for manufacturing.

In this research paper, an approach is proposed for dealing with uncertainty measurement and analysis for quality control and reliability in manufacturing. In the first step key quality and reliability, the characteristic is identified. Secondly, uncertainty factors or sources and uncertainties are determined for the manufacturing or production process and measurement or testing/ inspection process. Experimental data of manufacturing process parameters are obtained to develop a mathematical equation between each input and output parameters. Regression modelling and analysis is performed using MATLAB or MINITAB software. Similarly, data of measurement or testing process is obtained. The sensitivity coefficient for each input quantity and output is determined from the mathematical equation. The A-Type and B-Type uncertainties are estimated for the manufacturing process and measurement or testing process. Finally, the combined and expanded uncertainty is estimated. Based on the estimated uncertainties, the sensitivity analysis and control strategy is suggested. The proposed is applied to Selective Laser Melting (SLM) additive manufacturing technique.

2. Manufacturing process variations, uncertainties, quality and reliability

Manufacturing process variations and product quality and reliability have a close relationship. The reliability degradation in the final product is caused by quality variation in the manufacturing process [12]. The quality and reliability can be improved by targeting key characteristics and analysing critical process parameters in manufacturing [13]. These manufacturing process variations regarded as sources of errors and uncertainties in the system. During the manufacturing process, reliability can be assured by analysing and controlling the influencing factors or parameters, process variations and uncertainties [12,14].

Modern quality and reliability control systems are based on predictive and prognostic techniques. Accurate measurement and monitoring is an essential need for these predictive quality control and prognostic system. In manufacturing the key process, parameters are monitored and tracked for variation detection and subsequent action by predictive quality or prognostic system [5,6,15]. The uncertainty is an essential part of these measurements and monitoring.

Uncertainty exists in every measurement being performed during manufacturing. It is a critical factor which is significant as accuracy and precision. The realization and requirement for uncertainty measurement and analysis are emerging in manufacturing systems. Therefore, new approaches are required for quantification, measurement and analysis of manufacturing uncertainties.

3. Methodology

The methodology is based on a proposed approach which is explained below. The first general scheme of the proposed approach is defined. Then the quantitative treatment for uncertainty measurement and, sensitivity analysis and control for influence or contributing process parameters, sources, and uncertainties is explained.

3.1. General approach

The proposed approach is described in Figure 1. In general, the proposed approach has four steps. The first step includes identification of key characteristics or features which is critical for product or process quality and reliability. Secondly, determine contributing uncertainty factors and their uncertainties of both the manufacturing process and measurement process. Thirdly estimation of combined and expanded uncertainty, and finally perform sensitivity analysis for controlling and improving the uncertainties.

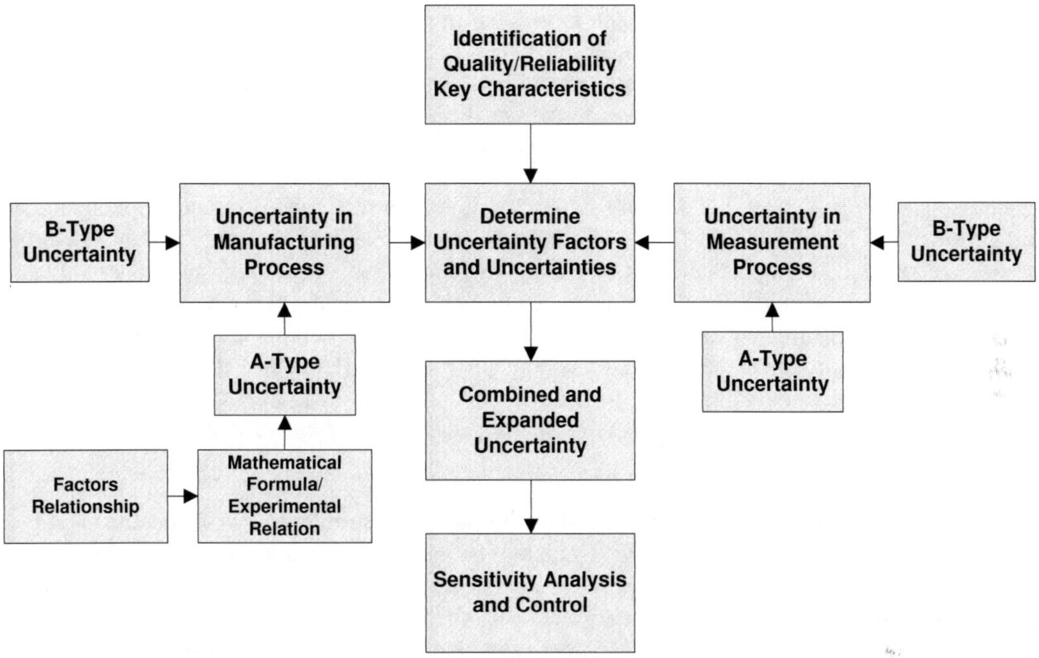

Figure 1. General approach

3.2. Identification of key characteristics

The product or process quality and reliability are associated with certain key characteristics. To improve quality and reliability, the key characteristics are needed to be identified, monitored and controlled during manufacturing, i.e., strength, stiffness, dimensional accuracy, surface finishing, etc. There are different originating sources of these key characteristics which include design and operational requirements, environmental conditions, field performance data, validation data, maintenance, quality, and reliability analysis, i.e., FMMEA, Failure Analysis, DOE, QFD, etc. In the first step, we need to identify and define the key characteristic(s) which is needed to be controlled during the manufacturing process.

3.3. Determination uncertainty factors and uncertainties

In this step, we need to determine sources of error and factors contributing to the uncertainty in the identified key characteristics. In general, different sources of errors and uncertainties originate from man, machines, materials, environment, procedures/ techniques, etc. We have divided the overall uncertainty into two parts, one includes uncertainties in the manufacturing process, and second are uncertainties in the measurement process. The numerical values of identified uncertainties are determined by formulas as defined in Guide for Uncertainty Measurement (GUM) and other reference guidelines developed for uncertainty measurement [16–20]. The uncertainties of the factors which are controlled, in a limit that they do not affect output, are not included in the uncertainty budget. The uncertainty can be overestimated and should not be underestimated.

3.3.1. Uncertainty in the manufacturing process: There are different manufacturing or production process parameters which affect the quality and reliability. The variation in these input process parameters contributes significantly to the uncertainty of the output quantity. We have two possibilities to estimate the uncertainties. In the first case, if we have a mathematical formula or relationship between process input and output quantities, then we can estimate the uncertainty in output due to variation or uncertainty in input quantities using this relation. In the second case if we do not have any formula or relationship between inputs and output, then we have to do experimentation in a systematic way, such as the design of experiment (DOE). We take different input setting values and determine output response against change in inputs. From data collected we develop a mathematical model or equation which gives the relationship between each input and output by using MATLAB or MINITAB software. The regression and correlation analysis can be done for correlated input quantities. Using developed equations, we determine the sensitivity coefficients, which are used to transform the numerical value of each input quantity into the output.

3.3.2. Uncertainty in the measurement process: The measurement process is an integrated part of manufacturing which exits in the form of testing, inspection, and calibration. All the measurable quantities are realized, verified, validated and qualified through this process. This process is used to compare the process or product output against standards or some references. The uncertainties originate from the measurement process, testing, inspection or calibration, itself. The major sources, but not limited these, include measuring or testing instruments, reference standards and materials, calibrations, traceability, environmental conditions, methods and procedures, sampling, instrument properties, i.e., resolutions, drift, stability, etc., maintenance and personnel performing measurement tasks [16–20].

3.3.3. A-Type and B-Type uncertainties: The uncertainties are determined as A-Type and B-Type. A-Types uncertainties are those which are estimated from data set values of input quantities and estimated using equation (1). The B-Type is estimated for those quantities which do not have data values. For such quantities, we estimate their standard uncertainty from a single available value. For each uncertainty component, we need its numerical value of uncertainty with the unit, distribution (normal, rectangular, triangular or U-shape) with respective divisor ($1, \sqrt{3}, \sqrt{6}$ or $\sqrt{2}$) and sensitivity coefficient. Examples of A-Type include repeatability or reproducibility performed by using a set of data values. The B-Type includes instrument resolution (taken as half value of resolution), calibration, drift, standard reference certificate value, etc. The numerical treatment and methodology for A-Type and B-Type uncertainties are addressed in detail in different uncertainty measurement standards and guides [16–20].

$$u(X) = \frac{\sigma}{\sqrt{n}} = \sqrt{\frac{\sum_{i=1}^{n}(X_i - \bar{X})^2}{n(n-1)}} \quad (1)$$

Where *u(X)* is the A-Type uncertainty, *Xi* is an ith measured value, *X-bar* is the mean value, σ is the standard deviation and *n* is the number of measurements or values taken for quantity *X*.

3.4. Combined and expanded uncertainty

The combined uncertainty $u_c(y)$ included estimated A-Type and B-Types uncertainties which are resulted from both manufacturing and measurement process. It is calculated by using equation (2) or (3) for uncorrelated and correlated input quantities respectively [15,16]. Equation (4) is useful when the uncertainties are required to be measured in fractions or percentage.

$$u_c(Y) = \sqrt{\sum_{i=1}^{n}\left(\frac{\partial Y}{\partial X_i}\right)^2 u^2(X_i)} \quad (2)$$

Where $u_c(Y)$ is the combined uncertainty in output quantity (Y), $\partial Y/\partial Xi = C_{xi}$ is the sensitivity coefficient of ith quantity, and *u(Xi)* is the standard uncertainty in i^{th} quantity.

$$u_c(Y) = \sqrt{\sum_{i=1}^{n}\left(\frac{\partial Y}{\partial X_i}\right)^2 u^2(X_i) + 2\sum_{i=1}^{n-1}\sum_{j=i+1}^{n}\frac{\partial Y}{\partial X_i}\frac{\partial Y}{\partial X_j}u(X_i, X_j)} \quad (3)$$

$$\frac{u_c(Y)}{Y} = \sqrt{\sum_{i=1}^{n}\frac{u^2(X_i)}{X_i}} \quad (4)$$

The estimated uncertainties have a 68% confidence level. The final expanded uncertainty U is determined, at 95% confidence level (coverage factor k=2), by using equation (5) [16,20].

$$U = k\, u_c(Y) \quad (5)$$

3.5. Sensitivity analysis and control

The sensitivity analysis of estimated uncertainties will provide a way to study the impact and degrees of variation in influencing inputs parameters and the resulting variation in the uncertainty of the output quantity. It helps in targeting, controlling and improving the different uncertainty contributors or sources. We can reduce the uncertainty by either reducing or eliminating the sources having a higher contribution in overall uncertainty. A control strategy can be developed to control and improve target uncertainties.

4. Case application

The additive manufacturing or 3D printing process case is taken to demonstrate the proposed approach. For this purpose, data of Selective Laser Melting (SLM) is taken for study. In this technique, the object is built by using a laser beam which targets powder material and solidifies it layer by layer by using an automated programme. This technique is being used to manufacture different complex parts and rapidly expanding in different sectors including in the aerospace sector.

There are different process parameters which affect the product quality and reliability. The key quality or reliability characteristics of the product, e.g., surface finishing, party density, hardness, strength, etc., depends on the properties of the material used and input process parameters settings. In order to get the best results, these input parameters settings are optimized.

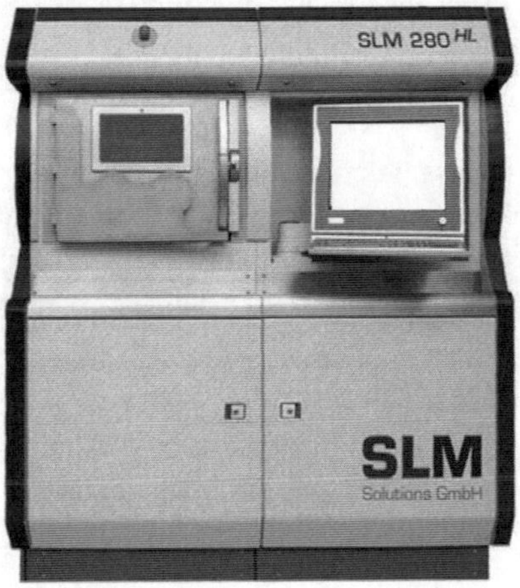

Figure 2. Selective laser melting system SLM 280 HL

In this study results of 27×3 samples of 8×6×10 mm size are analyzed which were fabricated on SLM 280 HL system, as shown in Figure 2, using gas-atomized AlSi10Mg powder material provided by SLM solutions. The surface roughness of the samples measured using Mitutoyo surface roughness tester Surftest SJ-210.

4.1. Identification of key characteristics and critical parameters
The manufactured parts have key characteristics which are linked with the quality and reliability of the final product. In this case, we have taken front surface roughness (Ra) as a key characteristic or critical output parameter. The input manufacturing process parameters laser power, scan rate, hatch distance, beam angle and layer thickness are considered as critical input parameters. The surface roughness measurement or testing process parameters, i.e., repeatability of roughness tester are taken as critical.

4.2. Determine uncertainty factors and uncertainties
The uncertainty factors and uncertainties based on critical parameters are determined both in manufacturing process and measurement or testing process. The known or available factors and uncertainties are included in the uncertainty budget.

4.2.1 Uncertainty factors and uncertainties in the manufacturing process: There are different input parameters in the SLM manufacturing process which includes laser power, scan rate, hatch distance, beam angle, and layer thickness. The relationship between these input factors and output parameter, the surface roughness is not available. Therefore, we need to determine the relationship between these inputs and output parameters. We have taken laser power, scan speed and hatch distance as three uncertainty factors and beam angle and layer thickness are fixed. First, considering independent input quantities, we perform experimentation and develop a linear equation as defined in equation (6). Then the sensitivity coefficient for each input is determined using relation (6).

$$Y = a + C_{xi}\, X_i \tag{6}$$

Where Y is the output, "a" is the y-intercept of the line, C_{xi} is the slop, which is called the sensitivity coefficient, corresponding to the ith parameter and X_i is the ith input.
The sensitivity coefficient will be determined with the following relation by using equation (6);

$$C_{xi} = \frac{\partial Y}{\partial X_i} \tag{7}$$

The layer thickness fixed as 30μm, vertical orientation or building direction and scanning strategy at 67°. There were twenty-seven settings of input parameters and three samples were developed at each setting. The laser power, scan speed and hatch distance (inputs) were varied in range as mentioned in Table 1, and average surface roughness (output) of the samples was determined at each setting.

Table 1. Manufacturing process parameters value selection for determination of the relationship.

Inputs (X_i)			Output (Y_i)
Laser Power (kW)	Scan Speed (m/s)	Hatch Distance (μm)	Surface Roughness (μm)
LP	*SS*	*HD*	R_a
0.32 ~ 0.40	0.60 ~ 0.90	81 ~ 116	3 ~ 10

If the surface roughness Ra is output, then the corresponding equations with each input will be in the following form;

$$R_a = a + C_{LP}\, LP \tag{8}$$

$$R_a = a + C_{SS}\, SS \tag{9}$$

$$R_a = a + C_{HD}\, HD \tag{10}$$

The mathematical equations are developed and sensitivity coefficients are determined, as defined in equations (8) – (10), using MATLAB or MINITAB software. The equations and estimated coefficients are mentioned in Table 2.

Table 2. Mathematical relationships and sensitivity coefficients.

Inputs	Equation	Sensitivity Coefficients C_{xi}		
Laser Power	R_a= -8.088+37.36 *LP*	C_{LP}	37.36	µm / kW
Scan Speed	R_a= 9.887-6.031 *SS*	C_{SS}	6.031	µm / m/s
Hatch Distance	R_a= -6.497+0.1285 *HD*	C_{HD}	0.1285	µm / µm

The optimum value of surface roughness is observed at laser power value 0.32 kW. In the samples developed at laser power 0.32kW setting, the average surface roughness is R_a=4.23 µm. The A-Type standard uncertainty in surface roughness is estimated as 0.48 µm (11.39 %) from repeatability which is due to variation in surface roughness in these samples. For A-Type uncertainty, the formula defined as equation (1) is used. The variation in surface roughness at the same parameters is due to variations in the manufacturing process that may be caused by some other random or assignable factors. The B-Type uncertainties are due to system resolution of laser power, scan speed, and hatch distance. The value of these uncertainties is taken as half of the resolution.

Table 3 shows the uncertainty factors and uncertainties in the SLM manufacturing process. For each factor, the numerical value of uncertainty with its type, distribution, and the divisor is identified.

Table 3. Uncertainty factors and uncertainties in SLM manufacturing process

Component of Uncertainty	Unit	Type	Uncertainty U_{xi}	Distribution, Divisor
Repeatability-Variation in Surface Roughness	µm	A	0.48	Normal, 1
Resolution of Laser Power	kW	B	5×10^{-7}	Rectangular, $\sqrt{3}$
Resolution of Scan Speed	m/s	B	5×10^{-7}	Rectangular, $\sqrt{3}$
Resolution of Hatch Distance	µm	B	0.05	Rectangular, $\sqrt{3}$

4.2.2. Uncertainty factors and uncertainties in the measurement process: The surface roughness of the samples is measured by using surface roughness tester of Mitutoyo Surftest SJ-210. The A-Type uncertainty in surface roughness tester is determined by repeatability by using precision reference specimen provided by Mitutoyo (Precision Reference Specimen No. 178-602). The value of precision reference specimen is the value of 2.91 µm or 115 µin. The resolution of surface roughness tester is taken as B-Type uncertainty factor. Half value of resolution is taken. The both A and B Type uncertainties in measurement/ testing process are mentioned in Table 4.

Table 4. Uncertainty factors and uncertainties in measurement process

Component of Uncertainty	Unit	Type	Uncertainty U_{xi}	Distribution, Divisor
Repeatability of Surface Roughness Tester using Precision Reference Specimen	µm	A	0.031	Normal, 1
Resolution of roughness meter	µm	B	0.0005	Rectangular, $\sqrt{3}$

4.3. Combined and expanded uncertainty

The combined uncertainty by considering both manufacturing process uncertainties and measurement process uncertainties is determined by using equation (3). The results are mentioned in Table 5. The standard uncertainty of each factor is determined by multiplying its uncertainty value with its sensitivity coefficient.

Table 5. Combined uncertainty estimations

Component of Uncertainty	Type	Uncertainty U_{xi}	Distribution	Divisor d	Sensitivity Coefficient C_{xi}	Standard Uncertainty $u_{xi} = (U_{xi}/d)\times C_{xi}$	Unit
Manufacturing Process							
Repeatability -Variation in Surface Roughness	A	0.48	Normal	1	1	0.48	µm
Resolution of Laser Power	B	5 x 10^{-7}	Rectangular	$\sqrt{3}$	37.36	0.0000032	µm
Resolution of Scan Speed	B	5 x 10^{-7}	Rectangular	$\sqrt{3}$	6.031	0.0000005	µm
Resolution of Hatch Distance	B	0.05	Rectangular	$\sqrt{3}$	0.1285	0.0111284	µm
Measurement Process							
Repeatability of Surface Roughness Tester using Precision Reference Specimen	A	0.031	Normal	1	1	0.0310	µm
Resolution of roughness meter	B	0.0005	Rectangular	$\sqrt{3}$	1	0.000866	µm
			Combined Standard Uncertainty u_c =			± 0.523	µm
			Coverage Factor k =			2	
			Expanded Uncertainty $U = (u_c \times k)$			± 1.05	µm

The estimated combined standard uncertainty in surface roughness is ± 0.523 µm which is at 68% confidence level. The estimated expanded uncertainty in surface roughness is ± 1.05 µm which is at 95% confidence level (k=2).

4.4. Sensitivity analysis and control

The sensitivity coefficient of laser power has the highest value of 37.36 µm/kW as compared to all three input parameters. It means a change of 1 kW in power can change surface roughness value to 37.36 µm. Similarly, one unit change in scan speed of 1 m/s and hatch distance 1 µm can vary the surface roughness value to 6.031 µm and 0.1285 µm respectively.

The estimated combined uncertainty provided overall uncertainty which has a cumulative effect of both manufacturing process uncertainties and measurement or testing process uncertainties. The manufacturing process uncertainties (93.91%) have a higher contribution as compared to measurement process uncertainties (6.09%) in combined uncertainty.

Figure 3 shows the percentage contribution of each factor in overall uncertainty. The variation in surface roughness – repeatability due to manufacturing process variation has the highest contribution of 91.78% in overall uncertainty value. These process variations are due to some random or systematic errors and assignable causes of other factors, i.e., material, laser beam properties, gas properties, etc. The second major contributor is the repeatability of surface roughness tester using a precision reference specimen which contributes 5.93% in overall uncertainty. The third is the resolution of hatch distance.

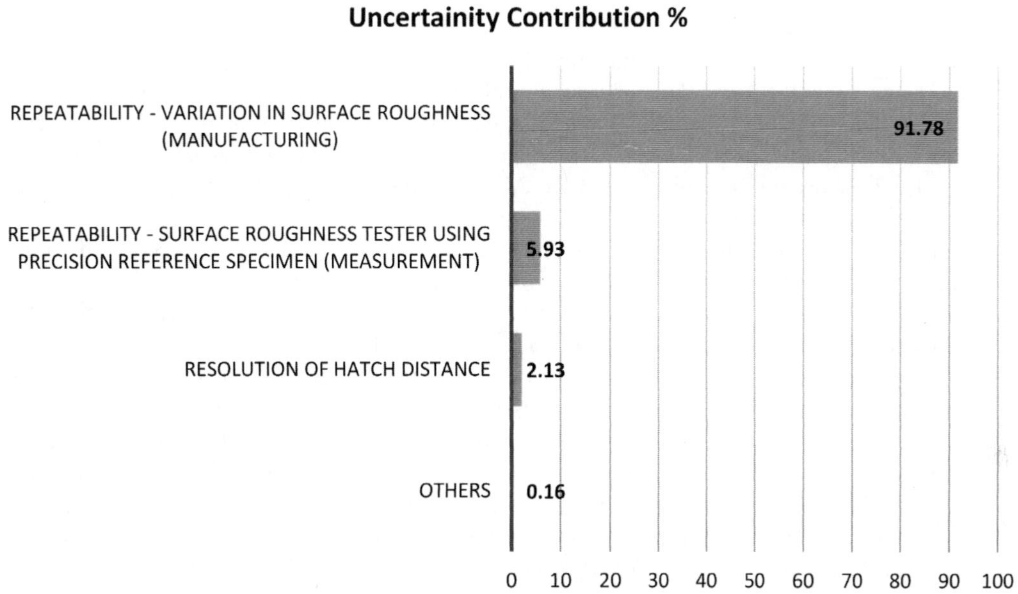

Figure 3. Contribution (%) of sources to the overall uncertainty

To achieve higher accuracy and minimize the value of uncertainty, we need to control and improve the significant uncertainty contributors. The SLM manufacturing process further needs to be analyzed, and assignable causes can be eliminated to improve the uncertainty in the surface roughness. Utilizing a higher resolution of scan speed can also reduce uncertainty.

5. Conclusions

The quality and reliability are one of the essential concerns of modern manufacturing. The key characteristics of quality and reliability are linked with different manufacturing process parameters. The estimation and analysis of uncertainty in the output value of key characteristic value and input process parameters need a quantitative approach.

The proposed approach for uncertainty measurement and analysis has found to be very practical and useful in implementation. It provides an insight view of critical parameters or factors and their uncertainties affecting product or process quality and reliability.

The uncertainties are estimated by considering both manufacturing process parameters and measurement process parameters. The uncertainty contribution from the manufacturing process is much higher as compared to the measurement process. Higher accuracy can be achieved by minimizing and targeting the primary uncertainty contribution sources.

6. References

[1] Xu L Da, Xu E L and Li L 2018 Industry 4.0: state of the art and future trends *Int. J. Prod. Res.* **56** 1–22
[2] Zhong R Y, Xu X, Klotz E and Newman S T 2017 Intelligent Manufacturing in the Context of Industry 4 . 0 : A Review *Engineering* **3** 616–30
[3] Arfan Majeed, Jingxiang Lv T P 2018 A framework for big data driven process analysis and

optimization for additive manufacturing *Rapid Prototyp. J.* **25** 308–21
[4] Imkamp D, Berthold J, Heizmann M, Kniel K, Manske E, Peterek M, Schmitt R, Seidler J, Sommer K and Vde-gesellschaft V D I 2016 Challenges and trends in manufacturing measurement technology – the " Industrie 4 . 0 " concept *J. Sensors Sens. Syst.* **5** 325–35
[5] Vogl G W, Weiss B A and Helu M 2019 A review of diagnostic and prognostic capabilities and best practices for manufacturing *J. Intell. Manuf.* **30** 79–95
[6] Atamuradov V, Medjaher K, Dersin P, Lamoureux B and Zerhouni N 2017 Prognostics and Health Management for Maintenance Practitioners-Review , Implementation and Tools Evaluation *Int. J. Progn. Heal. Manag.* **8** 060
[7] Abollado J R, Shehab E, Rose M and Schröter T 2017 Uncertainty Assessment for Measurement Processes in the Aerospace Manufacturing Industry *Procedia CIRP* **60** 326–31
[8] José Horst D, Adriano Duvoisin C and De Almeida Vieira R 2018 Additive Manufacturing at Industry 4.0: a Review *Int. J. Eng. Tech. Res.* **8** 3
[9] Dilberoglu U M, Gharehpapagh B, Yaman U and Dolen M 2017 The Role of Additive Manufacturing in the Era of Industry 4.0 *Procedia Manuf.* **11** 545–54
[10] Hoejin Kim, Yirong Lin T-L B T 2018 A Review on Quality Assurance in Additive Manufacturing *Rapid Prototyp. J.* **24** 645–69
[11] Berumen S, Bechmann F, Lindner S, Kruth J P and Craeghs T 2010 Quality control of laser- and powder bed-based Additive Manufacturing (AM) technologies *Phys. Procedia* **5** 617–22
[12] Yao J, Xiong B, Zhou Y and He Y 2016 Research on relationship between manufacturing process quality variation and product reliability *Proceedings of 2016 Prognostics and System Health Management Conference, PHM-Chengdu 2016* pp 1–6
[13] He Y and Chang W 2009 Systematical Design Quality Control model based on Key Quality Characteristics *2009 16th International Conference on Industrial Engineering and Engineering Management* pp 1208–12
[14] Wang W, Ren Y and Lv J 2016 Research on the assurance of reliability during the manufacturing process of mechanical products *Proceedings of 2016 Prognostics and System Health Management Conference, PHM-Chengdu 2016* pp 1–5
[15] Gu H, Zhang S and Ma L 2012 Process Analysis for Performance Evaluation of Prognostics Methods Orienting to Engineering Application *2012 International Conference on Quality, Reliability, Risk, Maintenance, and Safety Engineering* pp 681–6
[16] ISO 2017 *ISO/IEC 17025:2017 General requirements for the competence of testing and calibration laboratories*
[17] ISO 2012 ISO/IEC 17020:2012 Conformity assessment -- Requirements for the operation of various types of bodies performing inspection
[18] BIPM, IEC I and ILAC, ISO, IUPAC I and O 2008 *JCGM 200:2008 International vocabulary of metrology — Basic and general concepts and associated terms (VIM)*
[19] BIPM, IEC, IFCC, ISO, IUPAC, IUPAP O 2008 *JCGM 100:2008 Evaluation of measurement data — Guide to the expression of uncertainty in measurement*
[20] UKAS 2012 *M3003 The Expression of Uncertainty and Confidence in Measurement 3rd Ed.*

Vibration Characteristics Analysis and Optimization of Trawler Propulsion Shafting System

T R Sial[1], **Y Jin**[1,*] **and Z Juan**[1]

[1] School of Mechanical and Power Engineering, Guangdong Ocean University, Zhanjiang 524088, Guangdong, China.

Abstract. The propulsion shafting system enables power delivery from the main engine to the propeller and plays a vital role in promoting the motion of the ship. The increase in demand for trawlers in coastal areas has led to the increase in the need for an improvement in its performance. Therefore, while trawling at sea, vibration characteristics are essential in analyzing for the safe and efficient functioning of the propulsion shafting of trawler ship. The primary objective of this paper is to model, analyze and reduce the vibration characteristics of trawler propulsion shafting. A computational model has been modeled and analyzed by Modal Analysis using Finite Element Method, focusing on the Lateral Vibrations. The natural frequencies, mode shapes, and the maximum deformation with affected zones of trawler propulsion shafting are evaluated. The attempt has been made to reduce the vibration characteristics of trawler propulsion shafting by the optimization process considering the design parameters whereby, it comes in the acceptable range for normal operation. The results of analysis and optimization have been computed and discussed.

1. Introduction

Propulsion shafting system playing an important role is considered to be the major component to transfer the power from the main engine to propeller blades to drive the ship [1], [2]. It suffers all kinds of exciting forces during operation. Each point in the propulsion shafting generates a response in all the directions of disturbing force. The vibration aspects include torsional, longitudinal, lateral and the mutual coupling vibrations which affect adversely on the performance of propulsion shafting [3], [4]. It alters its dynamic as well as static response parameters [5]. Lateral vibration has a significant impact on the propulsion shafting which becomes the source of bending and whirling of the shaft causing an imbalance of the operational rotating part and the periodic motion of varying bending deformation due to the non-uniform flow where a propeller is operating in [6].

Researchers have paid much attention to the vibration analysis of propulsion shafting. Vibrations can be initiated by a propeller which is considered by Wang *et al.* [7] that can produce periodically varying exciting forces and can be transferred to the hull by the propulsion shafting system and fluids, significantly increasing the underwater noise and vibration of the hull. Jalali *et al.* [8] presented the model identification and ship propulsion shaft lines. The Finite element modeling and experimental modal parameters by updating propulsion shaft lines are considered in a ship structure such as natural frequencies and mode shapes. The corrected Finite Element models were used for the prediction of the dynamic response of the shaft lines. Due to the complexity of the system, the calculation of free and forced vibrations on the whole ship is a challenging issue by which vibration characteristics are governed at the local and global level. Yucel *et al.* [9] presented the free and forced vibration analyses of ship structures using the Finite Element method. Their research revealed that the undesired effects of the ship might result in the failure of local structural members or the malfunctioning of machinery and its component. Thus, excessive ship vibration has to be avoided.

The operating condition of propulsion shafting is different with respect to the kind of the ship. Generally, the focus of researchers is the commercial ships having larger sizes. By considering the study of large size ships, Tian *et al.* [10] investigated the dynamic interactions between the propulsion shafting system and ship hull in large scale ships. A simplified hull-propulsion system mathematical model was presented to investigate the dynamic interactions between the ship propulsion and hull deformations. As the trawl nets dragged behind the Trawler ship, it causes instability and oscillations in the propulsion shafting system which leads to the cause of vibration. Therefore, in enhancing the growing need of Trawler ship performance, the vibration characteristics need to be looked at.

Vibration analyses using Finite Element Method (FEM) have become a standard practice during various phases of structure design. As a result, subsequent potential issues are being analyzed and remedied before construction begins. In FEM, the selection of material, element type, mesh size, connections and boundary conditions should be well defined which are the influencing factors in acquiring accurate results [11]. It is essential to know the natural frequencies at which the excitation is likely to occur, mode shapes, and maximum deformation to figure out the dangerous and affected zones to prevent the failure of the structure. Therefore, the current study is aimed at adding to knowledge the modeling, analysis, and vibration reduction through engineering optimization, for the safe and efficient functioning of the propulsion shafting of Trawler ship.

2. CAD Modelling

2.1. Material Properties

The material used to model the Trawler propulsion shafting is Structural Steel. The material properties are listed in Table 1.

Table 1. Material Properties of Structural Steel

Sr. No.	Property (Symbol)	Value (Unit)
1.	Density (ρ)	7850 kg / m^3
2.	Young's Modulus (E)	2e+11 Pa
3.	Poison Ratio (v)	0.3

2.2. Model with Dimensions

Propulsion shafting of 35.5m Trawler ship is modeled and assembled in SolidWorks. The CAD model depicted in Figure 1, was built according to exact dimensions of shaft and propeller shown in Table 2 and 3 respectively.

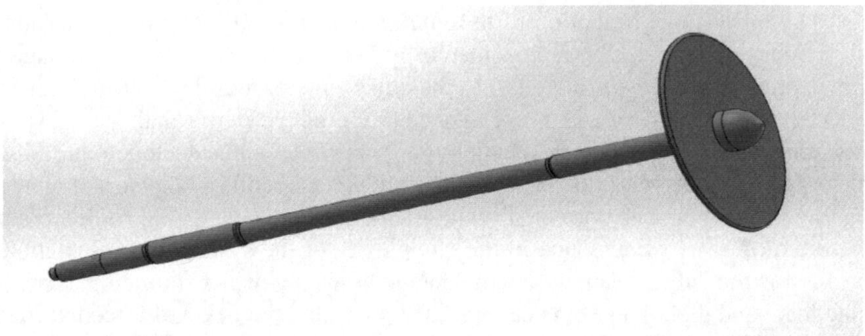

Figure 1. CAD model of shaft and propeller assembly

Table 2. Dimensions of Shaft

Sr. No.	Parameter (Symbol)	Value (Unit)
1.	Length (L_s)	4310 mm
2.	Mean Diameter (D_s)	148 mm

Table 3. Dimensions of Propeller

Sr. No.	Parameter (Symbol)	Value (Unit)
1.	Number of blades (Z)	4
2.	Diameter of propeller (D)	1450 mm
3.	Length of propeller hub (L_k)	315 mm
4.	Mean Thickness of blade ($t_{0.5R}=t_p$)	34.8 mm
5.	Diameter of propeller hub (d)	261 mm
6	Maximum width of blade (b_{max})	327.70 mm

3. Vibration Analysis of Trawler Propulsion Shafting

Trawler propulsion shafting was analyzed in ANSYS Workbench v16.0, and the analysis process is explained briefly.

3.1. Boundary Conditions

The boundary conditions are essential in defining the working condition of shafting operation. Here, one end of the shaft is attached with propeller and the other with the engine. The two sliding bearings are connected at the specific positions. The model with boundary conditions is depicted in Figure 2.

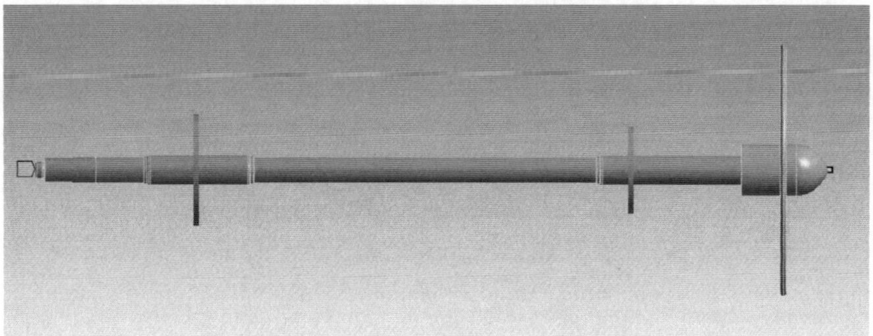

Figure 2. Model with boundary conditions

3.2. Modal Analysis

Modal analysis is one of the essential tools in ANSYS Workbench to find the vibration characteristics of structures, i.e., Natural frequency, Mode shape and Maximum deformation [12]. In this section, the natural frequencies with their respective mode shapes and the maximum deformation with affected zones of Trawler propulsion shafting are presented and analyzed.

According to the theory of classical mechanics, a simple machine may be represented by the general equation of motion as having Mass, Damping, and Stiffness:

$$[M]\ddot{X}(t) + [C]\dot{X}(t) + [K]X(t) = F(t) \qquad (1)$$

Where: $[M]\ddot{X}(t)$ is the Mass-Inertia, $[C]\dot{X}(t)$ is the Damping-Velocity, $[K]X(t)$ is the Stiffness-Displacement, and F(t) is the Force Vector.

When a structural system is set in motion due to inherent forces and allowed to vibrate freely, the state of free vibration occurs. In this state, the system vibrates at its natural frequency.

In the modal analysis, the Force vector F(t) is set to zero which acts as free vibrations. Usually, the Damping matrix [C] is ignored.

$$[M]\ddot{X}(t) + [K]X(t) = 0 \tag{2}$$

Here, the first six mode shapes are illustrated in Figure 3, evaluating the modal parameters.

(a) Mode 1

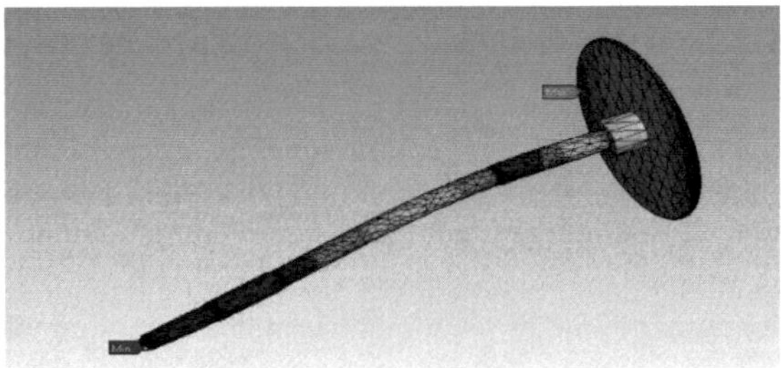

(b) Mode 2

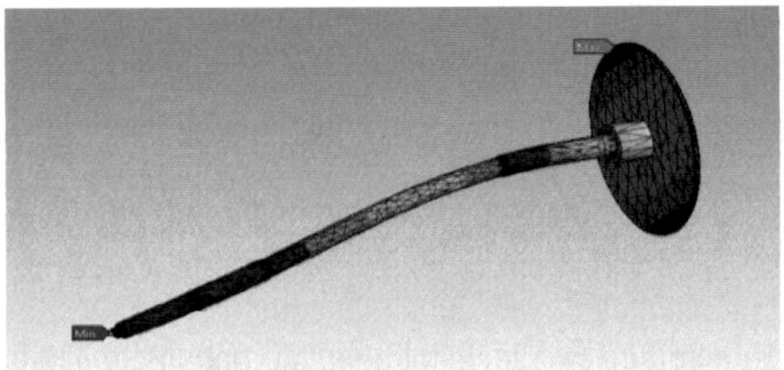

(c) Mode 3

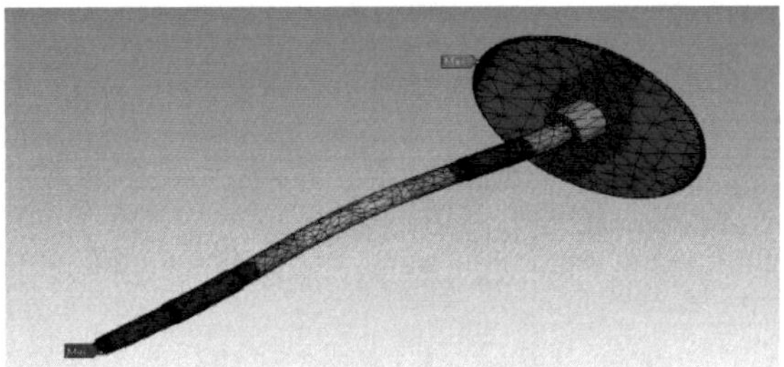

(d) Mode 4

(e) Mode 5

(f) Mode 6

Figure 3. First 6 order mode shapes

3.3. Analysis Results and Discussions

The maximum deformation and natural frequency values obtained from the modal analysis are listed in Table 4.

Table 4. Maximum Deformation and Natural frequencies for first 6 modes

Mode	Maximum Deformation (m)	Natural Frequency (Hz)
1.	6.6474e-002	15.941
2.	5.0557e-002	17.999
3.	5.056e-002	18.021
4.	0.1001	76.064
5.	0.1014	76.353
6.	7.9051e-002	81.842

The vibration characteristics of Trawler propulsion shafting can be analyzed from the illustration of mode shapes. Mode 1 is representing torsional whereas Mode 6 is representing longitudinal vibration. All the other modes from Mode 2 to Mode 5 are representing lateral vibrations (bending modes). The analyses of Trawler propulsion shafting made to notice that the shaft line is in a safe condition with having minor deformation, but the maximum deformation is occurring on the propeller. The cause of the maximum deformation on the propeller is because it is attached at the free end and the system is acting as a cantilever beam. It can also be seen that Mode 4 and Mode 5 are in maximum deformation among all the other modes, which can be the severe cause of vibration and must be avoided.

4. Optimization Scheme

Optimization is the process of maximizing or minimizing the desired objective function while satisfying the prevailing constraints [13]. Optimization can be applied to evaluate any engineering problem.

From the analysis of Trawler propulsion shafting, considerable deformation at the propeller as well as a minor deformation at the part of shaft line in between the bearing shells is found which indicates that these are the affected zones. Optimization is aimed at minimizing the deformation of affected zones which leads the system to the reduction in vibrations. Two Optimization schemes are adopted by using ANSYS Workbench v16.0 to reduce the vibration of these zones.

4.1. Goal Driven Optimization

Goal Driven Optimization (GDO) is a multi-objective technique in which the best possible designs are obtained from a sample set given the purposes as set for parameter [14]. The distance between the bearing shells has been changed by changing the position of tail shaft aft bearing, but the forward bearing remains at the same location.

While using the Response Surface Optimization in ANSYS Workbench which is one of the categories of GDO, aft bearing is considered to be the input parameter, and the Total deformation of the first six modes considers being the output parameters. The goal set at the input is to adjust the position of aft bearing, and at the output is to minimize the total deformation of all the first six modes. The actual value of the aft bearing from propeller was 1.015m, but by using the Response Surface Optimization, ANSYS Workbench finds the best-suited candidate point which is 0.9483m, reducing the maximum deformation. The optimized maximum deformation values at the new candidate point are listed in Table 5.

Table 5. Effect of tail shaft aft bearing position on deformation

Mode	Actual Maximum Deformation (m)	Optimized Maximum Deformation (m)
1.	6.6474e-002	6.6375e-002
2.	5.0557e-002	4.9893e-002
3.	5.056e-002	4.9897e-002
4.	0.1001	9.7481e-002
5.	0.1014	9.8893e-002
6.	7.9051e-002	7.8691e-002

4.2. Design Optimization

Design Optimization is the process to find the best design parameters that satisfy the project requirements [15]. The thickness of the propeller blade is increased by 1mm, and the radius of the shaft between bearing shells is increased by 2mm (diameter 4mm). The design parameters varied manually in SolidWorks to analyze the effect of variation in ANSYS Workbench. Several results obtained by altering the parameters, and finally selecting the most appropriate value. The effect of blade thickness and shaft diameter between bearings on deformation can be seen from Table 6 and 7 respectively.

Table 6. Effect of propeller blade thickness on deformation

Mode	Actual Maximum Deformation (m)	Optimized Maximum Deformation (m)
1.	6.6474e-002	6.5553e-002
2.	5.0557e-002	4.9143e-002
3.	5.056e-002	4.9134e-002
4.	0.1001	9.6536e-002
5.	0.1014	9.6623e-002
6.	7.9051e-002	7.2903e-002

Table 7. Effect of shaft diameter between bearing shells on deformation

Mode	Actual Maximum Deformation (m)	Optimized Maximum Deformation (m)
1.	6.6474e-002	6.6442e-002
2.	5.0557e-002	4.9788e-002
3.	5.056e-002	4.9761e-002
4.	0.1001	9.9946e-002
5.	0.1014	9.8947e-002
6.	7.9051e-002	7.8088e-002

4.3. Optimization considering parameters collectively

Effects of the parameters by variation have been analysed one by one (Table 5, 6 and 7). The values of all the parameters found after Optimization schemes applied. Modify the design of Trawler propulsion shafting, according to these values in SolidWorks and analyze the optimized maximum deformation values in ANSYS Workbench, which are listed in Table 8.

Table 8. Effect on deformation when all parameters considered collectively

Mode	Actual Maximum Deformation (m)	Optimized Maximum Deformation (m)
1.	6.6474e-002	6.5574e-002
2.	5.0557e-002	4.9829e-002
3.	5.056e-002	4.9837e-002
4.	0.1001	9.4285e-002
5.	0.1014	9.4319e-002
6.	7.9051e-002	7.4949e-002

4.4. Optimization Results and Discussions

The comparison has been made by altering the parameters separately and then collectively. The effect on the maximum deformation of Trawler propulsion shafting in the first six modes can be seen from Figure 4.

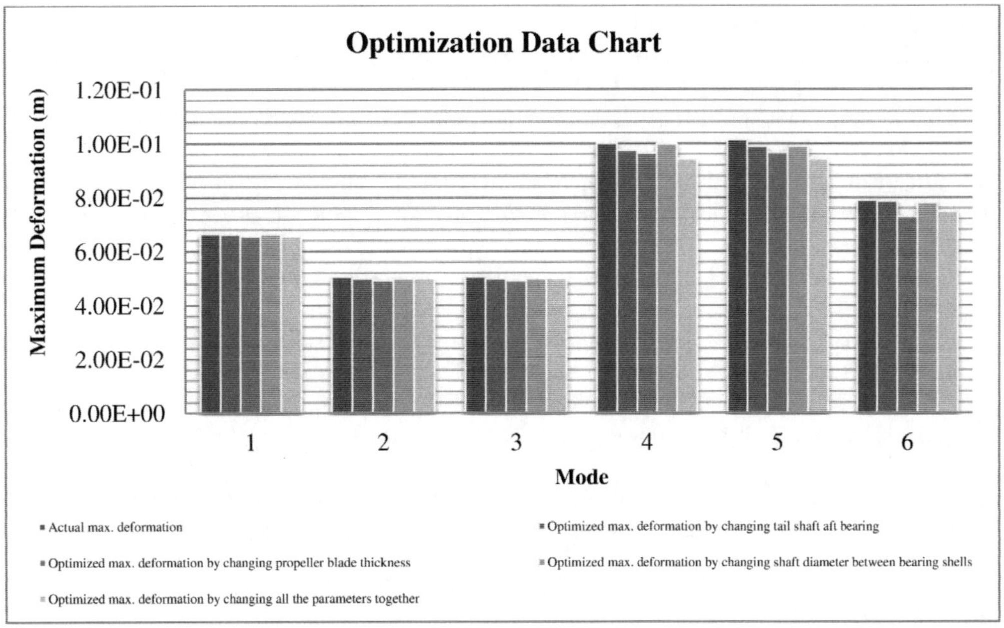

Figure 4. Comparison Optimization data chart

It is clear from Figure 4 that by a slight change in the parameters, the deformation of all the modes reduced and Trawler propulsion shafting comes in the acceptable range for normal operation and ultimate results in the reduction of vibration.

In GDO, it generates 1000 samples and finds three best-fitted candidate points. Among them, one is verified which is used as the appropriate position of aft bearing. This method uses a simple approach based on sampling and sorting. In Design Optimization, it has come to our realization that, as long as the thickness of the blade and the shaft diameter between the bearings shells increases, the deformation decreases. But these parameters cannot be increased much because it also increases the mass of the Trawler propulsion shafting system which can affect the efficiency of the system. Notably, the slight change in feasible limit has been done to optimize the design.

5. Conclusion

The study has shown the Modeling, Analysis, and Optimization of trawler propulsion shafting system with the ultimate aim of reduction in vibrations. The computational model has been modeled with accurate dimensions and analyzed by Modal Analysis using Finite Element Method. It is essential to know the natural frequencies, mode shapes, and maximum deformation to figure out the affected zones to prevent failure.

To analyze the effect of parameters on trawler propulsion shafting, the position of the aft bearing, the thickness of propeller, and the diameter of the shaft between the bearing shells have been optimized. By Optimization of these parameters, the maximum deformation is noticeably reduced leading to the reduction of vibration characteristics.

6. References

[1] Khande A G, Sable S A and Bidwai V R 2015 Vibration Analysis of Propeller Shaft using FEM *Int. J. Mech. Eng. Robot.* vol. 3, no. 4, pp. 70–75.

[2] Huang Q, Zhang C, Jin Y, Yuan C and Yan X 2015 Vibration analysis of marine propulsion shafting by the coupled finite element method *J. Vibroengineering*, vol. 17, no. 7, pp. 3392–3403.

[3] Murawski L 2004 Axial vibrations of a propulsion system taking into account the couplings and the boundary conditions *J. Mar. Sci. Technol.* vol. 9, no. 4, pp. 171–181.

[4] Ruíz F J D 2013 Analysis of most frequent cases of vibration in propulsion systems *Sh. Sci.*

Technol. vol. 7, no. 13, pp. 65–74.
[5] Sial T R, Jin Y and Juan Z 2018 Crack identification in Beams by Vibration based analysis techniques – A Review *Int. J. Sci. Eng. Technol. Res.* vol. 07, no. 10, pp. 766–771.
[6] Warikoo R and Haddara M R 1992 Analysis of propeller shaft transverse vibrations *Mar. Struct.*, vol. 5, no. 4, pp. 255–279.
[7] Wang C, Sun S, Li L and Ye L 2016 Numerical prediction analysis of propeller bearing force for full-scale hull-propeller-rudder system *Int. J. Nav. Archit. Ocean Eng.*, vol. 8, no. 6, pp. 589–601.
[8] Jalali H and Ahmadian H 2015 Model identification and dynamic analysis of ship propulsion shaft lines *J. Theor. Appl. Vib. Acoust.*, vol. 1, no. 2, pp. 85–95.
[9] Yucel A and Arpaci A 2013 Free and forced vibration analyses of ship structures using the finite element method *J. Mar. Sci. Technol.*, vol. 18, no. 3, pp. 324–338.
[10] Tian Z, Yan X, Zhang C, Xiong Y and Yang P 2016 Vibration characteristics analysis on ship propulsion system taking hull deformations into account *Chinese J. Clin. Oncol.*, vol. 33, no. 10, pp. 783–790.
[11] Hutton D V 2004 *Fundamentals of Finite Element Analysis* 1st ed. (McGraw-Hill).
[12] Gutierrez E S 2003 Modal Analysis of Rotating Machinery Strucrutes (Imperial College London).
[13] Belligundu A D and Chandrupatla T R (Optimization Concepts And Application in Engineering).
[14] Introduction to ANSYS DesignXplorer - Goal Driven Optimization (GDO) 2012.
[15] Belgasmia M 2018 *Optimization of Design for Better Structural Capacity* (Premier Reference Source).

Acknowledgement
The work is supported by The Project of Enhancing School with Innovation of Guangdong Ocean University (No.GDOU2016050258) and Guangdong Natural Science Foundation (No.2017A030313275).

A New Type of Energy Absorbing Structural Component: Spiral Thin-Walled Tube

Qiyang Zuo[1, 2, *], Feifei Zhang[1, 2] and Kai He[1, 2]

[1]Shenzhen Institutes of Advanced Technology, Chinese Academy of Sciences, Shenzhen, Guangdong, China
[2]Shenzhen Key Laboratory of Precision Engineering, Shenzhen, Guangdong, China

Corresponding author's e-mail: qy.zuo@siat.ac.cn

Abstract. Various thin-walled (TW) tubular components have been widely applied to absorb impact energy due to their sound crashworthiness and long stroke. This paper presents a novel spiral TW tubular component used for absorbing energy. The presented TW structural component is easily prepared and cost-effective because it can be handily manufactured by sheet metal rolling. Its geometry structure on cross section is spiral so it is called spiral TW tube in this work, which was studied by formulation method. Furthermore, its dynamic crushing and energy absorbing characteristics under impact loading was investigated by means of simulation analysis. The analysis result shows that the presented spiral TW tube has a superior energy absorbing performance and is promising used in passive safety area.

1. Introduction

Thin-walled (TW) tubular parts or components have big advantages in lightweight and crashworthiness and have been widely used in automotive, aerospace, transportation and defense industries. On one hand, wide application of lightweight TW tubular components in transport industry can save weight and reduce CO_2 emissions. On the other hand, sound crashworthiness of the TW tubular components can effectively absorb impact force of a serious crash and reduce death or injury risk of occupants. Therefore, studies concerning the TW tubular components have become increasingly attractive in recent years.

The structures of the traditional TW tubular components used for absorbing energy are various, mainly including TW circular structure, square structure, corrugated tubular structure, tapered circular structure and so on. Enormous investigations into the crashworthiness and energy absorbing performance of all kinds of the traditional TW tubular components have been carried out in the past decades [1]. Especially, as the most common structure, the axial crushing behaviors of the TW circular tubes were studied early [2]. It was revealed that the deformation mode of the TW circular tubes principally depends on their geometric parameters, which includes the so called concertina or ring mode, diamond mode, Euler mode and so on [2, 3]. By contrast, as another common TW structural tube, the crashworthiness and energy absorbing capacities of the TW square tubes have been also investigated in recent years. The studies about the TW square tubes with various cross-sectional shapes were carried out, including multi-corner [4, 5], multi-cell [6, 7], foam-filled [8] and so on. Meanwhile, the TW structural tubes with other cross sectional shapes such as multi-cell hexagonal tubes were also studied [9].

Furthermore, as far as axial direction structure is concerned, some investigations involved the TW tubes with functionally graded thickness (FGT), such as TW square tubes with FGT [10, 11] and TW tapered circular tubes FGT [12-14]. In addition, TW corrugated tubes also have non-uniform thicknesses in axial direction which were studied by some researchers [15-17].

Besides the traditional TW structural tubes, a kind of folded square tube which can be fabricated by sheet metal bending has recently excited interest [18, 19]. This kind of folded tube is easily prepared, cost-effective and flexible in sectional shape, so is promising to be used in many engineering fields. Especially, the crashworthiness of the folded tube is also a concern. For example, Zhang et al. investigated side bending collapse [18] and axial crushing [19] of the folded square tubes, respectively. However, it is a pity that the related studies are barely concerned with the folded tubes with square structure at present.

By contrast, this paper presents an investigation into a new kind of energy absorbing structural component, called spiral TW tube in this work, which is concerned with its manufacturing process, geometry structure, dynamic crushing and energy absorbing characteristics undergoing impact loading. As a lightweight and promising energy absorbing structural tube, this novel spiral TW tube can be potentially used for passive safety field. Furthermore, the fabricating cost of this kind of spiral TW tube is relatively lower because it can be easily obtained by sheet metal bending (or called sheet metal rolling, more exactly). The geometry structure of the spiral TW tube was studied via formulation method in this work. In addition, the axial crushing and energy absorption characteristics was investigated by means of simulation analysis. The analysis results show that the spiral TW tubes have a sound energy absorbing performance and a great potential applied to act as energy absorber. So to say, this work provides a new idea for passive safety study to some extent.

2. Manufacturing process and geometry structure

2.1. Manufacturing process

The spiral TW tube can be easily manufactured by sheet metal rolling. Especially, just one cylindrical bar with a certain diameter can serve as the mold, thus it is easily prepared and cost-effective. The manufacturing process is simply as follows: at first, roll the blank sheet metal round the mold with desired turns into the spiral TW tube, as shown in Fig. 1, and then, draw the mold out from the rolled spiral TW tube. It should be noted that, however, elastic springback would be very likely to occur if the rolled spiral TW tube is loosened immediately. So some measures should be taken to deal with the springback issue this moment.

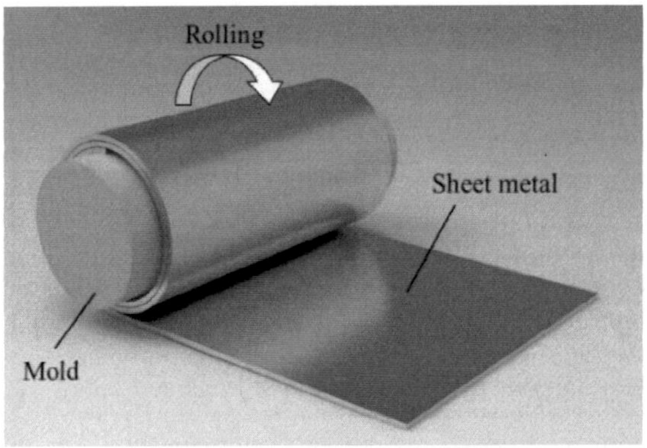

Figure 1. Manufacturing process of the spiral thin-walled tube by sheet metal rolling.

In fact, the springback can be prevented effectively if the sheet metal end is fixed by welding or gluing before loosening. Accordingly, there are three contact modes at the sheet metal end after rolling the sheet metal, i.e. welding, gluing and free contact, as shown in Fig. 2. In this work, laser welding is recommended for welding the sheet metal end due to simplification. The laser focus only need to aim at a point of the end and move long the end line of the sheet metal to weld the end fixedly. Therefore, there is a contact line along the sheet metal end if laser welding is used, as shown in Fig. 2(a). On the other hand, if the sheet metal end is fixed by gluing, there will be an adhesive layer at the contact area, as shown in Fig. 2(b). In this work, the springback is thought to be prevented effectively if the sheet metal end is welded or glued together.

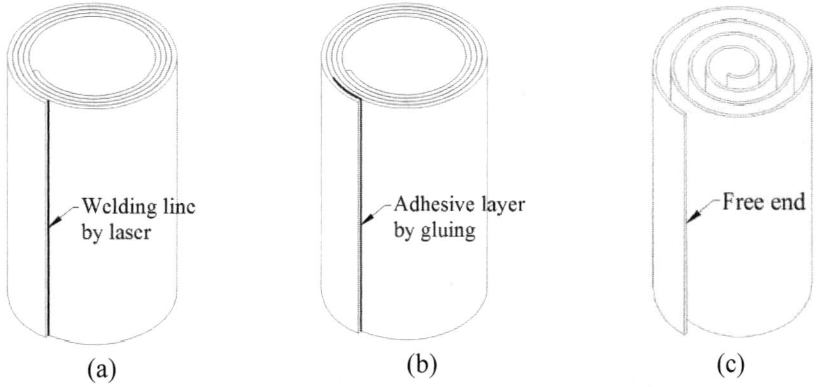

Figure 2. Contact modes of the sheet metal end after rolling.

With regard to the free contact mode of the sheet metal end (Fig. 2 (c)), the effect of the elastic springback cannot be neglected. The springback issue will inevitably lead to a certain structural or dimensional error in the final spiral TW tube. The extent of springback is related to the material of the sheet metal. Once the material is determined, the springback property should be considered and some measures must be taken to reduce the extent of springback as far as possible. For example, roll the tube round the mold repeatedly before pull the mold out from the rolled spiral TW tube. Or even, since the size (e.g. the outer diameter) of the spiral TW tube would be enlarged after springback, it would be better to appropriately decrease the diameter of the mold in advance so as to achieve the desired size of the final spiral TW tube after springback.

2.2. Geometry structure

From the view of cross section, the center line of the spiral TW tube is an Archimedean spiral, as shown in Fig. 3. Obviously, this Archimedean spiral center line is actually equal to the length of the sheet metal used for manufacturing the spiral TW tube. Therefore, in order to manufacture a desired spiral TW tube, the required length of the sheet metal can be worked out only need to calculate the length of the Archimedean spiral center line on the cross section. However, the calculation of the real length of the Archimedean spiral is rather complicated because an integral formula for calculating arc length in polar coordinates has to be used. Hence, for simplification, the required length of the sheet metal used for manufacturing the spiral TW tube is calculated by approximate method in this work and discussed as follows:

As we know, the mathematical equation of the Archimedean spiral in polar coordinates is $r = a + b\theta$, where θ is polar angle which indicates the total rolled degree of the Archimedean spiral; a is polar radius when $\theta = 0^{\circ}$; b is Archimedean spiral coefficient which indicates the increment of the polar radius when the spiral rolls 1 degree.

At first, let's discuss the case without springback. As shown in Fig. 3, the spiral TW tube is obtained by rolling the sheet metal with n turns plus a remnant length of k. The length of the spiral generated by every turn is approximately equal to the center line length of the annulus which envelop this corresponding spiral. For instance, the length of the spiral generated by ith turn, i.e. $\widehat{ABCDE}$ shown in Fig. 4, is approximately equal to the center line length of the annulus between the circles of which the radii are r_{i-1} and r_i, respectively. Accordingly, the total length of the sheet metal used for manufacturing the spiral TW tube can be calculated approximately by summing all the lengths of the center line of the annulus after n turns, of course, plus a remnant length of k.

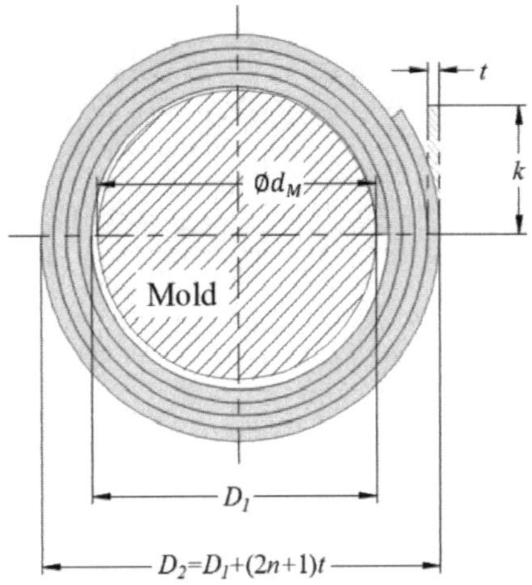

Figure 3. Cross sectional geometry structure of spiral thin-walled tube ($\Delta t = 0$).

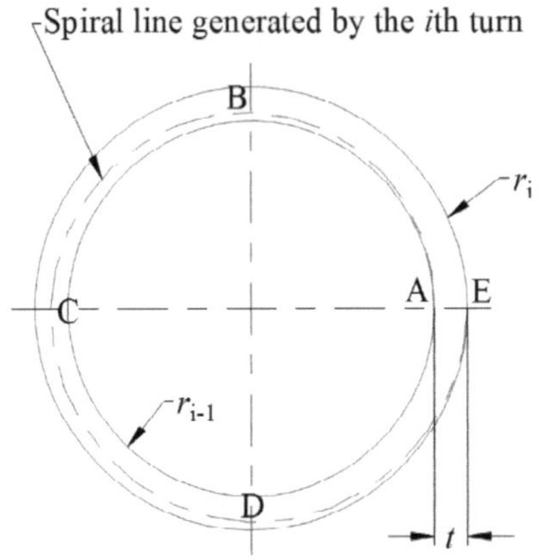

Figure 4. Schematic calculation method of spiral length of the ith turn.

Concretely speaking, according to the Archimedean spiral equation in polar coordinates, the radii of the circles which encircle the annulus enveloping the spiral of the ith turn (Fig. 4) can be respectively obtained by

$$r_i = a + 2\pi ib \tag{1}$$

$$r_{i-1} = a + 2\pi(i-1)b \tag{2}$$

Thus, the radius of the center line of the annulus enveloping the spiral of the ith turn can be calculated by

$$\overline{r_i} = \frac{r_i + r_{i-1}}{2} = a + 2\pi ib - \pi b \tag{3}$$

As the above mention, the length of the spiral generated by ith turn is approximately equal to the center line length of the corresponding annulus, which is calculated as

$$l_i \approx 2\pi\overline{r_i} = 2\pi(a + 2\pi ib - \pi b) \tag{4}$$

Therefore, the total spiral length of the spiral TW tube on the cross section is obtained by

$$L \approx \sum_{i=1}^{n} l_i = 2\pi\left(an + 2\pi b\sum_{i=1}^{n} i - \pi bn\right) + k \tag{5}$$

where, n is the total turn number of the spiral TW tube; k is the extra length after the nth turns. In addition, since $\sum_{i=1}^{n} i = n(n+1)/2$, the above equation can be modified as

$$L \approx 2\pi n(a + \pi bn) + k \tag{6}$$

From Fig. 3, it can be deduced that there are three relational expressions as follows:

$$D_1 = d_M + \pi b \tag{7}$$

$$a = \frac{d_M}{2} \tag{8}$$

$$b = \frac{t}{2\pi} \tag{9}$$

where, D_1 is the inner diameter of the spiral TW tube; d_M is the diameter of the mold; t is the thickness of the sheet metal.

Based on the above relational expressions (7), (8) and (9), the Eq. (6) can be modified as

$$L \approx \pi n\left(D_1 + nt - \frac{t}{2}\right) + k \tag{10}$$

Since the outer diameter of the spiral TW tube D_2 can be obtained by $D_2 = D_1 + (2n+1)$ shown in Fig. 3, the Eq. (10) also can be expressed as

$$L \approx \pi n\left(D_2 - nt - \frac{3t}{2}\right) + k \tag{11}$$

Both the equations (10) and (11) can be used to approximately calculate the required length of the sheet metal for manufacturing the spiral TW tube. Generally speaking, using Eq. (11) is more convenient because the outer diameter of the spiral TW tube D_2 can be measured directly and conveniently. It should be noted that the Eq. (10) or (11) was derived by the approximate method which the length of the spiral on every turn is approximately equal to the center line perimeter of the corresponding annulus, which means every turn would generate a calculating error. Therefore, the greater the turn number, the larger the cumulative calculating error will be.

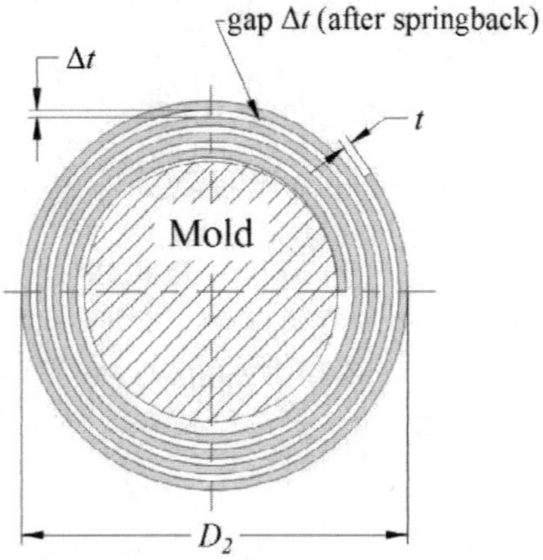

Figure 5. Cross sectional geometry structure of spiral thin-walled tube ($\Delta t \neq 0$).

If springback occur, there will be a certain gap among the coils, as shown in Fig. 5. Supposing that the gap among the spiral wall after springback is even and marked by Δt, the total spiral length of the spiral TW tube after springback on the cross section also can be calculated by using Eq. (10) or (11) only need to replace the parameter t with $t + \Delta t$ in the equation, such as:

$$L \approx \pi n\left[D_2 - n(t+\Delta t) - \frac{3(t+\Delta t)}{2}\right] + k \tag{12}$$

With regard to the value of the gap among the spiral wall after springback, a calculation method is given as follows:

In the light of Eq. (9), the Archimedean spiral coefficient after springback can be obtained by

$$b = \frac{t+\Delta t}{2\pi} \tag{13}$$

Based on Eqs. (7) and (13), the inner diameter of the spiral TW tube can be calculated by

$$D_1 = d_M + \frac{t+\Delta t}{2} \tag{14}$$

According to the relation of the inner and the outer diameters (shown in Fig. 3), the outer diameter of the spiral TW tube D_2 can be worked out by

$$D_2 = d_M + (2n+1)t + (2n-1)\Delta t + \frac{\Delta t}{2} \tag{15}$$

Therefore, based on the above equation, the value of the gap among the spiral wall after springback can be obtained by

$$\Delta t = \frac{D_2 - d_M - (2n+1)t}{2n - 0.5} \tag{16}$$

3. Numerical simulation

3.1. Evaluation methods of energy absorption

Acting as a novel energy absorber, the energy absorption capacity of the spiral TW tube is the most important property that should be paid attention to. There are several indicators to evaluate this capacity, e.g. total absorbed energy during crushing process (W_{total}), absorbed energy per unit mass (W_m) and average crash force (F_{avg}).

As the key indicator used for assessing the capacity of absorbing impact energy, the total absorbed energy (W_{total}) can be determined mathematically as

$$W_{total} = \int_0^{\delta} F(\delta)d\delta \tag{17}$$

where δ is the crushing displacement and $F(\delta)$ is the instantaneous crashing force with a function of the displacement δ.

The absorbed energy per unit mass (W_m), which is also called specific energy absorption (SEA) in some literatures [11], can be worked out by

$$W_m = \frac{W_{total}}{m} = \frac{\int_0^{\delta} F(\delta)d\delta}{m} \tag{18}$$

where, m is the total mass of the spiral TW tube. Since the total length of the sheet metal used for manufacturing the spiral TW tube L can be obtained by Eq. (11) or (12), the total mass m can be worked out if the material density and the height of the spiral TW tube are given.

The average crashing force (F_{avg}) is also one of the parameters to assess the energy absorption capacity of a structure, which can be calculated by W_{total} divided by the crushing displacement δ as

$$F_{avg} = \frac{W_{total}}{\delta} = \frac{\int_0^{\delta} F(\delta)d\delta}{\delta} \tag{19}$$

3.2. Finite element modeling

In this work, the crashworthiness and the energy absorption characteristics of the spiral TW tubes undergoing impact loading were investigated by means of numerical simulations. Those numerical simulations were carried out by using commercial explicit finite element code LS-DYNA.

For simulating impact loading situation, two rigid planes were applied to crush specimen (i.e. the spiral TW tube). One rigid plane was stationary fixed at the bottom for supporting the specimen, while another one moved toward the specimen with a certain impact speed from the top of the specimen. A typical finite element meshed model is shown in Fig. 6 where the moving rigid plane is concealed for better illustration.

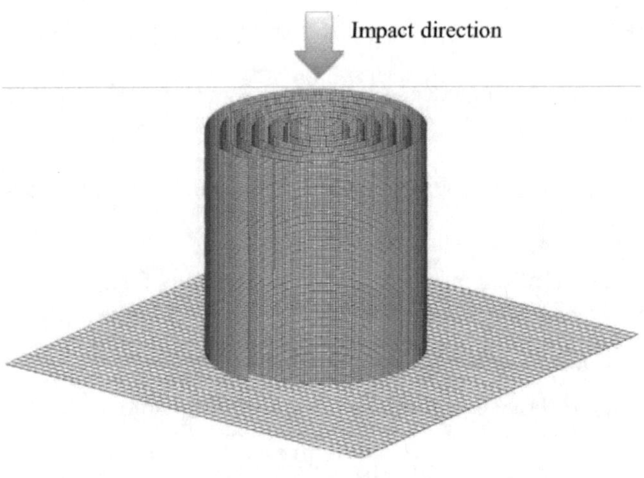

Figure 6. A typical finite element meshed model for spiral thin-walled tubes.

In this work, the selection of the specimen materials referenced that of the literature [6], namely aluminum extrusion AA6061 T4, with Young's modulus $E = 70$ GPa, Poisson's ratio $\nu = 0.28$, yield strength $\sigma_y = 110.3$ MPa and ultimate strength $\sigma_u = 213$ MPa. Likewise, the tensile stress–strain curve and power law expression were also the same as that of the literature [6].

3.3. Results and discussions

At first, collapsing deformation of the spiral TW tube during impacting process is discussed. Let's take the case shown in Fig. 7 for example, where a specimen was simulated to be crushed by the moving rigid plane with an impact speed 10 mm/s. The specimen was obtained by rolling the sheet metal round the mold of which the diameter is 20 mm (i.e. $d_M = 20$) with 6 turns (i.e. $n = 6$), and the sheet metal end was free (as shown in Fig. 2 (c)). The thickness of the sheet metal for manufacturing the specimen was 0.2 mm (i.e. $t = 0.2$ mm) and the height of the specimen was 100 mm.

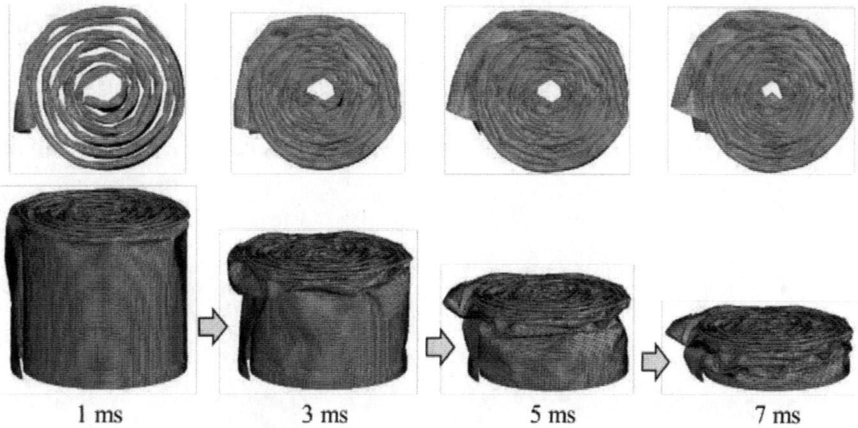

Figure 7. Collapsing deformation process of the spiral thin-walled tube ($d_M = 20, n = 6$).

The whole collapsing deformation process of this specimen is shown in Fig. 7. It can be seen that the sheet metal of every turn incrementally collapsed in the manner of so called diamond collapse mode, which means that the collapsing deformation mode of some spiral TW tubes during impacting process is similar to that of some TW circular aluminum tubes with certain sizes [2, 3].

Besides the above case, more simulation analyses were carried out in order to investigate the influence of the turn number (n) and the mold diameter (d_M) on the energy absorption performance of the spiral TW tubes. The cases with the turn numbers (n) in the range of 2 to 10 and the mold diameters (d_M) in the range of 0 to 60 were compared. For the sake of comparative analysis, the heights of the specimens were all selected as 100 mm and the thicknesses of the sheet metal using for manufacturing specimens were 0.2 mm (i.e. $t = 0.2$ mm) uniformly.

The simulation results are shown in Fig. 8 and Fig. 9. It can be found that the force-deformation curves of the specimens are rather similar to those of the most of cellular materials such as honeycomb [20], metal foams [21] and some periodic cellular metals [22-25]. Concretely speaking, at the impacting moment the crashing force of every specimen rapidly increased to the first peak point, but after that the crashing force decreased and then vibrated a little around an approximate constant value with the collapsing deformation until reaching densification stage. Obviously, the most impact energy will be absorbed in the stage that the crashing force slightly fluctuates around the constant value with the collapsing deformation. The absorbed energy can be evaluated by Eq. (17) and Eq. (18). Specifically, the Eq. (17) can be calculated by the method of trapezoid rule of NewtonCotes formula, which can refer to the literatures [24, 25]. Likewise, the constant value that the crashing force vibrates around is approximately equal to the average crashing force (F_{avg}) which can be calculated by Eq. (19).

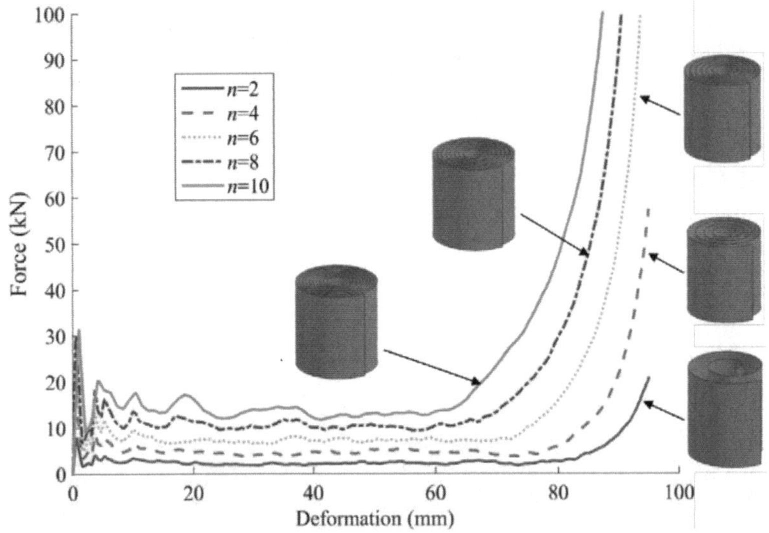

Figure 8. Force-deformation curves of spiral thin-walled tubes obtained by the mold of $d_M = 20$.

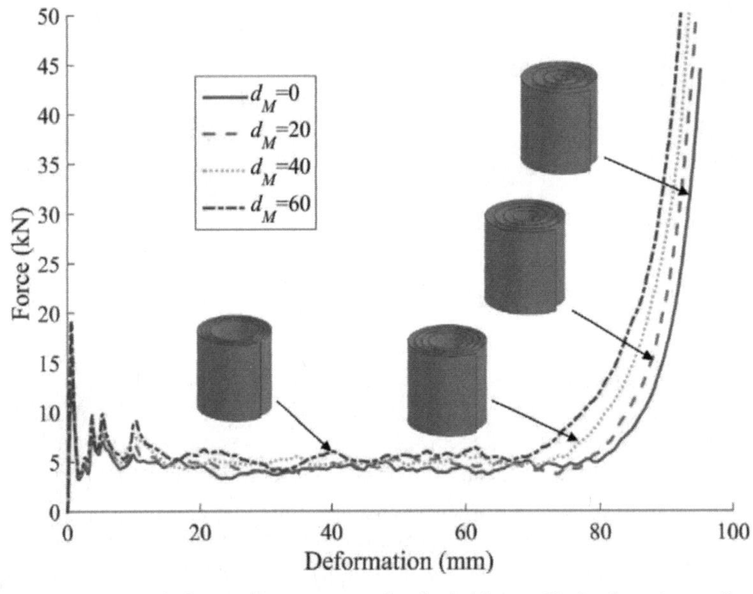

Figure 9. Force-deformation curves of spiral thin-walled tubes ($n = 4$).

From the simulation results shown in Fig. 8 and Fig. 9, it can be found that the spiral TW tube presented in this work indeed has the same energy absorbing performance as other superior cushioning materials such as honeycomb, metal foams and some periodic cellular metals. That is to say, the spiral TW tubes are very suitable for energy absorption and have a great potential used in passive safety area.

4. Conclusion

As a novel and promising structural component used for energy absorption, the spiral TW tube was investigated in this work, concerning its manufacturing process, geometry structure, dynamic crushing and energy absorbing performance under impact loading. The conclusions can be drawn as follows:

Firstly, the spiral TW tube can be easily fabricated by sheet metal rolling, meanwhile, the production cost is rather low because the manufacturing process can be carried out by just one cylindrical bar serving as the mold.

Secondly, the geometry structure of the spiral TW tube was studied by formulation method. The approximate equation for calculating the required length of the sheet metal used for manufacturing the spiral TW tube was derived. Furthermore, if springback occur, the value of the gap among the spiral wall after springback can be worked out by the calculation method given in this work.

Finally, the dynamic crushing and energy absorbing performance of the spiral TW tubes were investigated by means of simulation analysis. The analysis results show that the spiral TW tube has a similar energy absorbing performance as other superior cushioning and energy absorbing materials, so that is to say, has a great potential used for passive safety area.

5. References

[1] Baroutaji A, Sajjia M and Olabi A 2017 *Thin Wall Struct.* **118** 137-163
[2] Andrews K, England G and Ghani E 1983 *Int. J. Mech. Sci.* **25** 687-696
[3] Guillow S, Lua G and Grzebieta R 2001 *Int. J. Mech. Sci.* **43** 2103-2123
[4] Abramowicz W and Wierzbicki T 1989 *J. Appl. Mech. Trans. ASME* **56** 113–120
[5] Tang Z, Liu S and Zhang Z 2012 *Thin Wall Struct.* **51** 112–120
[6] Chen W and Wierzbicki T 2001 *Thin Wall Struct.* **39** 287-306
[7] Zhang X, Cheng G and Zhang H 2006 *Thin Wall Struct.* **44** 1185-1191
[8] An X, Gao Y, Fang J, Sun G and Li Q 2015 *Thin Wall Struct.* **91** 63–71
[9] Qiu N, Gao Y, Fang J, Feng Z, Sun G and Li Q 2016 *Thin Wall Struct.* **102** 111-121
[10] Zhang X, Wen Z and Zhang H 2014 *Thin Wall Struct.* **84** 263–274
[11] Sun G, Xu F, Li G and Li Q 2014 *Int. J. Impact Eng.* **64** 62-74
[12] Li G, Zhang Z, Sun G, Huang X and Li Q 2015 *Thin Wall Struct.* **94** 334–347
[13] Zhang X, Zhang H and Wen Z 2015 *Int. J. Mech. Sci.* **92** 12–23
[14] Li G, Xu F, Sun G and Li Q, 2015 *Int. J. Impact Eng.* **77** 68–83
[15] Salehghaffari S, Tajdari M, Panahi M and Mokhtarnezhad F 2010 *Thin Wall Struct.* **48** 379-390
[16] Singace A and El-Sobky H 1997 *Int. J. Mech. Sci.* **39** 249–268
[17] Eyvazian A, Habibi M, Hamouda A and Hedayati R 2014 *Mater. Des.* **54** 1028–1038
[18] Zhang X, Zhang H and Ren W 2016 *Int. J. Mech. Sci.* **117** 67–78
[19] Zhang X, Zhang H and Ren W 2018 *Thin Wall Struct.* **122** 252-263
[20] Hönig A and Stronge W 2002 *Int. J. Mech. Sci.* **44** 1665–1696
[21] Banhart J 2001 *Prog. Mater. Sci.* **46** 559–632
[22] Dang X, Du R, He K and Zuo Q 2016 *ASME IMECE2016-65362* **9** V009T12A057
[23] Dang X, He K and Du R 2017 *Int. J. Adv. Manuf. Technol.* **91** 2561–2569
[24] Zuo Q, He K, Mao H, Dang X and Du R 2018 *Mater. Des.* **153** 242-258
[25] Dang X, He K, Zuo Q, Li J and Du R 2015 *ASME IMECE2015-50952* **9** V009T012A041

Acknowledgements

This work was supported by the Shenzhen Basic Research Project [grant number JCYJ20160331190103609].

A Calculation Method for Dressing Amount of Monolayer Electroplated Grinding Wheel Based on Zernike Moment

Li Bin[1, 2, a], Wang Shuai[1,2], Bao Hua[1,2], Chen Xiaoqiang[2]

[1] State Key Laboratory of Superabrasives, China
[2] Zhengzhou Research Institute for Abrasive & Grinding Co., Ltd., China

E-mail: [a]zhifeng920920@126.com

Abstract. Aiming at the difficulties in accurate prediction of the dressing amount of monolayer electroplated grinding wheel, a new calculation method based on image processing was proposed. For the purpose of increasing the measurement precision, all images were captured by CCD camera and a sub-pixel level accuracy edge detection approach based on Zernike moment was used. Two dressing amount of electroplated grinding wheel were calculated by protrusion height and run-out respectively, and the larger number of them was chosen as the optimal dressing amount. Finally, this method is applied to calculate the optimal dressing amount of 30 gothic-arc profile electroplated grinding wheels, and they were dressed by optical contour grinder. The results showed that, form precision of grinding wheels and roughness of workpiece were fine.

1. Introduction

Electroplated grinding wheel is made of a single layer of abrasive which have 60 to 80% of the grit height are covered by the galvanic embedding layer for effective grit retention[1]. With more complex and more precise of profile of grinding wheel, dressing is an essential process after electroplate [2, 3]. Moreover, due to small protrusion height, the using life reduces sharply with excessive dressing amount, and if dressing amount is inadequate, the accuracy cannot meet the requirement. In order to dress the monolayer grinding wheel, many inadequate presented various methods, such as cup dresser dressing [4], electrical discharge dressing [5], laser dressing [6], however, the majority of them were contrast experiment, which did not provide a method for calculating the dressing amount of grinding wheel. In recent research, an image processing method was used to measure wheel wear in curve grinding, and the contour error was less than 2 μm [7, 8]. In this paper, a new calculation method for prediction of dressing amount of monolayer electroplated grinding wheel was proposed. The figure of the grinding wheel was captured by CCD camera and the edge of it is extracted with a sub-pixel level edge detection algorithm based on Zernike moment operator. The optimal dressing amount was determined by reasonable procedure, and then, a precision dressing experiment of gothic-arc profile grinding wheel was carried out.

2. Image processing and calculation

2.1. Zernike Moment and Sub-pixel Level Edge Detection

The kernel of Zernike moment is the set of orthogonal Zernike polynomials defined over polar coordinates inside a unit circle [9]. The Zernike basis function of order n repetition m is

$$V_{nm}(\rho,\theta) = R_{nm}e^{im\theta} \tag{1}$$

where n is a positive integer or zero, and m is an integer subject to the following constraints: n-|m| is even and . Rnm is given by

$$R_{nm}(\rho) = \sum_{s=0}^{(n-|m|)/2} \frac{(-1)^s (n-s)!\rho^{n-2s}}{s!(\frac{n+|m|}{2}-s)!(\frac{n-|m|}{2}-s)!} \tag{2}$$

Let f be a complex-valued function on the unit disc. The Zernike moment for f of order n repetition m is

$$Z_{nm}^{(f)} = \frac{n+1}{\pi} \iint_{x^2+y^2=1} f(x,y)V_{nm}^*(\rho,\theta)dxdy \tag{3}$$

According to the rotational invariant property of Zernike moment, the Zernike moment of the original image Znm and that of the rotated image Z'nm has following relationship:

$$Z_{nm}^{'} = Z_{nm}e^{-im\phi} \tag{4}$$

It is clear that Zernike moment merely acquire a phase shift on rotation and their magnitudes remain constant. This rotational invariant property is useful for accuracy edge detection.

(a) original image (b) rotated image

Figure 1. Ideal model of the sub-pixel edge detection.

Figure 1 shows that the ideal model of the sub-pixel edge detection. There are four parameters of every edge point: h is the background gray level, k is the step height, l is the perpendicular distance from the center of the circular kernel to the edge, and φ is the angle between l and x-axis. Figure. 1 (b) is generated by rotating Figure 1 (a) with φ. The most important parameters can be calculated as

$$l = \frac{Z_{20}}{Z_{11}^{'}}; k = \frac{3Z_{11}^{'}}{2(1-l^2)^{\frac{3}{2}}} \tag{5}$$

According to the invariant properties of the Zernike moment, angle φ can be expressed as

$$\varphi = \arctan(\frac{\mathrm{Im}[Z_{11}]}{\mathrm{Re}[Z_{11}]}) \tag{6}$$

Therefore, the sub-pixel level edge position of the image is,

$$\begin{bmatrix} x_s \\ y_s \end{bmatrix} = \begin{bmatrix} x \\ y \end{bmatrix} + l \begin{bmatrix} \cos\varphi \\ \sin\varphi \end{bmatrix} \tag{7}$$

Moreover, in discrete form, the calculating of the Zernike moment can be simplified to the mask calculating. In this paper, the unit circle is divided into 7*7 homogeneous grids as shown in Figure. 2. Convolving these masks with image points can get Zernike moment Z_{00}, Z_{10}, Z_{20}, Z_{31}, Z_{40} [10].

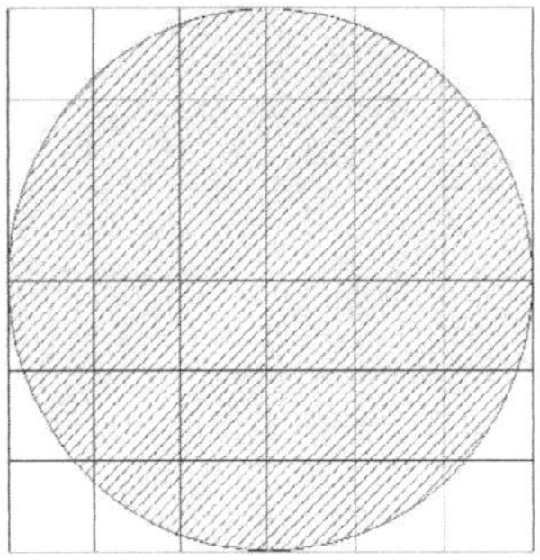

Figure 2. 7*7 mask.

Considering with the model effect, the sub-pixel level edge position of N*N mask can be improved as

$$\begin{bmatrix} x_s \\ y_s \end{bmatrix} = \begin{bmatrix} x \\ y \end{bmatrix} + \frac{Nl}{2}\begin{bmatrix} \cos\varphi \\ \sin\varphi \end{bmatrix}. \tag{8}$$

2.2. Calculate Dressing Amount Based on Protrusion Height

The protrusion height of electroplated grinding wheel is the distance between tip of abrasive and galvanic embedding layer. In theory, the grain size of abrasive follows the normal distribution, and the galvanic embedding layer has same level, as a result, the protrusion height follows normal distribution as well. However, such as 120/140 Cubic Boron Nitride, the extreme size of it is from 75 to 195 μ m and range is 122 μ m. Therefore, it is difficult to reduce the standard deviation by selecting grain. Chip space decreases with reduction of mean protrusion height, and amount of grains in the contact zone cuts down when standard deviation increases, which leads to increase of grinding force and surface roughness. As a result, one of the function of dressing is reducing the standard deviation while keeping the mean protrusion height.

Due to the random distribution of abrasive attached on abstract, the highest protrusion contacts the workpiece earlier than others. Figure 3 shows that the grinding wheel is divided to 8 homogeneous areas along the width direction, and the largest protrusion height of each area can represent the situation of this area. As a consequence, the mean protrusion height and standard deviation of grinding wheel can be calculated as long as superposing the largest protrusion height of each area into one image.

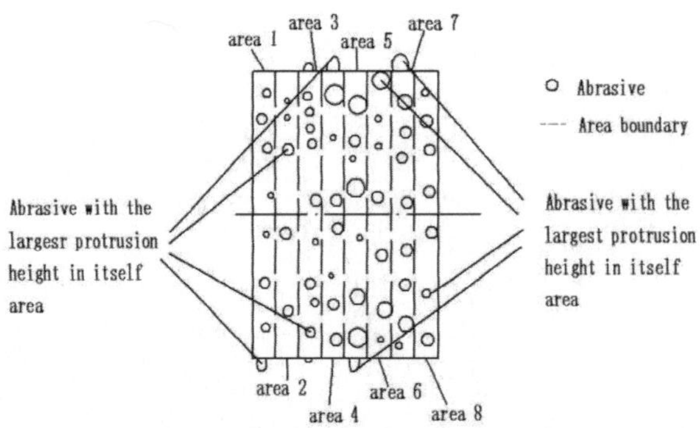

Figure 3. Largest protrusion height.

In this paper, optical contour grinder is chosen as dressing tool, and one of its properties is that the dresser can only feed along one direction while tool track is the nominal profile. The dressing amount, which means the reduction height of grains, of each area of complex profile grinding wheel contained line and arc is different, although the feed amount is same. According to the Figure 4, profile of grinding wheel is arc and dresser (not shown in the figure) feeds along X direction. The area painted black represents dressing amount and the volume of them shows that the dressing amount increases from area 1 to 5 and decreases from 5 to 8. Therefore, dressing amount can be calculated if a reasonable algorithm is proposed to process the superposition image, which can meet the function of dressing mentioned below.

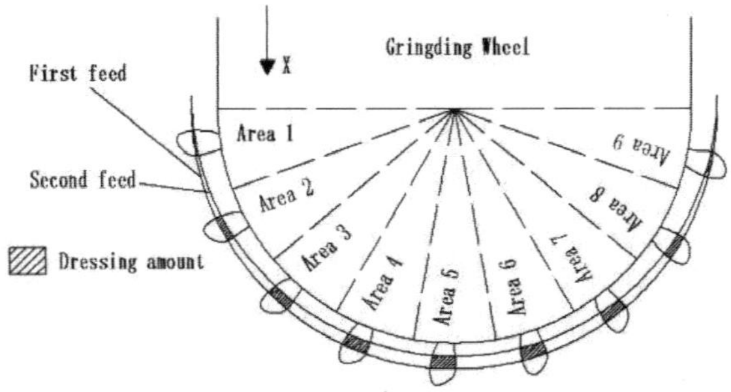

Figure 4. Reduction height of grains with same feed amount.

2.3. Calculate Dressing Amount Based on Run-out.

Another function of dressing is reducing the run-out of grinding wheel. Due to uneven surface, non-contact measurement is the most common method for measuring run-out of grinding wheel. For complex profile, such as the arc shown in the Figure 4, radius of fitting circle according to all grains of each section is different. Consequently, the range of those radius can be chosen as dressing amount.

3. Experimental set-up and step

3.1. Experimental Facilities

In this paper, gothic-arc chosen as experimental profile is a combined curve of two arcs with the same radius and shifted centers, and its fundamental requirements shown in the Table.1. Thereinto, contact angle is the angle between the line, which is connected between the tangent point of the gothic-arc and 3mm radius circle named A and center of A, and the vertical direction. In addition, the roughness Ra of workpiece grinded by grinding wheel after dressing is less than 0.6 μ m.

Table 1. Fundamental requirements of gothic-arc profile grinding wheel.

Size	Grit designation	Contact angle(Left)	Contact angle(right)
30*10*8-R3.118	120/140	$42\pm3°$	$42\pm3°$

Image measuring instrument with CCD camera and contourgraph contains function of measuring roughness were chosen as detecting instrument and optical contour grinder was used in dressing process. Table.2 shows the parameters of these facilities.

Table 2. Parameters of experimental facilities

CCD camera	Type	Port View 300
	Pixels in image output[pixel]	1280(h) x 1024(v)
	Vision field[mm]	5.69(h) x 4.55(v)
Optical contour grinder	Type	SPG-W
	Rotational accuracy[μm]	1
Dresser	Size[mm]	3F1 125*31.75*6
	Bonding	V
	Revolution[rpm]	6500
Contourgraph	Type	T8000RC200-400

3.2. Experimental Step

As shown in Figure 5, CCD camera captured a mass of images while the grinding wheel was rotated slowly by hand. After the edge position of each image shown in Figure 6 (a), of the grinding wheel is detected based on Zernike moment, which is shown in Figure 6 (b), and the threshold was subject to the following constraints:

$$l \le l_t = \bar{l} \cap k \ge k_t = \bar{k} \tag{9}$$

Considering pixels in image out and vision field in Table.2, the vertical and horizontal distance represented by each pixel could be calculated as

$$\begin{bmatrix} v \\ h \end{bmatrix} = \begin{bmatrix} 4.55/1024 \\ 5.69/1280 \end{bmatrix} = \begin{bmatrix} 0.0044 \\ 0.0044 \end{bmatrix} \tag{10}$$

Figure 6 (c) was the profile of one section of grinding wheel after translating sub-pixel edge position to actual size. And then, this profile was divided in two parts, named left arc and right arc, with the highest point as demarcation point. Furthermore, radius of right and left least-square circle of each image were denoted by set {R1，R2 ...Rn} and set {L1，L2 ...Ln}. After calculating the range of two sets, the larger number was chosen as the dressing amount, named V1.

Figure 5. Appearance of image capture system.

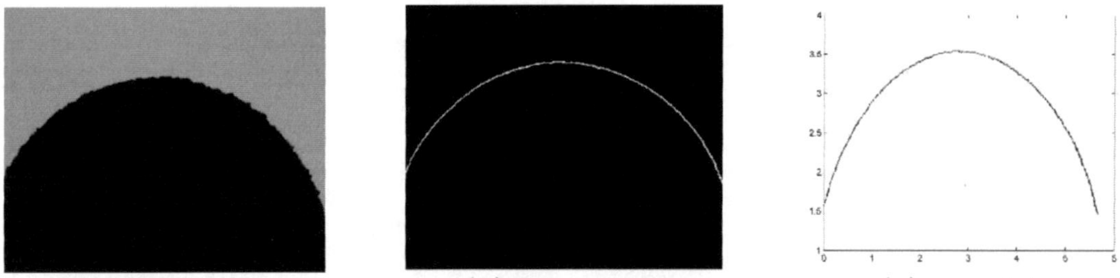

(a)Original image （b）Edge position （c）Actual size

Figure 6. Image processing based on Zernike moment.

Moreover, superposing all images which were grayed before into one image, and edge position was calculated based on Zernike moment. Similarly, radius and center of left and right least-square circle were calculated, and the difference between data points and radius were calculated as the protrusion heights. After that, using the algorithm to calculate mean protrusion height and standard deviation under increased feed amount, and optimal feed amount was chosen as the dressing amount based on protrusion height, named V2. At last, larger number of V1 and V2 was chosen as optimal dressing amount, and carbon sample was grinded by the dressed grinding wheel, as shown in Figure 7.

Figure 7. Carbon sample.

4. Experimental result

According to above-mentioned set-up and step, 30 gothic-arc were dressed in the experiment. Dressing amounts and measuring results of carbon sample were shown in Figure 8.

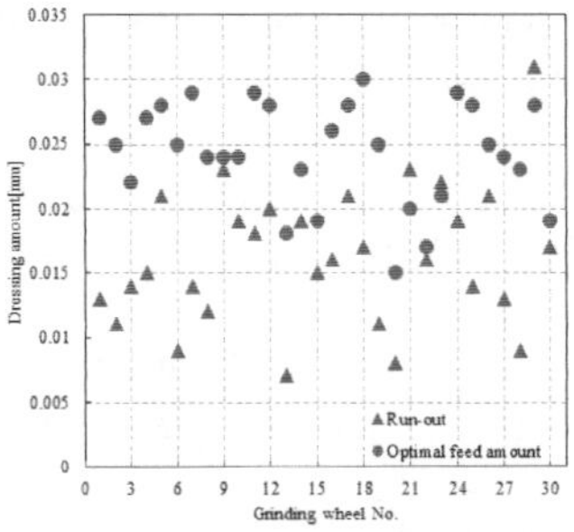

(a) Dressing amount

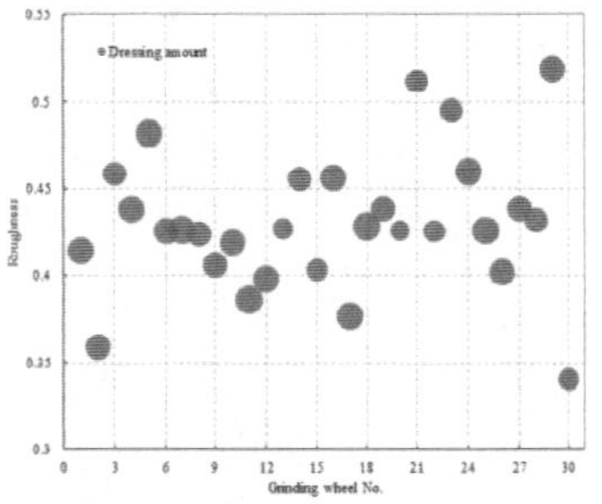

(b) Roughness vs dressing amount

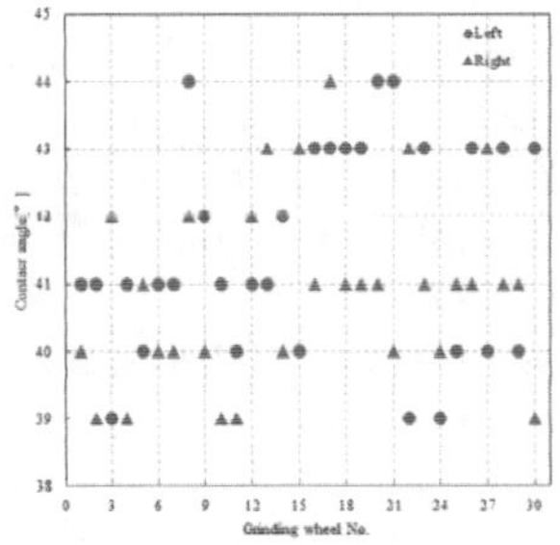

(c) Contact angle vs grinding wheel No

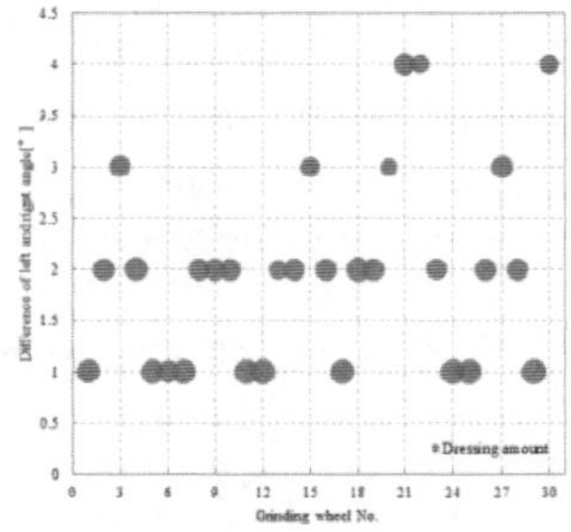

(d) Contact angle vs dressing amount

Figure 8. Dressing amounts and measuring results.

Figure 8 (a) shows that the majority optimal dressing amount were calculated based on protrusion height, besides No.21, 23, 29 were in opposite, which meant radius of each section was similar. Moreover, the number of experimental grinding wheel which led to greater roughness value of workpiece were 5, 21, 23, 29, dressing amount of three of them were calculated based on run-out. In a word, surface roughness of workpiece was better if the dressing amount was calculated based on protrusion height than based on run-out, and roughness value increased while dressing amount was bigger than V2.

In addition, Figure 8 (c) shows that left and right contact angle could meet requirement shown in Table.1, and Figure 8 (d) shows that the volume of data point representing dressing amount when the difference of left and right angle was equal to 1 degree was less than that was equal to 4 degree. In other words, the difference of left and right contact angle decreased while dressing amount increased.

5. Conclusion

Aiming at the difficulties in accurate prediction of the dressing amount of monolayer electroplated grinding wheel, a new calculation method based on image processing was proposed. Conclusions were drawn as follows:

1. With the sub-pixel level edge detector based on Zernike moment, the measurement accuracy is reached about 5 μ m.
2. Two calculation method for dressing amount are proposed. One is calculated based on protrusion height, the other is calculated based on run-out. Surface roughness of workpiece is better if the former one is chosen as optimal dressing amount.
3. The proposed image processing system and calculation method are valid. The experimental results show dimensional accuracy of grinding wheel and roughness of workpiece can meet requirements.

6. References

[1] A. Chattopadhyay, L. Chollet and H. Hintermann 1990 On performance of chemically bonded single-layer CBN grinding wheel.*CIRP Annals* **39** 309-312

[2] A. Ghosh and A. Chattopadhyay 2007 Experimental investigation on performance of touch-dressed single-layer brazed cBN wheels. *International Journal of Machine Tools and Manufacture* **47** 1206-1213

[3] C. Dold, R. Transchel, M. Rabiey, P. Langenstein, C. Jaeger, F. Pude, F. Kuster and K. Wegener 2011 A study on laser touch dressing of electroplated diamond wheels using pulsed picosecond laser sources.*CIRP Annals-Manufacturing Technology* **60** 363-366

[4] X. Lin, Z. Wang, Y. Guo, Y. Peng and C. Hu 2014 Research on the error analysis and compensation for the precision grinding of large aspheric mirror surface.*The International Journal of Advanced Manufacturing Technology* **71** 233-239

[5] E. Weingärtner, R. Roth, F. Kuster, M. Boccadoro and F. Fiebelkorn 2012 Electrical discharge dressing and its influence on metal bonded diamond wheels.*CIRP annals* **61** 183-186

[6] E. Westkämper 1995 Grinding assisted by Nd: YAG lasers.*CIRP annals* **44** 317-320

[7] K.-C. Fan, M.-Z. Lee and J.-I. Mou 2002 On-line non-contact system for grinding wheel wear measurement.*The International Journal of Advanced Manufacturing Technology* **19** 14-22

[8] Y. M. Luo and D. J. Hu 2008 Real-Time Measurement and Compensation for Wheel Wear in Curve Grinding Based on Image Processing Method.*Key Engineering Materials* **359-360** 479-483

[9] M. R. Teague 1980 Image analysis via the general theory of moments.*JOSA* **70** 920-930

[10] J.-q. Li 2003 Improved algorithm of subpixel edge detection using Zernike orthogonal moments.*Optical Technique* **29** 500-503

Design and Developed of High Performance Rotary Head Module for Core Drill Rig

Muhammad Sajjad and Dong Won Jung

Department of Mechanical Engineering, Jeju National University, 102 Jejudaehak-ro, Jeju-si, Jeju Special Self Governing Province, 63243, Republic of Korea

Abstract. Fossil fuel, especially oil, represents the engine of nowadays life. Oil companies work hard in order to extract it from subsurface hydrocarbon reservoirs. The operations at which oil is being extracted from these reservoirs are referred to as drilling operations. Drilling operations are per-formed with the means of drilling rigs. However, due to the high demand for oil, drilling for oil becomes more challenging and oil companies are always looking for the most effective extraction methods and are they moving toward the precious control of the drilling operation for different type of requirement and conditions. This paper aimed for the manufacturing of High-performance Rotary Head module of Core Drill equipment which is used for geological purposes. The fallow target criteria will be going to achieve which includes the rotation speed, rotation torque, gripping force, and speed change at different stages. In our study we find out that the rotational speed of head is greater than 1200 rpm and maximum torque generated is greater than 160 kgf.m also the drill rod breaking power is equal or higher than 6 ton. After the complete study it is certain that the rotary head module will be used for future use.

1. Introduction

Rotary drilling rig is a kind of pile drilling machinery which is suitable for deep-hole piling in building foundations, and is also widely used in different foundation constructions. A typical structure of a drilling rig is shown in 'figure 1'. It consists of rotary head, gear box, mast, swivel body, sand reel, rotary table, hydraulic system and electrical system [1]. Core drilling is used in a variety of fields to identify underground exploration. The equipment used for such core drilling has different shapes and functions depending on the manufacturer, but it is similar to the general drilling equipment and it is distinguished from general core drilling.

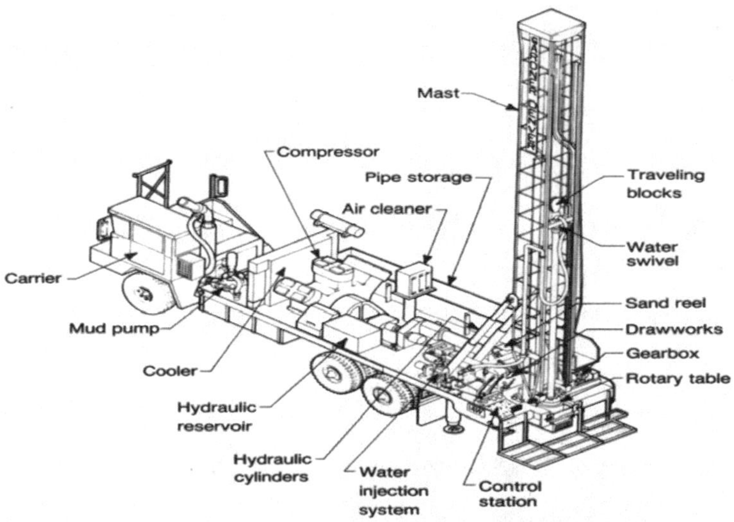

Figure 1. Structure diagram of the rotary drilling rig

The geologic drilling rig is a device used for a general purpose drilling rig, used to take samples (core) of rock by using high rotational speed. Therefore, it is necessary to perform a high-speed rotation of 1,000 rpm or more, to continuously grasp (clamp) the drill rod while working, and to develop a device capable of changing the number of revolutions according to working conditions.

In order to increase the work efficiency, we design and manufacture the high speed rotating structure so that we can build up the foundation for commercialization by accumulating know-how about core drill. Torque of the Core the Drill, and the tensile Force are key performance indicators. For the performance of the Core Drilling machine we check the performance of the "Rotary Head" module are indicated by number 5 in the 'figure 2' which is installed in the drilling rig. The main performance listed above depends on the "Rotary Head" module.

Figure 2. Rotatory head module of drilling rig

Archaeological records show that as early as 3000 B.C., the Egyptians may have been using a similar technique. Leonardo Di Vinci, as early as 1500, developed a design for a rotary drilling mechanism that bears much resemblance to technology used today. Despite these precursors, rotary drilling did not rise in use or popularity until the early 1900's [2, 3].

IOP Conf. Series: Materials Science and Engineering **576** (2019) 012011 doi:10.1088/1757-899X/576/1/012011

Although rotary drilling techniques had been patented as early as 1833, most of these early attempts at rotary drilling consisted of little more than a mule, attached to a drilling device, walking in a circle! It was the success of the efforts of Captain Anthony Lucas and Patillo Higgins in drilling their 1901 'Spindletop' well in Texas that catapulted rotary drilling to the forefront of petroleum drilling technology [4].

2. Designing and Mathematical Calculation of Rotatory Head Components

In the designing process two main component are design which is rotatory chuck and the transmission system for efficient rotation and power transmission of drilling rig. For rotatory chuck we Increase the number of jaws and the size of the ground plane to increase the gripping force of the rod. To maintain the rotational force of the rotatory head thrust bearing has been used which is the ideal for our designing condition [5, 6]. The 2D and 3D of rotatory chuck with swivel body are shown in 'figure 3'.

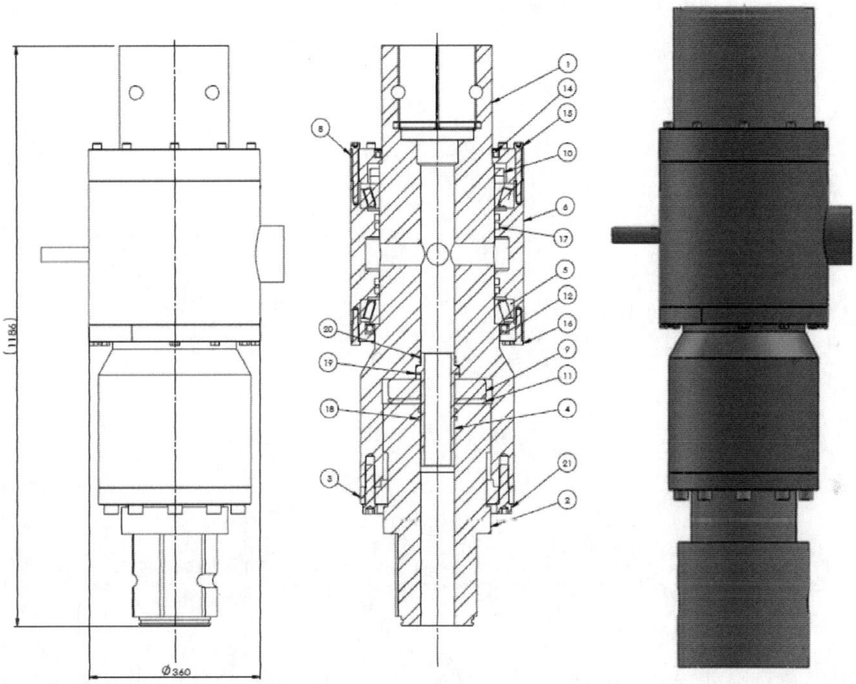

1	Swivel_Body	12	TC Oil-Seal OD300_ID260_20T
2	Shock Absorbor Shaft	13	Grease nipple-1_8inch
3	Shock Absorbor cap	14	TC_Oil Seal_ID230_OD260_16T
4	shock absorbor tube	15	Hex Bolt M12 x 110L
5	Taper Roller Bearing 32948	16	Bolt M12 x 65L
6	Swivel Body	17	SKY OD260_ID240
7	Swivel_Lower Cap	18	SKY OD80_ID70
8	Swivel_Upper Cap	19	D90 Snap Ring
9	Shock absorbor Rubber	20	G-70 O-Ring
10	M230 Nut	21	Hex Bolt_M20 x 80L
11	shock absorbor steel plate		

Figure 3. 2D and 3D diagram of rotatory chuck with swivel body

Force Calculation for chucking is to perform in which we measure the gas spring minimum initial force from the diagram through formulas and the value of forces are shown in the Table 1. Where Fc represent the chuck force of the drill rig. The complete model of rotary head module was initially design in solid-works for the required amount of rotational speed and gripping force. The model consist of many parts which is indicated in Figure. 3 above with numbering from 1 to 21. One end of the rotary head module is screw type which will connected to the Kelly Bars and there dimesons are 150*150*250 (mm), the other end of module is connected to hammer and it's a box type connection as we can see from the assembly diagram in 'figure 4'.

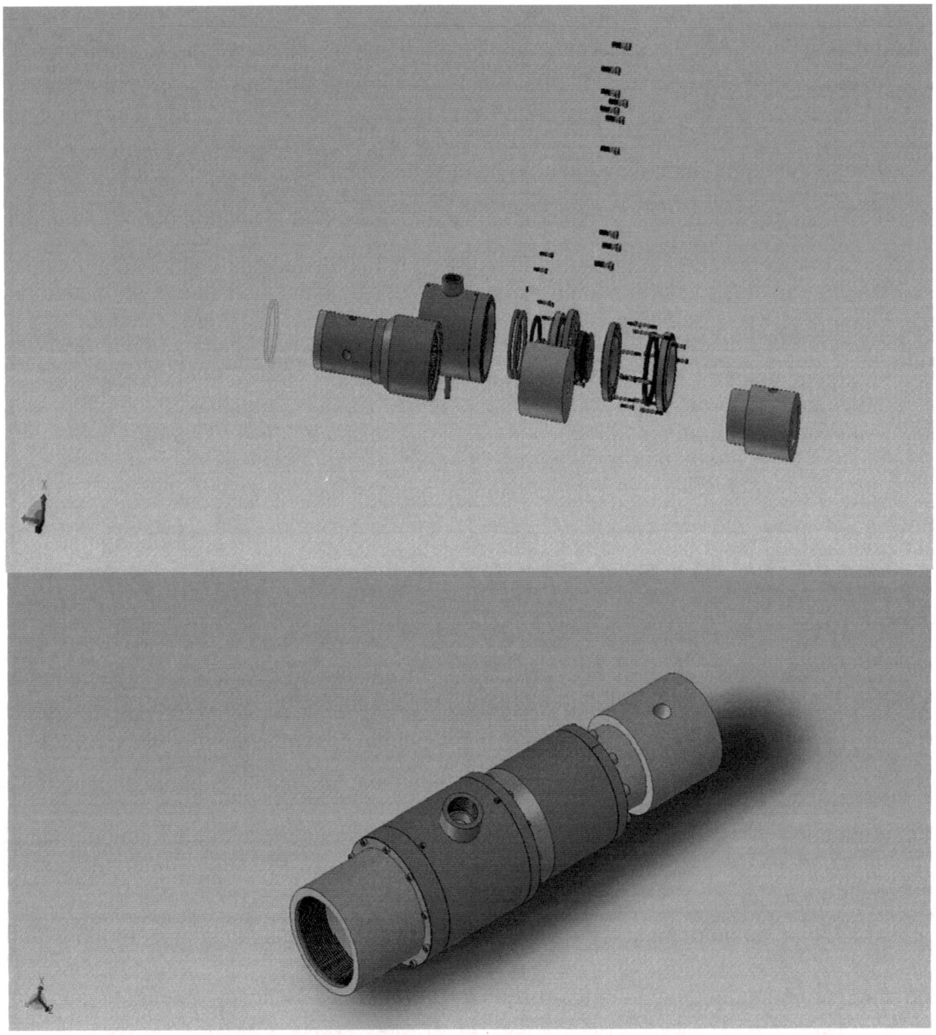

Figure 4. Mathematical model of rotary drill module in solidworks

3. Transmission Design for Improving Rotary Head Operating Force

Geological drilling Core drill is operated at a constant speed, which often causes a decrease in work speed or breakage of drill when faced rocks or obstacles in the ground. In this task, we have developed and installed a manual transmission that can control the output manually according to the condition of the object for sampling, so as to enhance the commerciality of the rotary head.

It is designed to increase the performance by controlling the output by designing and installing the 4-speed gearbox as a model of the manual transmission for the vehicle. The core drill transmission is considered to be suitable for non-clutch type manual transmission, and the basic type is a basic model of 4-speed manual transmission for vehicle. For the experiment three types of gear shifting has been used to find out the best suitable option for our design which are, selective shifting type, continuously shifting type and synchronized type.

Selectable folding type (Perturbation gear type) in which shift is possible by simply pushing gear shown in Figure. 5. These type are good for speed control but it has a big disadvantage when there is a lot of noise and the stage is changed that gears are difficult to fit.

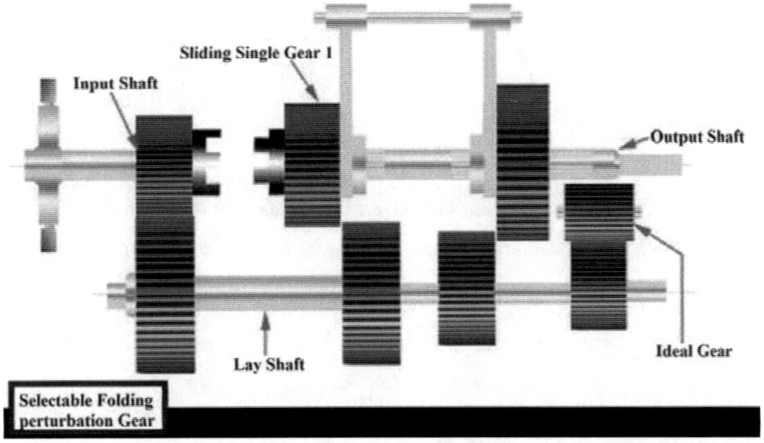

Figure 5. Selectable folding Perturbation gear transmission system

Always-on (Regular combination) Improved type of Selective-fold Gearbox. The gears are always, in engagement with is the auxiliary gear, which is powered by the Input Shaft, Since the Gears are not Fixed to the main shaft, they are Free to The Rotate and need of "Dog Clutch" for the Engagement of Gear which are mostly used in high-torque and high-speed vehicles. Figure. 6 shows the working mechanism of transmission type.

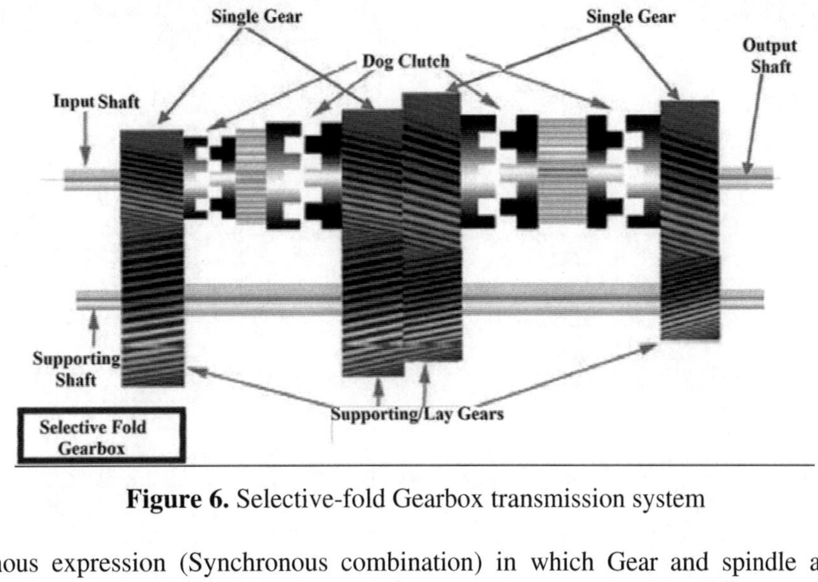

Figure 6. Selective-fold Gearbox transmission system

Synchronous expression (Synchronous combination) in which Gear and spindle are connected smoothly by sleeve and synchronizer ring and the power is transmitted in different gears shown in 'figure 7'.

IOP Conf. Series: Materials Science and Engineering **576** (2019) 012011 doi:10.1088/1757-899X/576/1/012011

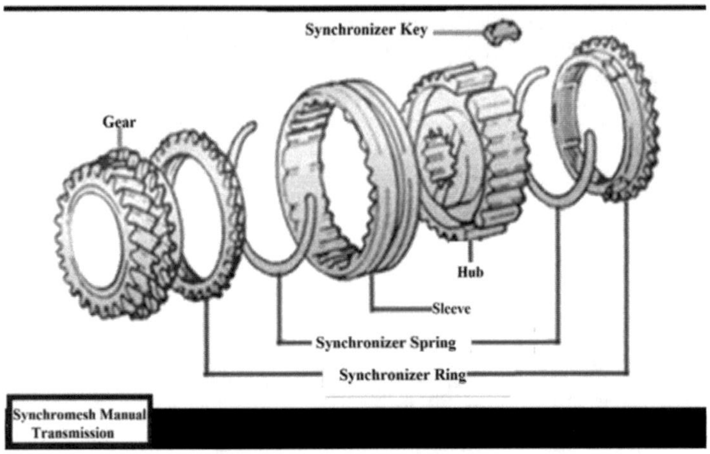

Figure 7. Synchromesh manual transmission system components

The final design of transmission module is shown in the 'figure 8', which consist of total of 29 components along with the numbering and names in the below figure.

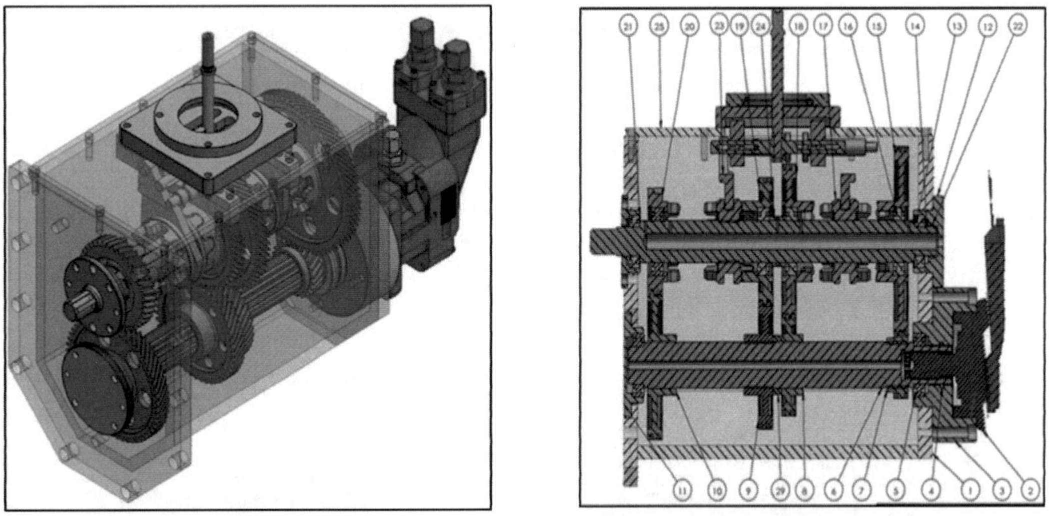

1	Transmission Case
2	Parker Motor F12 60
3	Motor Cover
4	Quad Ring : AS568 - 328
5	Tapered Bearing 45 x 75 x 16
6	Driving Shaft
7	Driving 1st Gear
8	Driving 2nd Gear
9	Driving 3rd Gear
10	Driving 4th Gear
11	Driving Shaft Cover
12	Driven Shaft Cover
13	Roller Bearing 35 x 62 x 14
14	Driven Shaft

15	Driven 1st Gear
16	Ball Bearing 50 x 80 x 16
17	Dog Clutches
18	Driven 2nd Gear
19	Driven 3rd Gear
20	Driven 4th Gear
21	Output Cover
22	Quad-Ring AS568-221
23	Dog Pusher
24	Distance Ring
25	Change Lever Assy
26	Axial Stop Ring S-55
27	Axial Stop Ring S-50
28	Lever distance R-80
29	Driving Shaft Distance Ring

Figure 8. Final Design of transmission system

4. Experimentation and Results

The fallowing equipment's are used to find the different parameters during the test.

a. Rotational speed - Measurement using tachometer Stable fixed structure when rotary head is mounted Measure the rotational force of the lower ground surface when rotating the rod on the rotary head.

b. Rotational torque - Pressure measurement with hydraulic actuator Maximum rotational torque: 160 kgf * m (45.3 bar pressure when achieving torque cylinder target)

c. Gripping force - Measurement of gripping force by drawing test through linear hydraulic drawing device Gripping force: 6,000 kgf or more (pressure when gripping force cylinder target is reached 101.9 bar)

d. Checking the speed change stage - Four speed shift mode.

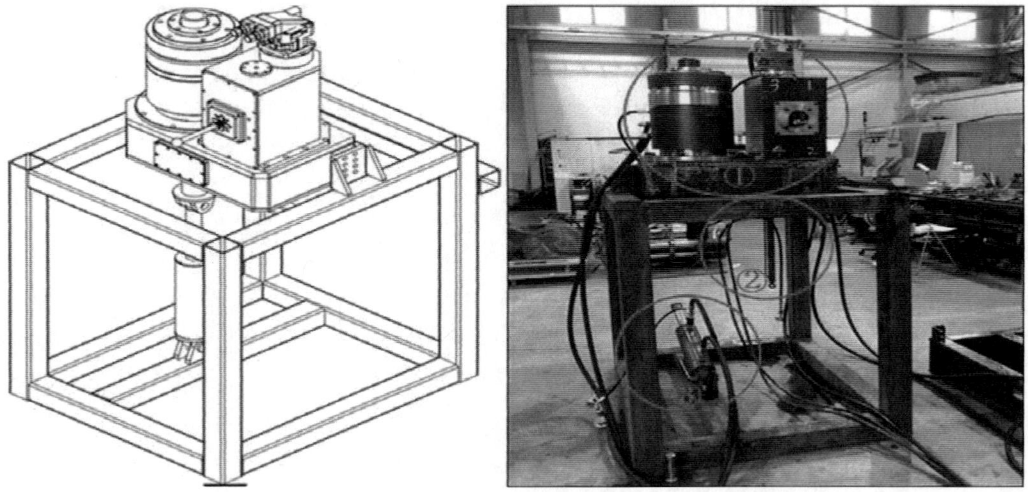

Figure 9. Assembly of Rotary Head for Performance Test

Complete assembly of Rotary Head for Performance Test and there setup are shown in the Figure 9. In the whole setup there are three main components which are indicated through numbers from 1 to 3.

① Rotary head (Body) - equipped with the high-speed motor and transmission performed the performance test after mounting the test rod.

② Cylinder for measuring rotational torque - the performance instrument measured torque during rotational motion.

③ Cylinder for gripping force measurement – Measure this values in case of maximum torque the chuck (Jaw) rod is pulled out.

5. Summary of Results

From the study we find out the fallowing results that the drilling rotation speed of the Core Drill is about 100 rpm. However, the Rotary Head of the Core Drill has been developed in this study to exhibit a rotational speed comparable to that of overseas competitors at 1,200 rpm. In order to realize the high rotational speed, it is necessary to design the mechanism that can maintain the speed stably, and it is necessary to design the structure that can cancel the rotational force and the centrifugal force.

By integrating hydraulic system design technology into the design technology of the drilling rig that we possess, we have accomplished the results through this development task and it is an opportunity to reduce the technology gap with related overseas technology. The design of Chuck (Jaw) for rod grip, which is a major performance part of Rotary Head, is designed as Nitro Gas spring type that is used overseas, so that it can overcome the compatibility part when exporting from abroad.

We also find out our final requirement of the study which is the rotational speed of head is greater than 1200 rpm and maximum torque generated is greater than 160 kgf.m also the drill rod breaking power is equal or higher than 6 ton. Our key performance indicators compare to other world class company are compared in the below Table 1. So by using our study In future we can manufacture the drilling module with in Korea which will fulfill the requirement of world top class companies.

Table 1. Study achievements and there comparison with world top class companies

Key Performance Indicators	Unit	Development Target		World-class (Company/ Country)	weight (%)
		Target Value	Result		
1. Rotation Speed	Rpm	1,200 More than	1,342	1,200 (Bo Art Longyear / USA)	30
2.Maximum Rotational Torque	kgf * m	160 More than	215.985	165 (Bo Art Longyear / USA)	30
3.Drill Rod Gripping force	ton	6 More than	7.658	-	20
4.Shift number	Only	4 More than	4	4 (Bo Art Longyear / USA)	20

6. References

[1] Q.F. Li, J.L. Li, Overall Design Of TR250D Rotary Drilling Rig, *Engineering Machinery*, **5** (2008) 41-45.

[2] J. Luo, Working Performance Analysis and Optimization Design of Rotary Drilling Rig under On Hard Formation Conditions, *Procedia Engineering*. (2014) 23-28.

[3] J.G. Zhang, X.L. Li, G.J. Tang, Y.G. Liu, B. Jiang, The Application Discussion And Generalization Of The Rotary Drilling Rig In Guizhou, *Construction Mechanization*, **8** (2012) 41-45.

[4] Pei Ran. The Market Analysis of the Rotary Drilling Rig under the New Situation [J]. T*oday Engineering Machinery*, 2010(6).

[5] H.M. Kang, Q.H. He, J.X. Zhu. Dynamics Simulation on Installation Angle of Mast Link Frame System of Rotary Drilling Rig, *International Conference on Measuring Technology and Mechatronics Automation*. (2009) 221-225.

[6] R.W. Tucker, C. Wang, Torsional Vibration Control and Cosserat Dynamics of a Drill Rig Assembly, *Meccanica*. **38** (2003) 145–161.

IOP Conf. Series: Materials Science and Engineering **576** (2019) 012012 doi:10.1088/1757-899X/576/1/012012

Steady-state temperature field of stator bar of double water internal cooling synchronous condenser

Fuchun Sun[1], Andong Wang[1], Gang Ji[2], Junqing Li[2], Jun Yong[3] and Jihao Wang[3]

[1]State Grid Shandong Electric Power Research Institute, 250002 Jinan, China
[2]School of Electric and Electronics Engineering, North China Electric Power University, 071003 Baoding, China
[3]State Grid Shandong, 250002 Jinan, China

Abstract. Synchronous condenser is a key part of ensuring stable operation of the extra-high voltage AC and DC transmission grid. In order to optimize and ensure the reliable operation of the synchronous condenser cooling system, it is necessary to master the temperature distribution in the synchronous condenser. In the double water internal cooling synchronous condenser, the fluid flow and heat transfer are quite complicated in the water-cooling system of the stator and the rotor. In this paper, the steady-state temperature field of stator coil of TTS-300-2 300MVar double water internal cooling synchronous condenser is calculated by using the finite volume method. The influence of empirical parameters on the calculation results is reduced. The calculation results show that the water temperature and copper temperature increase linearly along the direction of water flow.

1. Introduction

As China's Direct current ultra-high voltage (DC UHV) transmission continues to deepen, the capacity required for reactive power compensation continues to increase, and the static var compensator can no longer meet its capacity requirements. In order to meet the requirements of reactive power compensation, China has invested in a batch of 300Mvar synchronous condenser. The problem of heat generation and cooling of synchronous condensers has become one of the major concerns of the design, manufacturing and operation departments. Therefore, monitoring the temperature and temperature rise during its operation is an important part of ensuring its normal operation. For large steam turbine generators, the main cooling methods are air cooling [1], hydrogen cooling [2], water cooling [3], and evaporative cooling [4]. Large capacity synchronous condenser is improved on the basis of turbine synchronous generators. As one of the key indicators of generator design, temperature rise is directly related to the life and operational reliability of synchronous condenser. An excellent cooling system plays a vital role in controlling the temperature rise of the unit. At present, the large capacity synchronous condenser has mainly two cooling modes including in total-air-cooling and double-water internal cooling systems. All of components are cooled by air in the total-air-cooling synchronous condenser. The stator coil and rotor winding are seperately cooled by water and the iron core is cooled by air in the dual-water internal cooling synchronous condenser. The large-scale synchronous condensers have only emerged in recent years, however, the current researches are mainly directed to the stator winding temperature field of large generator sets[5].In order to make the synchronous condenser run stably, it is very necessary to study the temperature field of the synchronous condenser stator winding.

The scholars have studied the calculation of the stator temperature field of large steam turbines. The calculation method of temperature field in large capacity generators includes mainly finite element method[6], finite volume method[7] and heat net method[8], all of these methods obtained the precise result in partly. The finite volume method has good conservativeness, could solve complex engineering problems well, and has good adaptability to the grid. It has good adaptability to the grid. In the fluid-solid coupling analysis, it can be perfectly integrated with the finite element method. For example, the QFSN-220-2 turbo generator is used as an example to compare the calculation results with the experimental data and the two are in good agreement, which verified the reliability of the finite element method [9]. As far as the temperature field calculation of large generators is concerned, it can be found that the research on the the stator coil is mainly carried out by the method of finite element, and only local analysis can be performed due to the limitation of model size and meshing [10]. The synchronous condenser is optimization designed on the basis of turbine generator, so the finite element method is also suitable for the temperature field analysis of the synchronous condenser.

The paper studied the steady state temperature field of stator bar in the dual-water internal cooling synchronous condenser. Since 97 percent of the heat loss generated by the stator coil is taken away by water [11,12], it is assumed that the losses generated by the stator coil are all carried away by water, and the conjugate transfer between copper and water over a water path length is performed by the finite volume method. The temperture field is directly analyzed and calculated to improve the accuracy of the calculation results.

2. Stator bar physical model and meshing

2.1. Physical model of stator bars

The TTS-300-2 synchronous condenser is taken as an example for analysis and calculation of temperature field. The cooling method of TTS-300-2 synchronous condenser is double-water internal cooling, that is, the stator and the rotor windings are cooled by passing cooling water into the hollow strands. This paper mainly studies the cooling system of its stator windings. The synchronous condenser stator bar consists of 30 strands, in which there are 24 solid strands and 6 hollow strands, cooling water is passed through the hollow strands to achieve cooling of the bars. A temperature sensor is installed at the outlet end of each bar to detect the temperature of the cooling water there. The stator bar has an effective length of 12,000 mm, a height of 58.75 mm, and a width of 33.9 mm. The wire rod structure is shown in figure 1.

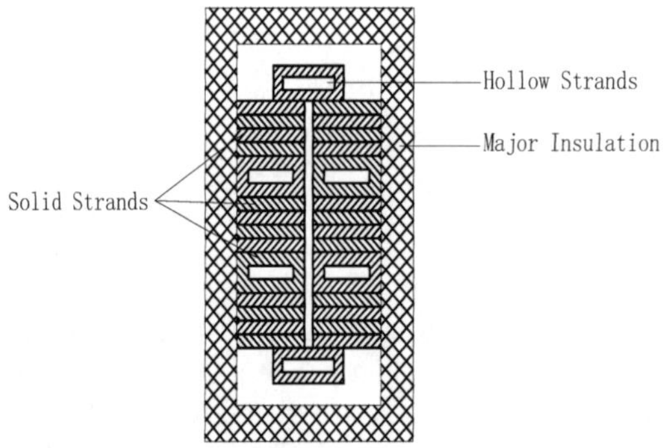

Figure 1. Cross-sectional view of the stator bar.

2.2. Meshing of stator bars

The ICM CFD software is used for calculating the temperature field of synchronous condenser, the fluid control equations are discrete in the spatial domain [13]. Since the axial length of stator bar is much larger than its equivalent diameter and the stator bar belongs to the elongated pipe fluid model, the hexahedral meshing strategy is most suitable when dividing the Solution domain. Because the length of stator bar is large, its grid partition sketch is divided into two visual angles: radial and axial. The partition results are shown in figure 2 and figure 3.

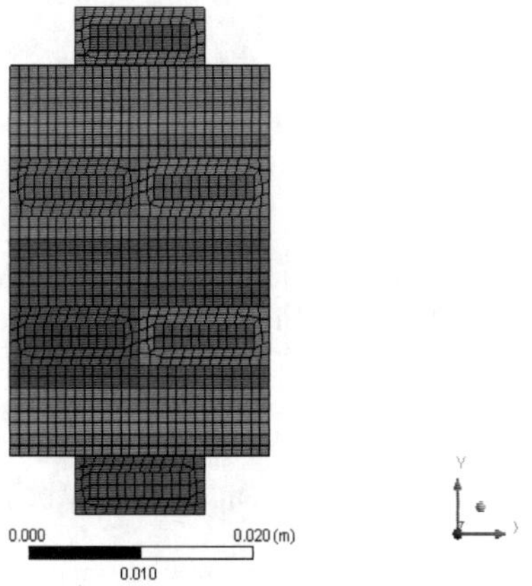

Figure 2. Radial meshing of stator bars.

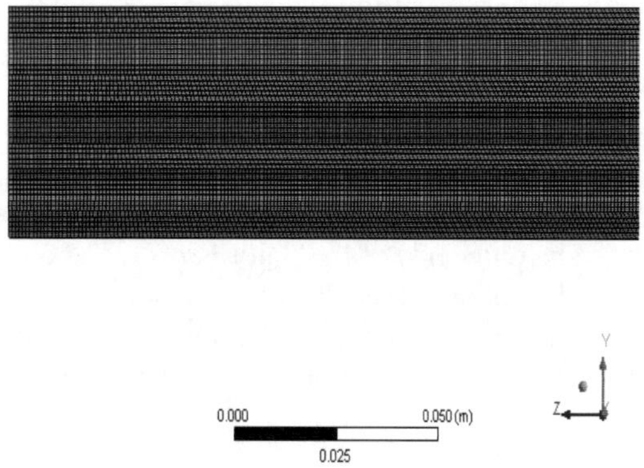

Figure 3. Axial meshing of the part of stator bars.

3. Mathematical model of three-dimensional fluid field and temperature field of stator bars

In order to simplify the model, the fluid field and temperature field in the stator bar are analyzed and calculated under the following preconditions according to the structure of the stator bar, fluid flow and heat transfer characteristics [12].

- Since the Mach number of the cooling water in the stator bar is small, the fluid is considered as an incompressible fluid.
- The Reynolds number of the cooling water in the stator bar is much larger than 2300, so it is in the turbulent state, and the turbulence model is used for calculation and analysis of internal cooling water.
- Ignore the effects of buoyancy and gravity.
- The fluid flow and fluid solid heat transfer in steady state are studied only.

The cooling water in the stator bar is incompressible and in a stable state. The mass conservation equation and the momentum conservation equation are as following [9][13]:

$$\nabla(\rho\vec{V}) = 0 \tag{1}$$

$$\nabla(\rho\vec{V}\vec{V}) = -\nabla p + \nabla\tau + F \tag{2}$$

Where, ρ is the density; p is the static pressure on the fluid micro-body; τ is the viscous stress generated by the viscous action; F is the volume force on the micro-body; ∇ is the divergence, ie , $\nabla(\rho\vec{V}) = div(\rho\vec{V})$.

For the fluid in the stator bar body and the cooling channel, the general energy conservation equation is [5]:

$$\frac{\partial(\rho T)}{\partial\tau} + \nabla\cdot(\rho\vec{V}T) = \nabla^2\left(\frac{\lambda}{C}T\right) + S_T \tag{3}$$

Where, $\vec{V}$ is the velocity vector; T is the temperature; λ is the thermal conductivity; C is the constant pressure specific heat capacity ; λ / C is the diffusion coefficient; S_T is the ratio of the sum of the heat source per unit volume and the mechanical energy converted to thermal energy due to viscous action to C, for the fluid in the cooling channel, $S_T = 0$.

The heat generated by the stator winding is mainly takes away by cooling water in the hollow wire with convection heat transfer, and satisfies the third type of boundary condition at the fluid-solid coupling boundary. Which is:

$$-\lambda\left(\frac{\partial T}{\partial n}\right)_{\omega} = h\left(T_{\omega} - T_{f}\right) \tag{4}$$

Where, n is the outer normal of the heat exchange surface; h is the surface heat transfer coefficient; T_{ω} and T_{f} are the temperature of the interface and the fluid around it, respectively. For the conjugate heat transfer problem, the solid domain and the fluid domain need to be defined first, and the subdomain loading heat source is set in the solid domain, the heat source is given in the form of heat generation rate[14][15]. The boundary conditions of the stator bar solution domain are as following:

- The outer side of the stator bar is an adiabatic boundary;
- The cooling water inlet is a mass flow inlet, and the cooling water outlet is a pressure outlet;
- The wall surface is smooth and has no slippage;
- The water channel of the hollow copper wire is the coupling heat exchange boundary.

4. Results and analysis

The basic parameters of TTS-300-2 synchronous condenser are: Q_N=300MVar, U_N=20kV, I_N=8660A, .The coupling calculation of the stator winding fluid field and temperature field was completed on the FLUENT software platform using the above conditions.

The boundary condition setting values are shown in table 1.

Table 1. Boundary condition setting

Waterway entrance flow（kg/s）	**Inlet water temperature（°C）**	**Water outlet pressure（MPa）**
0.02652	45	1.01

4.1. Determination of heat source of material

The main heat source of the stator is the copper loss caused by the stator winding and the iron core loss caused by the stator core. In the stator winding solution domain, the copper loss constitutes a part of the main heat source in the stator winding temperature field. The stator copper loss includes basic copper loss and additional eddy current loss, and the additional eddy current loss is calculated by the classic formula of skin effect [16]. Below is the copper consumption formula:

$$q = q_{Cu} + q_{ps} = I^2 r_0[1+0.004(T-15)] + I^2 k_f r_0[1+0.004(T-15)] \tag{5}$$

Where I is the strand current and T is the strand temperature, r_0 is DC resistance of strand wire at 15°C; k_f is the fielder coefficient that accounts for the increase in resistance caused by eddy current effect.

When the current phase of the upper and lower bars is the same, the average field coefficient k_{f1} of the upper bars is 7 times of the average field coefficient k_{f2} of the lower bars, that is, $k_{f1} = 7k_{f2}$. When the upper and lower bars are in different phase, the average field coefficient of the upper bars is 5.5 times of that of the lower bars, that is, $k_{f1} = 5.5k_{f2}$. The field coefficient of the lower stator bar is derived from document [17]. This paper considered the case when the current phase of upper and lower rods is the same. After calculation, the upper bar copper loss is 953886W/m3, and the lower bar copper loss is 820349W/m3.

IOP Conf. Series: Materials Science and Engineering **576** (2019) 012012 doi:10.1088/1757-899X/576/1/012012

4.2. Analysis of the fluid field and temperature field of stator bars

When the synchronous condenser is operated under the rated load, the velocity distribution of the fluid in the hollow tube is as shown in figure 4.

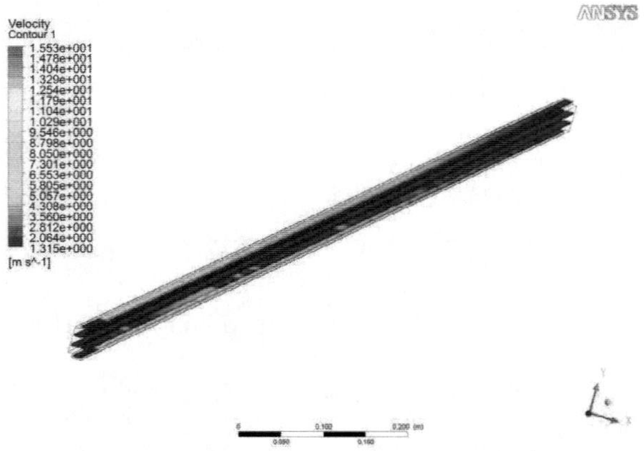

Figure 4. Velocity profile.

It can be seen from the velocity profile in figure 4 that the velocity distribution in the hollow tube is low at the inlet and outlet, and the velocity near the inlet of the hollow tube is slightly higher. From the axial section of the pipe, because of the viscosity of the fluid, the intermediate speed of the section is the highest, the peripheral speed is gradually reduced and the fluid is evenly distributed near the wall.

Figure 5 and figure 6 show the temperature distribution of the upper and lower stator bars respectively:

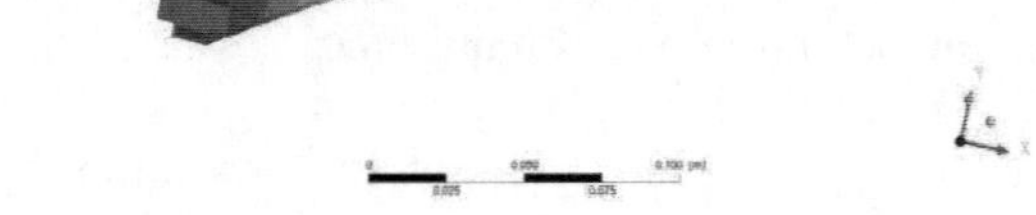

Figure 5. Upper stator bar temperature distribution cloud map.

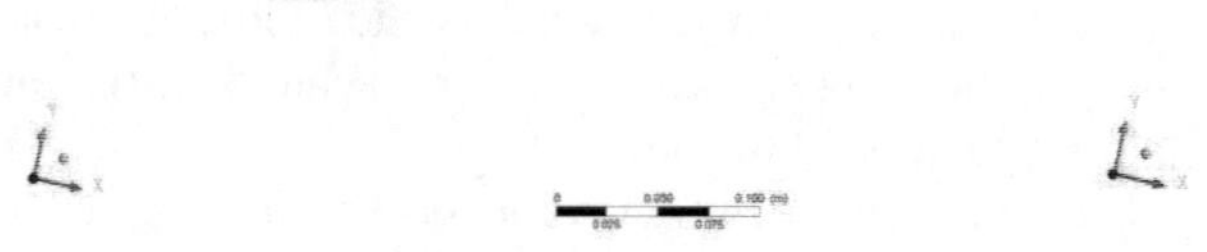

Figure 6. Lower stator bar temperature distribution cloud map.

It can be seen from figure 5 and figure 6 that the outlet temperature of the upper stator bar is 329.8K, and the outlet temperature of the lower stator bar is 328.2K, so that the average temperature of the outlet is 329K, and the experimental data is 328.1K. The calculation value is compared with the experimental one by 0.9K, which can explain that the model meets the requirements of the condenser state simulation system. The cooling water flows at a relatively stable speed in the hollow strand, the temperature distribution inside the strand is relatively uniform, and the cooling effect is sufficient to meet the temperature requirements for stable. operation of synchronous condenser.

The axial distribution of the cooling water temperature and the temperature of the copper wire portion in the hollow strand in the stator bar is shown in figure 7.

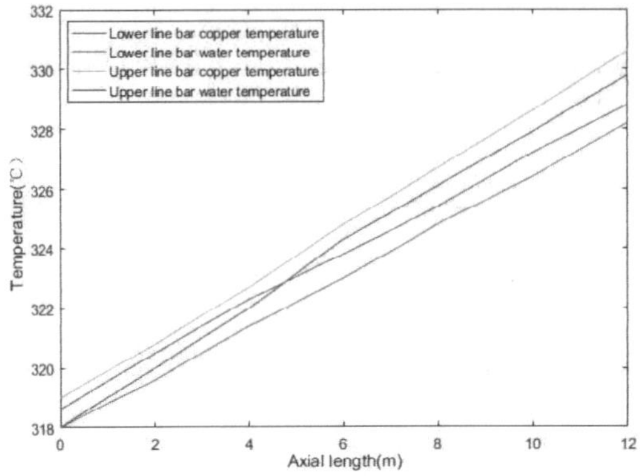

Figure 7. Axial temperature distribution of cooling water and copper wire.

It can be seen from Fig. 7 that the variation of the temperature of the cooling water and the copper wire in the stator bar in the axial direction is approximately linear. The temperature of the cooling water in the upper strand hollow strand is higher than the temperature of the cooling water in the lower strand hollow strand by an average of nearly 0.99K. The temperature of the upper copper wire corresponding to each portion of the cooling water in the axial direction is 0.72 K higher than the cooling water; the temperature of the lower copper wire corresponding to each portion of the cooling water in the axial direction is 0.73 k higher than the cooling water.

5. Conclusion

In this paper, FLUENT is used to calculate the water cooling conjugate heat transfer of stator bars. By directly solving the coupled heat transfer between water and copper, the influence of empirical parameters on the calculation is reduced. Through calculation and analysis, the following conclusions can be drawn:

- The flow velocity of water is larger in the middle of the strand than on both sides of the strand, and the water temperature and copper wire temperature in the hollow copper strand increase gradually along the flow direction.
- The water temperature of each section in the hollow strand is approximately linear with the water inlet temperature corresponding to the section.
- Due to the skin effect, the additional loss of the upper layer stator bar is higher than that of the lower layer stator bar, so the average temperature of the upper bar is higher than the lower bar, but the difference is small, which indicates that the cooling effect of the cooling water in the hollow copper wire is better. It can meet the temperature control requirements of the normal operation of the unit.

The new-type synchronous condenser with large capacity is put into operation for a short time, and research on its condition monitoring and fault diagnosis has not been carried out in depth. This paper could provide the early basis of the health fault research in synchronous condenser. In the next work, the heat fault characteristic of synchronous condenser will be studied using finite element method.

6. References

[1] Wen Jiabin, Meng Dawei, Zhou Meilan, *Transactions of China Electrotechnical Society* **15**(6), 1-4(2000)
[2] Li Deji, Cao Guoxuan, *Transactions of China Electrotechnical Society* (3), 3-10(1986)
[3] Wu Guizhen, Meng Dawei, Xu Mingyu, *EXPLOSION-PROOF ELECTRIC MACHINE* **43**(142), 1-4(2008)
[4] Dong Haihong, Gu Guobiao, *Proceedings of the CSEE* **28**(20), 137-141(2008)
[5] Zhou Feng, XiongBin, Li Weili, etc, *Proceedings of the CSEE* (24), 128-132(2005)

[6] Daud H.A., Li Q., Bég O. Anwar, Abdulghani S.A.A.,*WIT Transactions on Modelling and Simulation* **51**, 279-292(2011)
[7] Bian Xin, Li Tao, Jiang Liang, Zhang Ruigang, Huang Hongyan, *Applied Mechanics and Materials* **614**, 128-132(2014)
[8] Meroni Andea, Andreasen Jesper Graa, Persico Giacomo, Haglind Fredrik, *Applied Energy* **209**, 339-354(2018)
[9] Li Junqing, Xu Yanping, Liu Yanfeng, *Large Electric Machine and Hydraulic Turbine* (04), 30-34+46(2012)
[10] Li Junqing, Hu Jiwei, *IEEE Power and Energy EngineeringConference(APPEEC)*, pp 1-4(2010)
[11] Li Weili, Zhao Zhihai, Hou Yunpeng, *Transactions of China Electrotechnical Society* (03), 1-6+92(2002)
[12] Zhou Feng, XiongBin, Li Weili, etc, *Proceedings of the CSEE* (24), 128-132(2005)
[13] ZHAO Wei-duo, LIU Yu-xin, *Electric Machines and Control* **18**(04), 39-44+59(2014)
[14] Li Junqing, Li Heming, *Proceedings of the CSEE* (9), 122-125(2005)
[15] Shen Wenhua, *Energy and Energy Conser-vation* (5), 89 —90(2011)
[16] Tong Wenming, Shu Shenglang, *Proceedings of the CSEE* **37**(37), 2131-2142(2007)
[17] Ding Shunnian, *Heating and cooling of large motors*, (1992)

IOP Conf. Series: Materials Science and Engineering **576** (2019) 012013 doi:10.1088/1757-899X/576/1/012013

Investigation of Gear Dynamic Characteristics under Stochastic External Excitations

Yining Fang[1,2], Ming J Zuo[2] and Yue Li[1]

1 Laboratory of Science and Technology on Integrated Logistics Support, National University of Defense Technology, Changsha, China

2 Department of Mechanical Engineering, University of Alberta, Edmonton, Canada

E-mail: liyue@nudt.edu.cn

Abstract. Engine drive gear systems are widely used in lots of mechanical equipment. In engine drive gear systems, the external excitations, including the driving speed and external load, are generally stochastic due to manufacturing error, energy loss, variation of operating environmental, and so on. The reported works usually considered them as deterministic. In this paper, a stochastic nonlinear gear dynamic model considering the stochastic external excitations is presented. Then, the effects of the stochastic driving speed and external load on gear dynamic characteristics are investigated by Monte Carlo (MC) simulation. By the simulation results, some insightful conclusions, which are important for the gear random dynamics, are obtained.

1. Introduction

Gear systems, especially engine drive gear systems, are widely used in lots of mechanical equipment [1]. For an engine drive gear system, there are generally two kinds of excitations, internal excitations and external excitations. Internal excitations include time-varying mesh stiffness (TVMS), backlash, transmission error, and so on [2]. External excitations contain the external load [3] and driving speed of the pinion [4]. In existing studies, the effects of internal excitations have been well investigated while the research about the effects of external excitations achieve less attention, especially under stochastic domain. However, the dynamic responses of an engine drive gear system are significantly affected by the external excitations [4].

Under deterministic domain, external excitations have been investigated in some existing works. The mathematical models and the solving approaches for the nonlinear dynamics of gear systems under deterministic load was reviewed in [5]. Qiu et al. [6] considered the influence of deterministic input rotating speed on stiffness and introduced a velocity modulated stiffness model. In reality, however, the stochastic feature of external excitations is ubiquitous due to manufacturing error, energy loss, possible change of the engine status, and variation of operating environmental [7]. For example, the stochastic feature of dynamic loads for high speed gears was observed in [8]. Accordingly, to obtain more realistic representation of gear systems, the stochastic anticipated load profile and speed profile should be considered when modelling engine drive gear systems.

In recent works about modelling gear systems, attention of researchers has transferred from the deterministic domain to the stochastic domain. Tobe et al. modelled the randomness in the load as Gaussian white noise and validated their model through experiment results [9]. The random dynamic response of a gear pair under constant mesh stiffness, constant damping coefficient, and stochastic external load by adding Gaussian white noise was investigated in [10]. Fang et al. [2] investigated the effects of friction on the dynamic characteristics of a spur gear pair under stochastic external load.

However, in existing works, the effects of stochastic driving speed, especially the coupled effects of stochastic driving speed and stochastic external load, are still short of investigation. In this paper, we model an engine drive gear system under stochastic external excitations, i.e., stochastic external load and stochastic driving speed. Then, the dynamic characteristics focusing on the effects caused by the stochastic external excitations are analyzed, and thus, some insightful conclusions are obtained.

The remaining parts of this paper are organized as follows. In Section 2, the proposed stochastic dynamic model for a spur gear pair is described. The effects of the stochastic external excitations are illustrated in Section 3. Section 4 draws the conclusion.

2. Nonlinear stochastic gear dynamic model

In this section, the proposed stochastic dynamic model for a spur gear pair is described. A spur gear pair is given in Fig. 1. Thus, the equations of motion for the system is given as

$$I_p\ddot{\theta}_p = T_p - (F+G)R_p - \mu(F+G)X_1 \tag{1}$$

$$I_g\ddot{\theta}_g = T_g + (F+G)R_g + \mu(F+G)(l-X_1) \tag{2}$$

where I_p, T_p, $\ddot{\theta}_p$, R_p are the moments of inertia, driving torque, angular acceleration, and base circle radius of the pinion (similarly the subscript g denotes the gear), respectively, F is the total elastic force, G is the damping force, μ is the friction coefficient, X_1 denotes the tangent distance in the action line, $l=(R_p+R_g)\sin\alpha_0$ means the length of the action line, and α_0 denotes the pressure angle. According to Eq. (1) and Eq. (2), we have the following expression

$$\ddot{\theta}_g = \left[I_p\ddot{\theta}_p + T + (K\delta + \alpha\dot{\delta})L\right]/I_g \tag{3}$$

where $\delta = R_p\theta_p - R_g\theta_g$ denotes the relative angular displacement of the gear pair, $F = Kg(\delta)$, $L = R_g - R_p + \mu(R_g+R_p)\sin\alpha_0$, $T = T_g - T_p$, K is the effective mesh stiffness, $g(\delta)$ denotes the function of backlash, L is the equivalent length of the force arm, c is the damping coefficient, and T is the equivalent torque (i.e., the external load in this paper) that reflects the combined fluctuation of T_g and T_p. In this model, we consider the classic backlash function and the mesh stiffness function [2], i.e., $g(\delta)=\begin{cases}\delta-b, & \delta>b\\ 0, & -b\le\delta\le b\\ \delta+b, & \delta<-b\end{cases}$ and $K(\theta)=\begin{cases}k_1, & (n-1)\phi_n\le\theta<(n-1)\phi_n+\phi_m\\ k_2, & (n-1)\phi_n+\phi_m\le\theta<n\phi_n\end{cases}$, where b denotes the backlash, ϕ_n and ϕ_m represent a mesh period of gear teeth and the double pairs of teeth mesh duration in a mesh period, respectively.

Accordingly, the second-order ordinary differential equation (ODE) of the considered gear model is given as

$$\ddot{\theta}_g = \left\{I_p\ddot{\theta}_p + T + [Kg(\theta_pR_p - \theta_gR_g) + c(\dot{\theta}_pR_p - \dot{\theta}_gR_g)]L\right\}/I_g \tag{4}$$

Note that, the inputs (excitations) in (4) are $T, \theta_p, \dot{\theta}_p, \ddot{\theta}_p$ while the outputs are $\theta_g, \dot{\theta}_g, \ddot{\theta}_g$ (or called the responses of the gear dynamic model).

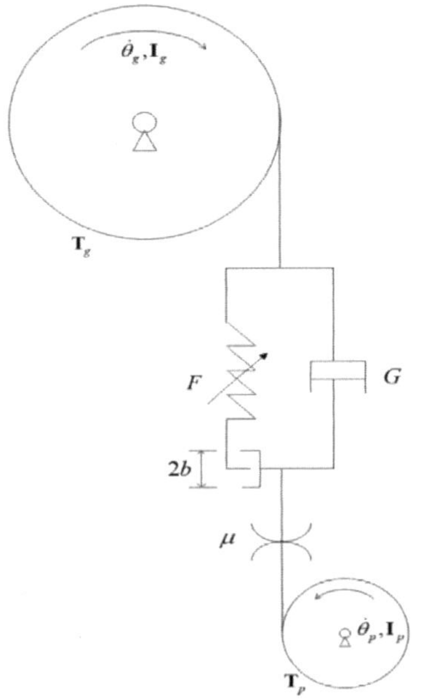

Figure 1. Dynamic model of a gear pair

Then, we need to consider the stochastic external load and driving speed formulation. According to the study in [11], the randomness in the external load and the driving speed is generally set as Gaussian white noise $\xi_i(t)$,

$$E(\xi_i(t)) = 0 \tag{5}$$

$$E[\xi_i(t)\xi_i(t+\tau)] = r_i\Theta(\tau) \tag{6}$$

where $\xi_1(t)$ denotes the noise of the driving speed with variance r_1, $\xi_2(t)$ denotes the noise of the external load with variance r_2, and Θ is the Dirac Delta function. Accordingly, the expression of $\dot{\theta}_p(t)$ can obtained as

$$\dot{\theta}_p(t) = \chi_0 + \sum_n \chi_n \sin(\rho_n t + \psi_n) + \xi_1(t) \tag{7}$$

Similarly, the stochastic external load T can be expressed as

$$T(t) = f_0 + \sum_j f_j \cos(v_j t + \beta_j) + \xi_2(t) \tag{8}$$

where f_0 is a constant, j denotes the harmonic of the torque, v_j, f_j, and β_j are the corresponding angular frequency, amplitude, and initial phase, respectively.

It is known that Gaussian diffusion process $\Phi(t)$ is the integration of Gaussian noise, and thus, $\Phi(t)$ can be approximated by Itô stochastic differential equation (SDE) [11][12], which is given as follow.

$$d\Phi(t) = \lambda dt + \sigma dW(t) \tag{9}$$

where λ is the drift scalar, σ is the diffusion scalar, and $W(t)$ is a standard Wiener process. In addition, $dW(t)$ can be approximated as $dW(t)$: $\sqrt{\Delta t}$ [11]. For Gaussian white noise $\xi_1(t)$, $\lambda = 0$ and $\sigma = \sqrt{r_1}$. Accordingly, Eq. (7) can be approximated as

$$\theta_p(t) = \theta_p(t_0) + \chi_0 \Delta t + \sum_n \chi_n \frac{\sin(\rho_n t) - \sin(\rho_n t_0)}{\rho_n} + \sqrt{r_1 \Delta t} \tag{10}$$

Then, $\ddot{\theta}_p(t)$ can be defined as

$$\ddot{\theta}_p(t) = -\sum_n \chi_n \rho_n \sin(\rho_n t) + \frac{\xi_1(t) - \xi_1(t_0)}{\Delta t} \tag{11}$$

Note that the proposed stochastic model (i.e., Eq. (4) combined with Eqs. (7), (10), and (11)) is considered in the pinion rotational motion for the first time. Under the proposed model, there are two stochastic factors in the gear dynamic model, and thus, the complexity in solving the model and the analysis is largely increased.

3. Results and discussion

In this section, the numerical simulation results are given, and then, we analyze the simulation results. The gear pair parameters adopted in the simulation are summarized in Table 1.

Table 1. Gear pair parameters.

Inertia moment	$I_p = 2.6 \times 10^{-4}\ Kg \cdot m^2$ $I_g = 0.0045\ Kg \cdot m^2$
Base circle	$r_p = 0.034\ m$ $r_g = 0.052\ m$
Backlash	$b = 1 \times 10^{-5} m$
Pressure angle	$\alpha_0 = 20^\circ$
Stiffness	$k_1 = 1.6 \times 10^8 N/m$ $k_2 = 0.9 \times 10^8 N/m$
Friction coefficient	$\mu = 0.04$

3.1. Validation of the proposed model

In [4], a model of a spur gear pair under deterministic load and deterministic driving speed was validated with a finite element model. It was concluded that the model in [4] captures the salient behaviour of the actual system. Accordingly, we take that model as a sample to validate our proposed model. The simulation results are given in Fig. 2. Note that Liu's Model denotes the model in [4].

From Fig. 2, the relative displacement of our proposed model is almost the same with that in Liu's model. Thus, it validates that our proposed model can capture the characteristics of the actual systems.

3.2. Dynamic responses of the proposed model

In this section, we analyze the dynamic responses of the proposed model. Let $r_1 = 25$, and we have the expression of $\dot{\theta}_p(t)$ as follow.

$$\dot{\theta}_p(t) = 800 + 80\cos(100t + \pi/2) + 20\cos(80t + \pi/2) + \xi_1(t) \tag{12}$$

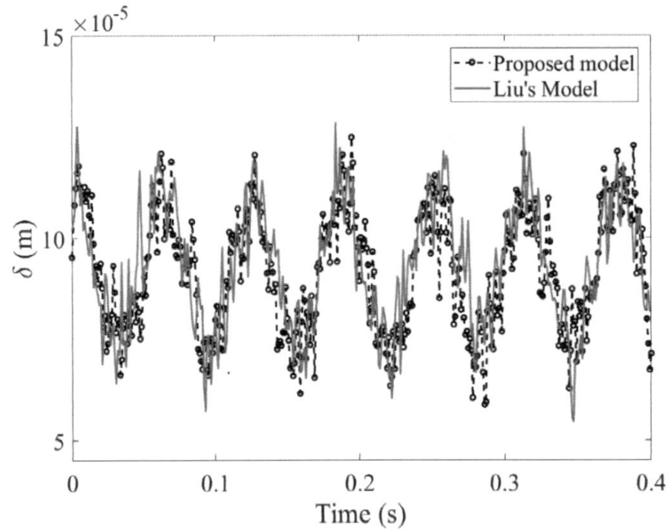

Figure 2. Relative angular displacement with different models

Let $r_2 = 2500$, and we have the expression of $T(t)$ as follow.

$$T(t) = 500 + 50\cos(100t) + \xi_2(t) \tag{13}$$

Therefore, the gear model is excited by two stochastic excitations, i.e., Eq. (12) and Eq. (13). To demonstrate the effects of the stochastic driving speed, we take the case with deterministic driving speed as a comparison. Fig. 3 presents two instantaneous probability density functions (PDFs) with two profiles (i.e., stochastic driving speed and deterministic speed) of the driving speed at 0.2 s. Fig. 4 presents the standard derivation of δ with time variation under two cases. The difference of the PDF and standard derivation of δ under two cases is caused by the stochastic term in the driving speed. From Fig. 3, we can see that the responses excited by stochastic driving speed have more dispersion than those excited by deterministic driving speed. The variance of responses under the stochastic driving speed case is larger than that under the deterministic driving speed case. From Fig. 4, it illustrates that the standard deviation of δ obtained by stochastic driving speed is always greater (about twice greater) than that obtained by deterministic driving speed during the simulation period. Accordingly, it can further prove that the case with stochastic driving speed has a larger variance than the case with deterministic driving speed.

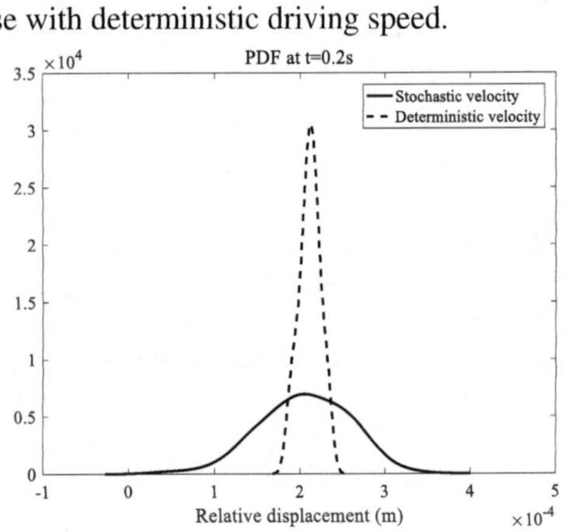

Figure 3. PDF under two cases

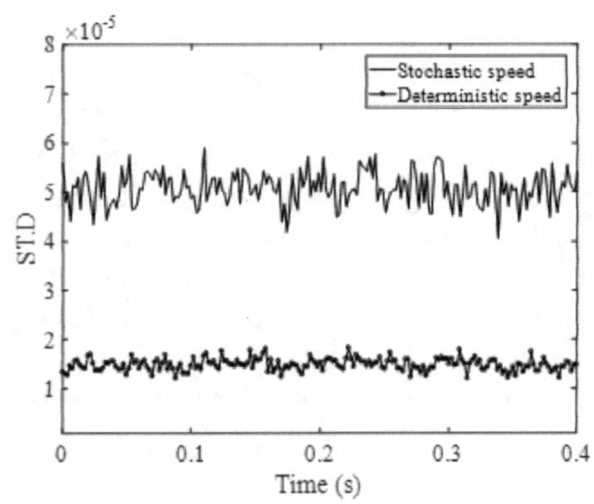

Figure 4. Standard derivation of with time variation under two cases

3.3. Coupled effects of the stochastic external load and stochastic driving speed

In this section, we analyse the coupled effects of the stochastic external load and stochastic driving speed. First, two sensitivity parameters η_1 and η_2 for driving speed and load are defined as follows.

$$\eta_1 = \frac{\sqrt{r_1}}{E(\dot{\theta}_p)} \tag{14}$$

$$\eta_2 = \frac{\sqrt{r_2}}{E(T)} \tag{15}$$

Keep the mean values $E(\dot{\theta}_p)$ and $E(T)$ invariant and adjust the values of r_1 and r_2 to change the values of η_1 and η_2. Table 2 summarizes the standard deviation of δ under different η_1 and η_2. From Table 2, we can see that the standard deviation of δ increases with the increase of η_1 and η_2. In addition, it also can be concluded that the standard deviation of δ increases more with the variation in η_1.

Table 2. Standard deviation of δ under different η_1 and η_2.

η_2 \ η_1	0.5%	0.65%	0.8%
0.5%	1.0665×10^{-5}	2.0038×10^{-5}	3.7009×10^{-5}
0.65%	1.0715×10^{-5}	2.1316×10^{-5}	4.3268×10^{-5}
0.8%	1.1444×10^{-5}	2.4168×10^{-5}	4.5109×10^{-5}

To further analyse the results, we define the average increase rates of the driving speed and external load as follows.

$$\kappa_1 = \frac{1}{6}\sum_{i=1}^{3}\left(\frac{<i,2>}{<i,1>} + \frac{<i,3>}{<i,2>}\right) \tag{16}$$

$$\kappa_2 = \frac{1}{6}\sum_{i=1}^{3}\left(\frac{<2,i>}{<1,i>} + \frac{<3,i>}{<2,i>}\right) \tag{17}$$

where κ_1 and κ_2 are the average increase rates of the driving speed and external load, respectively, $<i,j>$ represents the element in i th row and j th column in Table 2 without counting the header. Under the same η_1, the average increase rate κ_2 is 1.09. For the same η_2, the average increase rate κ_1 is 1.97. Thus, the standard deviation of δ is more sensitive to the variation of η_1. In other words, the stochastic feature in the driving speed $\theta_p(t)$ is much more dangerous than that in the external load $T(t)$. To some degree, the randomness in the $\dot{\theta}_p(t)$ may cause more vibration and failure. The control of its randomness is very important in real applications.

Then, we consider the phase diagram ($\delta-\dot{\delta}$ relationship) with the stochastic driving speed. The simulation results are given in Fig. 5. Noted that δ and $\dot{\delta}$ are the mean value of the total samples. Recalling the phase diagram in [2], the orbit of $\delta-\dot{\delta}$ is smooth and periodic. However, in our proposed model, the orbit of $\delta-\dot{\delta}$ in the phase portrait is non-smooth and non-periodic, which is caused by the stochastic driving speed. This phenomenon could be affected by the coupling of the stochastic driving speed and stochastic external load.

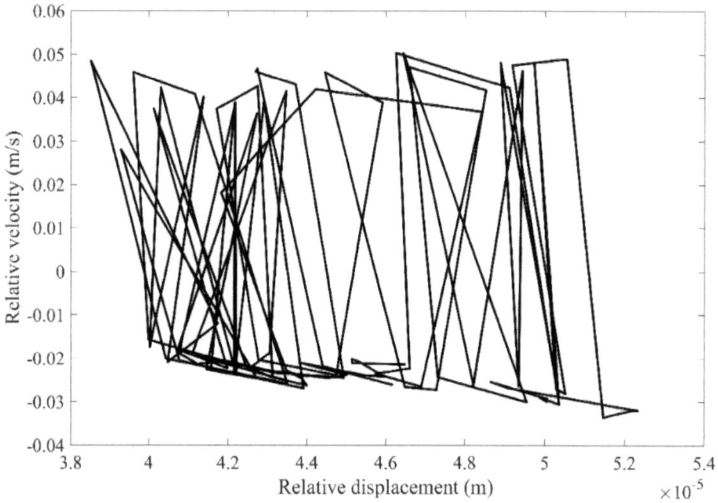

Figure 5. Phase Diagram

3.4. Discussion

According to the simulation results, we can obtain some insightful conclusions.

(1) The established model is more realistic by considering the randomness in the driving speed and external load,

(2) Under the same stochastic load, a small ratio of randomness in the driving speed will greatly increase the dispersion in responses,

(3) The dispersion in responses is more sensitive to the uncertainty in the driving speed than that in the external load.

Therefore, our proposed gear dynamic model and numerical results can be used as a useful tool to investigate the gear random dynamics.

4. Conclusions

In this paper, a gear stochastic dynamic model for a spur gear pair considering TVMS, gear mesh damping, backlash, and friction under stochastic external excitations is established. Then, the stochastic external excitations, i.e., the driving speed and the external load, are modeled in the gear dynamic model. After that, the dynamic responses of this model are investigated using numerical simulation and compared with previous works.

5. References

[1] X. Liang, M.J. Zuo, Z. Feng, Dynamic modeling of gearbox faults: A review, *Mech. Syst. Signal Process*. **98**, 852–876 (2017).

[2] Y, Fang, X. Liang, M. J. Zuo. Effects of friction and stochastic load on transient characteristics of a spur gear pair, *Nonlinear Dyn.* **93**, 599-609 (2018).

[3] W. Bartelmus, F. Chaari, R. Zimroz, M. Haddar, Modelling of gearbox dynamics under time-varying nonstationary load for distributed fault detection and diagnosis, *Eur. J. Mech. - ASolids*. **29**, 637–646 (2010).

[4] F. Liu, H. Jiang, S. Liu, X. Yu, Dynamic behavior analysis of spur gears with constant & variable excitations considering sliding friction influence, *J. Mech. Sci. Technol.* **30**, 5363–5370 (2016).

[5] J. Wang, R. Li, X. Peng, Survey of nonlinear vibration of gear transmission systems, *Appl. Mech. Rev.* **56**, 309–329 (2003).
[6] X. Qiu, Q. Han, F. Chu, Analytical investigation on unstable vibrations of single-mesh gear systems with velocity-modulated time-varying stiffness, *J. Sound Vib.* **333**, 5130–5140 (2014).
[7] S. Wei, Q. Han, F. Chu, Review on research of uncertain dynamics for gear system, *J. Mech. Eng.* **52**, 1–19 (2016).
[8] M. Utagawa, T. Harada, Dynamic loads on spur gear teeth having pitch errors at high speed, *Bull. JSME.* **5**, 374–381 (1962).
[9] T. Tobe, K. Sato, N. Takatsu, Statistical analysis of dynamic loads on spur gear teeth : experimental Study, *Bull. JSME.* **20**, 1315–1320 (1977).
[10] Y. Wen, J. Yang, P. Yang, Random vibration response of a spur gear pair with periodic stiffness and backlash, in: *Proc. ASME 2013 Int. Mech. Eng. Congr. Expo.*, ASME, San Diego, California, USA, (2013).
[11] R.B. Randall, Basic Signal Processing Techniques, in: Vib. Based Cond. Monit., John Wiley & Sons, Ltd, 63–141 (2011).
[12] K. Itô, Diffusion Processes, in: Encycl. Stat. Sci., John Wiley & Sons, Inc., (2004)

Acknowledgments
This research is supported by the National Natural Science Foundation of China (Grant No. 51675523), Natural Sciences and Engineering Research Council of Canada (NSERC), and Future Energy Systems under Canada First Research Excellent Fund (CFREF).

IOP Conf. Series: Materials Science and Engineering **576** (2019) 012014 doi:10.1088/1757-899X/576/1/012014

FSI simulation and experimental validation of nonlinear dynamic characteristics of a gas-pressurized hydraulic shock absorber

XU Wen-xue[1], LÜ Zhen-hua[1] and ZHOU Ping-zhang[1]

[1]Department of Automotive Engineering, Tsinghua University, Beijing 100084, China

Corresponding author: xuwx1509@163.com

Abstract. The structure, fluid and gas dynamic models of a gas-pressured hydraulic shock absorber are established to form a 3-D FSI (Fluid-Structure Interaction) simulation model, which can be used to predict the working characteristics and kinematical responses of the valve plates, as well as the flow characteristics, transient fluid pressure and velocity distribution of the shock absorbers. The working characteristics obtained from the simulation model are compared with the experimental results to prove the effectiveness of the simulation model. It is shown that the dynamic characteristics of the shock absorber can be predicted quite accurately by the 3-D FSI simulation model, which is meaningful to the understanding of the inner working mechanism of the shock absorber system.

1. Introduction

Ride comfort and handling of a car is directly influenced by the performance of automobile shock absorbers. There are many kinds of automotive shock absorbers, such as twin-tube, monotube, ER and MR shock absorbers. Monotube gas-pressurized hydraulic shock absorber is a new type of liquid-resistance shock absorber, which is developed in the 1960s [1]. A typical monotube gas-pressurized hydraulic shock absorber is composed by adding a floating piston in the compression chamber to form a gas chamber with high pressure inside, which is used to compensate the volume change caused by the piston rod [2].

In the traditional lumped-parameter models, simulation of shock absorbers' dynamic characteristics is carried out using semi-empirical formulas of fluid mechanics [3-6], which owes limited accuracy and can't accurately describe the dynamic characteristics of the shock absorber system. The FSI simulation model can describe the structure and fluid of the shock absorber more accurately [7-10]. Moreover, information of flow field and structure responses that can't be obtained from the traditional models can be obtained to help with the understanding of the inner working mechanism of the shock absorber.

By using the finite element method, an FSI simulation model of a monotube gas-pressurized hydraulic shock absorber designed for a special heavy-duty vehicle is established and analysed. Direct FSI method and Leader-Follower parameter moving mesh control strategy are used in the simulation model. The working characteristics, kinematical response of the valve plates, flow characteristics, transient fluid pressure and velocity distribution of shock absorbers are obtained from the model, which is helpful to the understanding of the inner working mechanism of the monotube gas-pressurized hydraulic shock absorber

2. Mathematical model of the shock absorber

2.1. Fluid mathematical model

The motion of the oil and gas in the shock absorber is governed by the Navier-Stokes equations. The conservative forms for mass, momentums and energy equations in a fixed Cartesian coordinate frame of reference can be expressed as follows [11],

$$\begin{aligned} &\frac{\partial \rho}{\partial t}+\nabla\cdot(\rho\mathbf{v})=0 \\ &\frac{\partial \rho\mathbf{v}}{\partial t}+\nabla\cdot(\rho\mathbf{v}\mathbf{v}-\boldsymbol{\tau})=\mathbf{f}^{\mathbf{B}} \\ &\frac{\partial \rho \mathrm{E}}{\partial t}+\nabla\cdot(\rho\mathbf{v}\mathrm{E}-\boldsymbol{\tau}\cdot\mathbf{v}+\mathbf{q})=\mathbf{f}^{\mathbf{B}}\cdot\mathbf{v}+\mathrm{q}^{\mathrm{B}} \end{aligned} \tag{1}$$

, where, ρ is the density, $\mathbf{v}$ is the velocity vector, $\boldsymbol{\tau}$ is the stress tensor, $\mathbf{f}^{\mathbf{B}}$ is the body force vector of the fluid medium, E is the specific total energy, $\mathbf{q}$ is the heat flux and q^B is the specific rate of heat generation.

Additional equations, i.e. the state equations, must be provided to obtain a closed system for solution variables.

Oil in the shock absorber can be modelled by a slightly compressible fluid model. Its state equations are expressed as follows,

$$\begin{aligned} &\rho=\rho_0\left(1+\frac{p}{\kappa}\right) \\ &e=C_v\theta \end{aligned} \tag{2}$$

, where, ρ_0 is the density at $p=0$, κ is the fluid bulk modulus of elasticity, C_v is the specific heat at constant volume, θ is temperature.

Gas in the shock absorber can be modelled by a compressible fluid model. Its state equations are expressed as follows,

$$\begin{aligned} &p=(\mathrm{C}_p-\mathrm{C}_v)\rho\theta \\ &e=\mathrm{C}_v\,\theta \end{aligned} \tag{3}$$

, where, C_p is the specific heat at constant pressure.

2.2. Solid mathematical model

The motion of the solids of the shock absorber is governed by motion equations,

$$\rho_s\frac{\partial^2\mathbf{u}}{\partial t^2}=\nabla\cdot\boldsymbol{\tau}_s+\mathbf{f}_s^B \tag{4}$$

, where, ρ_s is the density, $\mathbf{u}$ is the displacement vector, $\boldsymbol{\tau}_s$ is the stress tensor, $\mathbf{f}_s^B$ is the body force vector of the solid medium.

2.3. FSI mathematical model

In FSI analyses, the fluid boundaries are influenced by the deformation of the solid domain, while the fluid forces are applied onto the solid domain through the FSI boundaries.

Direct method is used in the simulation model, in which the fluid and solid solution variables are fully coupled. The fluid and solid equations are combined in one system, and linearized in a matrix system. The matrix system can be written as,

$$\begin{bmatrix} \mathbf{A}_{ff} & \mathbf{A}_{fs} \\ \mathbf{A}_{sf} & \mathbf{A}_{ss} \end{bmatrix} \begin{bmatrix} \Delta \mathbf{X}_f^k \\ \Delta \mathbf{X}_s^k \end{bmatrix} = \begin{bmatrix} \mathbf{B}_f \\ \mathbf{B}_s \end{bmatrix} \tag{5}$$

, where, X is the solution vector of the coupled system, **A** and **B** are the linearized matrixes.

The convergence of the coupled system is controlled by the stress and displacement tolerances, relaxation factors and convergence criteria.

3. FSI model of the shock absorber

The monotube gas-pressurized hydraulic shock absorber used in this paper is designed for a special heavy-duty vehicle, which needs to meet the large stoke and damping force requirements. A simplified model of the damper is shown in Figure.1. The valve system includes a piston, two valve plates and two compression springs. As the piston is forced to move in and out along the fluid-filled cylinder, a pressure difference is developed across the piston causing the fluid to flow though the valves on the piston. The piston moves in the cylinder during the compression stroke, causing the volume decrease and pressure increase in the compression chamber, as well as the volume increase and pressure decrease in the rebound chamber. The piston moves out the cylinder during the rebound stroke, causing the volume increase and pressure decrease in the compression chamber, as well as the volume decrease and pressure increase in the rebound chamber. The volume change caused by the piston rod is compensated by the air chamber.

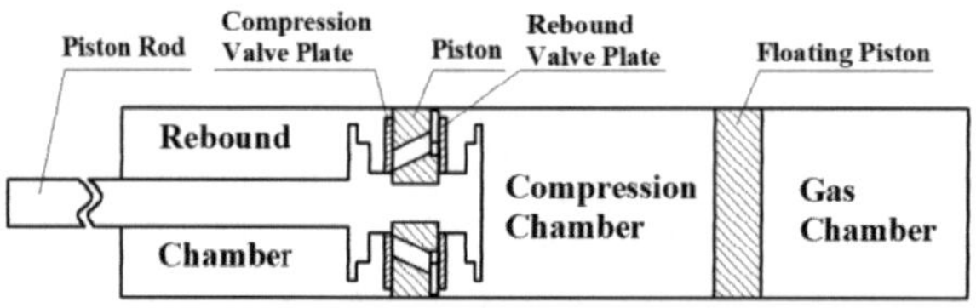

Figure 1. Simplified structure of the shock absorber.

The fluid geometry model is obtained from Boolean operation of the structure. In order to improve the FEA model accuracy and calculation efficiency, 1/6 FEA models of structure and fluid are established for FSI dynamic simulation analysis. Hexahedral meshes are used for the structure model except the piston body, whose structure is too complex to be meshed with hexahedral elements, so the tetrahedral meshes are used. In order to limit the number of finite elements, only structure associated with the FSI surfaces of the piston body is meshed. Hexahedral meshes are used for the fluid model. Throttling gaps are generated when the valve plates are opened, which should be reserved between the valve plates and the piston in the FEA model. In order to capture the drastically changed pressure and velocity in the gap, the gap is discretized into four layers of meshes to meet the deformation here. The average element size is only 0.05mm at the gap. Pressure and velocity in the gas, compression and rebound chamber is relatively uniform, so elements with relatively large size are adopted at these regions. The established structure and fluid FEA models are shown in Figure.2.

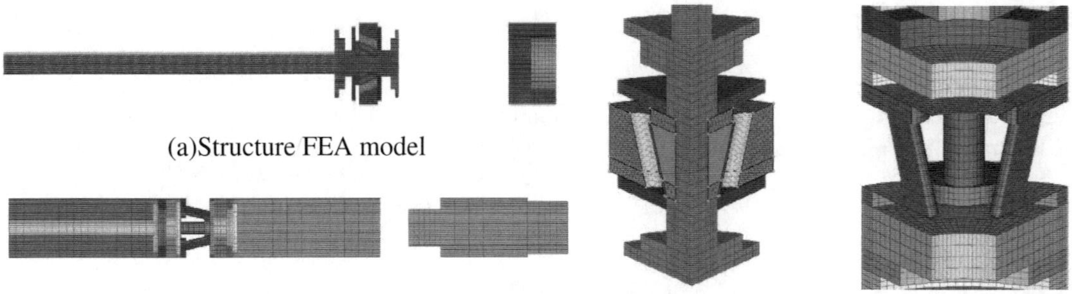

(a)Structure FEA model

(b)Fluid FEA model

(c)Valve system FEA model

(d) Part of the fluid FEA model

Figure 2. FEA mesh models.

In the structure model, a displacement excitation of the sine wave form with the frequency 5 Hz and amplitude 32 mm is applied to the piston rod, and a sinusoidal displacement excitation of 0.1Hz is applied to the piston rod in the beginning 2.5 seconds of the simulation process to ensure the formal simulation's starting excitation speed is zero, which is helpful to the convergence of the FSI analysis. The displacement excitation is shown in Figure.3. The spring finite elements are used to replace the structures of the compression springs. The designed stiffness characteristics and the experimental results are shown in Figure.4, which shows great agreements with each other. There are two contact pairs defined in the structure model, i.e. the valve plates surfaces and the piston surfaces. FSI interfaces include surfaces of the valve plates, flow holes' surfaces of the piston and both side surfaces of the floating piston.

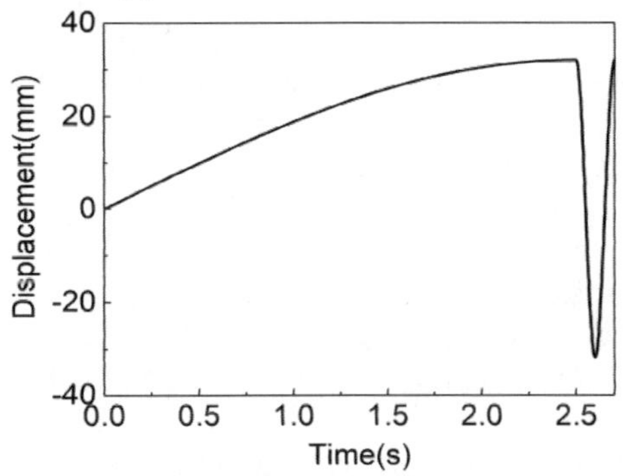

Figure 3. Input displacement excitation.

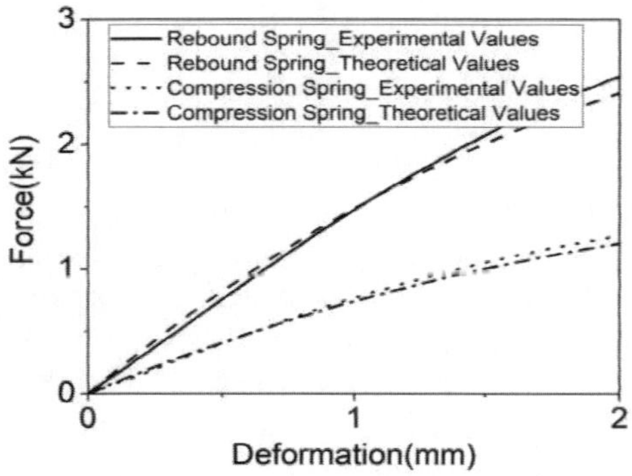

Figure 4. Spring stiffness characteristics.

In the fluid model, the oil is modelled as a slightly compressible fluid, and the gas is modelled as a low-speed compressible fluid. The material parameters needed in the FSI analysis are shown in Table 1. What's more, Leader-Follower method is used for parametric dynamic moving mesh control, initial pressure of 2 MPa is applied to the fluid model and FSI interfaces corresponding to the structural model's FSI interfaces are defined.

Table 1. Material parameters.

Materials	**Solid**	**Oil**	**Gas**
Density/(kg/m^3)	7800	832	/
Viscosity/(Pa · s)	/	0.017	22
Elastic modulus /(GPa)	210	1.5	/
Poisson's ratio	0.3	/	/
C_p **/(J/(kg· K))**	/	/	968.5
C_v **/(J/(kg· K))**	/	/	764.84

4. Results and discussion

4.1. Model validation

To verify the reliability of the results of the FSI simulation model, the experimental and simulation work diagrams are presented in Figure.5.

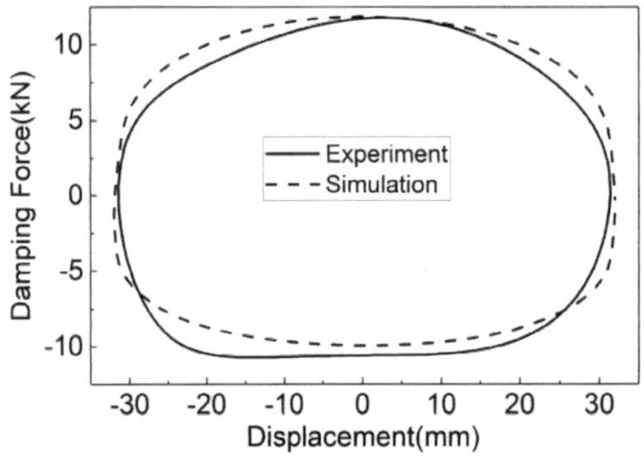

Figure 5. Work diagrams of the shock absorber

As can be observed in Figure.5, the simulation result shows good agreements with the experimental result. The friction between the pistons and the cylinder inner surface is neglected in the FSI model. Moreover, the experimental result is affected by the machining error and measurement error. Above factors make the simulation work diagram a fuller, better symmetric curve.

4.2. Flow characteristics

It is hard to obtain transient fluid pressure and velocity distribution in the absorber by experiments. However, FSI dynamic analysis can help get the fluid pressure and velocity distribution at anytime and anywhere. Fluid pressure and velocity distribution around valve system at the maximum velocity points of the compression and rebound strokes are shown in Figure.6 and Figure.7.

Figure.6 shows the pressure gradient is mainly concentrated in the throttling gap between the piston and the valve plates. Negative pressure zone appears in the compression stoke, which means cavitation happens in this area. It should be noted that if the pressure were negative, the oil in the actual shock absorber would become gas medium. Pressure distribution far away from the valve system shows small pressure gradient, the variation in the pressure of the compression and rebound chambers are shown in Figure.8. As can be observed in Figure.8, pressure in the rebound chamber changes drastically, pressure in the compression chamber changes little.

As can be observed in Figure.7, the maximum velocity value in the compression stroke is 65m/s, which occurs at the throttling gap between the piston and compression valve plate. The maximum velocity value in the rebound stroke is 71.8m/s, which occurs at the throttling gap between the piston and the rebound valve plate. Therefore, an excitation of 1m/s at the piston rod can lead to nearly 70 times velocity magnitude in the shock absorber.

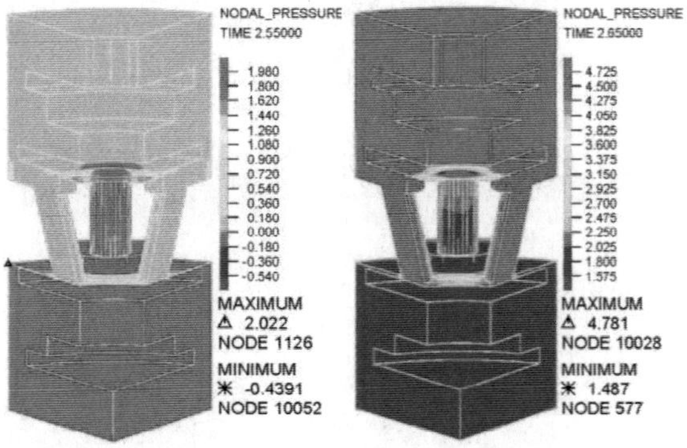

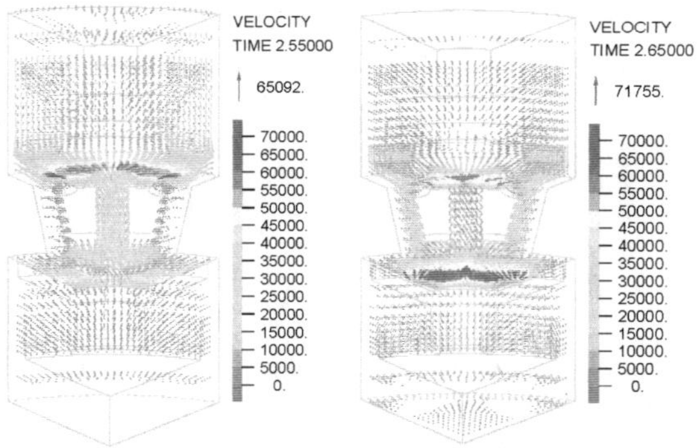

(a)Compression stroke (b)Rebound stroke (a)Compression stroke (b)Rebound stroke

Figure 6. Fluid pressure distribution. **Figure 7.** Fluid velocity distribution.

Figure.9 shows the variation in the flow rate through the compression and rebound flow holes. There is small amount of flow rate through the rebound flow holes during the compression stroke and small amount of flow rate through the compression flow holes during the rebound stroke, which are caused by the reserved gap between the piston and valve plates. This amount of flow rate can replace the amount of flow rate for constant flow channels and piston leakage.

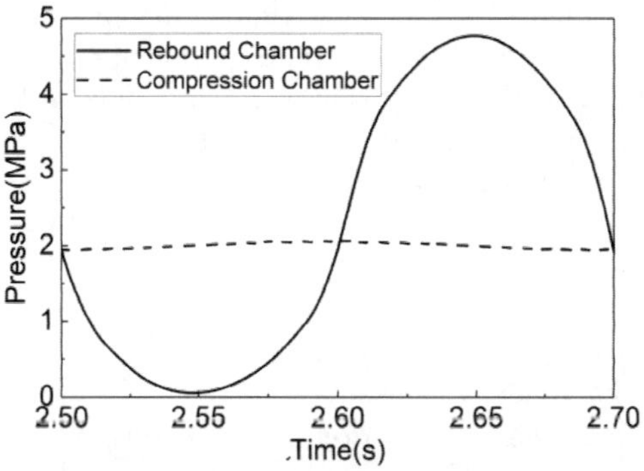

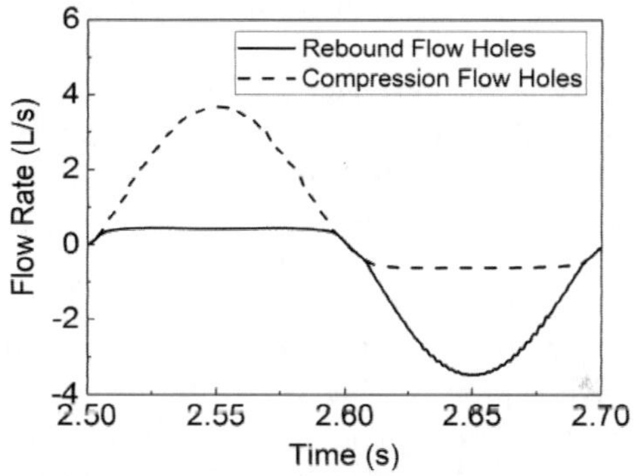

Figure 8. The variation in the pressure of the compression and rebound chambers.

Figure 9. The variation in flow rate.

4.3. Nonlinear dynamic response of the structure

Figure.10 shows the displacements of the valve plates, the maximum displacement of the compression valve plate is greater than the maximum displacement of the rebound valve plate, which is caused by the stiffness difference of the compression springs.

Figure.11 shows the displacements of the floating piston and the piston. As can be observed in Figure.11, there is a small phase difference between the displacement of the floating piston and the displacement of the piston, as the mass of the floating piston and oil compressibility are considered in the FSI model.

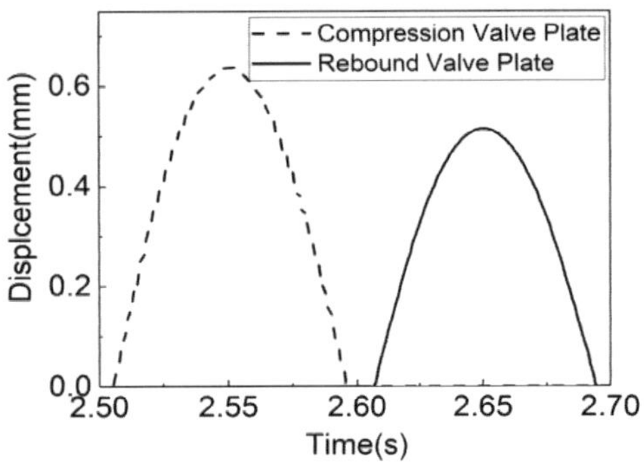

Figure 10. Displacement of the valve plates.

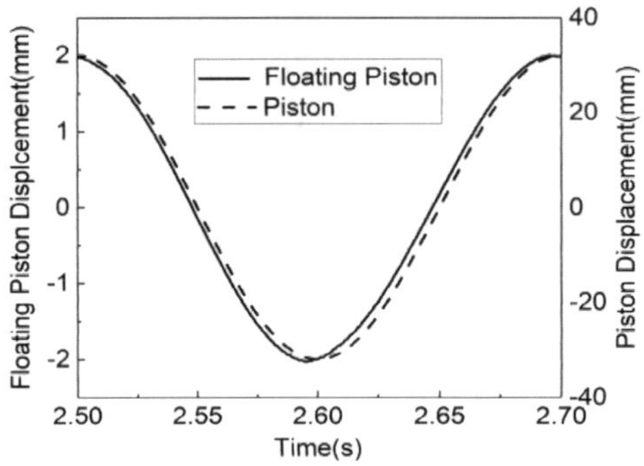

Figure 11. Displacement of pistons.

5. Conclusion

An FSI simulation model of a monotube gas-pressurized hydraulic shock absorber is built, in which the slightly compressible characteristic of the oil and the compressible characteristic of the gas are considered, direct FSI method and Leader-Follower moving mesh control strategy are used. The simulation work diagram shows very similar tendency with the experimental result, which verifies the reliability of the FSI model.

The following results can be obtained from the FSI model: the working characteristics, the flow characteristics which include transient fluid pressure and velocity distribution, as well as the variation in flow rate, the kinematical responses of the valve plates and floating piston, which are helpful to the understanding of the inner working mechanism of the monotube gas-pressurized hydraulic shock absorber system. This work is meaningful to the guiding and optimization of the design of shock absorbers.

6. References

[1] Chen Jia-rui. *Automobile Construction* (Beijing: Mechanical Industry Press, 1995)

[2] Dixon, John C. *The Shock Absorber Handbook* (John Wiley & Sons, 2008)

[3] Lee K. *Numerical modelling for the hydraulic performance prediction of automotive monotube dampers,* **28(1)**:25-39 (Vehicle System Dynamics, 1997)

[4] Alonso M, Comas Á. *Modelling a twin tube cavitating shock absorber*, **220(8)**: 1031-1040 (Proceedings of the Institution of Mechanical Engineers, Part D: Journal of Automobile Engineering, 2006)

[5] Alonso M, Comas Á. *Thermal model of a twin-tube cavitating shock absorber*, **222(11)**: 1955-1964 (Proceedings of the Institution of Mechanical Engineers, Part D: Journal of Automobile Engineering, 2008)

[6] Castellani F, Scappaticci L, Bartolini N, Astolfi, D. *Numerical and experimental investigation of a monotube hydraulic shock absorber*, **87(12)** : 1929-1946 (Archive of Applied Mechanics, 2017).

[7] Shams M, Ebrahimi R, Raoufi A, Jafari B. J. *CFD-FEA analysis of hydraulic shock absorber valve behavior,* **8(5)** :615-622 (International journal of automotive technology, 2007).

[8] LU Zhen-hua, JIANG Li-quan. *FSI FEA simulation of liquid-supplement valves in gas-pressurized hydraulic dampers*, **23(11)**: 163-169 (Engineering Mechanics, 2006)

[9] LI Ming, LU Zhen-hua. *FSI dynamic response analyses of a conical orifice valve during working process with several major influences*. **34(9)** :239-247 (Engineering Mechanics, 2017).

[10] CHEN Qi-ping, et al. *Fluid structure interaction for circulation valve of hydraulic shock absorber*. **20(3)**: 648-654 (Journal of Central South University, 2013)

[11] Bathe K J. *Finite element procedures,* 485-640 (USA: Prentice Hall,1996)

IOP Conf. Series: Materials Science and Engineering **576** (2019) 012015 doi:10.1088/1757-899X/576/1/012015

Effect of roundness and coaxiality error on the lubrication performance of ICE main bearings

Peirong Ren[1], Zhengxing Zuo[1], Ying Cheng[1] and Guoxi Jing[2]

[1]School of Mechanical Engineering, Beijing Institute of Technology, China
[2]Chine North Engine Research Institute, China

E-mail: 1120120759@bit.edu.cn

Abstract. Nowadays, with higher power density trend and lightweight design requirements in internal combustion engine, more factors are considered in bearing lubrication analysis to improve the lubrication performance and to decrease the risk of failure. The effect of roundness and coaxiality on the lubrication performance of a V-type diesel engine main bearings were investigated systematically and comprehensively using numerical simulation method based on elasto-hydrodynamic lubrication theory. The average flow model, the asperity contact theory and the multi-body dynamics based on component mode synthesis were used to improve simulation accuracy and efficiency. The results showed that out-of-roundness has an apparent negative effect on the lubrication performance of main bearings and the out-of-roundness wave height and the wave number should not exceed 2μm and 10 respectively. In addition, the lubrication performance increased slightly when the bearing house coaxiality tilt angle is the same as the journal load bending direction. The study has some guiding significance for improving the lubrication performance and tolerance requirements design of main bearings.

1. Introduction

Plain bearings are machine elements with widespread application, which are used in internal combustion engine (ICE) due to their simple structure and high load carrying capacity. Engine main bearing is one of the main sources of power loss in ICE, and its lubrication performance has an important impact on engine's operation. Nowadays, with higher power density trend and lightweight design requirements in ICE, more factors are considered in the lubrication analysis to improve lubrication performance and to decrease the risk of failure. Due to deviation of machining and assembly, the wear and environmental changes during work, etc., the journal and house of main bearing has some shape and position tolerance which may have an impact on the lubrication performance. The shape and position tolerance of main bearings mainly include roundness and coaxiality error because both journal and shell are cylindrical rotary parts.

Some work were dedicated to the study of the influence of journal out-of-roundness and misalignment on the bearing lubrication performance. D Vijayaraghavan [1] analysed the effect of out-of-roundness on the dynamic performance of a diesel engine connecting-rod bearing. The journal profiles of circular, elliptical, semi-elliptical and three lobe epicycloid were analysed and compared. D S Mehenny [2] theoretically analysed the influence of circumferential waviness of a rigid journal on the Lubrication of Dynamically Loaded Journal Bearings. The influence of lobe number and size on operating parameters such as maximum lubricant pressure and minimum film thickness is examined. Xiang [3] analysed the effect of oval and tooth circularity errors of the journal on the lubrication performance of sliding bearings, which show that the tooth errors would bring a periodic fluctuation of lubricant film at the load bearing region and the profile of film pressures has a multi-peak distribution. Based on assumption of smooth surface and rigid bearing, P Maspeyrot [4-5] analysed the influence of journal misalignment on a big-end bearing lubrication performance including minimum film thickness, the friction torque and the axial flow. A comparison of the dynamic performance characteristics of two misaligned crankshaft bearings belonging to a four-stroke petrol engine was carried out by M Lahmar [6] which showed that the presence of even a small skew of the main journal axis can cause asperity contact at one of the bearing edges.

However, few studies have been done on the effect of roundness and coaxiality error of bearing house on lubrication performance of ICE main bearings. In addition, early lubrication calculation accuracy is not high because roughness and flow factor were not considered. In this paper, the effect of roundness and coaxiality error on the lubrication performance of a V-type diesel engine main bearings were investigated systematically and comprehensively using numerical simulation method based on elasto-hydrodynamic lubrication theory. The average flow model, the asperity contact theory and the multi-body dynamics based on component mode synthesis were used to improve simulation accuracy and efficiency. The study has some guiding significance for improving the lubrication performance and tolerance requirements design of main bearings.

2. Problem formulation and theory

The two-dimensional modified Reynolds equation is used as lubrication governing equation. Neglecting the centrifugal effects for an incompressible Newtonian lubricant the Reynolds equation can be expressed as in equation (1):

$$\frac{\partial}{\partial\theta}\left(\frac{1}{12\eta}\gamma\phi_\theta h^3\frac{\partial p}{\partial\theta}\right)+\frac{\partial}{\partial x}\left(\frac{1}{12\eta}\gamma\phi_x h^3\frac{\partial p}{\partial x}\right)=\frac{V_j-V_s}{2}\frac{\partial}{\partial\theta}\left(\gamma h+\gamma\sigma_s\phi_s\right)+\frac{\partial(\gamma h)}{\partial t} \quad (1)$$

Where θ and x are circumferential and axial coordinates, respectively. η is the oil film viscosity, and γ is the fill ratio which describes the oil percentage in the gap between shell and journal. p and h are hydrodynamic pressure and oil film thickness, respectively. ϕ_θ and ϕ_x are circumferential and axial pressure flow factor, respectively. ϕ_s is the shear flow factor and σ_s is integrated surface roughness. V_j and V_s are circumferential velocity components of journal and shell.

According to Patir/Chang average flow model [7], the pressure flow factor and shear flow factor are determined when the surface direction parameter is 1. The total friction of bearing is composed of hydraulic friction force and summit friction force. The isotropic surface summit asperity contact pressure is determined using Greenwood/Tripp contact model [8]. Finite difference method and JFO mass conservation boundary are used to solved equation (1). The oil film thickness h is affected by elastic deformation and surface roughness of journal and shell which can expressed as in equation (2) :

$$h=h_m+\Delta h+h_\delta+h_\sigma \quad (2)$$

Where h_m is the minimum oil film thickness under the assumption of rigidity, $h_m+\Delta h$ is the oil film thickness under the assumption of rigidity, h_δ and h_σ are changes in oil film thickness caused by elastic deformation and surface roughness, respectively. The condensation of degree of freedom method in the dynamic substructure reduction technique is used to improve the multibody dynamic calculation efficiency and the finite element model of crankshaft and bearing house are condensed using Guyan method.

3. Simulation model

A numerical simulation model of main bearing of a V-type diesel engine was established. Figure 1 shows the finite element model of bearing house and crankshaft. The condensation of degrees of freedom of bearing house and crankshaft retain the first 30 and the first 60 modes respectively. The global parameters, nonlinear connection joint parameters and main bearing parameters were defined firstly. The combustion gas pressure as a function of crank angle was predicted using a cycle simulation code which the peak gas pressure is 22MPA. Inertia load and output torque were calculated by mass information, geometry information and combustion gas pressure. Using Gear prediction emendation multi-step algorithm and Newton-Raphson to solve flexible multi-body dynamic equations. Elastic deformation equations and modified Reynold equation were solved by iterative calculation using the finite element method and the finite difference method respectively. The node pressure is the boundary of the equilibrium equation, and the node displacement is the boundary of the Reynold equation. The global and bearing parameters are shown in table 1 and numerical methodology is shown in figure 2 which was based on a program named EXCITE, developed by AVL.

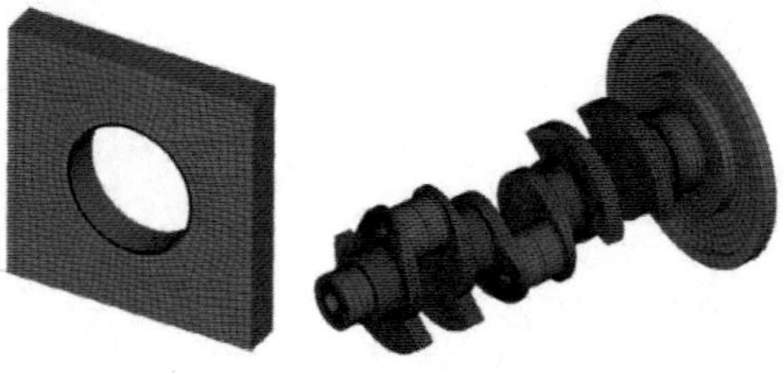

Figure 1. Finite element model.

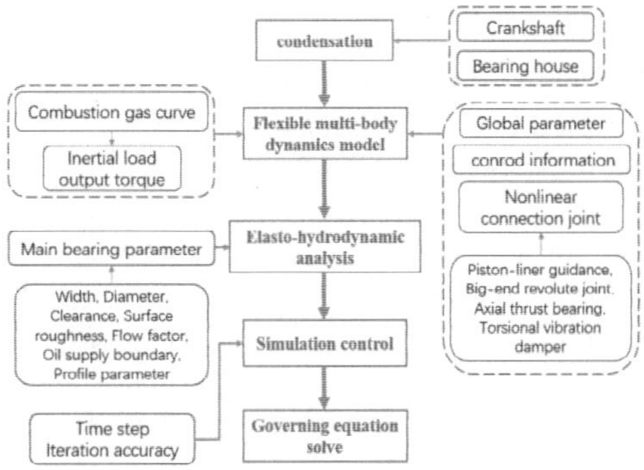

Figure 2. Numerical methodology.

Table 1. Global and bearing parameter

Parameter name	Valve	Parameter name	Valve
Engine speed/(r/min)	4200	Ambient pressure/(MPA)	0.1
Bore/(mm) & Stroke/(mm)	110&110	Cavitation pressure/(MPA)	0.098
Bearing width/(mm) & diameter/(mm)	30&98	Supply pressure/(MPA)	0.5
Radial clearance/(μm)	35	Oil type	SAE-5W30
Journal/shell Young's modules/(MPA)	210000/206000	Groove position/(deg) & width/(mm)	330~30& 7
Journal/shell Poisson's ration	0.3/0.33	Bore position/(deg) & diameter/(mm)	270, 90&8
Journal/shell surface roughness/μm	0.1/0.3		

4. Results and discussion

4.1. Effect of roundness error

Due to the machine shake during machining process and the deformation of the workpiece, the journal and shell of the main bearing may form roundness error of tooth profile. Measurement data shows that some engine crankpins are machined with up to 21 circumferential lobes and some lobes have amplitudes even exceeding 5μm [2]. Bearing are subjected to cyclic circumferential load, so tooth roundness error could decrease lubrication performance which may lead to premature damage. The radial height change of the bearing surface caused by the roundness error can be expressed by the Fourier formula, as shown in equation (3):

$$\delta(\theta) = {a_0}/{2} + \sum_{n=1}^{n=k}\left(a_n \cos\theta + b_n \sin\theta\right) \tag{3}$$

δ(θ) is the radial height change of different circumferential angle *θ*, a_n and b_n are the Fourier series. The frequency and amplitude of machine shake during machining process usually close to constant, so the tooth profile is usually close to the regular wave shape shown in figure 3, which can be represented by few Fourier series. The height between wave peak and trough and wave number are two parameters of the waveform roundness error. Generally, the wave height is about 3μm and the wave number is between 5 and 50. Two sets of numerical simulations were carried out to investigated the influence of these two parameters separately, and each set includes five cases. In the first set, the wave number (*n*) is 20, and the wave height (*h*) varies from 0μm to 5μm. In the second set, the wave height is 3μm, and the wave number varies from 0 to 40. Details of the two sets are shown in table 2.

Table 2. The details of set 1 and set 2.

	Set 1, *n*=20						Set 2, *h*=3μm				
	Case1	Case2	Case3	Case4	Case5		Case1	Case2	Case3	Case4	Case5
h/μm	0	2	3	4	5	*n*	0	10	20	30	40

The peak total pressure (PTP), the minimum oil film thickness (MOFT) and the load uniformity were selected as evaluation index of lubrication performance in this paper. The most prominent types of wear in diesel engine bearings are abrasive wear, adhesive wear and surface fatigue wear [9]. The abrasive wear is mainly caused by oil contamination with foreign particles and wear products which not been considered here. The adhesive wear is caused by oil film breakdown which results in poor lubrication conditions like boundary lubrication even dry friction which can be reflected by the minimum oil film thickness. The surface fatigue wear is mainly influenced by the total pressure, and too large total pressure will accelerate surface fatigue wear. The load uniformity includes circumferential and axial uniformity, which is important for bearing reliability.

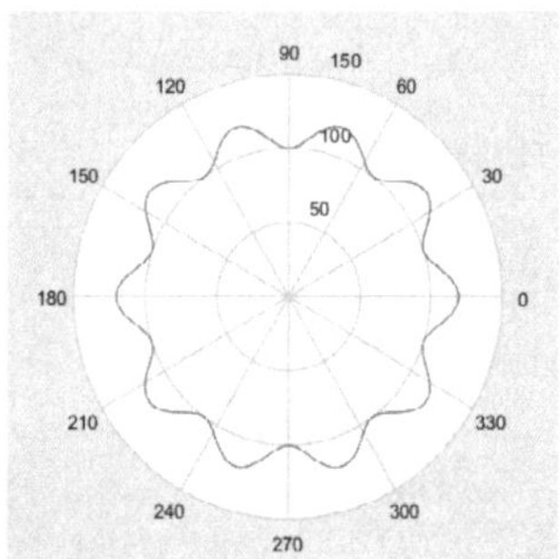

Figure 3. Regular wave shape.

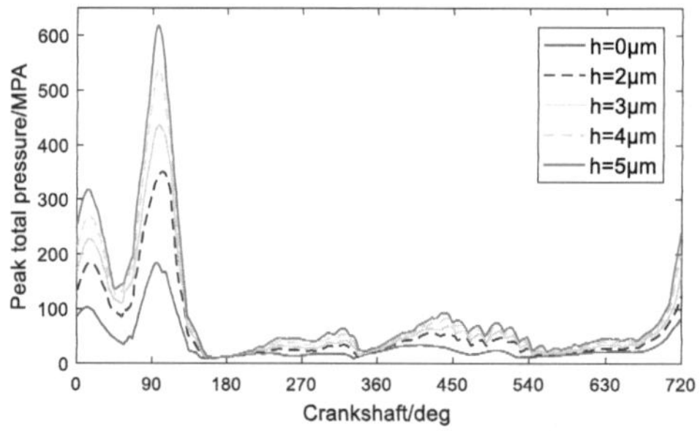

Figure 4. Variation in the PTP with different height.

4.1.1. Effect of the weight height. The first main bearing has the worst lubrication performance, so the performance of the first main bearing was analyzed here. The PTP of the bearings with roundness error is much higher than cylindrical bearing at the ignition moment of first two cylinders (about 10 degrees and 100 degrees in cycle) shown in figure 4. The PTP increased significantly with increasing wave height and it is even higher than 300MPA when the height is greater than 3μm.

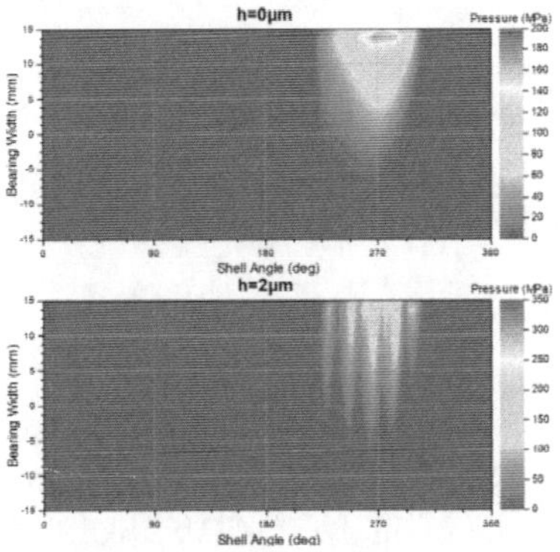

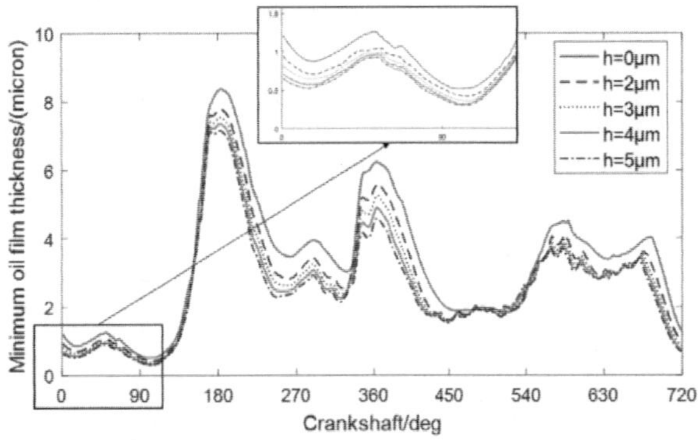

Figure 5. The total pressure distribution with *h*=0μm and *h*=2μm.

Figure 6. Variation in the MOFT with different height

The figure 5 shows the total pressure distribution of the cylindrical bearing and the bearing with roundness error (*h*=2μm) at 100° in cycle. It indicates that the wave trough area was not loaded, and the wave peaks area was subjected to most of the load, so the loaded area of bearing with roundness error significantly reduced and the total pressure distribution is multi-peak which approximate strip. The total pressure in wave peak area is very high, and it even exceeded the limits in some regions. As the wave height increase, the MOFT decreased by different extent within the cycle shown in figure 6. The MOFT is less than 1μm within 100 degrees of cycle, so the film thickness ratio is less than 1 and the main bearing is in the boundary lubrication condition which will reduce the bearing efficiency and reliability. In addition, the load density in wave peak area is extremely high and asperity contact occurs in peak area which decreased the lubrication performance. The total pressure distribution of different roundness error bearings at 100 degree in cycle are shown in figure 7. As the wave height increase, the total pressure at the wave peak increased significantly and the area subjected to load reduced, so the circumferential load uniformity decreased obviously with the wave height.

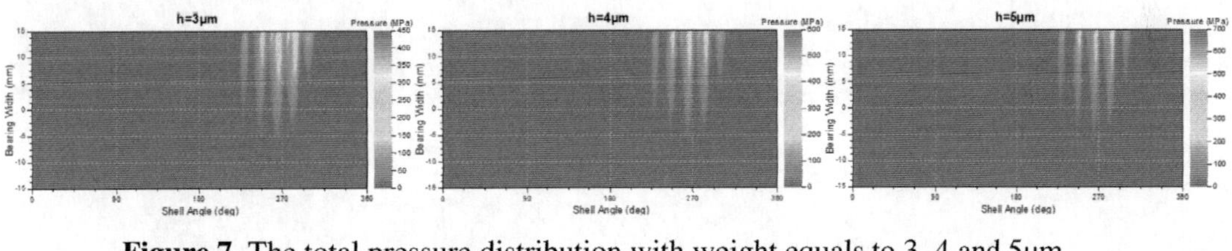

Figure 7. The total pressure distribution with weight equals to 3, 4 and 5μm.

In conclusion, the roundness error of bearing has a significant negative impact on lubrication performance of main bearing. As the wave height increase, the peak total pressure increased sharply and the minimum oil film thickness decreased, so the risk of adhesive wear and surface fatigue wear increased with wave height. In addition, the circumferential load uniformity decreased obviously with the wave height which reduce the bearing reliability.

4.1.2. Effect of the wave number. The PTP of bearings with roundness error is much higher than that of cylindrical bearing within 150 degree in cycle and it increased significantly with the wave number shown in figure 8. The reason for the increase of the PTP is that the metal-to-metal contact occurs in bearing peak region when the wave number is greater than 10, so the peak asperity contact pressure (PACP) increased sharply with the number shown in figure 9. Moreover, the PACP even greater than the limit value of 200MPA when the number is greater than 20.

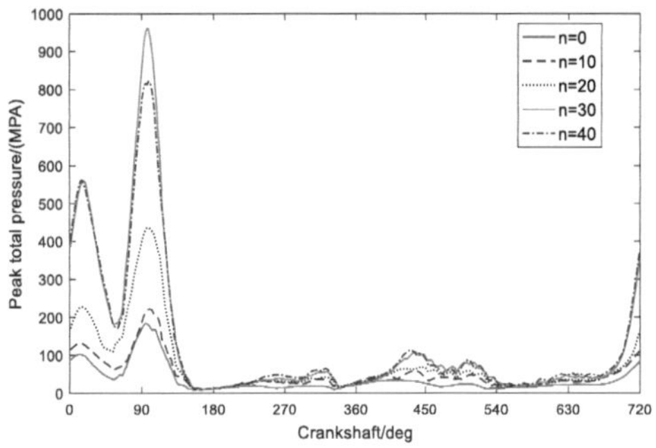

Figure 8. Variation in the PTP with different number.

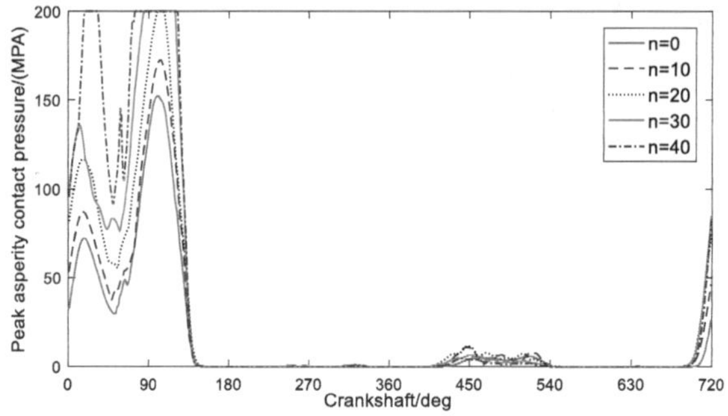

Figure 9. Variation in the PACP with different number.

The MOFT decreased with the wave number within 150 degree shown in figure 10 which is corresponding to the PACP. When the wave number is greater than 40, the MOFT is 0 in some moment which mean the lubrication condition is dry friction and the bearing reliability will greatly reduce. The enlargement of the total pressure distribution of bearings with different wave number roundness error at 100 degree in cycle are shown in figure 11. It shows that the area subjected to load decreased sharply with the wave number and the peak total pressure increased significantly. When the wave number is greater than 10, the circumferential load is in the shape of multi-peak and the circumferential load uniformity decreased obviously with the wave number.

In conclusion, the roundness error wave number has a significant negative effect on main bearing lubrication performance in terms of the peak total pressure, the minimum oil film thickness and the load uniformity. When the wave number is greater than 10, the bearing reliability and performance decrease significantly so the wave number should not exceed 10.

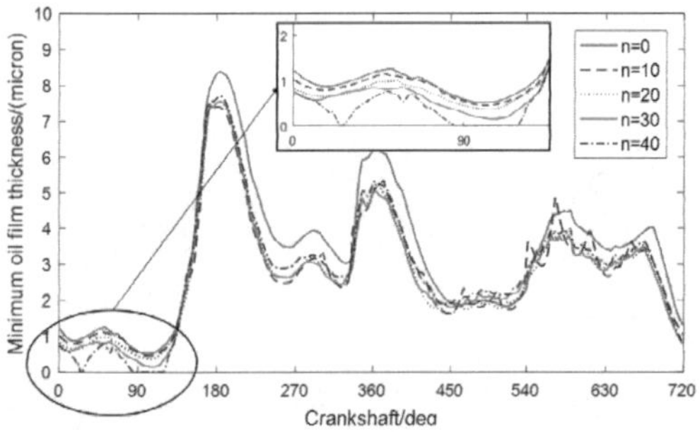

Figure 10. Variation in the MOFT with different number.

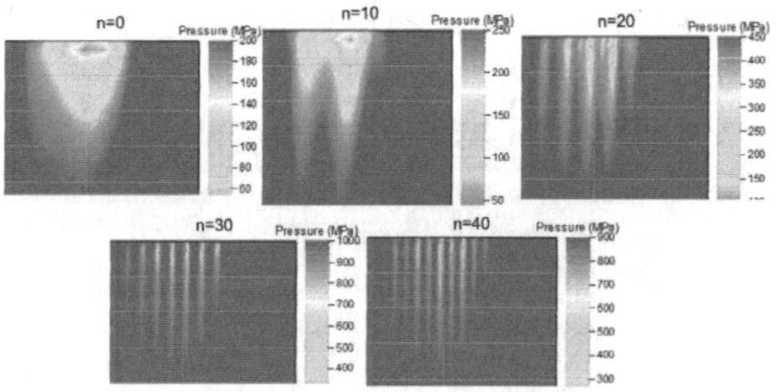

Figure 11. The enlargement of the total pressure distribution of bearings with different number.

4.2. Effect of coaxiality error

The main journal is bent due to load during operation, resulting a higher pressure at the edge of the bearing. If the main bearing house has a certain coaxiality error due to machining assembly error, the coaxiality error causes the bearing clearance at different axial position to be different, which may increase the asperity contact and total pressure in some bearing areas, resulting in severe friction and wear of bearing. Therefore, the influence of coaxiality error on main bearing lubrication performance was analysed in this part. Five cases of numerical simulations with different bearing house tilt angle were performed. The tilt angle (β) varies from -0.03° to 0.03° which is positive in the clockwise direction around the Y-axis, and e is the maximum bearing clearance change of each case shown in figure 12. The details of each case are shown in table 3.

Table 3. The details of each case.

	Case1	Case2	Case3	Case4	Case5
β/°	-0.03	-0.015	0	0.015	0.03
e/μm	7.85	3.93	0	3.93	7.85

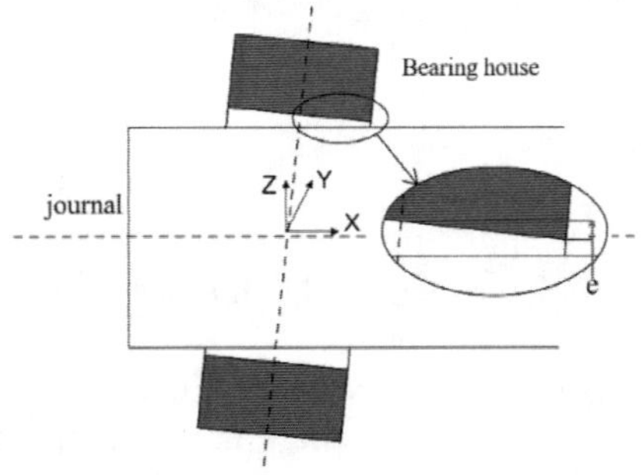

Figure 12. Bearing house tilt coaxiality error.

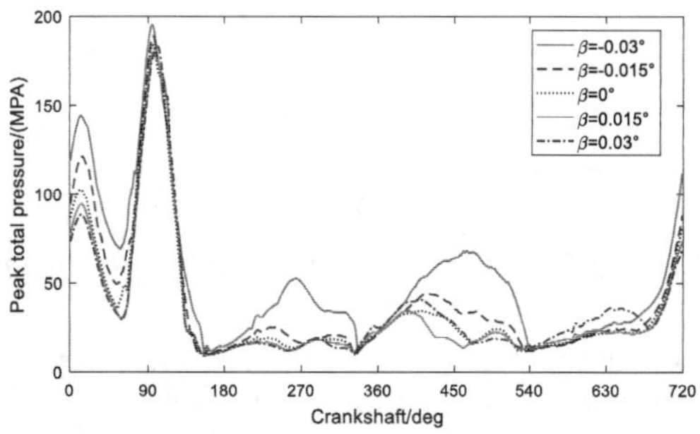

Figure 13. Variation in the PTP with different tilt angle.

The PTP of each bearing with different tilt angle are shown in figure 13. When the tilt angle is negative, the PTP increased significantly in the ignition moment of the left side cylinder like 10 degree, 250 degree and 480 degree. The reason of the phenomenon above is that the left cylinder piston moves in Z direction, so the asperity contact occurs in the bearing edge area in ignition moment. By contrast, when the tilt angle is positive, the tilt direction is consistent with the crankshaft journal bending direction, so the area subjected to load increased and the PTP decreased at 10 degree moment in cycle. Therefore, as the tilt angle increase from negative to positive, the peak asperity contact pressure (PACP) and the asperity contact percentage (ACP) decreased steadily within 100 degree in cycle shown in figure 14 and figure 15. In addition, both the PACP and the ACP increased slightly at ignition moment of the left second and third cylinder when the tilt angle is 0.03°.

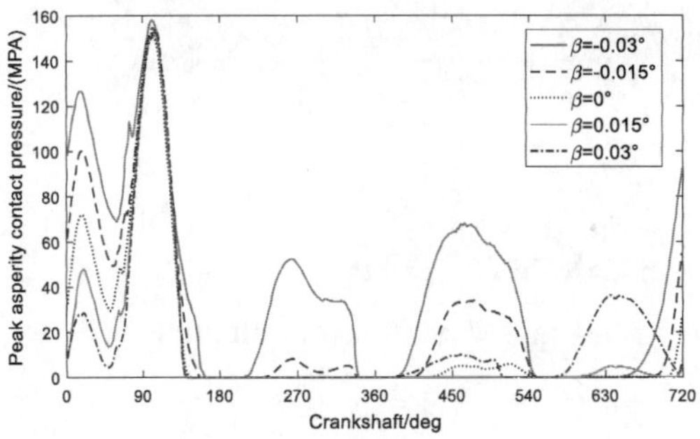

Figure 14. Variation in the PACP for various tilt angle.

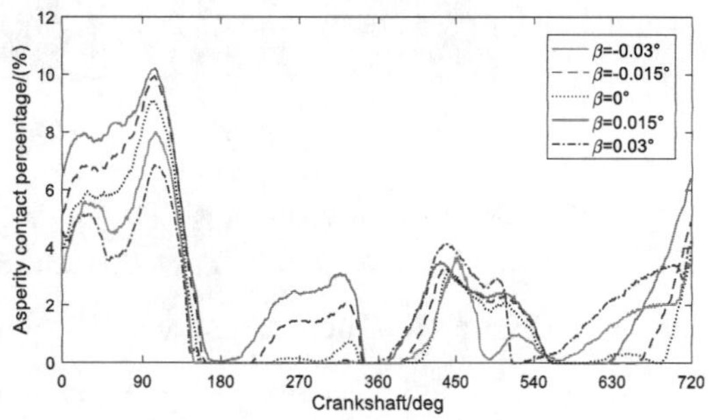

Figure 15. Variation in the ACP for various tilt angle.

Affected by the above factors, the average peak total pressure (APTP) decreased sharply with tilt angle firstly, reaching its lowest value when tilt angle is 0.015° (drop by 4.5% compared to no tilt bearing), then increased slightly with tilt angle, which is approximately a concave quadratic curve shown in figure 16. As the tilt angle increase from negative to positive, the MOFT increased within 90 degree in cycle from about 0.8μm to 1.6μm shown in figure 17, so the lubrication condition change from boundary lubrication condition to mixed lubrication condition which will raise the lubrication efficiency.

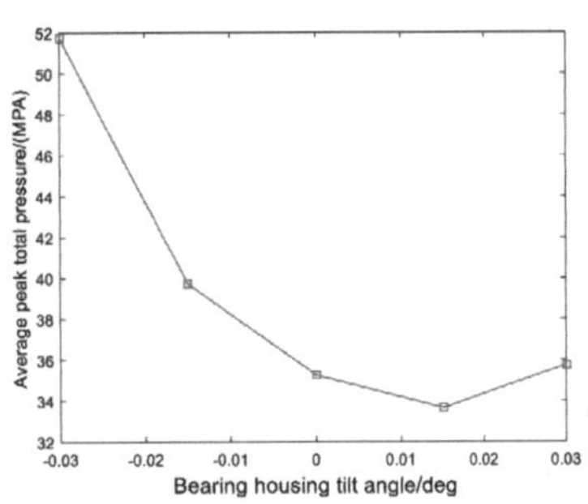

Figure 16. Variation of the APTP for various tile angel.

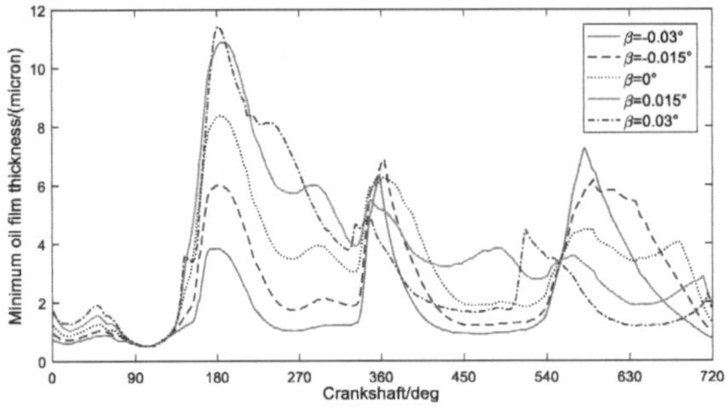

Figure 17. Variation of the MOFT for various tile angel.

The figure 18 shows the total pressure distribution of main bearings with different tilt angle at ignition moment of left first cylinder (about 10 degree in cycle). It indicates that the area subjected to load increased with the tilt angle and the peak total pressure decreased with the tilt angle. In addition, the central of load translates from edge region to central with the increase of tilt angle, so the edge asperity contact pressure dropped with the tilt angle. The average total pressure (ATP) of each axial section are shown in figure 19 which indicates that the ATP increased with tilt angel at left side of bearing and it decreased with tilt angle at right side, so the load percentage in central area increased and the axial load uniformity raised with the tilt angle.

In conclusion, when bearing house tilt angle is opposite to the journal bent direction, the lubrication performance decreased significantly, however, the performance increased slightly when the tilt angle is the same as the bend direction. Therefore, for V type 6 cylinder diesel engine, the first main bearing house and the fourth main bearing house could be tilted by a small angle clockwise and counterclockwise, respectively. In addition, the tilt angle should not be too large which is recommended to be within 0.02°.

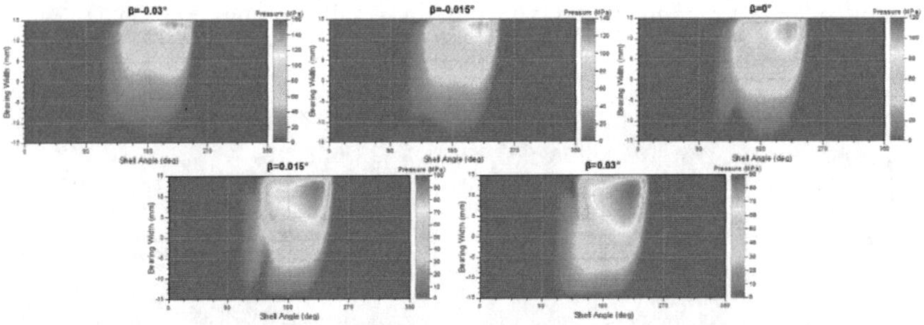

Figure 18. The total pressure distribution of bearings with different tilt angle.

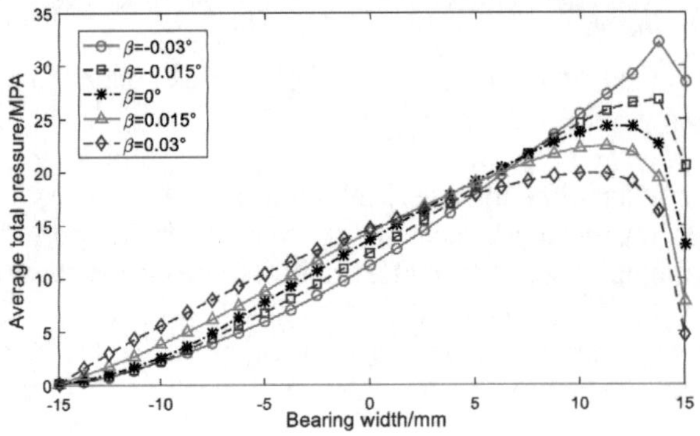

Figure 19. The ATP of each axial section of bearings with different tilt angle.

IOP Conf. Series: Materials Science and Engineering **576** (2019) 012015 doi:10.1088/1757-899X/576/1/012015

5. Conclusion

Out-of-roundness error has an apparent negative effect on the lubrication performance of the main bearings. Lubrication performance such as the peak total pressure, the minimum oil film thickness and the circumferential load uniformity decreased significantly with the wave height and the wave number. The reliability of main bearings reduced significantly when the roundness error wave number is greater than 10 and the wave height is greater than 2μm, which can be used as a basic machining tolerance limit requirement. When bearing house coaxiality error tilt angle is opposite to the journal load bending direction, the main bearing lubrication performance decreased significantly, however, the performance increased slightly when the tilt angle is the same as the bending direction. Therefore, for a V type six-cylinder diesel engine, the first and the fourth main bearing houses can be tilted a small angle clockwise and counterclockwise to improve lubrication performance, respectively.

6. Reference

[1] Vijayaraghavan D, Brewe D E and Keith T G 1993 Effect of Out-of-Roundness on the Performance of a Diesel Engine Connecting-Rod Bearing[J] *J. Tribol.* **115(3)**:538-43

[2] Mehenny D S, Taylor C M and Jones G J 1997 *Sae Internal Congress & Exposition.* **970216**

[3] J H Xiang, L F Zou and R D Liao 2012 *Transactions of Beijing Inst of Tech.* **32(12)**:1241-46

[4] Maspeyrot P and Frene J 1989 Shape defects and misalignment effects in connecting-rod bearings[J]. *Tribology Series.* **14**:317-22

[5] Maspeyrot P and Frene J 1991 *Tribology Series.* **18**:19-26.

[6] Lahmar M, Harouadi F and Frihi D 2000 *Proceedings of the Institution of Mechanical Engineers Part D: Journal of Automobile Engineering.* **214(8)**:991-97

[7] Patir N and Cheng H S 1978 An Aver Flow Model for Deter Effects of Three-D Roughn on Partial Hydrod Lubrica [J]. *J. of Lubrication Tech.* **100(1)**:12-17

[8] Greenwood J A 1970 The Contact of Two Nominally Flat Rough Surfaces[J]. *Proceedings of the Institution of Mechanical Engineers.* **185**:625-34

[9] Vencl A and Rac A 2014 Diesel engine crankshaft journal bearings failures: Case study [J] *Engineering Failure Analysis.* **44(9)**:217-28

IOP Conf. Series: Materials Science and Engineering **576** (2019) 012016 doi:10.1088/1757-899X/576/1/012016

Research on Tool Wear Monitoring and Turning Simulation

Liao Wei

Guangxi Radio and TV University, Nanning 530022, China

Abstract. The condition of cutting tool has a direct effect on processing quality, productivity and produce cost. So, it is very important to monitor tool condition in cutting process. The on line monitoring of tool wear is an important topic in the research of flexible manufacturing system. Cutting force is one of the important physical parameters in cutting process, which affects tool wear and machined surface quality. The correlation between cutting force and tool wear is experimentally analyzed. It is proposed that the change of tool cutting force is consistent with the change of wear. The simulation study of turning processing based on finite element analysis software Deform 3D. The size distribution of cutting force was simulated, the simulation results and the test results were compared and analyzed, which provided guiding significance for the determination of process parameters.

1. Introduction

The cutting tool can withstand higher temperature and mechanical shock during machining, which will cause the tool to wear, and the residual stresses will accelerate tool wear and influence the surface machining quality of the workpiece[1-3].

As an important part of advanced manufacturing technology, tool state intelligent monitoring technology is an emerging technology developed on the basis of modern sensor technology, signal processing technology, computer technology and manufacturing technology[4]. Many scholars at home and abroad are devoted to the research of on-line monitoring technology for tool wear, among them, R Teti et al. [5], JT Roth et al. [6] and JV abellan-nebot et al. [7] summarized the development status of tool wear monitoring technology from the perspectives of sensor signal selection, signal processing and pattern recognition.

304 stainless steel is a kind of universal stainless steel, which is widely used in the production of equipment and parts requiring good comprehensive performance. Cemented carbide has high hardness and good abrasion resistance. It is one of the commonly used tool materials for processing 304 stainless steel. Wc-co cemented carbide is widely used in cutting tools, mining tools and wear-resistant parts due to its high hardness, high red hardness, high strength and toughness[8-10].

There are two methods to monitor tool wear: direct measurement and indirect measurement. The direct method is to directly measure the size of the tool wear surface or the change of cutting edge shape when the tool is damaged. The advantage is high accuracy, but common direct measurement needs to stop monitoring, which can not achieve the purpose of online monitoring, thus affecting the processing efficiency[11]. The indirect method can obtain the working condition information indirectly by detecting the signal change between one or several parameters which have a strong internal connection with tool wear and damage. The advantage is that it can realize on-line real-time monitoring. Indirect measurement method is closer to the requirement of online monitoring[12], and cutting force monitoring method is adopted in this paper.

With the successful application of finite element method in the field of machining, the process of tool wear can be predicted by studying tool wear mechanism, and the information of tool stress, strain and temperature distribution can be obtained through simulation, which is directly related to tool wear[13]. The simulation study of turning processing based on finite element analysis software Deform 3D. The size distribution of cutting force was simulated, the simulation results and the test results were compared and analyzed, which provided guiding significance for the determination of process parameters.

2. Experiment

2.1. Experimental equipment and materials

The equipment used in the experiment includes numerical control lathe, image measuring instrument and dynamometer. The experimental tool is YW1 carbide blade and the workpiece material is 304 stainless steel. Cutting conditions: dry cutting. The mechanical properties of YW1 tool are shown in Table 1. Mechanical properties of 304 stainless steel are shown in Table 2. The cutting parameters are listed in Table 3.

Table 1. Mechanical properties of YW1 tool

Mechanical property	YW1
Hardness /HRA	91
Bending strength /MPa	1290
Tensile strength /MPa	≥1300
Density（g/cm^3）	12.6~13.5

Table 2. Mechanical properties of 304 stainless steel

Mechanical property	304 stainless steel
Hardness /HRB	≤92
Tensile strength /MPa	≥520
Yield strength /MPa	≥205
Elongation /%	≥40
Area reduction /%	≥60

Table 3. Cutting parameters

Technics	Cutting speed（m/min）	Cutting depth（mm）	Feed rate（mm/r）
NA	50	0.8	0.3
NB	60	0.8	0.3
NC	70	0.8	0.3

2.2 Experimental process

The cutting force sensor adopts the Kistler 9257B dynamometer manufactured by Swiss company. During the turning test, the cutting force was collected during the external round cutting process. The force in X, Y and Z direction was collected using Kistler 9257B piezoelectric dynamometer. The charge signal generated by the dynamometer was amplified by Kistler 5070 charge amplifier and transmitted to the computer through the data acquisition card. The block diagram of cutting force system design is shown in Figure 1.

IOP Conf. Series: Materials Science and Engineering **576** (2019) 012016 doi:10.1088/1757-899X/576/1/012016

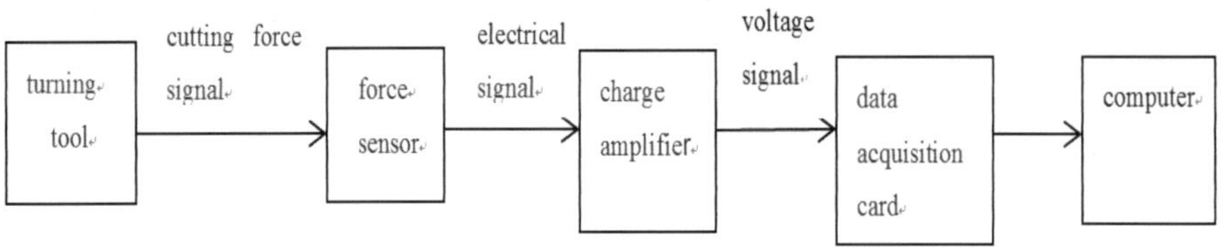

Figure 1. Design block diagram of cutting force system

The cutting tool is used for turning the outer circle on the numerical control lathe. During the experiment, after cutting the cutter for a certain distance, the blade was taken off, and the abrasion of the tool surface was measured with an image measuring instrument. Meanwhile, the cutting force was measured with dynamometer.

3. Results and discussion

3.1. Analysis of cutting force and wear

Cutting force is one of the important physical parameters in the process of cutting. Studying the rule of cutting force is helpful to analyze the process of cutting, which is of great guiding significance for determining reasonable cutting amount and improving machining efficiency.

The variation curve of the cutter's main cutting force and wear amount can be obtained by changing cutting speed without changing cutting depth and feed amount, as shown in Figure 2 and Figure3.

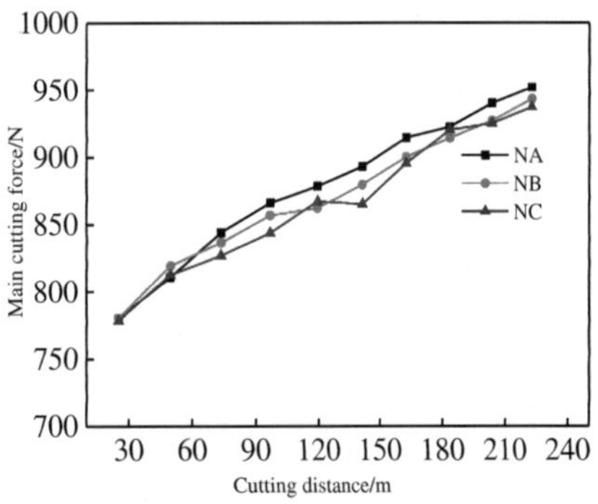

Figure 2. The effect of cutting speed on main cutting force

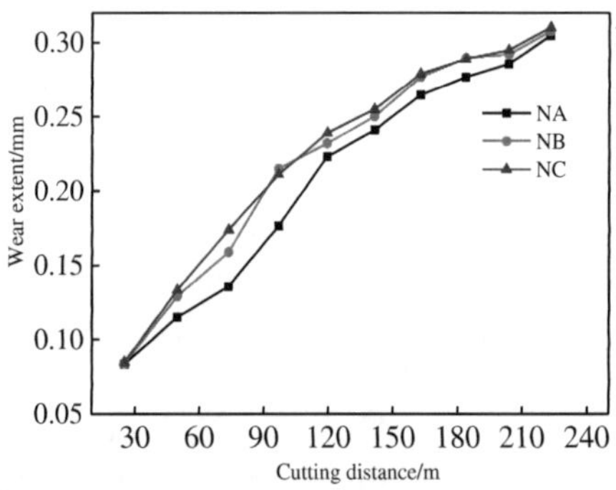

Figure 3. The effect of cutting speed on wear extent

It can be seen from Figure 2 that between different cutting speeds, the main cutting force in each segment decreases slightly as the cutting speed increases. Generally speaking, the cutting depth has the greatest impact on the cutting force, followed by the amount of feed and finally the cutting speed.

As can be seen from Figure 3, as the cutting speed increases, the amount of wear increases slightly.

It can be seen from Figure 2 and Figure 3 that under the same process conditions, the main cutting force and wear curve of the cutter change consistently. The trend of subsection change in each process is also generally consistent. Therefore, the change of cutting force to monitor the change of wear has certain research value and scientific basis.

In order to test whether the result of tool state change is reliable by monitoring the change of cutting force signal, take process NA as an example, the cutting depth ap=0.8mm, the feed rate f=0.3mm/r, and the cutting speed Vc=50m/min. Figure 4 shows the change curve of main cutting force and wear.

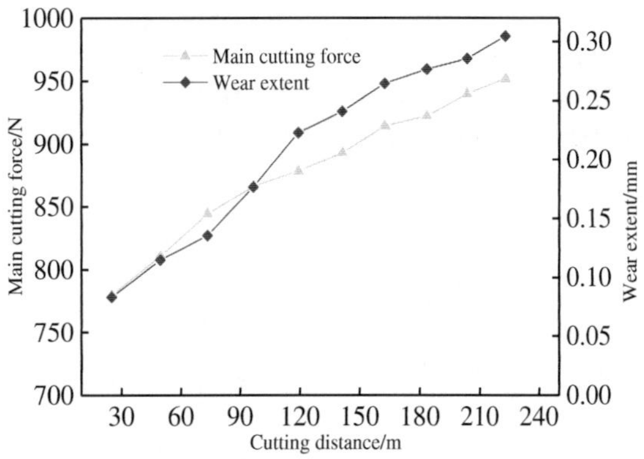

Figure 4. Change curve of main cutting force and wear extent

It can be seen from Figure 4 that the tool wear goes through three stages, namely the initial wear, normal wear and severe wear. With the extension of the cutting process, the cutting force increases with the increase of cutting distance (time), mainly due to the sharp wear of the cutter and the increase of cutting edge radius in the cutting process, so the cutting force increases with the increase of cutting distance.

As the most stable and reliable signal in machining process, cutting force signal is closely related to tool wear and damage. In addition, the successful application of force sensor in industry makes the cutting force monitoring technology most widely used in the research field of tool wear monitoring, which is also the most advantageous method.

3.2. Analysis of cutting force simulation

In the process of metal cutting, the cutting force includes the main cutting force, feed force and back force. The main cutting force is to calculate the strength of the lathe, design the machine parts and determine the power of the machine tool. Therefore, the main cutting force is selected as the research object.

Figure 5 showed the diagram of the simulation analysis of turning process in Deform-3D environment.

Figure 5. The diagram of the simulation analysis of turning process

The cutting model is used to describe the machining process by using feed rate(f), cutting speed(Vc)and cutting depth(ap). The influence of these parameters on the machining process is analyzed by analyzing the degree of simulation and simulation results.

The orthogonal experimental parameter and numerical comparison of cutting force simulation and experimental results was shown in Table 4.

Tab.4 Orthogonal experimental parameter and simulation and experimental results

Number	(ap)	(f)	(Vc)	simulation values (N)	experiment values (N)	Error (%)
1	0.5	0.15	50	806	883.6	9.6
2	0.5	0.2	60	841	879.2	4.5
3	0.5	0.25	70	874	905.5	3.6
4	0.8	0.15	60	862	854.8	-0.8
5	0.8	0.2	70	883	913.2	3.4
6	0.8	0.25	50	958	946.1	-1.2
7	1.1	0.15	70	894	935.6	4.7
8	1.1	0.2	50	956	981.3	2.6
9	1.1	0.25	60	982	1012.2	3.1

It can be seen from Table 4 that the error is within 10%, considering that the failure and friction conditions of the material in the actual cutting are very complicated, so the simulation results are acceptable. The DEFORM simulation is under ideal conditions, there are various influencing factors in actual processing, and it can be seen that there is a certain error between the simulated value of cutting force and the experimental value. But the overall error is small, within acceptable limits.

In order to verify the reliability of the cutting results obtained by simulation, the same experimental parameters are used to compare the results of the cutting force measured with the simulation results. Such as ap=0.5mm, f=0.2mm/r and Vc=60m/min.

Figure 6 showed the experimental curve of the cutting force measured by the dynamometer. Figure 7 showed the 3D simulation curve of cutting force.

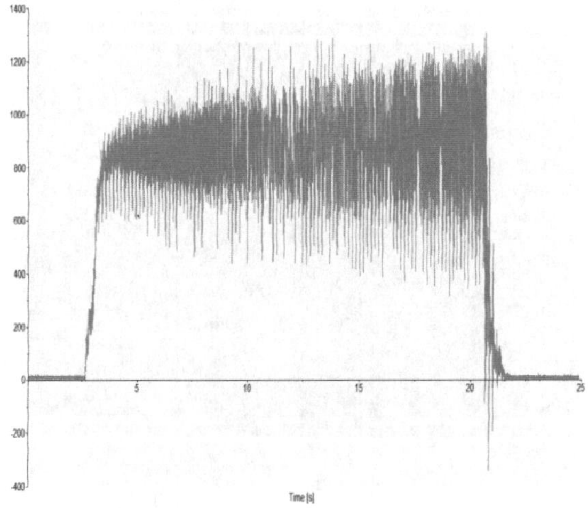

Figure 6. The experimental curve of the cutting force

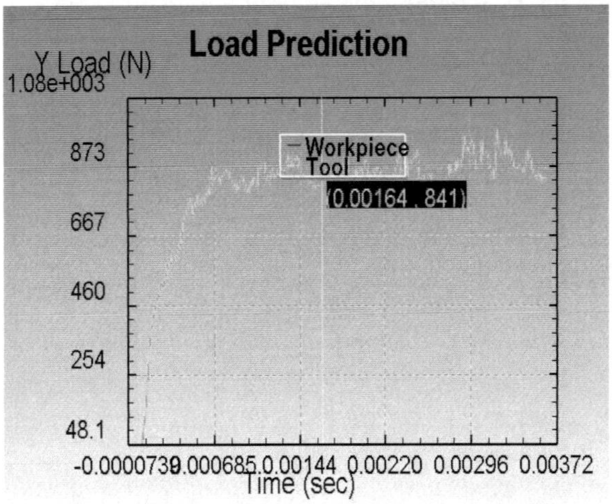

Figure7. The 3D simulation curve of cutting force

It can be seen from the Figure 6 and Figure 7 that when the tool cuts the workpiece, cutting force over a period of time will reach the steady value, and the value within a certain range, in the actual experiment, the image characteristics obtained by the dynamometer and software are consistent.

4. Conclusions

The on line monitoring of tool wear is an important topic in the research of flexible manufacturing system. Cutting force is one of the important physical parameters in cutting process, which affects tool wear and machined surface quality. During the turning process, the amplitude of the cutting force signal changes with the change of the cutting state. It is an ideal monitoring method to monitor the tool wear state by using cutting force.

In this paper, The correlation between cutting force and tool wear is experimentally analyzed, and it is suggested that the change of cutting force is consistent with the change of tool wear. The simulation study of turning processing based on finite element analysis software Deform 3D. The size distribution of cutting force was simulated, the simulation results and the test results were compared and analyzed, which provided guiding significance for the determination of process parameters.

5. References

[1] J Xu, H Hu, W Fu, et al 2014. *Journal of Steel Research*, vol 26, p 60
[2] J Lin, H Zhao, Z Cai, et al 2007. *Tsinghua Univ(Sci &Tech)*, vol 47, p 161
[3] Z Jiang, L Wang, L Shi, et al 2014. *Journal of mechanical engineering*, vol 50, p 178
[4] Y Liu, Y Chen, W Yang, et al 2004. *Machine Tool & Hydraulics*, vol 42, p 174
[5] R Teti, K Jemielniak, G O'Donnell, D Dornfeld 2010. *Cirp Annals-Manufacturing Technology*, vol 59, p 717
[6] JT Roth, D Djurdjanovic, X Yang, L Mears, T Kurfess 2010. *Journal of Manufacturing Science and Engineering*, vol 132, p 575
[7] JV Abellan-Nebot, FR Subirón 2010. *The International Journal of Advanced Manufacturing Technology*, vol 47, p 237
[8] Bounhoure V, Lay S, Coindeau S et al 2015. *International Journal of Refractory Metals and Hard Materials*, vol 52, p 21
[9] W Su, Y Sun, J Feng et al 2015. *International Journal of Refractory Metals and Hard Materials*, vol 48, p 369
[10] Rumman M R, Xie Z, Hong S J, et al 2015. *Materials & Design*, vol 68, p 221

[11] DAN L, MATHEW J 1990. *International Journal of MachineTools and Manufacture*, vol 30, p 579
[12] Dimla E. DimlaSnr 2000. *International Journal of Machine Tools and Manufacture*, vol 40, p 1073
[13] Y Chen, S Yang, Y FU, et al 2013. *Acta Aeronautica ET Astronautica Sinica*, vol 34, p 2230

Acknowledgments
Foundation item: Project (GXGZJG2017A070) supported by Guangxi Vocational Education Teaching Reform Research Project

Robust tracking control for Robotic Manipulators based on Super-twisting Algorithm

Muhammad Umer Javed[1], Haitao Liu[1], Jianhao Nie[1] and Jian Sun[1]

[1]Robotics Lab, School of Mechanical and Power Engineering, Guangdong Ocean University, Zhanjiang 524088, PR China

Abstract: Robotic manipulators are broadly used in industries for different kinds of specialized operational responsibilities and it is very complicated to design of a robust and stable controller for these systems. The performance of industrial robotic manipulators has augmented concerning stability and safety due to growths in robust controller design. However, designing a stable and reliable control method for a robot manipulator is important for real-world applications. This paper proposed a robust composite super-twisting sliding mode controller (STSMC) for robotic manipulators with uncertainties. To improve the robustness of robotic manipulators and reduce the chattering problem of sliding mode control, the Super-twisting algorithm is employed to design a continuous controller, and the stability is proved. Simulations on the 2-DOF robotic manipulator are shown to illustrate the effectiveness of the proposed method in which demonstrated that the proposed model has smoothness at 10^{-5}.

1. Introduction

Robotic manipulators are widely used in different kinds of industries for different types of specialized operational responsibilities. These systems are enthusiastically coupled, time-varying and nonlinear. Therefore, it is very difficult to design a stable and robust controller for these kinds of systems. In recent few years, Research control methodologies for industrial robots have increased. The performance of industrial robotic manipulators has augmented regarding safety and stability due to developments in robust controller design. However, it is important to design a reliable and stable control method for a robot manipulator for real-world applications [1], [2]. It is essential to develop and apply control algorithms to complicated tasks. Several types of controllers have been recognized for robot manipulators. These controllers can be separated into two main classes: linear control systems and nonlinear control systems. Non-linear controllers are recommended in this research because in linear control some challenges are needed to overcome the limitation of the velocity and acceleration of the system. Nonlinear control algorithms are sub-divided into three main types: model-based control algorithms, soft computing (artificial intelligence)-based control theory and hybrid control actions. The main benefit of a nonlinear model-free controller is structure information, which has been improved by researchers over the years [1–6]. While it has several advantages, this method has challenges associated with system reliability, precision and robustness. Model reference control algorithms are suggested to improve the stability, robustness, and reliability. The feedback linearization method (FLC), back-stepping control procedure (BSC), passivity-based control (PBC), Lyapunov-based algorithm (PBC) and sliding mode control (SMC) are general methods for designing model-based controllers. However, selecting a suitable control technique is a foremost challenge for many researchers.

In recent years, different kinds of methodologies have been used to control the systems and the Sliding Mode Control (SMC) [7], [8] is one of the efficient techniques to control a complicated dynamic system. This controller reduces the sensitivity to disturbances and plant parameters differences. Another kind of control for robotics systems as the adaptive control, adaptive fuzzy sliding mode [9], [10], fuzzy logic, neural adaptive [11], neural networks [12], sliding mode neuronal network [13], and neuro-fuzzy.

SMC became the main working mode for the situation of robustness in the class of control systems [14], [15]. For the nature of this control, the principal disadvantage is the chattering phenomenon caused by the "sign" function while forcing the states to zero. The chattering problem is dangerous for the physical systems high-frequency oscillations. A method to reduce the chattering is the Super-twisting Algorithm (STA) [16], this algorithm was developed to avoid the chattering effect. The Super-twisting Control (STC) can be applied to any system where it can obtain the first derivative of the sliding variable. The man advantage of STC is that it compensates uncertainties and perturbations and guarantees the system of finite time convergence to the origin of the sliding variable and its derivate [17].

The main idea of this paper is to propose a controller for robotic manipulator based on a super twisting algorithm. To improve the trajectory tracking is the priority of this paper using a continuous controller. MATLAB/Simulink was used to validate the dynamic model and simulate the proposed controller for the realization of a particular task.

The rest of the paper is designated as a mathematical model of robot manipulator in section 2, the proposed design is discussed in section 3, and the stability will be described in section 4. In section 5 Results and analysis presented. In the last conclusion of this paper is described.

2. A mathematical model of Robot Manipulator

The primary challenge in this study is to control the robot manipulator using Super-twisting controller in the presence of uncertainty and disturbance. The dynamics of serial n-link is described in [18].

$$M(q)\ddot{q} + V(q,\dot{q})\dot{q} + g(q) = \tau \tag{1}$$

Where $M(q) \in \mathbb{R}^{nxn}$ Inertial vector Matrix, $V(q,\dot{q}) \in \mathbb{R}^{nx1}$ Centripetal and Coriolis vector Matrix, $g(q) \in \mathbb{R}^{nx1}$ is the Gravitational vector Matrix, $q \in \mathbb{R}^{nx1}$ is the Position Vector Matrix, $\dot{q} \in \mathbb{R}^{nx1}$ is the Velocity vector Matrix and $\tau \in \mathbb{R}^{nx1}$ is the applied input torque Matrix. The two-dimensional state vector x is defined as:

$$x = \begin{bmatrix} x_1 \\ x_2 \end{bmatrix} = \begin{bmatrix} q \\ \dot{q} \end{bmatrix} \tag{2}$$

Then the non-linear system in (1) can be represented as a state space model as follows:

$$\begin{cases} \dot{x}_1 = x_2 \\ \dot{x}_2 = M^{-1}(x_1)[\tau - V(x_1,x_2)x_2 - g(x_1)] \\ y = x_1 \end{cases} \tag{3}$$

Where y is the output of the system.

3. Super-twisting Controller Design

3.1. Design of controller

For system (1) the desired homogenous linear time-invariant sliding surface can be written as follows in equation (4)

$$s = cx_1 + x_2 \tag{4}$$

where $s, (c > 0), x_1, x_2$ are sliding surface, sliding polynomial, position and velocity of the system respectively. In robotics, the main principal of designing super twisting control law is to move the error and derivative of error to zero. So, sliding surface can be written in the form of error as follows:

$$s = ce + \dot{e} \tag{5}$$

where $e, \dot{e}$ are error and derivative of error and can be described in (6) and (7):

$$e = x_1^d - x_1 = q_d - q = \begin{bmatrix} q_{d1} - q_1 \\ q_{d2} - q_2 \end{bmatrix} \tag{6}$$

$$\dot{e} = \dot{x}_1^d - \dot{x}_1 = x_2^d - x_2 = \begin{bmatrix} \dot{q}_{d1} - \dot{q}_1 \\ \dot{q}_{d2} - \dot{q}_2 \end{bmatrix} \tag{7}$$

where, x_1^d *and* $\dot{x}_1^d$ are the desired position and velocity of the system, respectively.
To designed super twisting control for this system, we need to take the derivative of the sliding surface. Equation (5) can be written in the following form:

$$\dot{s} = c\dot{e} + \ddot{e} \tag{8}$$

where, $\ddot{e}$ is the acceleration error. Now substituting the value of (6) and (7) in (8) we get

$$\begin{aligned} \dot{s} &= c\dot{e} + \dot{x}_2^d - \dot{x}_2 \\ &= c\dot{e} + \dot{x}_2^d - M^{-1}(x_1)[\tau - V(x_1, x_2)x_2 - g(x_1)] \end{aligned} \tag{9}$$

The total τ of the proposed system can be designed as follows:

$$\tau = M(x_1)\big[\big(c\dot{e} + \dot{x}_2^d + u_{STA} + u_s\big)\big] + V(x_1, x_2)x_2 + g(x_1) \tag{10}$$

where, u_s & u_{STA} are the sliding mode control and super twisting algorithm respectively, and can be defined in (11) & (12) respectively:

$$\begin{cases} u_{STA} = k_2|s|^{\frac{1}{2}} sign(s) + v \\ \dot{v} = -k_3 sign(s) \end{cases} \tag{11}$$

$$where, \qquad k_2, k_3 > 0$$

and

$$sign(s) = \begin{cases} -1, & s < 0 \\ 0, & s = 0 \\ 1, & s \geq 0 \end{cases}$$

$$u_s = -k_1 sign(s) \tag{12}$$

$$where, \qquad k_1 > 0$$

Where, k_1, k_2 & k_3 are positive constants.

3.2. Stability Analysis

From (9) to (12) we obtain the following closed loop system

$$\begin{cases} \dot{s} = -k_1 sign(s) - k_2|s|^{\frac{1}{2}} sign(s) + v \\ \dot{v} = -k_3 sign(s) \end{cases} \tag{13}$$

Proof:
Let

$$\chi^T = [|s|^{\frac{1}{2}} sign(s), v] = [\chi_1, \chi_2]^T$$

Select a candidate Lyapunov function [19]:

$$V = \chi^T P \chi \tag{14}$$

where P is the positive definite symmetric matrix and satisfies the following equation:

$$A^T P + PA = -Q \tag{15}$$

where A=$\begin{bmatrix} -\frac{1}{2}k_2 & \frac{1}{2} \\ -k_3 & 0 \end{bmatrix}$ is Hurwitz Matrix. Q is any positive definite symmetric Matrix

Since:

$$\begin{aligned} \dot{\chi} &= \begin{bmatrix} \frac{1}{2}\frac{1}{|s|^{\frac{1}{2}}}\left[-k_2|s|^{\frac{1}{2}} sign(s) + v - k_1 sign(s)\right] \\ -k_3 sign(s) \end{bmatrix} \\ &= \frac{1}{|\chi_1|}\left[A\chi - \begin{bmatrix} \frac{1}{2}k_1 sign(s) \\ 0 \end{bmatrix}\right] \end{aligned} \tag{16}$$

Differentiate (14) with time, we have

$$\dot{V} = \dot{\chi}^T P \chi + \chi^T P \dot{\chi}$$

$$= \frac{1}{|\chi_1|}\left(A\chi - \begin{bmatrix} \frac{1}{2}k_1 sign(s) \\ 0 \end{bmatrix}\right)^T P\chi + \chi^T P \frac{1}{|\chi_1|}\left(A\chi - \begin{bmatrix} \frac{1}{2}k_1 sign(s) \\ 0 \end{bmatrix}\right)$$

$$= \frac{1}{|\chi_1|}\left(A\chi - \begin{bmatrix} \frac{1}{2}k_1 sign(s) \\ 0 \end{bmatrix}\right)^T P\chi + \chi^T P \frac{1}{|\chi_1|}\left(A\chi - \begin{bmatrix} \frac{1}{2}k_1 sign(s) \\ 0 \end{bmatrix}\right)$$

$$= \frac{1}{|\chi_1|}\left(\chi^T A^T P \chi - \chi^T P A \chi - \begin{bmatrix} \frac{1}{2}k_1 sign(s) & 0 \end{bmatrix} P\chi - \quad \chi^T P \begin{bmatrix} \frac{1}{2}k_1 sign(s) \\ 0 \end{bmatrix}\right)$$

$$\leq \frac{1}{|\chi_1|}\chi^T (A^T P + PA)\chi - \frac{1}{|\chi_1|}k_1 sign(s)|s|^{\frac{1}{2}} sign(s)\lambda_{\min}(P)$$

$$\leq -\frac{1}{|\chi_1|}\chi^T Q \chi - k_1\lambda_{\min}(P)$$

$$\leq -\frac{1}{|\chi_1|}\chi^T Q \chi \leq -\frac{1}{\|\chi_1\|}\lambda_{\min}(Q)\|\chi\|^2 \tag{17}$$

$$\leq -\lambda_{\min}(Q)\|\chi\| \tag{18}$$

For equation (14) we have

$$\lambda_{\min}(P)\|\chi\|^2 \leq \chi^T P \chi \leq \lambda_{\max}(P)\|\chi\|^2 \tag{19}$$

Then according to (19)

$$V^{\frac{1}{2}} \leq \lambda^{\frac{1}{2}}{}_{max}(P)\|\chi\| \tag{20}$$

Combining (18) and (20), we obtain

$$\dot{V} \leq -\frac{\lambda_{\min}(Q)}{\lambda^{\frac{1}{2}}{}_{max}(P)} V^{\frac{1}{2}} \tag{21}$$

Then the system (13) can converges to zero in finite time i.e. when t→T, s→0. Since $s = ce + \dot{e}$ we can conclude $e \to 0$, $\dot{e} \to 0$ when t→∞. i.e. the robot system is asymptotically stable.

4. Simulations

The effectiveness of the proposed method a two-link robot manipulator described in [20] is considered, as shown in Figure 1. The dynamic model of the two-link robot manipulator and parameters selected from [17].

For STC the value of K2=0.8, K3=1.5 selected and constant reference is chosen as 0.2 Radian for simulations. Simulation Results of the considered robotic manipulator are shown below:

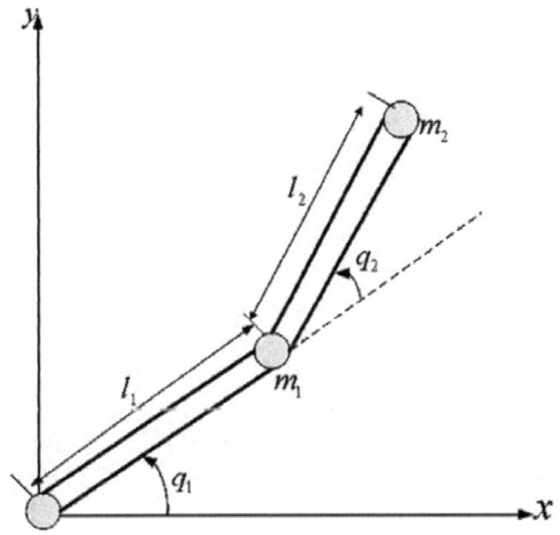

Figure 1. Two Link Robotic Manipulator.

Figure 2 and figure 3 shows the trajectory tracking of link 1 & 2 of proposed STC and [18] STC respectively. For the same parameters, the proposed STC shows the smooth trajectory tracking of the system while the trajectory tracking of the controller used in [18] is not smooth. The trajectory time for proposed system model is 0.04 seconds while the trajectory tracking time for the previously used controller is 0.02 seconds.

Figure 4 and figure 5 shows the Tau applied to link 1 & 2 of the proposed model and STC model used in [18] respectively. It is shown that the proposed controller is smoother and more robustness to the system while the [18] one is not smoother. From the zoomed view of figure 4 and figure 5 it is shown that the proposed controller has a minimal amount of changes rather than controller used in [18]. It is clearly shown that the proposed controller has more continuity and smoothness which proves that the system has more robustness while the STC used in [18] overshoot which can cause damage.

Tracking error of link 1 & 2 of proposed model and controller used in [18] is shown in figure 6 and figure 7, respectively. It is clear that the proposed model has less chattering effect than [18] model and error is smoothly going to zero. The Zoomed area of figure 7 and figure 7 shows the tracking error of the proposed model and it can be seen that the proposed model has very less error even in the zoomed section of figure 6 and figure 7 it can be seen that the controller used in [18] have an error in 10^{-4} while at this stage proposed controller has smoothness.

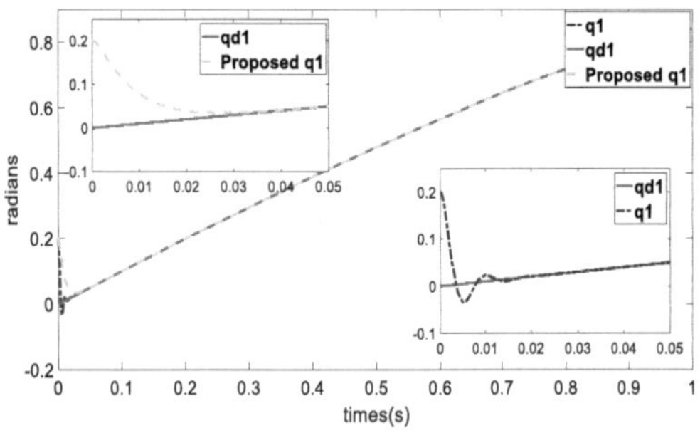

Figure 2. Trajectory tracking of link 1 using proposed STC vs. STC used in [18].

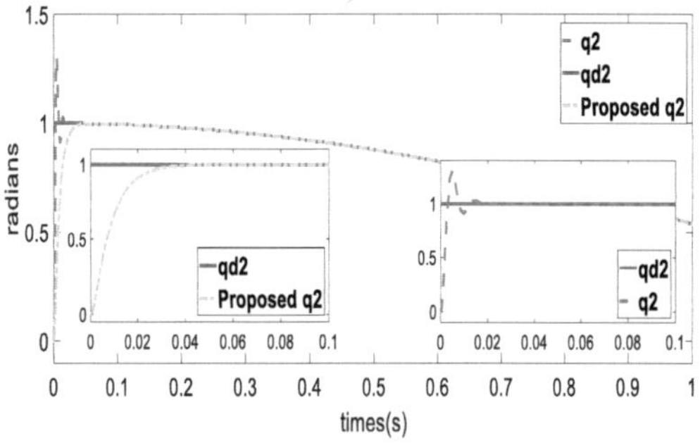

Figure 3. Trajectory tracking of link 2 using proposed STC vs. STC Used In [18].

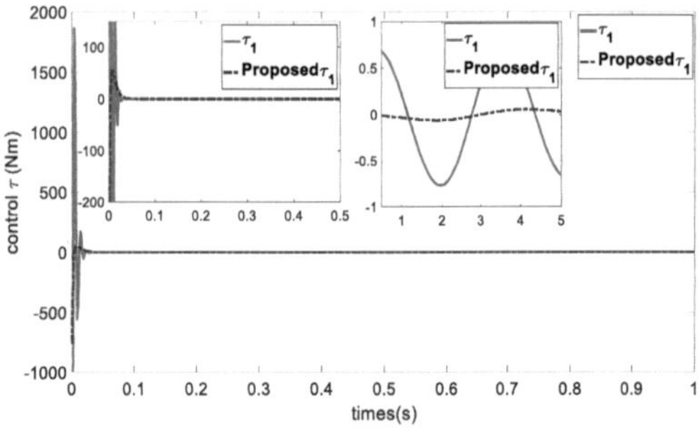

Figure 4. Tau applied of link 1 using proposed STC vs STC used in [18].

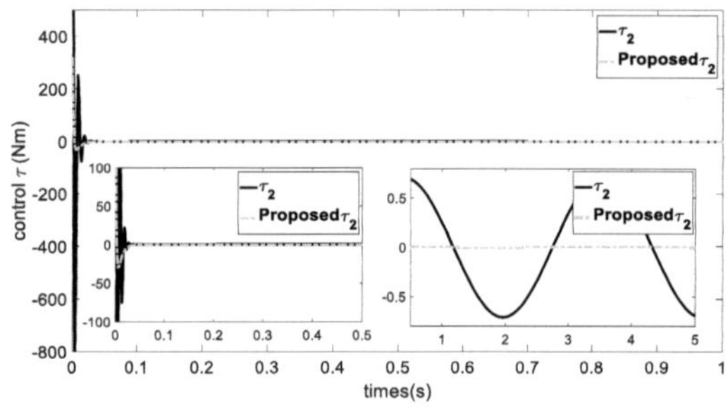

Figure 5. Tau applied of link 2 using proposed STC vs. STC used in [18].

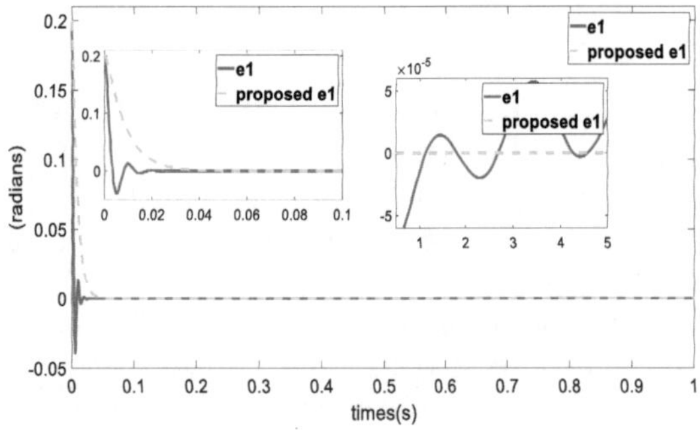

Figure 6. Error tracking of link 1 using proposed STC vs. STC used in [18].

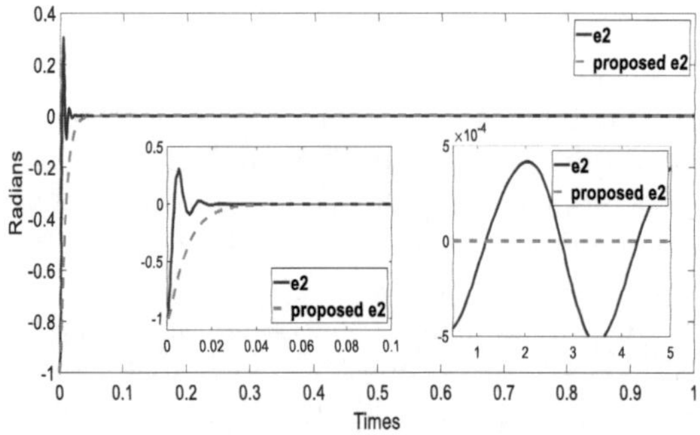

Figure 7. Error tracking of link 2 using proposed STC vs. STC used in [18].

5. Conclusions

This study aim was to propose a Robust tracking control for Robotic Manipulators based on super-twisting Algorithm. According to simulations, the proposed super-twisting controller approach shows the better performance in the term of reaching the object, in control sense and in the term of Error. The controller is smoother and continuous. The Result obtained from constant reference 0.2 radian for trajectory tracking shows that both controllers present good trajectory tracking for the stabilization of the position with the desired position, but the trajectory tracking error of the proposed STC is smaller than previously used STC. Additional types of controller and observer can be considered for Chattering attenuation In the future.

6. References

[1] Siciliano E B and Khatib O 2013 Springer handbook of robotics *Choice Rev. Online*, vol. **46**, no. 06, pp. 46-3272-46–3272.
[2] Son N N, Anh H P H and Nam N T 2016 Robot manipulator identification based on adaptive multiple-input and multiple-output neural model optimized by advanced differential evolution algorithm *Int. J. Adv. Robot. Syst.*, vol. **14**, no. 1, pp. 1–14.
[3] Alonge F, Cirrincione M, Pucci M and Sferlazza A 2016 Input-output feedback linearization control with on-line MRAS-based inductor resistance estimation of linear induction motors including the dynamic end effects *IEEE Trans. Ind. Appl.*
[4] Lascu C, Jafarzadeh S, Fadali M S and Blaabjerg F 2017 Direct Torque Control With Feedback Linearization for Induction Motor Drives *IEEE Trans. Power Electron.*
[5] Utkin V, Guldner J and Shi 2009 Sliding Mode Control in Electro-Mechanical Systems *CRC Press Boca Raton, FL, USA*, vol. **34**.
[6] Li H, Wang J and Shi P 2016 Output-Feedback Based Sliding Mode Control for Fuzzy Systems with Actuator Saturation *IEEE Trans. Fuzzy Syst.*
[7] Slotine J J and Sastry S S 1983 Tracking control of non-linear systems using sliding surfaces with application to robot manipulators *Int. J. Control.*
[8] Slotine J J E, Hedrick J K and Misawa E A 2009 On Sliding Observers for Nonlinear Systems *J. Dyn. Syst. Meas. Control.*
[9] Sun G F, F C 1996 Design of adaptive fuzzy sliding mode controller for robot manipulators *Proc. IEEE 5th Int. Fuzzy Syst.* ISBN: 0-7803-3645-3.
[10] N D and Medhaffar T D 2006 A decoupled fuzzy indirect adaptive sliding mode controller with application to a robot manipulator *Int. J. Model. Identif. Control*, vol. **1**, pp. 23–29.
[11] Yu W, Poznyak A S and Sanchez E N 2003 Neural adaptive control of two-link manipulator with sliding mode compensation.
[12] Ertugrul M and Kaynak O 1998 Neural computation of the equivalent control in sliding mode for robot trajectory control applications in *Proceedings - IEEE Int. Conf. on Rob. and Auto.*
[13] Uran S, Sabanovic A, Rodic M, Safaric R and Jezernik K 2002 Sliding-mode neural network robot controller.
[14] Utkin V I and Chang H C 2002 Sliding mode control on electro-mechanical systems *Math. Probl. Eng.*
[15] Edwards S K and Spurgeon C 1998 Sliding Mode Control Theory and Applications *NY: CRC Press.*
[16] Levant A 2007 Principles of 2-sliding mode design *Automatica.*
[17] Chalanga A, Kamal S, Fridman L M, Bandyopadhyay B and Moreno J A 2016 Implementation of Super-Twisting Control: Super-Twisting and Higher Order Sliding-Mode Observer-Based Approaches *IEEE Trans. Ind. Electron.*
[18] Cruz G L, Alazki H, and Hernández R G 2018 Super Twisting Control For Thermo's Catalyst-5 Robotic Arm *IFAC-PapersOnLine*, vol. **51**, no. 13, pp. 303–308.
[19] H. Liu. H. K. Khalil 2018 Output feedback stabilization using super-twisting control and high-gain observer *Int. J. Robust Nonlinear Control*, vol. **29**, no. 3.
[20] Tairen b H and Sun A 2011 Neural network-based sliding mode adaptive control for robot manipulators *Elsevier.*

Acknowledgment

This work was supported by the Natural Science Foundation of Guangdong Province in China [grant number 2018A0303130076], the Science and Technology Planning Project of Zhanjiang City [grant number 2017A02025], and the Innovation Strong School Project in Guangdong Province [grant number 2017KTSCX087].

IOP Conf. Series: Materials Science and Engineering **576** (2019) 012018 doi:10.1088/1757-899X/576/1/012018

The influence of the temperature of hydro-pneumatic spring into the state of tracked vehicle

Mang Jiang

Department of Automotive Engineering, Tsinghua University, Beijing 100084, China

Abstract. Based on the multi-body dynamics model, the influence of the temperature of the hydro-pneumatic spring on the state of a lighted tracked vehicle under several driving conditions is studied. A multi-body dynamics model of chassis and body of a tracked vehicle is established, and verified with simulation and experimental results in the relevant literatures. Typical driving conditions are studied including running on a flat road with constant velocity, passing a semi-round obstacle with the radius of 200 mm, and running on a sinusoidal surface with wavelength of 2 m and amplitude of 100 mm. Some responses of the vehicle are compared and analyzed, so as to obtain the influence of the temperature of hydro-pneumatic spring. The numerical analysis results show that the temperature of hydro-pneumatic spring has a significant effect on the vehicle's responses including height of centroid, vertical acceleration of centroid, pitch angle, track tension and gas pressure of the springs.

1. Introduction

When tracked vehicle is running on flat road surface, the stroke of the road-wheel is small, and small stiffness of the suspension can improve the ride comfort. While running on rough road surface, the stroke of the road-wheel is large, and large stiffness of the suspension is required to increase the vibration absorption ability. The stiffness of hydro-pneumatic increases as the piston stroke increases, and this characteristic is very suitable for such driving requirements of tracked vehicles. The hydro-pneumatic spring provides elastic force by compressing the high pressure gas in the closed container. The state equation of the ideal gas shows that the pressure of gas in the chamber is related to its volume and temperature. Due to long-term work of the hydro-pneumatic spring or changes in the ambient temperature, the gas temperature will change greatly, then the stiffness will change, which will eventually affect the state of the tracked vehicle.

When the oil passes through the orifice at high speed, the internal friction between its molecules will occur, then the temperature of the oil increases. This is the heat generation mechanism of the hydro-pneumatic spring. Then the heat is transferred to the gas chamber through the floating piston to make the gas temperature increases [1, 2]. Chen Yijie et al. [3] studied the law of temperature rise of hydro-pneumatic spring under the excitation of E and F level roads, but did not pay attention to the influence of temperature rise on vehicle state. Chen Yijie et al. [4] studied the influence of viscous heat of hydro-pneumatic spring on its damping force, the law of temperature rise when running on the C, D, E level roads at a constant speed, and optimize the factors affecting temperature rise. Els P. S. et al. [5] studied the effect of heat transfer on the temperature of hydro-pneumatic springs. It is mentioned that temperature has a significant influence on the stiffness of hydro-pneumatic springs. It is also pointed out that the heat source is affected by road roughness, vehicle speed and damping force.

Many researches in the literature are about the heat generation mechanism of hydro-pneumatic springs, the way of heat dissipation, the factors that affect the temperature rise, and the effect of temperature on the stiffness and damping characteristics. Temperature changes affect the stiffness of the hydro-pneumatic springs, which in turn affects the state of the vehicle, however there are few studies on this topic.

In this paper the multi-body dynamics model of a high-speed and high-mobility tracked vehicle with hydro-pneumatic spring suspension is established. This paper studies the influence of gas temperature on the vehicle state when running on several typical road surfaces, in order to reveal the importance of controlling the temperature rise of the hydro-pneumatic spring to the stability of tracked vehicles.

2. Establish and validation of the dynamic model

2.1. Multi-body dynamics model of a tracked vehicle

The research object of this paper is a high-mobility tracked vehicle. The multi-body dynamics model of the vehicle is built based on the dynamic software RecurDyn, as shown in figure 1.

This model has 1038 degrees of freedom and 265 rigid bodies, among which every track link is a rigid body. Each track subsystem shown in figure 2 has a sprocket, six road wheels, six road arms, three rollers, 83 track links, a crank-link mechanism of the hydro-pneumatic spring, and a track tensioner mechanism at the idler. The sprockets, rollers, idlers and road arms are connected to vehicle body through revolute joints. The road wheels are connected with road arms by revolute joints. The link for hydro-pneumatic spring is connected to crank and main piston by revolute joint, while the crank and the road arm are connected by a fixed joint. Translational joint is added between main piston and vehicle body, also between idler and one tensioner link. Double-pin type of track link is used, and contact is defined between track link and idler, sprocket, roller, road wheel, and the ground. The vehicle model is driven by applying angular velocity to sprockets, since no engine and transmission system model have been established.

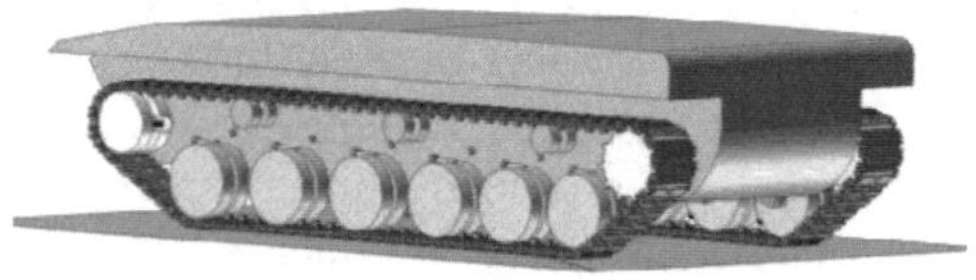

Figure 1. Multi-body dynamic model of a tracked vehicle

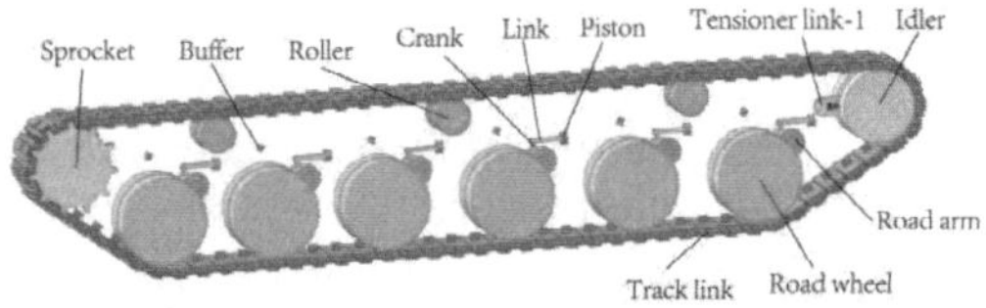

Figure 2. Track subsystem

Before performing dynamic simulation, it is necessary to determine the initial equilibrium state of the vehicle. All the suspensions of the tracked vehicle studied in this paper use hydro-pneumatic springs, so the stiffness is greatly affected by temperature. Here the temperature is set to be 30 °C. Elastic tensioners are used at the idlers. Some parameters of the initial equilibrium state are shown in table 1 and table 2.

Table 1. Parameters of the initial equilibrium state

Pitch angle/deg	Stiffness of tensioner /(N/mm)	Compression of tensioner /mm	Track tension around the idler/N
0.021	208.5	313	34024.4

Table 2. Parameters of the initial equilibrium state-continued

No. of road-wheel	Gas pressure/MPa	Incline angle of road-arm/deg	Contact of track and ground under road-wheel/N
1	14.2	24.5	10658.2
2	15.8	25.8	31317.9
3	17.0	25.8	33154.2
4	18.7	25.9	37374.5

5	18.5	25.7	35852.9
6	17.3	25.4	16678.3

2.2. Gas chamber model of hydro-pneumatic spring

Figure 3 shows the structure of the gas chamber, in which H_u is the depth of the floating piston chamber, d_u is the diameter of the floating piston chamber, d is the diameter of gas chamber, and l_g is the effective length of the gas chamber.

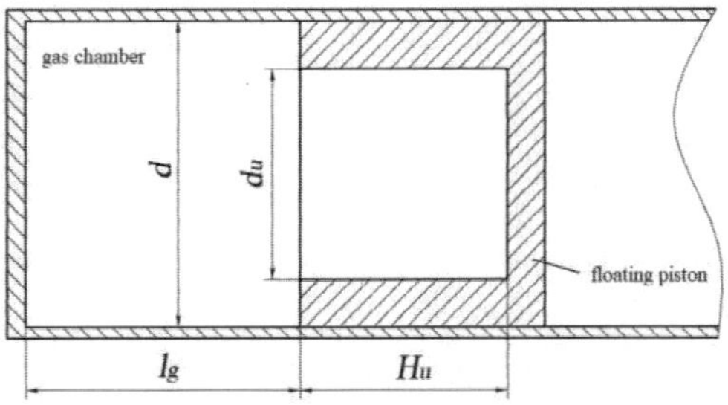

Figure 3. Gas chamber of a hydro-pneumatic spring

The oil and gas cylinder have the same inner diameter and their cross-sectional area is $A = \pi(d/2)^2$. The volume of the floating piston inner chamber is $V_u = \pi(d_u/2)^2 \cdot H_u$, and all the volume of gas is $V = V_u + A \cdot l_g$.

According to state equation of the ideal gas that is $pV/T = \text{const.}$, the relationship between the gas pressure and the displacement of the piston and also the temperature can be obtained by

$$p = p_0 \cdot \frac{(V_u + l_{g0} \cdot A) \cdot T}{(V_u + l_g \cdot A) \cdot T_0} . \quad (1)$$

Where p_0 is the initial gas pressure, l_{g0} is the initial length of the gas chamber; when the gas temperature is changed to T the gas chamber length is $l_g = l_{g0} - \Delta x$ and gas pressure is p, where Δx is the displacement of the floating piston.

After obtaining the gas pressure, the force of the high pressure gas on the floating piston can be obtained as

$$F_g = p \cdot A . \quad (2)$$

In the multi-body dynamics model, the elastic force of the hydro-pneumatic spring is built as the axial force acting on the piston, which is calculated by the equations (1) and (2). The multi-body dynamics model of the vehicle is simulated under different scenarios, and by modifying the temperature T in the equations, the influence of gas temperature on the vehicle state can be studied.

2.3. Validation of multi-body dynamics model

The multi-body dynamics model established in this paper has not been verified by field experiments due to some limited conditions. However, some literature has carried out similar modeling and experimental validation of high-mobility tracked vehicles. This section reproduces the same scenarios in the literature, and compares the same responses, in order to validate the rationality of the model indirectly.

In [6], the FFT amplitude spectrum of the vertical acceleration of the centroid is carried out when the vehicle is running on flat road at a constant speed of 10 km/h. Figure 4(a) shows the simulation and experimental results in the literature and figure 4(b) shows the simulation results of this paper. It can be seen that the curve shape is similar. The peak frequency in the literature is about 15 Hz, that of this paper is about 17.5 Hz, and the acceleration amplitude at the peak frequency is consistent in magnitude. The difference of amplitude values is due to the difference in vehicle parameters.

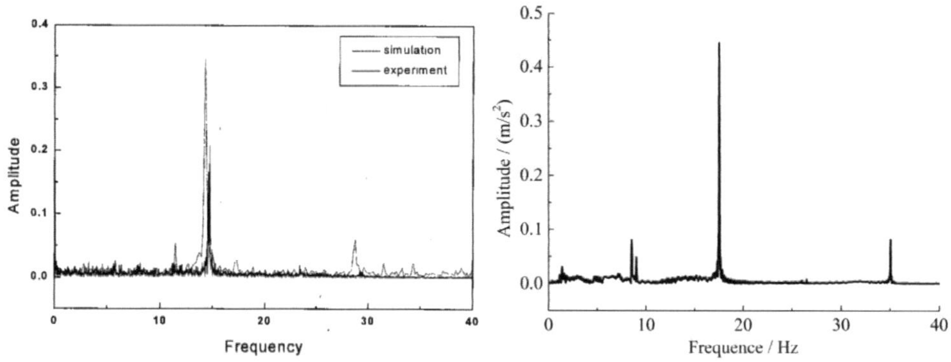

(a) Result in [6] (b) Simulation result of this paper

Figure 4. FFT of the vertical acceleration of the center of chassis

In [7], the vehicle is simulated on a flat road at a constant speed of 50km/h. This paper carries out the same scenario. Figure 5 shows the vertical displacement of a track link in literature [7] and this paper. It can be seen that the curve shape is quite similar.

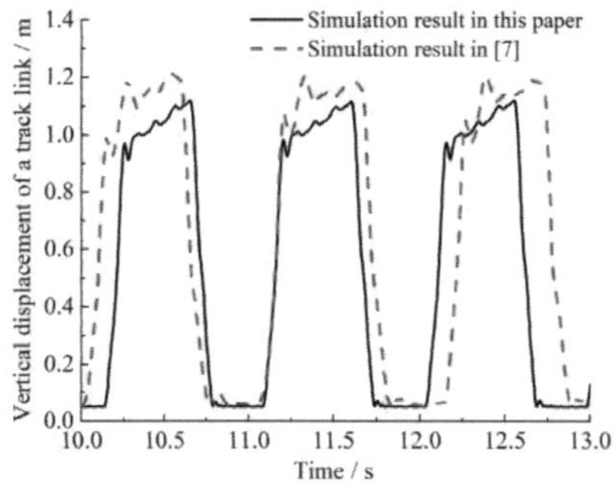

Figure 5. Vertical displacement of a track link

Through analysis of these results, the rationality of the multi-body dynamics model of the tracked vehicle built in this paper is validated indirectly.

3. Analysis of the simulation results

3.1. Driving on flat road

Changes in the ambient temperature of the vehicle or continuous driving for a long time will cause the temperature of all the hydro-pneumatic springs to change, and the magnitude of the change for each spring can be considered approximately the same. In this scenario, the temperature of all hydro-pneumatic springs is -45, -30, -15, 0, 30, 60, 90, 120, 140, 160 °C, and some responses are analyzed when driving on flat road at a constant speed. The input torque of the sprocket is a constant value of 8500 N•m.

Figure 6 shows the curves of several responses varying with the gas temperature. Figure 6(a) is the curve of the centroid driving velocity. During the steady driving phase, the velocity of the centroid decreases with the increase of gas temperature, and the fluctuation range is from 43.1 to 39.4 km/h. Figure 6(b) is a graph showing the height change of the vehicle centroid relative to normal temperature of 30 °C. The height of centroid increases with the increase of gas temperature. After the temperature is higher than 120 °C, the height remains almost unchanged.

Figure 6(c) shows the pitch angle of the vehicle body varying with the gas temperature. In this paper, the pitch angle of the vehicle body is positive when the tail of the car is low, and vice versa. The pitch angle of the vehicle body is positive at each temperature and decreases with the increase of gas temperature. The pitch angle decreases to 0° when the temperature reaches higher than 120 °C. Figure 6 (d) is a plot of the average value of the track tension. The average track tension is defined as the mean value of the track tension of all track links at the given time. It increases from 5.1×10^4 N to 6.5×10^4 N as the gas temperature increases. Figure 6(e) is the curve of the gas pressure of each hydro-pneumatic spring. All the gas pressure increases with the increase of temperature and the gas pressure of the 1st spring is always the lowest.

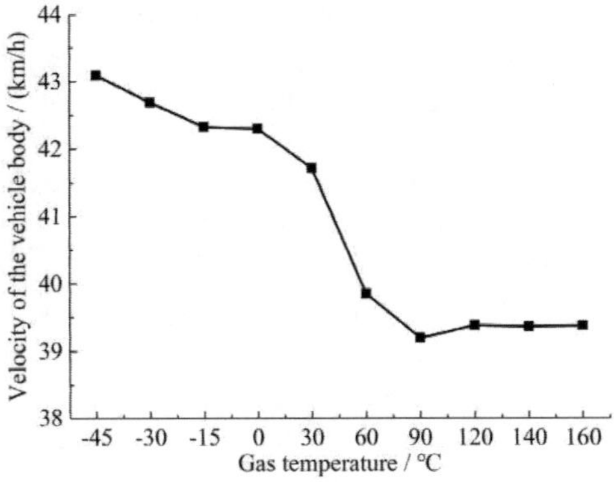

(a) Velocity of the vehicle body

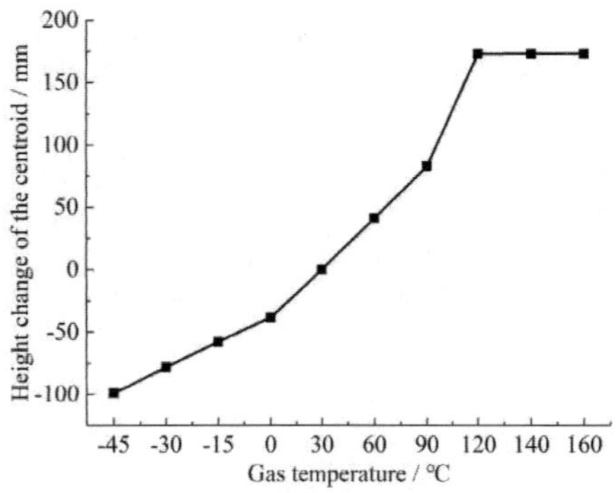

(b) Height change of the centroid

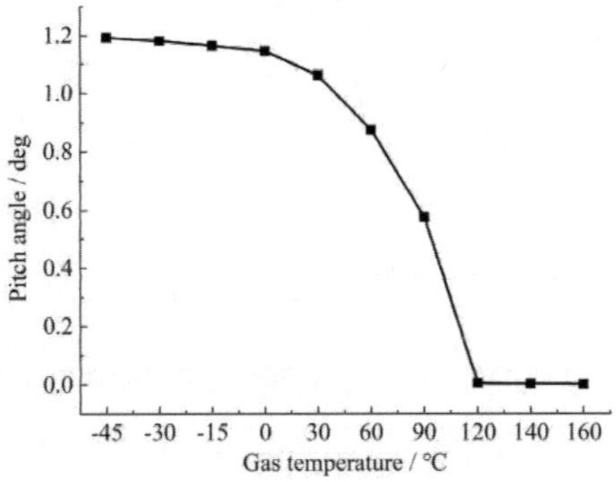

(c) Pitch angle

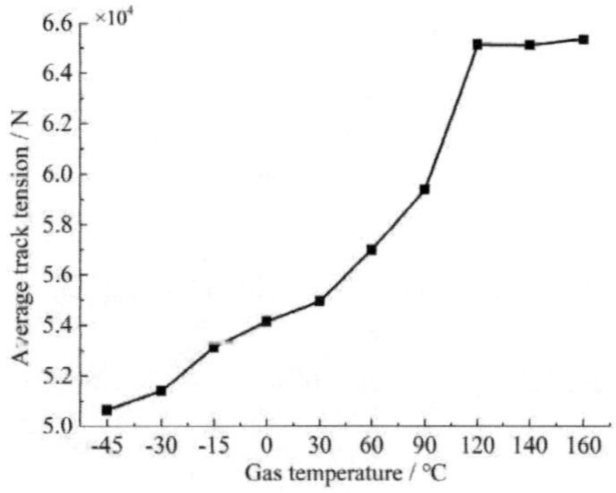

(d) Average track tension

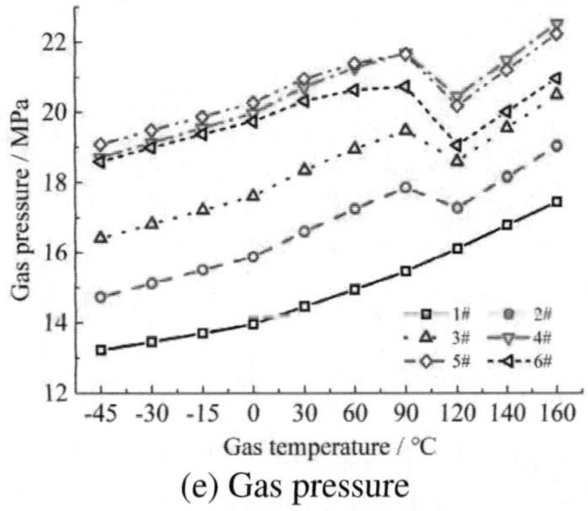

(e) Gas pressure

Figure 6. Responses of vehicle running on flat road with constant velocity

After the temperature is higher than 120 °C, the velocity of the vehicle body, height of the centroid, pitch angle and average track tension remain almost unchanged, because all the road-arms have hit their lower limiters, and this situation is undesired for tracked vehicle.

3.2. Driving through a single semi-round obstacle of radius 200 mm

When the gas temperature is -45, -30, -15, 0, 30, 60, 90, 120, 140, 160 °C, the vehicle drives through a single semi-round obstacle of radius 200 mm with constant input torque to the sprocket of 8500 N•m.

Some responses are analyzed of this scenario as shown in Figure7. Figure 7(a) shows the variation curve about the range of the centroid height. As the temperature increases, the height variation range shows a decreasing trend. When the temperature is lower than 30 °C, the speed of the decrease is faster; while the temperature is higher than 30 °C, the speed gets slower.

Figure 7(b) shows the time history of the vertical acceleration of vehicle centroid when passing the obstacle. The analysis is carried out at a normal temperature of 30 °C and a typical high temperature of 120 °C together with a low temperature of -30 °C. At the temperature of -30 °C, the vertical acceleration of vehicle centroid has a significant impact peak, which is due to the phenomenon that some road-arm hits its upper limiter. This situation is undesired for ride comfort. Figure 7(c) is the curve of the pitch angle displacement of the body. As the temperature increases, the maximum value decreases, the minimum value changes little, and the range decreases. Figure 7(d) is the curve of all the gas pressure range. As the temperature increases, the pressure range increases firstly and then decreases.

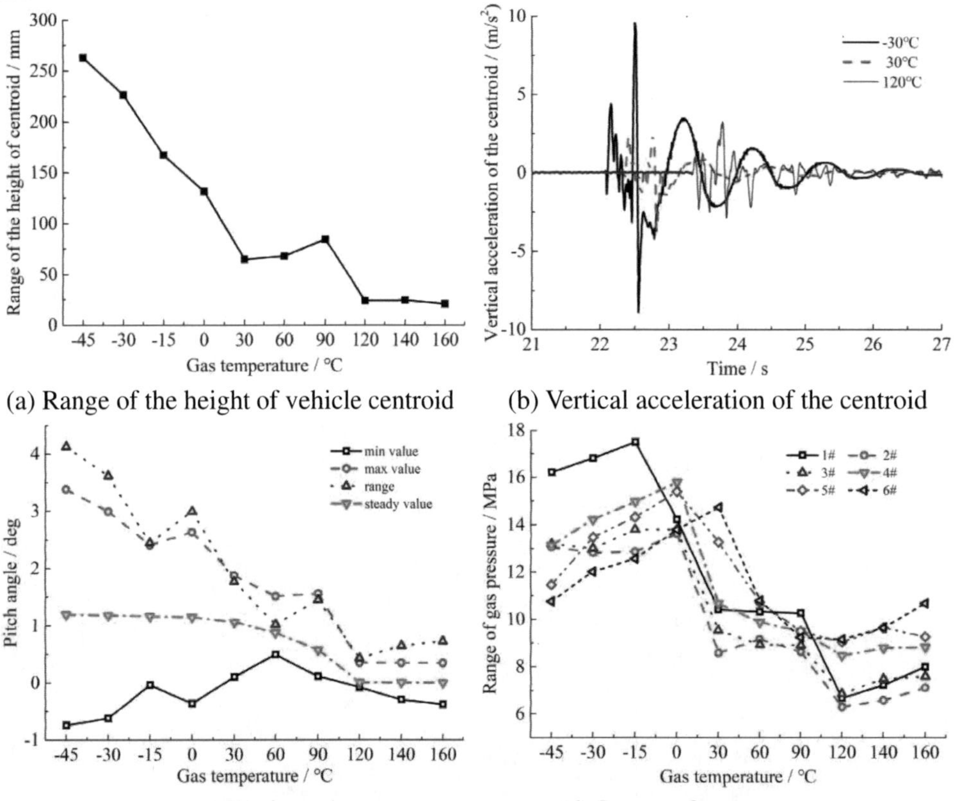

(a) Range of the height of vehicle centroid (b) Vertical acceleration of the centroid

(c) Pitch angle (d) Range of gas pressure

Figure 7. Responses of vehicle when passing a semi-round obstacle

3.3. Driving on a sinusoidal surface with the wavelength of 2 m and amplitude of 100 mm

When the temperature of all hydro-pneumatic springs is -45, -30, -15, 0, 30, 60, 90, 120, 140, 160 °C, the vehicle drives on a sinusoidal surface with the wavelength of 2 m and amplitude of 100 mm at a constant speed. The maximum, minimum, and range values of each response are statistical results of the stable driving phase on the sinusoidal road surface. The input torque of the sprocket is also a constant value of 8500 N•m just like the two scenarios above.

Figure 8 shows the typical responses varying with gas temperature under this scenario. Figure 8(a) shows the height variation of the vehicle centroid. With the increase of temperature, the maximum and minimum values increase, while the range of the height does not change much, from 4 mm to 14.5 mm. Figure 8(b) is the curve of the pitch angle displacement of the body. As the temperature increases, the maximum and minimum values of the pitch angle increase first and then decrease, and the turning point is 30 °C; the range of the pitch angle gradually increases. Figure 8(c) is the curve of all the gas pressure range. As the temperature increases, the range of gas pressure decreases. The pressure range of the 1st spring is the smallest at each temperature, while its change value with temperature is the largest.

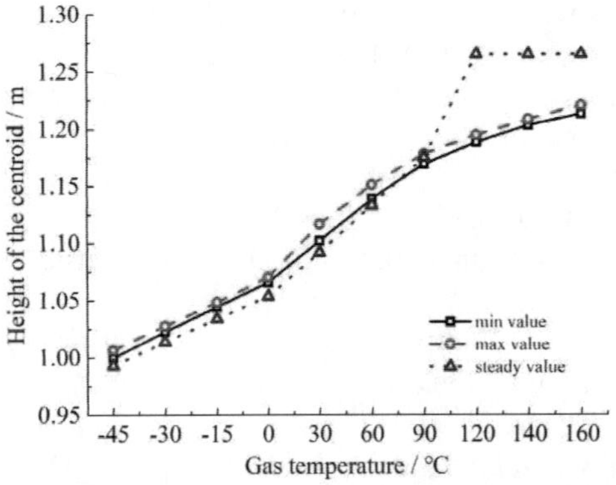

(a) Height of the centroid from the ground

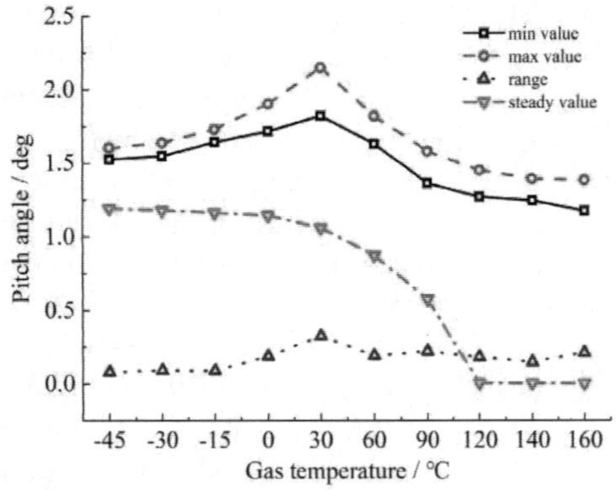

(b) Pitch angle

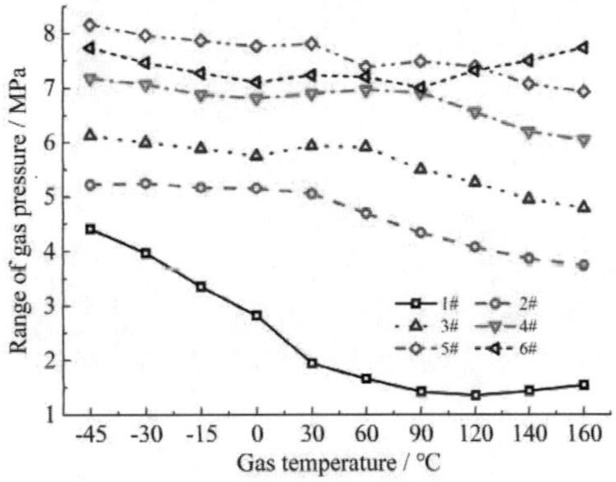

(c) Range of gas pressure

Figure 8. Responses of vehicle when running on sin wave road

4. Conclusions

Based on the dynamic model of a tracked vehicle, this paper studies the influence of the gas temperature on the state of the vehicle body. The typical driving scenarios include running on flat road at a constant speed and passing through a semi-round obstacle with radius of 200 mm, running on a sinusoidal road surface with wavelength of 2 m and magnitude of 100 mm. The responses indicating the running state of the vehicle include the height of the vehicle centroid, vertical acceleration of the centroid, pitch angle, gas pressure of the hydro-pneumatic springs, the track tension. The main conclusions drawn are as follows:

(1) The height of the vehicle centroid increases with the increase of gas temperature; when the temperature is higher than 120 °C, each road-arm will hit the lower limiter, which is undesired for tracked vehicle; in the low temperature environment such as -30 °C when passing the semi-round obstacle the centroid acceleration is large due to the impact of road-arm and its upper limit. In general the state of the vehicle body changes a lot with the change of gas temperature in all the three scenarios

displayed in this paper.

(2) Good heat dissipation performance of the hydro-pneumatic spring is very important, which can ensure that the gas temperature changes within a small range. Thus its working reliability can be increased, and the vehicle body posture and the driving stability can be maintained.

5. References

[1] Chen Yijie, Gu Liang, Guan Jifu. Research on the different methods thermodynamic model of hydro-pneumatic spring[J]. Machinery Design & Manufacture, 2008(06): 54-56.

[2] Chen Yijie, Gu Liang. Research on Natural Convection Thermodynamic Model of Hydro-pneumatic Spring [J]. Journal of System Simulation,2008(15): 4198-4201.

[3] Chen Yijie, Zhao Bo, Zhang Xu, et al. Research on the temperature rise of hydro-pneumatic suspension under the random power[J]. Journal of Mechanical Strength, 2010(06): 869-872.

[4] Zhao Jingkai, Gu Zhengqi, Zhang Sha, et al. viscous heating of hydro-pneumatic suspension and its optimization [J]. Journal of vibration and shock, 2015(20): 129-134.

[5] Els P S, Grobbelaar B. Heat transfer effects on hydro- pneumatic suspension systems[J]. Journal of Terramechanics, 1999,36(4):197-205.

[6] Ryu H S, Huh K S, Bae D S, et al. Development of a Multi-body Dynamics Simulation Tool for Tracked Vehicle Part I Efficient Contact and Nonlinear Dynamic Modeling[J]. JSME International Journal Series C, 2003,46(2):540-549.

[7] Ryu H S, Bae D S, Choi J H, et al. A Compliant Track Link Model for High-speed, High-mobility Tracked Vehicles[J]. International Journal for Numerical Methods in Engineering, 2000, 48 (10) :1481-1502.

IOP Conf. Series: Materials Science and Engineering **576** (2019) 012019 doi:10.1088/1757-899X/576/1/012019

Research on consistency check of natural storage and enhancement test for missile servo system

Xu Wang [1,2] and Quan Sun [1,3]

[1]National University of Defense Technology, Changsha, Hunan, 410073, China
[2]The Second Academy of China Aerospace Science and Industry Corporation, Beijing 100854, China,
[3]Hunan Ginkgo Reliability Technology Institute Limited Corporation, Changsha, Hunan, 410100, China

Abstract. In order to check the consistency of two mechanisms for a missile servo system, we conduct two enhancement tests and utilize two sets of test data to establish the accelerated degradation models of the servo system. Based on the above accelerated degradation models and natural storage profile of the servo system, we present and demonstrate a procedure to check the consistency of two mechanisms by checking the correlation of two sets of degradation data. The results indicate that two degradation mechanisms are significantly consistent with each other.

1. Introduction

The missile servo system is a single-use product with long storage periods in its lifecycle. Natural storage temperature of the missile servo system varies with the alternations of day-nights and four seasons, so temperature change rate is regarded as the main stress that influences the storage lifetime of the servo system during storage periods. To accelerate the degradation process of the servo system, we hastened performance degradation of the servo system by reliability enhancement testing(RET) and generate large amounts of performance degradation data. Based on these degradation data, we established the accelerated degradation models to describe the relationships between degradation characteristics and applied stresses. The life or reliability under normal stress conditions can be extrapolated based on accelerated degradation models estimated from various accelerated conditions[1,2]. However, this extrapolating method is based on such a basic assumption: the degradation mechanism under accelerated stress is the same as the one under normal stress[3,4]. It is an important theory basis for reliability enhancement testing. But sometimes accelerated stresses may change degradation mechanism.

Currently, the methods of checking mechanism consistency are classified into two categories: one identified the consistency of mechanisms and identify the stress boundary before accelerated tests, and the other was based on accelerated degradation models after accelerated tests. For example, Guo [5] provided a method to rapidly identify the consistency at the preliminary stage of accelerated test. Feng[6] proposed a method by measuring the difference of the degradation path based on the Spearman rank correlation coefficient. Wang [7] investigated a method based on degradation model after accelerated test. Based on the constant principle of acceleration coefficient, Xi [8] presented an identification method of failure mechanism consistency with taking Gamma process as study object. The literatures [9,10] utilize the grey theory to identify the consistency.

The remainder of this paper is organized as follows. Section 2 presents the stress profiles and degradation curves of these two tests. Section 3 develops the accelerated degradation models based on test data. Section 4 provides a method to check the consistency of natural storage and enhancement test. Finally, a brief summary is given in Section 5.

2. Enhancement Test and Degradation data

According to standard principles, reliability enhancement tests are usually conducted by applying several accelerated stress levels with simple profiles, such as step-temperature, rapid cyclic-temperature, step-vibration and combined stresses. We design two enhancement tests orderly: a cold step-temperature test and a rapid cyclic-temperature test. Only one sample is provided for testing. The sample goes through a cold step-temperature test and a rapid cyclic-temperature test two test orderly. A certain performance is chosen as the degradation metric which possesses significant degradation tendency. The stress profiles and performance degradation curves of these two tests are plotted as shown in figure 1. The solid curves represent the temperature stress profiles of enhancement tests, and the dashed represent performance degradation of the tested system. In the cold step-temperature test, as shown in figure 1(a), degradations are measured and recorded at the end of every temperature step. There is a remarkable increasing tendency for performance degradation below -35℃. In the rapid cyclic-temperature test, as shown in figure 1(b), the degradations are measured at the end of every temperature keep stage. Degradations steadily fluctuate in the first three cycles, and the amplitudes of degradation vibration become large from the forth cycle.

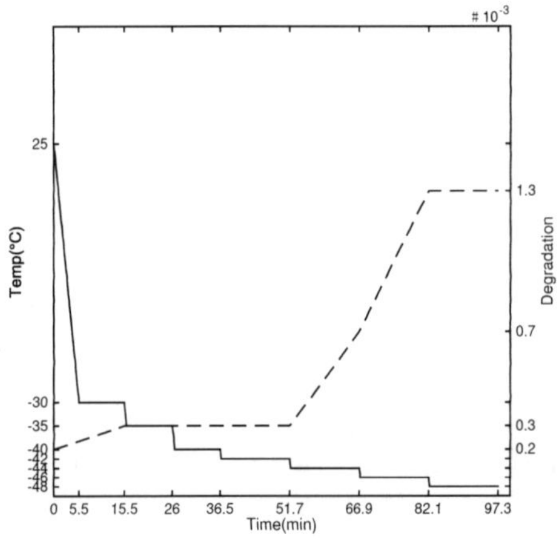

(a)

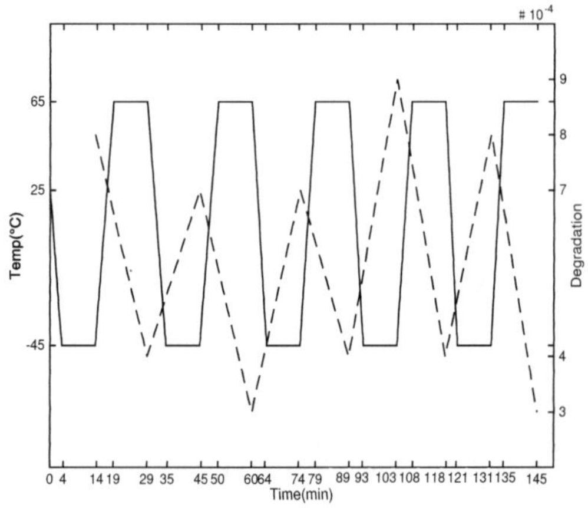

(b)

Figure 1. Stress profiles and performance degradation curves changing with time.

3. Degradation Modeling and Parameter Estimates

Let Y_i be the individual performance degradation of the sample at stress level i, q be the number of stress levels, p_i be an accelerated coefficient of the sample at stress level i; Δt_i be the duration of stress level i. then the cumulative degradation after the sample has experienced the qth stress level by time t can be expressed by

$$W_q(t) = \sum_{i=1}^{q} p_i \Delta t_i \quad (1)$$

For the cold step-temperature test, we suppose that the total amount of degradation observed at stress level i can be divided into two components: temporary damage and permanent damage

$$\begin{cases} W_{T_i} = W_{T_i}^{tem} + W_{T_i}^{per} \\ W_{T_i}^{tem} = \gamma \cdot T_i \\ W_{T_i}^{per} = \sum_{j=1}^{i} Y_j = \sum_{j=1}^{i} p_j^{per} \Delta t_j \end{cases} \quad (2)$$

where $W_{T_i}^{tem}$ and $W_{T_i}^{per}$ represent the temporary and permanent damage at stress level *i* respectively; T_i is the temperature value of stress level *i*; γ is an unknown coefficient on temporary damage; p_j^{per} is an accelerated coefficient on permanent damage at stress level *i*.

Then, the performance degradations per unit time at every stress level can be rewritten as

$$\tilde{Y}_i = p_i^{per} + \gamma \cdot \frac{\Delta T_i}{\Delta t_i}, \qquad i = 1,2, \dots, q \tag{3}$$

where $\Delta T_i = T_i - T_{i-1}$ denotes the temperature increment from stress level *i*-1 to *i*, therefore, $\frac{\Delta T_i}{\Delta t_i}$ is the temperature change rate of stress level *i*.

For the cold step-temperature test, temperature is the main stress, so we adopt a typical temperature accelerated model, i.e., the Arrhenius law. Then, the permanent degradation per unit time can be proposed as

$$p_i^{per} = a \cdot exp\left(-\frac{b}{T_i}\right) \tag{4}$$

where a is an unknown constant relevant to the material, shape, and machining process of the sample; and $b = E/k$, here k is Boltzmann's constant and E is the activation energy of the reaction.

Substituting (4) into (3) yields the relationships between $\tilde{Y}_i$ and T_i as follows:

$$\tilde{Y}_i = \gamma \cdot \frac{\Delta T_i}{\Delta t_i} + a \cdot exp\left(-\frac{b}{T_i}\right) + \varepsilon_i, \qquad i = 1,2, \dots, q \tag{5}$$

where ε_i denotes random error of $\tilde{Y}_i$, which follows a normal distribution, i.e., $\varepsilon_i \sim N(0, \sigma_{\tilde{Y}_i}^2)$. The parameters can be estimated by a non-linear fitting method based on the data $\{\tilde{Y}_i, \frac{1}{T_i}, \frac{\Delta T_i}{\Delta t_i}; i = 1,2, \dots, q\}$.

The rapid cyclic-temperature test profile can be divided into two categories of stages: temperature keep stages and temperature change stages. It is assumed that two categories are independent of each other. By the temperature accelerated equation (5), the degradation at the high and low temperature keep stages can be written as respectively

$$\begin{cases} Y_{\text{high}} = \hat{\gamma} \cdot T_{high} + \hat{a} \cdot exp\left(-\frac{\hat{b}}{T_{high}}\right) \cdot \Delta t^{keep} \\ Y_{\text{low}} = \hat{\gamma} \cdot T_{low} + \hat{a} \cdot exp\left(-\frac{\hat{b}}{T_{low}}\right) \cdot \Delta t^{keep} \end{cases} \tag{6}$$

where Δt^{keep} is the duration of temperature keep stage in the rapid cyclic-temperature test profile; T_{high} and T_{low} are the temperature values of high and low temperature keep stages respectively.

As a result of the main stress being temperature change rate, we eliminate the performance degradations generated at high and low temperature keeping stages from the total degradations of every temperature change cycle. The performance degradation increment due to temperature change rate can be obtained by eliminating the high and low temperature degradations, namely

$$Y_{V_i}^{change} = Y_{V_i} - Y_{\text{high}} - Y_{\text{low}} \tag{7}$$

where Y_{V_i} is the degradation of stress level V_i.

Furthermore, the degradation per unit time $\tilde{Y}_{V_i}^{change}$ can be obtained by

$$\tilde{Y}_{V_i}^{change} = \frac{Y_{V_i}^{change}}{\Delta t_{V_i}^{\text{change}}} \tag{8}$$

where $\Delta t_{V_i}^{\text{change}}$ is the duration of temperature change stage at stress level V_i.

Then, the mean and variance of performance degradation per unit time at stress level V_i are

$$\begin{cases} \tilde{\mu}_i^{change} = \dfrac{\sum_{k=1}^{n} \tilde{Y}_{V_{ik}}^{change}}{n} \\ \tilde{\sigma}_i^{2change} = \dfrac{\sum_{k=1}^{n} (\tilde{Y}_{V_{ik}}^{change} - \tilde{\mu}_i^{change})^2}{n-1} \end{cases} \tag{9}$$

According to the characteristics of temperature change rate, we choose a power law model as the form of the temperature change rate accelerated equation, i.e.,

$$\begin{cases} \tilde{\mu}_i^{change} = \alpha_1 \cdot V_i^{\beta_1} \\ \tilde{\sigma}_i^{2change} = \alpha_2 \cdot V_i^{\beta_2} \end{cases} \tag{10}$$

where $\alpha_1, \beta_1, \alpha_2$ and β_2 are the unknown parameters that need to be estimated.

Take logarithm on both sides of the equations (10), and we obtain the linear forms as follows:

$$\begin{cases} \log \tilde{\mu}_i^{change} = \log \alpha_1 + \beta_1 \cdot \log V_i \\ \log \tilde{\sigma}_i^{2change} = \log \alpha_2 + \beta_2 \cdot \log V_i \end{cases} \tag{11}$$

The estimates $\hat{\alpha}_1, \hat{\alpha}_2, \hat{\beta}_1, \hat{\beta}_2$ can be obtained by the least square estimation (LSE) method based on the data $\{\tilde{\mu}_i^{change}, \tilde{\sigma}_i^{2change}, V_i\}$ and the linear equations (11).

4. Consistency Check of Natural Storage and Enhancement Test

In the natural storage profile of missile servo system, there are 11 months in which temperature varies within the range of 15°C~25°C and 1 months in which temperature varies within the range of 5°C~40°C. Therefore, there are 11 months of temperature change rate $\frac{10°C}{12h*60min} = 0.0139°C/min$ and 1 months of temperature change rate $\frac{35°C}{12h*60min} = 0.0486°C/min$. The annual mean of temperature change rate is $\frac{0.0139\times11+0.0486\times1}{12}$=0.0168°C/min.

In practice, the missile servo system is tested every half a year and the period of storage is 5 years. The performance degradation data under natural storage condition as shown in table 1.

Table 1. Performance degradation data under natural storage condition

Measurements	Cumulative test time (a)	Critical performance degradation
1	0	0
2	0.5	0.0006
3	1	0.0013
4	1.5	0.0020
5	2	0.0026
6	2.5	0.0032
7	3	0.0038
8	3.5	0.0044
9	4	0.0052
10	4.5	0.0058
11	5	0.0065

We propose a procedure to check the consistency of two degradation mechanisms. The detailed procedure is recommended as follows:

Step 1: The performance degradation characteristics are obtained by the method presented in section 3.

The estimates $\hat{\gamma}, \hat{a}, \hat{b}$ and $\hat{\sigma}_{\tilde{Y}_i}^2$ of the formula(5) are obtained by based on the cold step-temperature test data are listed in table 2.

Table 2. The parameter estimations of the cold step-temperature test

$\hat{\gamma}$	$\hat{a}$	$\hat{b}$	$\sigma^2_{\hat{Y}}$
$\mathbf{2.1386 \cdot 10^{-6}}$	$1.1806 \cdot 10^{-5}$	-0.9957	$2.5054 \cdot 10^{-10}$

The mean $\tilde{\mu}_i^{change}$ and variance $\tilde{\sigma}_i^{2change}$ of degradation per unit time at three levels of temperature change rates can be calculated by the formulas (10). The results are listed in table 3.

Table 3. Means and variances of degradations per unit time due to temperature change rates

Temperature change rate(°C/min)	$\tilde{\mu}_i^{change}$	$\tilde{\sigma}_i^{2change}$
20 °C/min	$0.0850 \cdot 10^{-3}$	$1.4452 \cdot 10^{-8}$
25 °C/min	$0.1233 \cdot 10^{-3}$	$2.0702 \cdot 10^{-8}$
30 °C/min	$0.1548 \cdot 10^{-3}$	$4.7916 \cdot 10^{-8}$

Step 2: The accelerated degradation equations are established by using the method presented in section 3.

Based on the data listed in table 3, the parameter estimations $\hat{\alpha}_1, \hat{\alpha}_2, \hat{\beta}_1, \hat{\beta}_2$ are obtained by LSE as shown in table 4.

Table 4. Parameter estimations of the temperature change rate accelerated equations

$\hat{\alpha}_1$	$\hat{\beta}_1$	$\hat{\alpha}_2$	$\hat{\beta}_2$
$\mathbf{1.0067 \cdot 10^{-6}}$	1.4850	$2.1852 \cdot 10^{-12}$	2.9065

Therefore, the temperature change rate accelerated equations for the missile servo system can be written as:

$$\begin{cases} \mu_i^{change}(t) = 1.0067 \cdot 10^{-6} \cdot V^{1.4850} \cdot t \\ \sigma_i^{2change}(t) = 2.1852 \cdot 10^{-12} \cdot V^{2.9065} \cdot t \end{cases} \tag{12}$$

Step 3: According to the natural storage profile of the servo system, the mean and variance of performance degradation per half a year can be obtained by the accelerated equations (12). By the B-S cumulative damage theory [11], $\tilde{Y}_{V_0}^{\mathrm{half}year}$ follows a normal distribution, so $\tilde{Y}_{V_0}^{\mathrm{half}year}$ can be obtained by simulating the distribution. The cumulative degradation estimations under natural storage condition can be obtained by accumulating $\tilde{Y}_{V_0}^{\mathrm{half}year}$.

The transformed results from enhancement test to natural storage are shown in table 5.

Table 5. Transformed results from enhancement test to natural storage

Measurements	Cumulative test time (a)	Transformed critical performance degradation
1	0	0
2	0.5	0.0007
3	1	0.0013
4	1.5	0.0020
5	2	0.0026
6	2.5	0.0033
7	3	0.0040
8	3.5	0.0046
9	4	0.0053
10	4.5	0.0060
11	5	0.0066

Step 4: Compare the above transformed data $\{\widehat{W}_{01}, \widehat{W}_{02}, \dots, \widehat{W}_{0n_0}\}$ with the actual degradation data under natural storage $\{W_{01}, W_{02}, \dots, W_{0n_0}\}$, and calculate the Pearson correlation coefficient of two sets of data

$$\hat{\rho} = \frac{\sum_{j=1}^{n_0}(W_{0j} - \overline{W_0})(\widehat{W}_{0j} - \overline{\widehat{W}_0})}{\sqrt{\sum_{j=1}^{n_0}(W_{0j} - \overline{W_0})^2} \cdot \sqrt{\sum_{j=1}^{n_0}(\widehat{W}_{0j} - \overline{\widehat{W}_0})^2}} \quad (13)$$

where $\overline{W_0} = \frac{1}{n_0}\sum_{j=1}^{n_0} W_{0j}$, $\overline{\widehat{W}_0} = \frac{1}{n_0}\sum_{j=1}^{n_0} \widehat{W}_{0j}$.

According to two sets of data from Table 1 and Table 5, Pearson correlation coefficient during period of 5 years $\hat{\rho}_5 = 0.9996$.

Step 5: Determine the correlation criteria. If $\hat{\rho} \geq 0.7$, then two sets of data is strongly correlated; If $\hat{\rho} \leq 0.4$, then two sets of data have no apparent linear relationship, which means that some new mechanisms may be interpolated due to accelerated stress, hereby we need to review accelerated stress conditions; If $0.4 \leq \hat{\rho} \leq 0.7$, we are unable to draw final conclusions immediately and need to apply other methods to make further determination.

According to the correlation criterion, $\hat{\rho}_5$ is greater than 0.7, then there are strong correlations between two degradation mechanisms. Therefore, we can draw a conclusion that two degradation mechanisms are consistent with each other.

5. Concluding Remarks

In this paper, we utilize the degradation data obtained from the rapid cyclic-temperature test as well as the step-temperature test to develop the performance degradation models based on the theory of probabilistic physics, and apply a method based on Pearson correlation coefficient and differences analysis to check the consistency of degradation mechanisms of the missile servo system under the natural storage condition and enhancement test condition. The paper provides a way to validate the effectiveness of accelerated equations and the consistency check of various degradation mechanisms.

6. References

[1] Shang G Z, Fu G C and Wan B 2009 *Electronic Product Reliability and Environment Testing* **27** 32-36
[2] Gao L N and Zhao L 2014 *Electronic Components and Materials* **33** 72-76
[3] Nelson W 2014 *Accelerated Testing: Statistical Methods, Test Plans, and Data Analysis* (New York: Wiley–Interscience)
[4] Hu J M, Barker D and Dasgupta A 1992 *Proc. Annual Reliability and Maintainability Symp. IEEE* p 181
[5] Guo C S 2013 *Chinese J. of Phys.* **62** 068502
[6] Feng J 2011 *J. of Aerospace Power.* **26** 2439-44
[7] Wang H W, Xu T X and Wang W Y 2015 *J. of Aeronautics* **36** 889-97
[8] Xi W J, Wang H W and Wang R 2015 *J. of Beijing University of Aeronautics and Astronautics* **41** 2198-204
[9] Pan X Q and Kang R 2013 *J. of Beijing University of Aeronautics and Astronautics* **39** 787-91
[10] Li X G and Wang Y H 2014 *J. of Beijing University of Aeronautics and Astronautics* **39** 899-904
[11] Birnbaum Z and Saunders S 1969 *J. of Applied Probability* **6** 319-327

IOP Conf. Series: Materials Science and Engineering **576** (2019) 012020 doi:10.1088/1757-899X/576/1/012020

Operational Efficiency Improvement of Ply Cutting Machine for Aircraft Tires

Kulanit Eaknipitsari[1], Parames Chutima[1,2,*]

[1]Department of Industrial Engineering, Chulalongkorn University, Thailand
[2]Regional Centre for Manufacturing Systems Engineering, Chulalongkorn University, Thailand

E-mail: kulanit.eak@hotmail.com

Abstract. This research focuses on the operational efficiency improvement of ply cutting machines for aircraft tires by implementing the design of experiments (DOE) and Single Minute Exchange of Die (SMED) methods to decrease loss time caused by machine stoppages found at let-off and splicer processes. The problem comes from high loss time because the system's parameters are not set properly and high setup time of the machine. Therefore, the operations efficiency improvement has been conducted by setting a suitable flow rate at the splicer and changing activities of the operator in order to reduce waiting time. Improved efficiency is implemented in the Ply cutting machine from July 2018 to August 2018. The result shows the increase in the average production from 23 rolls/shift to 28 rolls/shift. Furthermore, a cost saving for efficiency improvement is equal to 36M baht per year.

1. Introduction

Ply is component in aircraft tire, it is produced by the semi-auto machine. A study is one of cutting machine, its production begins by receiving master roll until sent good component to tire building machine. Ply cutting machine consist of two let off, one cutter, two tables of the splicer and two wind up post. The flow of the ply cutting machine as shown in Figure. 1.

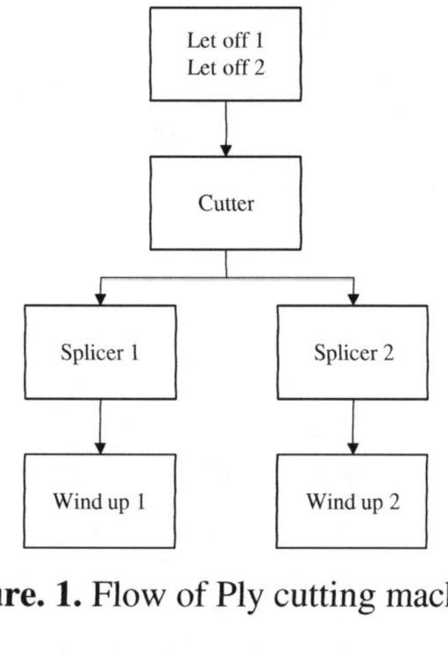

Figure. 1. Flow of Ply cutting machine.

Loss time was found at let off and splicer process that was impacted to production because they stopped the machine for change size at let off and waited on a wind-up process for folding ply out of the sensor. It made high waiting time. Often, production does not achieve production capacity's target per shift. Production data for Ply cutting machine from July 2017 to June 2018 shows the average of production per shift equal to 23 roll/shift as shown in Figure. 2

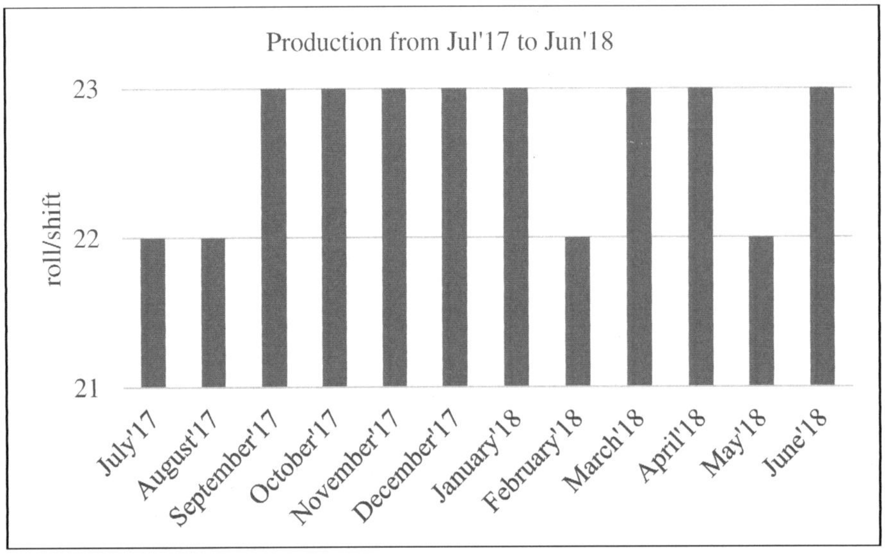

Figure. 2. Actual Production of Ply cutting machine.

2. Problems

Focusing on Ply cutting machine, Pareto chart of cutting ply makes researcher realize that the root cause of low production because of high waiting time in the current. Currently, the table of the ply cutting machine has two positions for load small roll, splicer and unload small roll but have no operator for load master roll at let off during the process. The average three main waiting times from July 2017 to June 2018 are as follows: waiting for loadmaster roll at let off, waiting for splicer and waiting time for machine breakdown. Their waiting times equal to 81%, 8%, and 4%, respectively, as shown in Figure. 3.

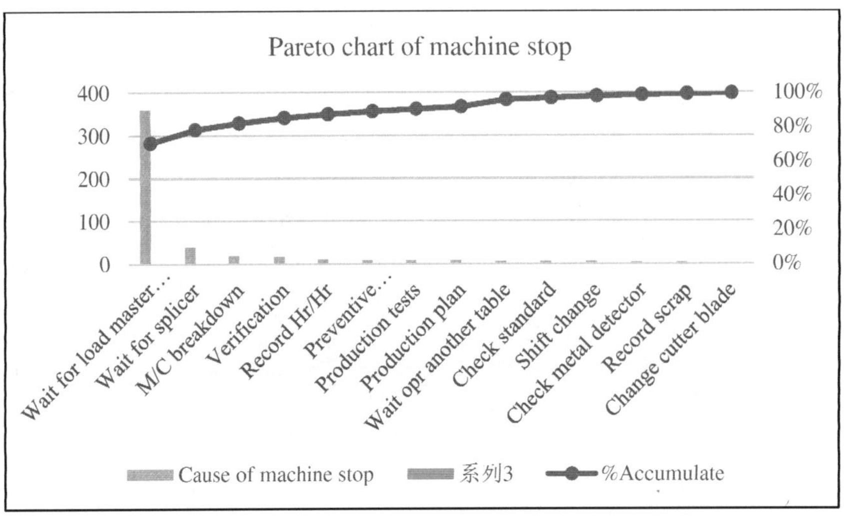

Figure. 3. Pareto chart of machine stop.

3. Process Improvements

Wastes removal tool based on Design of Experiment (DOE) and Single Minute Exchange of Die (SMED) were used to improve the efficiency of operation of Ply cutting machine. There were totally two approaches to improve efficiency as follow

3.1. Improving the Working Method

The objective was to reduce waiting time of machine from internal setup to external setup by applying Single Minute Exchange of Die method (SMED) [1] to separate internal and external setup, converting internal to external setup and streamlining all aspects of setup operation. All of the setup activity as shown in Table 1. This simulation result need to increase one man per shift for external setup activity in order to achieve customer demand in AC 2019 and compare opportunity cost and labor cost as shown in Table 2.Therefore reduce waiting time after improvement production are 27 roll/shift and average operator workload is 90.6% as shown in Table 3, increase from the original 23 rolls/shift.

Table 1. The activity of setup machine.

Activity	Total time (min/shift)
Bring master roll to post by forklift	8
Drive forklift to machine	16
Unpack master roll	5
Load master roll to let off	33
Unload liner master roll	9
Load empty core to let off	13
Unload core master roll	5
choose master roll from storage area	13
Keep empty container to area	10
Total	113

Table 2. Compare opportunity cost and labor cost.

Cost	M THB/year
Opportunity Cost	38
Labor cost	2

Table 3. Actual production for a shift.

Machine	Target (roll/shift)	Actual production (roll/shift)	Workload (%)
Splicer table1	14	14	93.79%
Splicer table2	13	13	87.43%
Service opr	-	-	95.12%

3.2. Improving Process of the Splicer

The objective was to optimize the parameter settings of splicer process for splicing Ply and returning to an original position with staffs have to manual move ply out of splicing path every time after splicing finish, which is a waste that occurs in the process. Therefore, the concept of the setting parameter was used to solve the problem by applying DOE. The experimental design method was adopted to find the factors that show the statistically significant effect on the cycle time. Three factors were experiments including splicer temperature, air pressure, and flow rate. The 2K full factorial design [2] with 3 factors, 2 levels and 2 replicates of each factor were adopted to test whether or not the main effects and their interactions were significant as shown in Table 4.

Table 4. Factors and levels in 2^K full factorial design.

Factors	Unit	Symbols	Levels of factor	
			Low (-1)	High (1)
Temperature	Celsius	A	80	100
Pressure	Bar	B	3	4
Flow rate	m^3/min	C	1	1.5

The experimental data were analyzed by using the Minitab software [3]. From the Pareto chart of effects as shown in Figure. 4, the factor effect with its value greater than 2.31 is a significant factor at 95% confident interval equal. The Pareto chart of effect showed that the main effects C have P-values less than 0.05 (significant level α), whereas factor A and B and interaction (AB, AC, BC, and ABC) were not significant. Therefore, it can be concluded that the main effects C are significant at 95% confident interval. As a result, all significant factors need further investigation to find their appropriate level setting.

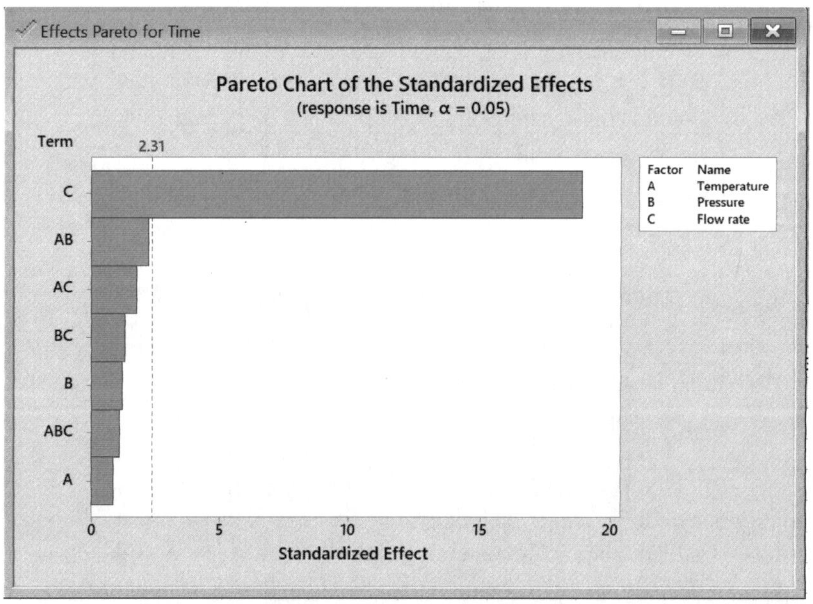

Figure. 4. Pareto chart of effects.

4. Results

The 2^K full factorial experiments were tested to find their appropriate setting parameter that can reduce the waiting time of the splicing process. The statistical analysis showed that the main effect obtained the p-values of less than 0.05 as shown in Figure. 5. Therefore, it can be concluded that the main effect affected the cycle time at 95% confident interval. Since the main effect was used as a mean to select the proper setting for factors, factor C was set at 1.5 m^3/min (high level of factor C) as shown in Figure. 6 and production increased 1 roll/shift.

Factorial Regression: Time versus Temperature, Pressure, Flow rate

Analysis of Variance

Source	DF	Adj SS	Adj MS	F-Value	P-Value
Model	7	73.5044	10.5006	53.00	0.000
Linear	3	71.3969	23.7990	120.12	0.000
Temperature	1	0.1406	0.1406	0.71	0.424
Pressure	1	0.2756	0.2756	1.39	0.272
Flow rate	1	70.9806	70.9806	358.26	0.000
2-Way Interactions	3	1.8819	0.6273	3.17	0.085
Temperature*Pressure	1	0.9506	0.9506	4.80	0.060
Temperature*Flow rate	1	0.6006	0.6006	3.03	0.120
Pressure*Flow rate	1	0.3306	0.3306	1.67	0.232
3-Way Interactions	1	0.2256	0.2256	1.14	0.317
Temperature*Pressure*Flow rate	1	0.2256	0.2256	1.14	0.317
Error	8	1.5850	0.1981		
Total	15	75.0894			

Model Summary

S	R-sq	R-sq(adj)	R-sq(pred)
0.445112	97.89%	96.04%	91.56%

Figure. 5. Result of ANOVA.

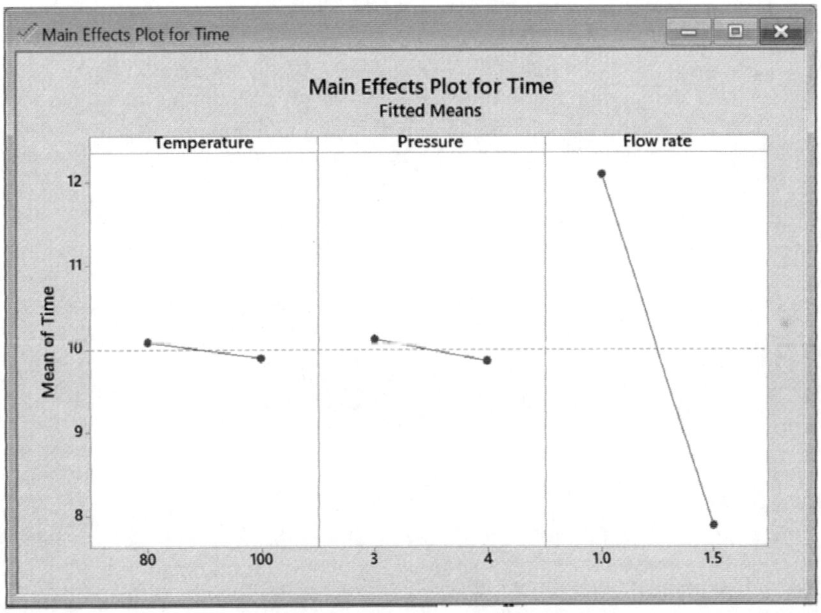

Figure. 6. Main effects plot.

5. Conclusions

The objective of this research was to apply the Design of Experiment (DOE) and Single Minute Exchange of Die (SMED) to improve the efficiency of ply cutting machine caused by waiting time. DOE and SMED were systematically utilized to find the appropriate factor which was needed to improve. Production's result after implementing the new organization for external setup machine and setting flow rate from July 2018 to August 2018 is increased as expected. The production increased from 23 rolls/shift to 28 roll/ shift and cost-saving equal to 36M Baht per year.

6. References

[1] Information on https://www.leanproduction.com/smed.html
[2] D.C. Montgomery 2008 Design and analysis experiments (America: John Wiley & Sons, Inc)

[3] D.C. Montgomery and G.C. Runger 2010 Applied statistics and probability engineers (America: John Wiley & Sons, Inc)

Acknowledgments
Authors wishing to acknowledge assistance or encouragement from colleagues, special work by technical staff or financial support from organizations should do so in an unnumbered Acknowledgments section immediately following the last numbered section of the paper.

IOP Conf. Series: Materials Science and Engineering **576** (2019) 012021 doi:10.1088/1757-899X/576/1/012021

Remaining Useful Life Prediction of Cutting Tools Based on Support Vector Regression

Y C Liu, X F Hu* and S X Sun

Shanghai Key Laboratory of Advanced Manufacturing Environment, School of Mechanical Engineering, Shanghai Jiao Tong University, Shanghai, China

E-mail: wshxf@sjtu.edu.cn

Abstract. Remaining useful life (RUL) prediction of cutting tools is critical to effective condition based maintenance for reducing downtime, ensuring quality and avoiding accidents. In this paper, a RUL prognostic method based on support vector regression (SVR) is proposed for predicting cutting tool's life. The proposed method consists of two main phases: an off-line phase and an on-line phase. In the first phase, the signal features are extracted from raw data, and then the SVR models with considering different length of signals at past times are established to reflect the relationship between monitoring data and tool life. In the second phase, the constructed models are used to predict cutting tool's RUL, and the best signal length for accurate prediction result is obtained. The proposed method is applied on experimental data taken from a computer numerical control (CNC) rotor slot machine in a factory. The result shows the validity and practicability of this method.

1. Introduction

As the major equipment in manufacturing system, tool health condition has a great effect on product quality. Tool failure in the machine may lead to unscheduled downtime, quality issues and even serious accidents. It is estimated that about 20% of the downtime is attributed to tool failure, which result in significant economic losses[1]. Thus it is necessary for prognostic and health management (PHM) of cutting tools.

The main task of cutting tool prognostics is to estimate its remaining useful life (RUL), which is defined as the length from the current service time until it fails[2]. Most of the existing RUL prediction models can be divided into two main categories: physics-based models and data-based models[3]. The physics-based models are more accurate than others, but they require a complete understanding of the mechanistic knowledge and theories and are therefore very difficult to build for practical applications. The data-driven models attempt to derive models directly from historical condition monitoring (CM) data instead of building complicated models, and it is widely used in many fields. The data-driven methods mainly include: artificial neutral network, Markov models, and support vector machine (SVM) et al. SVM is widely used for RUL prediction. Gokulachandran[4] compared two soft computing techniques, neuro fuzzy logic technique and SVR technique for cutting RUL prediction. Benkedjouh[5] and Shen [6] proposed the RUL prediction method based on SVR to assess bearing's degradations.

Benkedjouh[7] predicted the amount of wear and calculated the RUL of cutting tools by using SVR. Wu[8] predicted the RUL of aero-engine based on SVR.

Generally, the above approaches have achieved accurate RUL prediction results, which focused on establishing the relationship between the current monitoring data and machine health states. However, in reality, the machine health state is not only related to the current monitoring data, but also to the past times. Thus in this paper, a SVR based method with considering different length of signals at past times is proposed for RUL prediction. It consists of two main phases: an off-line phase and an on-line phase. In the first phase, the signal features are extracted from raw data, and then the SVR model with considering information at past times is established. In the second phase, the constructed models are used to predict cutting tool's RUL and the best signal length for accurate prediction result is obtained. The proposed method is validated by the experimental data taken from a computer numerical control (CNC) rotor slot machine in a factory.

2. RUL prediction model of cutting tools

2.1. SVR model

SVM is divided into two main categories: support vector classification (SVC) and support vector regression (SVR). SVR is the most common application form of SVMs. It has been proposed in 1997 by Vapnik[9]. The main objective of SVR is to map the data into a higher dimensional feature space and to estimate a relationship between input and output random variables under the assumption that the joint distribution of the variables is completely unknown.

Given a data set $X=\{(x_i,y_i),i=1,2,...,l\}$, where $x_i \in R^N$ is the signal feature vector of the ith cutter, $y_i \in R$ is the RUL of the ith cutter, l is the sample size. The regression function can be expressed as

$$y=f(x)=w\varphi(x)+b \tag{1}$$

where $\phi(x)$ denotes the high dimensional space transformation of AE features, w and b are the coefficients. The coefficients are estimated by minimizing the following regularized risk function:

$$R_{SVM}=\frac{1}{2}\|w\|^2+C\frac{1}{N}\sum_{i=1}^{N}L(d_i,y_i) \tag{2}$$

$$R_{SVM}=\frac{1}{2}\|w\|^2+C\frac{1}{N}\sum_{i=1}^{N}L(d_i,y_i) \tag{3}$$

$$L(d_i,y_i)=\begin{cases}|d_i-y_i|-\varepsilon & if\ |d_i-y_i|\ge\varepsilon \\ 0 & otherwise\end{cases} \tag{4}$$

where $\frac{1}{2}\|w\|^2$ is the regularized term, $L(d_i,y_i)$ is the ε-insensitive loss function, ε is the tube size of SVM, and C is the regularization constant determining the trade-off between the empirical error and the regularized term. Two variables ξ and ξ^* are introduced to representing the distance from desired values to the corresponding boundary values of the ε-tube, and the Eq. (2) can be rewritten as:

$$R_{SVM}=\frac{1}{2}\|w\|^2+C\sum_{i=1}^{l}(\xi_i+\xi_i^*)$$
$$s.t.\begin{cases} y_i-w^T\phi(x)-b\le\varepsilon+\xi_i \\ w^T\phi(x)+b-y_i\le\varepsilon+\xi_i^* \\ \xi_i,\xi_i^*\ge 0, i=1,2,...,l \end{cases} \quad (5)$$

By introducing Lagrange multipliers and kernel function $K(x_i,x_j)=\varphi(x_i)\cdot\varphi(x_j)$, the dual Lagrange form of Eq. (4) is given by:

$$R(a_i,a_i^*)=\frac{1}{2}\sum_{i=1}^{l}\sum_{j=1}^{l}K(x_i,x_j)(a_i-a_i^*)(a_j-a_j^*)-\sum_{i=1}^{l}y_i(a_i-a_i^*)+\sum_{i=1}^{l}\varepsilon(a_i-a_i^*)$$
$$s.t.\begin{cases} a_i,a_i^*\in[0,C], \quad i=1,2,...,l \\ \sum_{i=1}^{l}(a_i-a_i^*)=0 \end{cases} \quad (6)$$

Finally, the regression function Eq. (1) is transformed into:

$$y=f(x)=\sum_{i=1}^{l}(a_i-a_i^*)K(x_i,x_j)+b \quad (7)$$

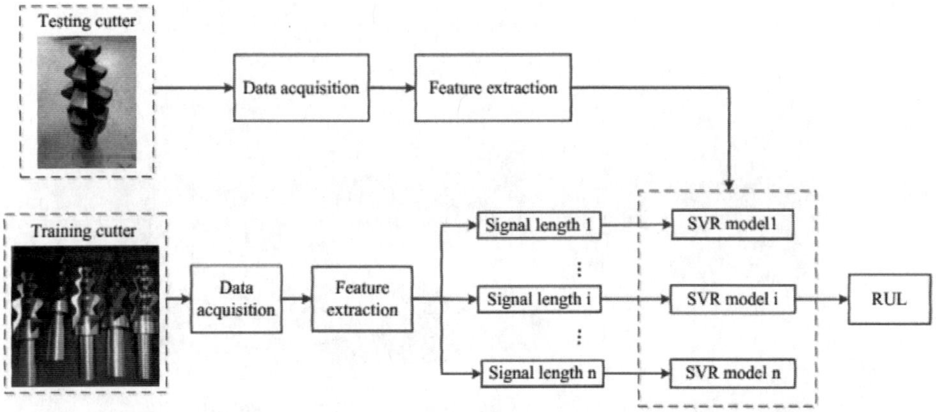

Figure 1. The framework for RUL prediction.

2.2. The Framework for RUL Prediction.

The framework for RUL prediction of cutting tools is shown in Figure 1. The principle of the proposed method relies on two main phases: an off-line phase and an on-line phase. In the first phase, the signal features are extracted from raw monitoring data, and then the SVR models with considering different length of data at past times are established to reflect the relationship between monitoring signals and tool life. In the second phase, the constructed models are used to predict cutting tool's RUL, and the best signal length for accurate prediction result is obtained.

IOP Conf. Series: Materials Science and Engineering **576** (2019) 012021 doi:10.1088/1757-899X/576/1/012021

3. Application and results

3.1. Experiment Setup.

The experiment was carried out on a special turbine rotor slotter of a factory, as shown in Figure 2. The workpiece material was high strength low alloy structural steel 30Cr2Ni4MoV and the material of J1-formed 4-flute slotting cutter was powder metallurgy high speed steel (PM HSS) M42-4. The experiment data were obtained under constant conditions. The cutting parameters were: the spindle speed of the slotting cutter was 200 rpm and the feed rate was 15mm/min.

The PCI-2 based AE system was adopted in this experiment and it was fixed on the rotor through the magnetic base, as show in Figure 2. Though it maybe a litter far away from the machining position than fixed on the machine tool, it can avoid the interference of external factors such as machine tool vibration, the impact of cutting fluid, and get more useful information.

The operating frequency of the sensor is from 100 to 450 KHz and the sampling frequency was set at 1 MHz. The flank wear value was measured at the end of each cutter's life by using an optical microscope XDC-10A-E630. The mean wear of the four flutes was considered as the wear status of slotting cutter. In this paper, a total of 7 J1-formed slotting cutters were used during the experiment. Among them, cutter no.3 is used to verify the proposed method. The wear threshold is set as 125μm considering the actual experience of the factory.

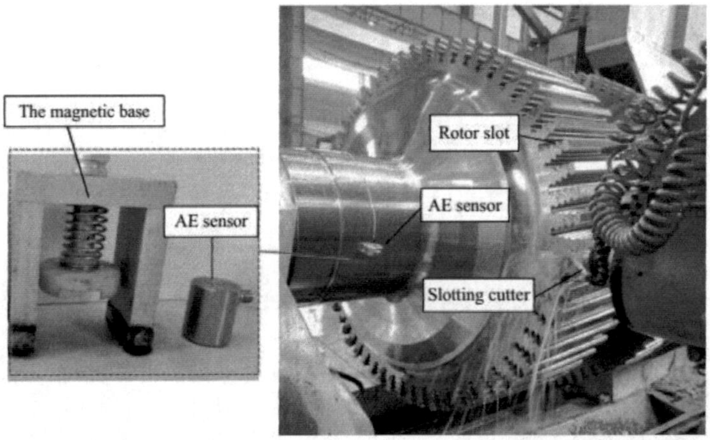

Figure 2. Experimental setup for RUL prediction of slotting cutter.

3.2. Signal Analysis.

AE is one of the most effective signals for tool wear monitoring, especially for the industrial applications. It can avoid the interference of external factors because the frequency of AE signals is much higher than that of machine vibration and environmental noises[10, 11]. Previous studies[12-14] have shown some AE features can indicate tool wear, in this work, 14 features are extracted: rise time(RT), counts(C), energy(E), amplitude(A), average frequency(AF), root mean square(RMS), average signal level(ASL), Counts To Peak(CTP), Reverberation Frequency(RF), Initiation Frequency(IF), Signal Strength(SS), Absolute Energy(AbE), Frequency Centroid(FC), and Peak frequency(PF).

Taken AbE as an example, the signal changes with time is shown in Figure 3 (a). It can be seen that The AbE values increase along with time, and each slotting processing cycle consists of two stages: slotting processing stage and no-load operation stage. For the convenience of analysis, the mean values of the above features change over each slot processing stage were calculated by using Matlab software. The feature AbE changes with the number of processed slots is shown in Figure 3 (b).

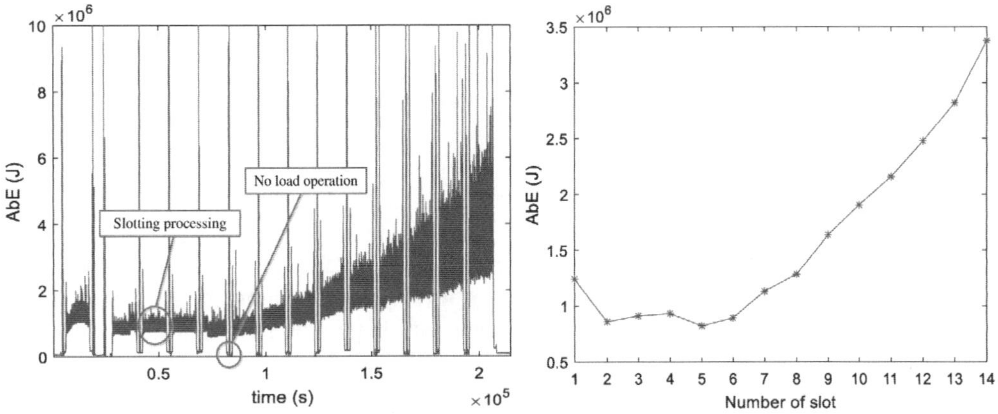

(a) (b)

Figure 3. AbE of AE signal features.

Although the above mentioned features are indicative of tool wear from different aspects, they have different important degrees. Some features are salient and closely related to the wear, but others are not. Thus, it is necessary to select the important and discard the irrelevant features. Correlation coefficient between two vectors x and y can be calculated by equation

$$\rho(x,y)=\frac{\sum(x-\overline{x})(y-\overline{y})}{\sqrt{\sum(x-\overline{x})^2}\sqrt{\sum(y-\overline{y})^2}} \tag{8}$$

where x and y are the mean values of x and y, respectively. The value of ρ ranges between −1 and +1. The larger the absolute value of ρ, the higher the degree of correlation.
According to Eq. (10), the correlation coefficient between the above 14 features and tool wear when the cutting tools were replaced are calculated and displayed in Table 1.

Table 1. Correlation coefficients between AE features and tool wear.

AE feature	Correlation coefficients	AE feature	Correlation coefficients
RT	0.1148	CTP	0.1640
C	0.1582	RF	0.2890
E	0.7566	IF	0.3111
A	0.6672	SS	0.7164
AF	0.2235	AbE	0.7089
RMS	0.8483	FC	0.2247
ASL	0.6689	RF	0.2507

It can be seen that these 6 features E, A, RMS, ASL, SS, and AbE is significant correlated with tool wear. They compose the input vector X= [E, A, RMS, ASL, SS, AbE] of the ARMA and SVR model. The extracted features are normalized to [0.1, 0.9] by the follow equation:

$$y = 0.1 + \frac{(0.9-0.1)}{(x_{max} - x_{min})} \times (x - x_{min}) \tag{9}$$

where x is the feature vector to be normalized; x_{min} is the minimum value of x; x_{max} is the maximum value of x; y is the normalization result.

3.3. RUL Prediction

In this paper, cutter no.3 is used to verify the proposed method, and the other six cutters are used for modeling. Assuming the RUL of cutting tools is only related to the condition monitoring signal of current slot. The AE features extracted from the six cutters during their whole working life compose the input vector X= [E, A, RMS, ASL, SS, AbE] of SVR model. Then the RUL of tool life-cycle are the output of this model.

The common kernel functions used for SVR approaches are linear, polynomial, and RBF. Literature demonstrates that the RBF kernel outperformed the others in most scenarios. Thus in this paper, RBF kernel $K(x,y) = \exp(-\gamma|x-y|^2)$ is employed to train SVR model. By using grid search method, the optimal values of the parameters are obtained ε=0.1, $C=4$ *and* $\gamma=0.5$. The grid search result is shown in Figure 4. After being trained, SVR model is used to predict the RUL of cutter no.3, and the prediction result is shown in Figure 5.

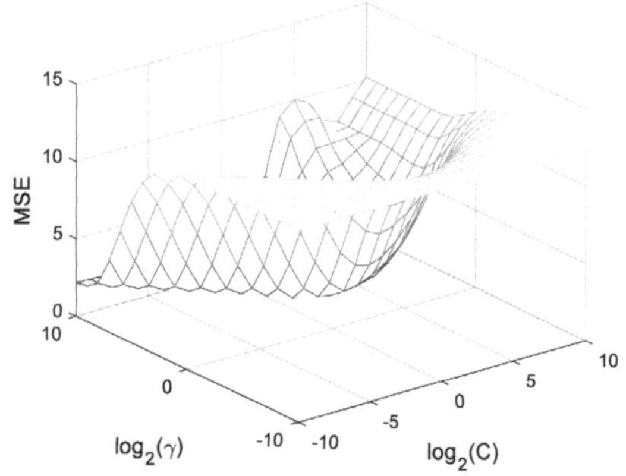

Figure 4. Grid search result.

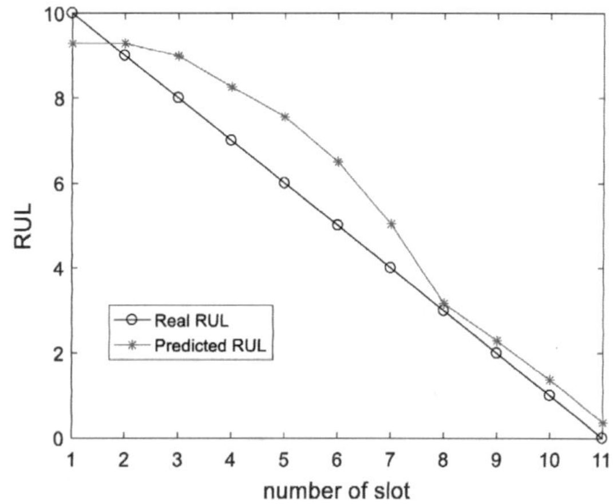

Figure 5. RUL prediction result.

In order to evaluate the performance of the prediction result, the following prognostic metrics are used in this paper: accuracy, mean absolute error and mean relative error. The three metrics can be expressed as:

$$\text{Accuracy} = \frac{1}{C}\sum_{c=1}^{C} e^{-\left|RUL_{real}(c) - RUL_{estimated}(c)\right| / RUL_{real}(c)} \tag{10}$$

$$MAE = \frac{1}{C}\sum_{c=1}^{C}\left|RUL_{real}(c) - RUL_{estimated}(c)\right| \tag{11}$$

$$MRE=\frac{1}{C}\sum_{c=1}^{C}\left|\frac{RUL_{real}(c)-RUL_{estimated}(c)}{RUL_{real}(c)}\right| \tag{12}$$

It can be calculated that the Accuracy=0.8399, MAE=0.7777 and MRE=0.1802. Similarly, assuming the RUL of cutting tools is not only related to the current slot monitoring signal but also the past N slots machining data. In this paper, the value of N is 0, 1, 2, 3 and 4. Repeat the above procedures, and calculate the RUL of cutter no.3 when N is 1, 2, 3 and 4, respectively. The RUL prediction results of different signal length are shown in Figure 6. The three prognostic metrics were calculated and shown in Table 2.

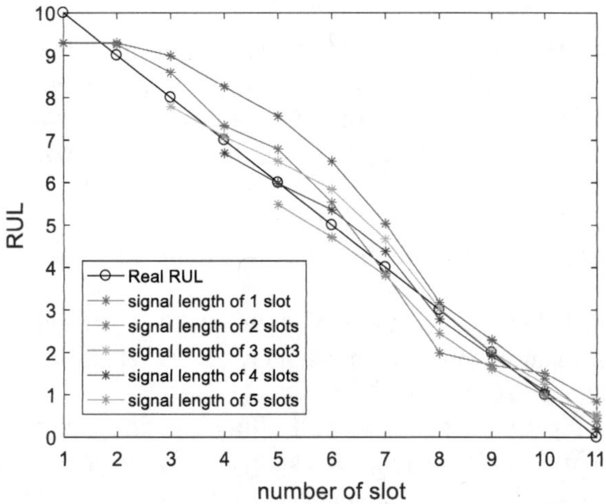

Figure 6. RUL prediction result of different signal length.

Table 2. Results of prognostic metrics.

N	Accuracy	MAE	MRE
0	0.8399	0.7777	0.1802
1	0.8635	0.5305	0.1576
2	0.9183	0.3374	0.0887
3	0.9435	0.2039	0.0586
4	0.9079	0.3550	0.0986

It can be seen that when the monitoring data of the past slot machining process is taken into consideration, the RUL prediction results are more accurate than those only used the current slot machining signals. The prediction results are the best when the machining process of the past 3 slots is considered and the best prediction accuracy is 0.9435.

4. Conclusion

In this paper, a RUL prognostic method based on support vector regression (SVR) is proposed to predict cutting tool life. The proposed method consists of two main phases: an off-line phase and an on-line phase. In the first phase, the signal features are extracted from raw data, and then the SVR models with considering different length of signals at past times are established to reflect the relationship between monitoring signals and tool life. In the second phase, the constructed models are used to predict cutting tool's RUL, and the best signal length for accurate prediction result is obtained. The proposed method is applied on experimental data taken from a computer numerical control (CNC) rotor slot machine in a factory. The best RUL prediction accuracy reaches 94.35%. The result shows the validity and practicability of this method.

Future work will include developing the RUL prediction method with considering different signal length of past times at different machining times, since the cutting tool condition at different times maybe depends on different length of condition monitoring readings. Furthermore, other algorithms like neural network will be investigated to compare with the proposed method.

5. References

[1] S. Kurada and C. Bradley 1997. A review of machine vision sensors for tool condition monitoring, *Computers in Industry,* **34**, 5572.

[2] X.S. Si, W. Wang, C.H. Hu, D.H. Zhou 2011. Remaining useful life estimation–A review on the statistical data driven approaches, *European Journal of Operational Research*, **213**, 0114.

[3] Y. Lei, N. Li, L. Guo, N. Li, T. Yan, J. Lin 2018, Machinery health prognostics: A systematic review from data acquisition to RUL prediction, *Mechanical Systems and Signal Processing*, **104**, 799834.

[4] J. Gokulachandran, K. Mohandas 2013. Comparative study of two soft computing techniques for the prediction of remaining useful life of cutting tools, Journal of Intelligent Manufacturing, **26**, 255268.

[5] T. Benkedjouh, K. Medjaher, N. Zerhouni, S. Rechak 2013. Remaining useful life estimation based on nonlinear feature reduction and support vector regression, *Engineering Applications of Artificial Intelligence*, **26**, 17511760.

[6] Z. Shen 2013. Remaining Life Predictions of Rolling Bearing Based on Relative Features and Multivariable Support Vector Machine, *Journal of Mechanical Engineering*, **49**,183189.

[7] T. Benkedjouh, K. Medjaher, N. Zerhouni, S. Rechak 2013. Health assessment and life prediction of cutting tools based on support vector regression, *Journal of Intelligent Manufacturing,* **26**, 213223.

[8] X.H. Wu, D. Wen, R.G. Li 2013. A model to predict the residual life of aero-engine based upon Support Vector Machine. In *Quality, Reliability, Risk, Maintenance, and Safety Engineering (QR2MSE)*, p1880-1882.

[9] V. Vapnik, S.E. Golowich, A. Smola 1997. Support vector method for function approximation, regression estimation, *Advances in Neural Information Processing Systems*, **9**, 281287.

[10] X. Li 2002. A brief review acoustic emission method for tool wear monitoring during turning, *International Journal of Machine Tools & Manufacture*, **42**, 157165.

[11] Y. Zhou, W. Xue 2018, Review of tool condition monitoring methods in milling processes, The *International Journal of Advanced Manufacturing Technology*, **96**, 25092523.

[12] Q. Ren, M. Balazinski, L. Baron 2014. Type-2 fuzzy tool condition monitoring system based on acoustic emission in micromilling, *Information Sciences*, **255**, 121134.

[13] Y. Liu, X. Hu and S. Yan 2016. Tool condition monitoring and degradation estimation in rotor slot machining process, *The International Journal of Advanced Manufacturing Technology*, **91**, 3948.

[14] H. Li, Y. Wang, P. Zhao, X. Zhang, P. Zhou 2014, Cutting tool operational reliability prediction based on acoustic emission and logistic regression model, *Journal of Intelligent Manufacturing*, **26**, 923931.

Acknowledgements
This research is supported by National Natural Science Foundation of China (Grant No. 51435009).

IOP Conf. Series: Materials Science and Engineering **576** (2019) 012022 doi:10.1088/1757-899X/576/1/012022

Effect of working environment on C/C composite finger seal performance

L N Wang, L C Sun, D H Meng, W Sun and L Qi

Beijing Institute of Spacecraft Environment Engineering, 104 Youyi Road, Beijing, China

Abstract. With the development of the technology, the working conditions, which include high temperature, assembly and working state, have important influence on the dynamic performance of finger seal. Therefore, it is necessary to consider above factors together when the finger seal performance is carried out. But so far, the finger seal performance under considering several factors together have not been investigated. Based on this, the material parameters, which are needed in the dynamic performance analysis process of C/C composite finger seal, are obtained by the tribology test of C/C composite. Then the equivalent dynamic model of finger seal based on distributed masses is established under considering the rotor incline, temperature, assembly and impact together. And the dynamic performance of finger seal is analyzed by the model. The results are shown that the above working states have complex influence on the finger seal performance. Meanwhile, the leakage performance of finger seal is compared with those of labyrinth seal in the bearing chamber working environment, and the advantage of finger seal is represented. The dynamic performance analysis of finger seal under considering complex working states is carried out, and the theory research system of finger seal is further improved.

1. Introduction

A seal device as an important component of aero-engine, has decisive influence on the performance of aero-engine. Finger seal as a novel seal device recently attracts many attentions of scholars[1-3].

Many scholars investigated the performance of finger seal under dynamic condition by equivalent dynamic model. Braun et al treated non-contact finger seal as a system of lumped mass-spring-damping, neglecting the frictional resistances between aft cover plate and finger element and among finger elements[4,5]. Marie treated non-contact finger seal as two degrees of freedom equivalent dynamic model based on lumped mass, only considering the frictional resistance between low pressure finger element and aft cover plate[6]. Su et al carried out the hysteresis performance analysis of finger seal by equivalent dynamic model based on lumped mass, and compared the finger seal leakage rates under dynamic condition with those under static condition[7]. Chen et al established equivalent dynamic model of finger seal based on distributed masses[8].

Recently, the performance analysis of C/C composite finger seal begins to attract many attentions of scholars. Chen et al analyzed the dynamic performance of the C/C composite finger seal by equivalent dynamic model based on the distributed masses[9]. Wang et al discussed the effect of temperature on the dynamic performance of C/C composite finger seal based on the distributed masses[10]. Wang et al also analyzed the effect of rotor incline on the dynamic performance of C/C composite finger seal[11], and discussed the effect of impact velocity and impact restitution coefficient on the dynamic performance of C/C composite finger seal[12].

But finger seal locates in the coupling effect among some working states in fact. Therefore it is necessary to carry out the dynamic performance analysis under complex working states. Based on this, the material parameters, which are needed in the dynamic performance analysis of C/C composite finger seal, are obtained by the tribology test of C/C composite. The equivalent dynamic model of C/C composite finger seal under complex working states is established, and the finger seal performance under single working state is compared with those under complex working states. Then the effect of rotor incline angle and impact velocity on the finger seal performance under complex working states is carried out. Meanwhile the advantage of finger seal is represented by the comparison between the leakage rates of finger seal and labyrinth seal in bearing chamber working environment. The dynamic analysis technology contributes to improve the practicality of finger seal engineering design.

2. Tribology test of C/C composite

Finger seal and rotor or aft cover plate are made of C/C composite and steel, respectively. Therefore the tribology performance analysis between finger seal and aft cover plate or between finger seal and rotor could be transformed for those between C/C composite and steel. Based on this, C/C composite sample and columnar steel dowel are machined, respectively. The tribology test is carried out by the UMT multifunction friction-abrasion testing machine. Then the effect of fiber orientation of C/C composite and working state on the tribology performance between C/C composite and steel is investigated. The experimental results provide some input parameters for the dynamic analysis of finger seal.

2.1. Structure of UMT multifunction friction-abrasion testing machine

The frictional coefficients between C/C composite and steel under different working states and fiber orientation are obtained by the UMT multifunction friction-abrasion testing machine, which is composed by the systems of hydraulic pressure, dynamical force and control.

2.2. Preparation of C/C composite sample

The schematic diagram of C/C composite sample is shown in Figure.1. To guarantee fiber integrality, machining precision and repetition tests of C/C composite sample, the C/C composite is made into a cube, whose length, width and height are 30mm, 20mm and 3mm by wire electrode cutting respectively, that is shown in Figure.2. And then the surface roughness of C/C composite sample is 1.5μm. In addition, columnar steel dowel is made of $45^{\#}$ steel in the test.

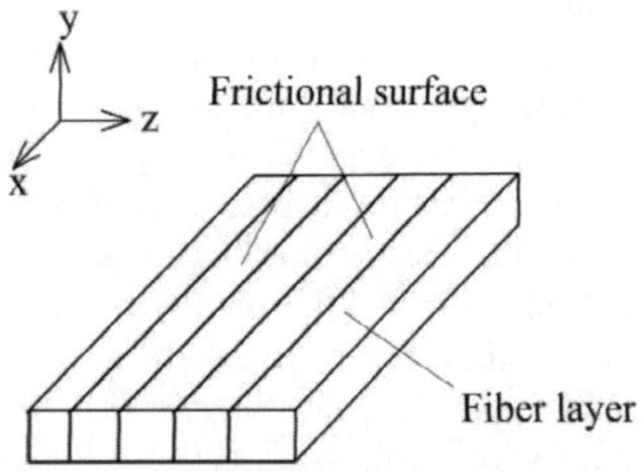

Figure 1. Schematic diagram of C/C composite sample.

Figure 2. Sample of C/C composite.

2.3. Tribology test principle of C/C composite

The working principle of UMT multifunction friction-abrasion testing machine is shown in Figure.3. The static friction sample or columnar steel dowel is perpendicular to the surface of C/C composite sample, and C/C composite sample begins to move backwards and forwards in a straight line. The friction-abrasion between C/C composite sample and columnar steel dowel is generated in the process. Then frictional coefficient and amount of abrasion could be measured by sensors in test device.

Figure 3. Working principle interpretation of UMT multifunction friction-abrasion testing machine.

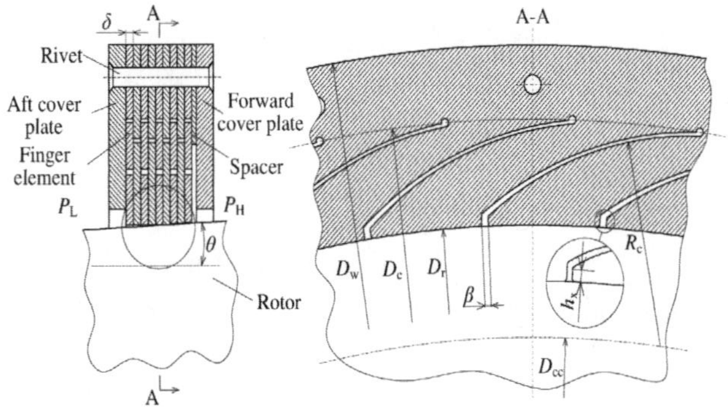

Figure 4. Schematic diagram of finger seal structure.

3. Equivalent dynamic model of finger seal

3.1. Finger seal structure

Finger seal is constituted by multiple finger elements, which are staggered, and every finger element includes some finger sticks (see Figure.4). The structural parameters of finger seal are introduced by the literatures [11] in the manuscript.

3.2. Equivalent dynamic model of finger seal based on distributed masses

When the fingers stick and rotor are in contact state, the forced vibration differential equations can be defined as

$$\begin{cases} m_1\ddot{x}_1 = -k_1x_1 + k_{c1}\left(y_1 - x_1\right) + F'_{f1} - F_{f1} \\ m_2\ddot{x}_2 = -k_2x_2 + k_{c2}\left(y_2 - x_2\right) + F'_{f2} - F_{f2} \\ \vdots \\ m_{n-1}\ddot{x}_{n-1} = -k_{n-1}x_{n-1} + k_{c(n-1)}\left(y_{n-1} - x_{n-1}\right) + F'_{f(n-1)} - F_{f(n-1)} \\ m_n\ddot{x}_n = -k_nx_n + k_{cn}\left(y_n - x_n\right) - F_{fn} \end{cases} \tag{1}$$

When the finger sticks and rotor are in separation state, the forced vibration differential equations can be defined as

$$\begin{cases} m_1\ddot{x}_1 = -k_1x_1 + P_1 + F'_{f1} - F_{f1} \\ m_2\ddot{x}_2 = -k_2x_2 + P_2 + F'_{f2} - F_{f2} \\ \vdots \\ m_{n-1}\ddot{x}_{n-1} = -k_{n-1}x_{n-1} + P_{n-1} + F'_{f(n-1)} - F_{f(n-1)} \\ m_n\ddot{x}_n = -k_nx_n + P_n - F_{fn} \end{cases} \tag{2}$$

where n is number of finger elements, $n \geqslant 2$ and n is an integer.

3.3. Equivalent parameters

The correlative parameters, which include the equivalent mass m_i and structural stiffness k_i of finger stick, contact stiffness coefficient k_{ci} between finger stick and rotor, frictional resistances F_{fi} and F'_{fi}

between finger element and aft cover plate and among finger elements, thermal deformation of finger stick, fluid pressure P_i, leakage rate and contact pressure, are introduced by the literature [11] in the manuscript. In addition, frictional coefficient, which is used to calculate frictional resistance, is obtained by C/C composite tribology test.

3.4. Displacement excitation of rotor

The amount of rotor incline corresponding to random finger element can be defined as[11]

$$\begin{cases} \Delta r_i = -(i_{\text{th}} - i)\delta \tan(\pi + \theta) & 1 \le i < i_{\text{th}} \\ \Delta r_i = 0 & i = i_{\text{th}} \\ \Delta r_i = (n - i)\delta \tan(\theta) & i_{\text{th}} < i < n \end{cases} \quad (3)$$

where i_{th} is serial number of reference finger elements.

The displacement excitation of the rotor under considering complex working states, which include the incline of the rotor, assemble condition, temperature effect and impact effect, can be defined as

$$y_i = a_c \Delta r \sin(\omega t) + \Delta r_i \sin(\omega t) \pm e + l_{ci} \quad (4)$$

where a_c is impact magnification coefficient; ω is angular velocity of rotor; Δr is the radial displacement amplitude of rotor; t is the time; Δr_i is the amount of rotor incline corresponding to finger element; e is the amount of interference or clearance, '+' and '-' express interference fit and clearance fit, respectively; l_{ci} is the thermal deformation of the finger stick.

3.5. Leakage rate Q

The leakage gap $h_i(t)$ between finger stick in the i_{th} finger element and rotor at time t could be defined as

$$h_i(t) = x_i(t) - y_i(t) \quad (5)$$

where $x_i(t)$ and $y_i(t)$ are displacement response of finger stick in the i_{th} finger element and displacement excitation of rotor at time t, respectively. And the schematic of finger seal leakage gap $h_i(t)$ are shown in Figure. 5.

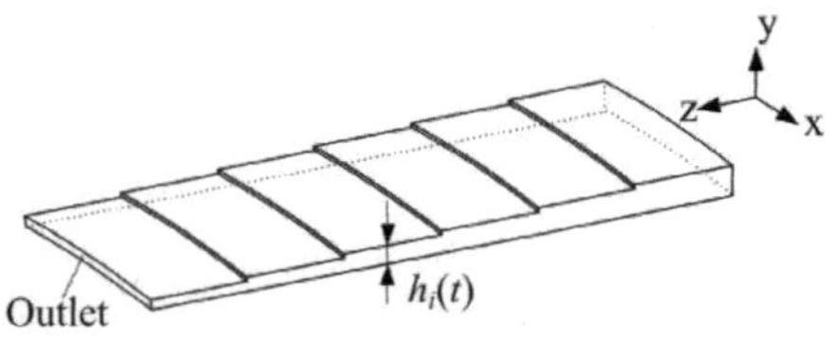

Figure 5. Schematic of finger seal leakage gap.

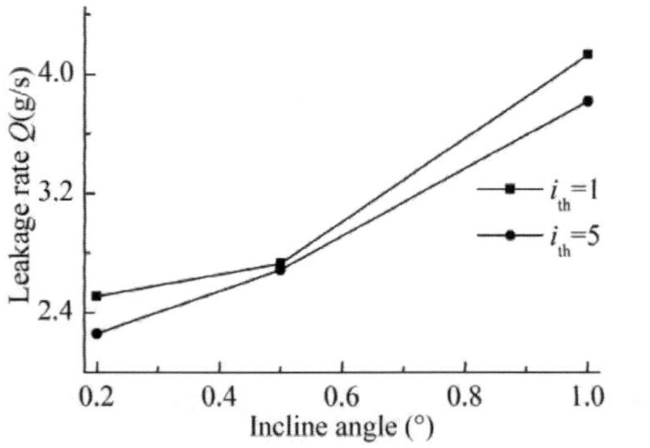

Figure 6. Effect of incline angle on leakage rate of finger seal under complex working states.

The outlet of leakage gap is chosen as the location of leakage rate calculation, and then the leakage rate $q(t)$ of finger seal at time t could be defined as

$$q(t) = \sum_{m=1}^{M}\sum_{k=1}^{K}\left[w_{mk}(t)\rho_{mk}(t)\Delta l_x(t)\Delta l_y(t) \right] \quad (6)$$

where $w_{mk}(t)$ is velocity of arbitrary fluid microunit in z the direction; $\rho_{mk}(t)$ is fluid density of microunit; M and K are total of grids in the x and y directions, respectively; $\Delta l_x(t)$ and $\Delta l_y(t)$ are sizes of fluid microunit in the x and y directions at time t, respectively.

Therefore, the average leakage rate of finger seal in a motion period could be defined as

$$\bar{q} = \frac{1}{f}\sum_{j=1}^{f} q(t + j\Delta t) \tag{7}$$

where f is time steps quantity.

4. Results and discussions

C/C composite finger seal, which is constituted by nine finger elements, is treated as research object, and the effect of complex working states, which include the incline of the rotor, assemble condition, temperature effect and impact effect, on the dynamic performance of finger seal is analyzed.

4.1. Leakage rate of finger seal in bearing chamber

The future application of C/C composite finger seal, which is due to its self-lubricating performance and lower structural stiffness, could be leaned to adopt by interference fit in bearing chamber working environment. Therefore the following analysis is carried out under temperature being 400℃ and interference fit.

4.1.1. Effect of rotor incline angle on finger seal leakage rate under complex working states. Under the conditions of the impact velocity being 0.05m/s, temperature being 400 ℃ and amount of interference fit being 0.01mm between finger seal and rotor, the leakage rates under different incline angles of rotor are calculated, that are shown in Figure.6. The leakage rates under precession incline i_{th}=1 are all larger than those under nutation incline i_{th}=5. It is consistent with the phenomenon of reference [11]. The main reason is that the displacement excitations of rotor acting on finger sticks are different under considering incline of rotor. In addition, the leakage rates of finger seal are all increased with increasing incline angles of rotor under precession incline i_{th}=1 and nutation incline i_{th}=5, that are also shown in Figure.6. It is different from the phenomenon of reference [11]. Therefore the complex working states makes the finger seal performance to become complicated.

4.1.2. Effect of impact velocity on finger seal leakage rate under complex working states. Under the conditions of the incline angle of rotor being 1°, temperature being 400℃ and amount of interference fit being 0.01mm between finger seal and rotor, the leakage rates under different impact velocities are calculated, that are shown in Figure.7. The leakage rates of finger seal under the complex working states are increased with increasing impact velocities. It is consistent with the phenomenon of reference [12]. The main reason is that the displacement excitation of rotor acting on finger stick is larger under considering impact effect. In addition, the leakage rates of finger seal under precession incline i_{th}=1 are all larger than those under nutation incline i_{th}=5. It is consistent with the phenomenon of reference [12]. The reason is also that the displacement excitations of rotor acting on finger sticks are different under considering incline of rotor

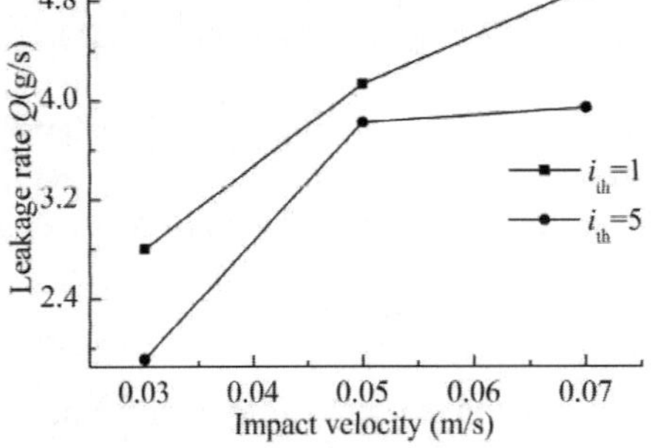

Figure 7. Effect of impact velocity on leakage rate of finger seal under complex working states.

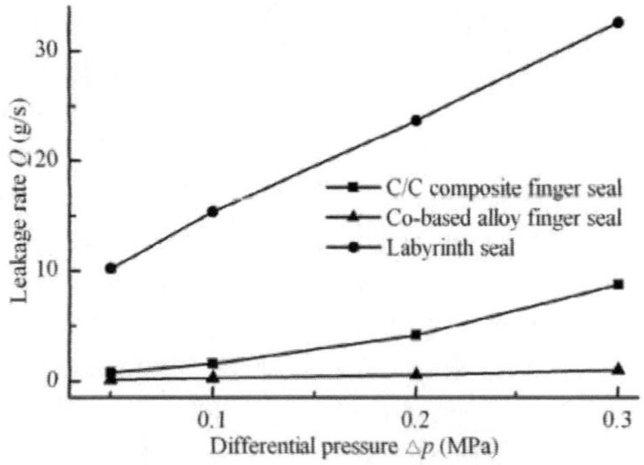

Figure 8. Comparison between leakage rates of finger seal and labyrinth seal with increasing differential pressure.

4.2. *Comparison between leakage rates of finger seal and labyrinth seal in bearing chamber.* Under rotational speed being 9000r/min, the comparison between the leakage rates of finger seal, which are made of C/C composite and Co-based alloy, and labyrinth seal with increasing differential pressure are shown in Figure.8. It is shown that the leakage rates of finger seals, which are made of C/C composite and Co-based alloy, are all remarkably smaller than those of labyrinth seal with increasing differential pressure. The feasibility of finger seal being used in bearing chamber is represented. In addition, it is also shown that the leakage rates of finger seal and labyrinth seal are all increased with increasing differential pressure. It is easy to understand the phenomenon of labyrinth seal. But for finger seal, the frictional resistances between low pressure finger element and aft cover plate and among finger elements are increased with increasing differential pressure, and the hysteresis degree of finger seal is also increased. When the finger stick is separated from rotor, the fluid pressure acting on the finger stick is increased with increasing differential pressure, and then the hysteresis degree is further increased. Therefore the leakage rate of finger seal is increased with increasing differential pressure. The leakage rate of Co-based alloy finger seal is remarkably smaller than those of C/C composite. The results are shown that the seal performance of Co-based alloy is excellent than those of C/C composite, but the shorter abrasion life-span is noticeable disadvantage.

5. Conclusions

The frictional coefficients, which are obtained by C/C composite tribology performance test, are treated as the main material parameters, and they are used to analyze the dynamic performance of C/C composite finger seal. Based on this, the equivalent dynamic model of C/C composite finger seal based on distributed masses is established, and the effect of complex working states on the dynamic performance of finger seal is carried out by the model. The main conclusions are as follow.

The effect of incline angle of rotor and impact velocity on the dynamic performance of finger seal under complex working states is discussed, respectively, and compared with those of references [11] and [12]. The prominent difference among the above results is represented. Meanwhile the complicated effect of complex working states on the finger seal dynamic performance is proved.

The leakage rates of finger seal under considering complex working states are remarkably smaller than those of labyrinth seal in bearing chamber. Therefore the advantage of finger seal is also represented.

6. References

[1] Arora G K, Proctor M P, Steinetz B M and Delgado I R. Pressure balanced, low hysteresis, finger seal test results. NASA TM 1999-209191, June 1999

[2] Arora G K and Glick D L. Pressure balanced finger seal[P]. Patent 6,196,550, USA, 2001

[3] Proctor M P, Kumar A and Delgado I R. High-speed, high-temperature finger seal test results. *J Propul Power*, 2004, **20**: 312-18

[4] Braun M J, Pierson H M, Deng D, Choy F, Proctor M P and Steinetz B M. Structural and dynamic considerations towards the design of a padded finger seal. In: 39th AIAA/ASME/SAE/ASEE joint propulsion conference, Huntsville, Alabama, 20-23 July 2003, paper no. AIAA 2003-4698. Reston: AIAA

[5] Braun M J, Pierson H M and Deng D. Thermofluids considerations and the dynamic behavior of a finger seal assembly. *Tribol Trans*, 2005, **48**: 531-47

[6] Marie H. A study of non-contacting passive-adaptive turbine finger seal performance. PhD Thesis, University of Akron, USA, 2005

[7] Su H and Chen G D. Dynamic behavior analysis of finger seal based on equivalent dynamic model. *Chin J Mech Eng*, 2010, **23**: 590-9

[8] Chen G D, Lu F, Yu Q P and Su H. Dynamic analysis of finger seal using equivalent model based on distributed mass method. *Proc IMechE, Part C: J Mechanical Engineering Science*. 2014, **288**: 2992-3005.

[9] Chen G D, Wang L N, Yu Q P and Su H. Dynamic analysis of C/C composite finger seal. *Chinese J Aeronaut*, 2014, **27**: 745-58.

[10] Wang L N, Chen G D, Su H, Lu F and Zhang Y C. Effect of temperature on the dynamic performance of C/C composite finger seal. *Proc IMechE, Part G: J Aerospace Engineering*. 2016, **230**(12): 2249-64.
[11] Wang L N, Chen G D, Su H, Lu F and Zhang Y C. Effect of work status on leakage and contact pressure of C/C composite finger seal. *Proc IMechE, Part C: J Mechanical Engineering Science*. 2017, **231**(5): 925-40.
[12] Wang L N, Chen G D, Su H, Lu F and Zhao H L. Impact dynamic analysis of C/C composite finger seal. *Proc IMechE, Part G: J Aerospace Engineering*. 2017, **231**(7): 1225-37.

Acknowledgments
The study was supported by the National Natural Science Foundation of China (No. U1537109) and National Defence Technique Foundation of China (No. JSJL2015203B019).

IOP Conf. Series: Materials Science and Engineering **576** (2019) 012023 doi:10.1088/1757-899X/576/1/012023

Finite element analysis of hot roll-forming process of local thickened U-rib for bridge

Xue-Feng Peng[1], Jing-Tao Han[2] and Jia-ShengLiu[1]

[1]Ninth department, 713th Research Institute of China Shipbuilding Industry Corporation, Zhengzhou 450015, China
[2]School of Materials Science and Engineering, University of Science and Technology Beijing, Beijing 100083, China

Abstract. To solve the problems of small welding area and poor welding quality between traditional U-rib and bridge decks, a new type local thickened U-rib (LTU-rib) proposed by bridge experts. In this paper, the simulation software ABAQUS is used to simulate a hot roll-forming process of LTU-rib. Finite element model of hot roll-forming process was established. The influence of the pass design schemes and temperature parameters on the hot roll-forming process was analyzed and discussed. The simulation results show that first scheme is more suitable for hot roll-forming process. Although forming temperature can reduce the roll-forming force and improve the metal fluidity, excessive temperature will lead to redundant deformation. The simulation results provide a reference for the parameters design of actual hot roll-forming process.

1. Introduction

U-ribs for bridge are widely used in modern orthotropic steel deck due to their large torsional stiffness and bending stiffness [1-3]. However, in recent years, with the rapid growth of traffic vehicles and long-term overload of bridges, some steel bridges which had been built and used in China have different degrees of fatigue cracks in the welded joints of U-rib and bridge deck [4].The reason is that the U-rib can only be welded from the outside. The industry often uses 75%~85% partial penetration welding and the welding area is small. The unfused part itself forms a natural initial crack. Under repeated loading, it causes fatigue cracking at the weld of the bridge deck and U-rib, reducing the life of the entire bridge. Recent years, some bridge experts have put forward a viewpoint of local thickened U-rib [5], which can improve the quality of welded joints by increasing the weld area. In this paper, taking the new LTU-rib as an example, the finite element analysis of the hot roll-forming process is carried out by ABAQUS software. The influence of various parameters on simulation results is analyzed to guide the hot roll-forming process design of the LTU-rib.

2. Finite element model

2.1. Hot roll-forming process finite element modeling

The commonly specification 280×300×8mm U-rib is selected as the object of study. And its cross-sectional dimension is shown in figure 1. The focus of this paper is to simulate the hot roll-forming process of the LTU-rib. Therefore, the cold bending of U-rib is not considered. The traditional U-rib is selected as the preformed material, and the hot roll-forming is carried out after local induction heating in the deformation zone.

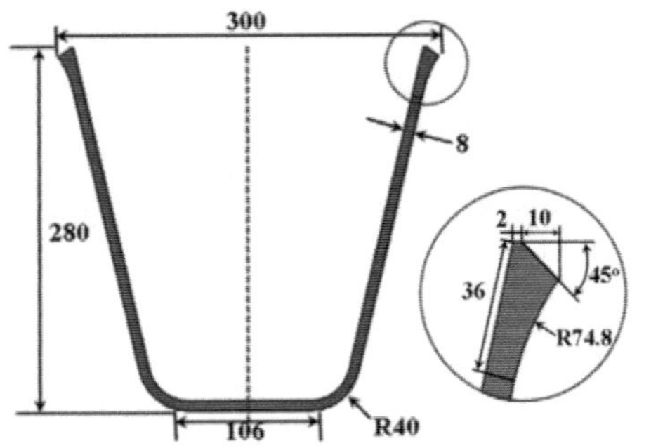

Figure 1. The section sizes of LTU-rib.

The simplified finite element model of hot roll-forming for LTU-rib is shown in figure 2. The two forming passes are selected. The first pass is mainly rough forming, consisting of upper roller and side rollers, and the second pass is finish forming, consisting of two side rollers. In the figure, the first group of flat rollers play a guiding role in order to ensure that the raw material enter the forming path steadily. Roller element types are selected as analytic rigid body. According to symmetry, 1/2 symmetry model is adopted. U-rib element is C3D8R. The mesh refinement at the forming end and the corner area at the bottom are refined appropriately. To ensure the U-rib bitten, bevel treatment is carried out at the feeding end. The friction coefficient between rollers and raw material is 0.35[6, 7]. And the forming speed is 6.0m/min.

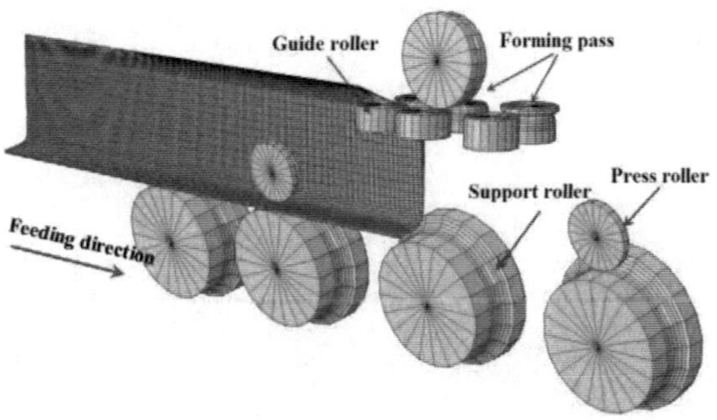

Figure 2. Finite element model of hot roll-forming process.

2.2. Material constitutive law

The investigated material, Q345qD, is a steel widely used in bridge engineering. Many research scholars [8-11]systematically studied the performance, high temperature thermal deformation constitutive model and dynamic recrystallization diagram. According to the production rate of U rib, the suitable deformation rate and temperature are selected.

Combining with the true stress-strain curve given in the reference [8], the hyperbolic sinusoidal function Arrhenius model is selected to characterize the Q345 thermal deformation constitutive model, and the relevant parameters are obtained by curve fitting. The constitutive equation of flow stress, deformation rate and deformation temperature of Q345 steel during high-temperature deformation is as follows:

$$Z=\dot{\varepsilon}\exp\left(\frac{450.1\times10^{3}}{8.314T}\right)=5.7\times10^{15}\left[\sinh\left(0.015\sigma\right)\right]^{5.71} \tag{1}$$

3. Simulation results and discussion

3.1. Effect of pass design on hot rolling process

In the overall pass design, two kinds of pass design schemes are given. The first one is direct forming method, i.e. the cross-sectional shape of two pass is similar with cross-sectional size of LTU-rib product. The second scheme is upsetting to thick type firstly, and then precision forming. The upsetting deformation is more in line with the metal flow law, and it is not easy to appear unstable behavior. The specific pass design， metal flow law and LTU-rib section size are shown in figures3 to 6.

Figure 3 shows the pass design of the first scheme. It can be seen that there is a sloped surface between the roller and sheet material, which is also the reason for instability in figure 4. However, the LTU-rib obtained by this scheme is close to the ideal product. The disadvantage is that the fillet radius of corner position is larger than the ideal size. Considering the subsequent welding, this value has a small impact. The black line range in figure is the strain area. The first pass satisfies the radius of the bottom part, and the second pass determines the radius of the upper part. The whole solution can meet the roll-forming technology of LTU-rib.

Figure 5 shows the pass design of the second scheme. It can be seen that the roller is symmetrical and the upper end surface is plane, which also makes no instability phenomenon. The size obtained is ideal. But there is a significant difference in the second pass. On the lateral displacement of U1, the displacement value is changed from 3.74mm in first scheme to 5.36mm in second scheme, as shown in figure 6. It is clear that the ideal sharp angle can be obtained by the second scheme, but it is found that the downward flow distance of metal is insufficient, which ultimately leads to a great difference between the connecting arc and the ideal value.

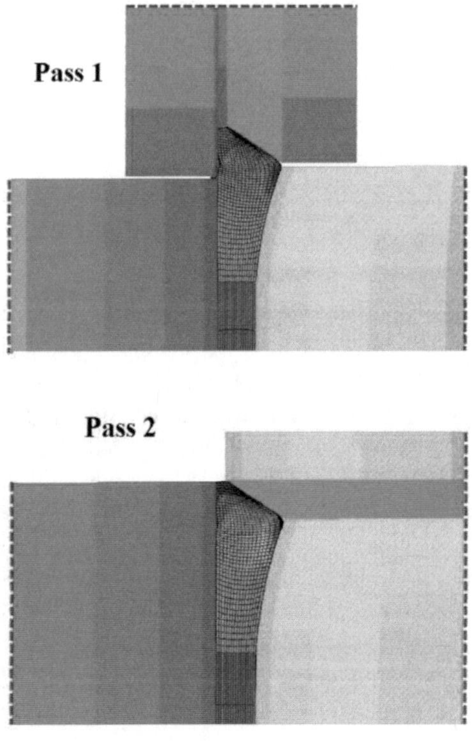

Figure 3. The pass design of the first scheme.

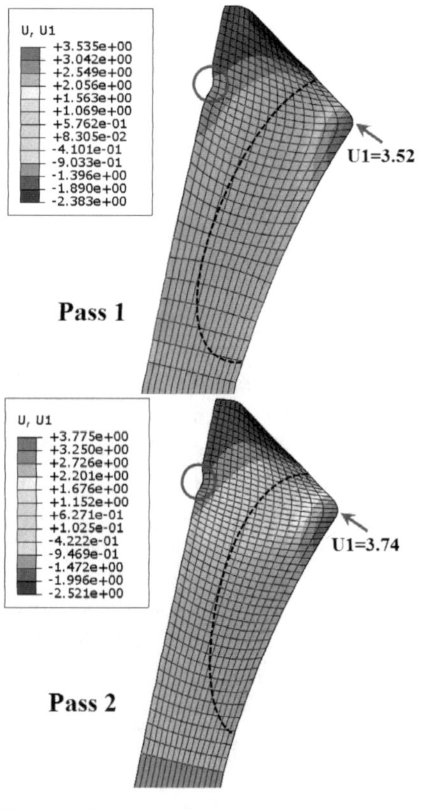

Figure 4. Metal flow at end of LTU-rib in first scheme.

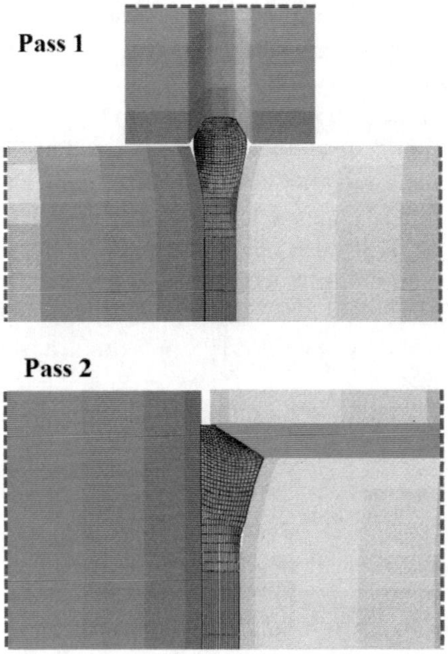

Figure 5. The pass design of the second scheme.

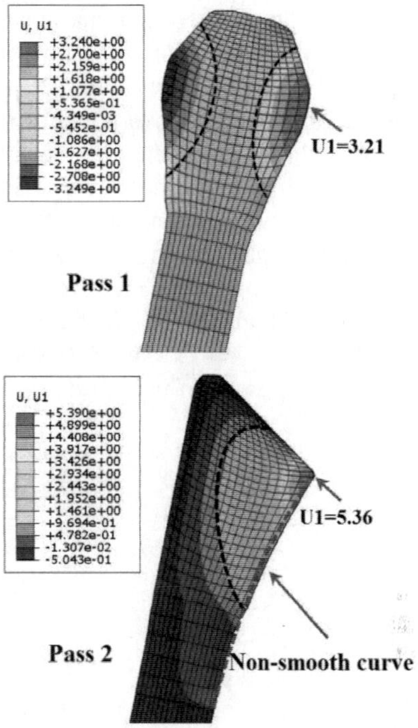

Figure 6. Metal flow at end of LTU-rib in second scheme.

In the above two design schemes, the rolling force and torque are similar because the same reduction amount is given at first pass. But the metal forming mechanism is different, which makes the stress of second pass different. Figure 7 shows the difference of the torque values of the secondary flat roller and profile roller. It can be seen that the values of flat roller and profile roller in second scheme are larger than those in first scheme. And the values of flat roller are larger than that of profile roller. This is because the large strain exists in second forming pass, while the flat roller and the profile roller in first scheme are similar in magnitude, about 508N•m.

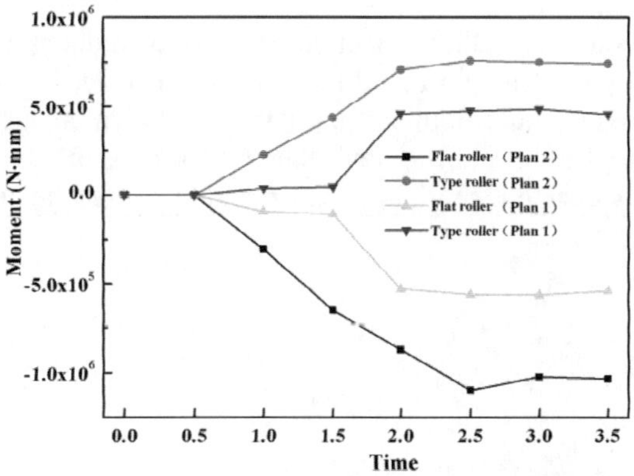

Figure 7. Roller torque of different process schemes.

3.2. Effect of temperature on hot rolling process

Temperature is one of the important technical parameters in hot roll-forming process, which is related to the forming equipment, rolling force and microstructure. According to critical transition temperature Ac1 and Ac3 of Q345qD sheet, forming temperatures of 700°C, 900°C and 1000°C are selected. And the forming at room temperature 25°C is taken as a comparison.

The first scheme is selected as forming pass design. Figure 8 gives the changing of rolling force upper roller at different temperatures. As shown in the figure, the variation trend of rolling force at different temperatures is similar. In the biting zone, the rolling force increases rapidly. The rolling force decreases slightly and then stabilizes into the stable zone. However, the forming temperature has a significant effect on the rolling force. As temperatures rise, the rolling force decreases sharply. When forming at room temperature, the maximum rolling force is 282.9kN. The maximum rolling force is only 110.9kN at 1000°C. The roll forming force is reduced by nearly 61%, which means that temperature has a great influence on selection of forming power.

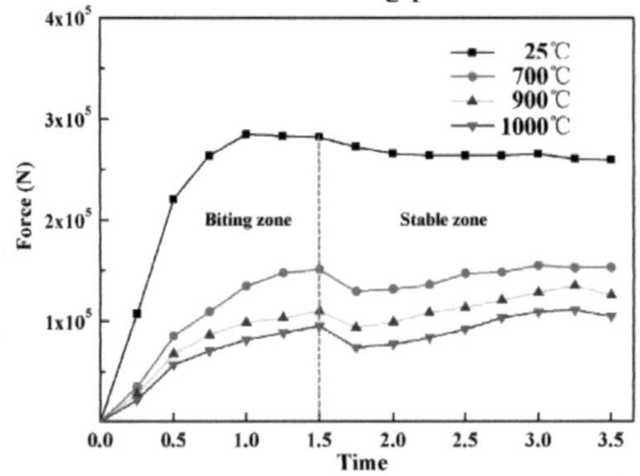

Figure 8. Rolling force at different temperatures.

Figure 9 is the section size and stress distribution of LTU-rib after roll-forming at room temperature. It can be seen that the cross-section size is different from the target product size. And the sharp corner is not formed, which will affect the welding penetration between U-rib and bridge deck, thereby affecting the welding quality. If the product size requirements are to be met, more forming passes and greater rolling force may be required. In addition, it is found that the residual stress of LTU-rib formed at room temperature is high, which will greatly increase the problem of cracking due to stress concentration after welding.

The simulation results show that the temperature not only reduces the forming force, but also improves the end metal fluidity. The end metal flows backwards due to the forming resistance. With the roll forming process, the metal accumulates gradually, as shown in figure 10 (a). The continuously accumulated metal is released at the trailing end, thereby forming the deformed tail shown in figure 10(b), which has a great influence on the yield of sheet, and increases the cutting process and cost.

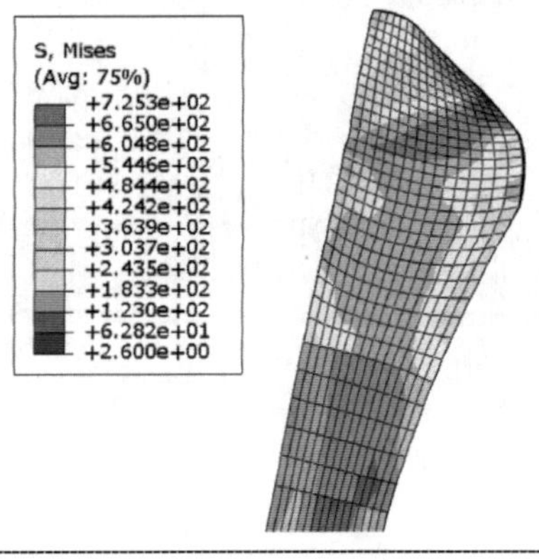

Figure 9. Stress distribution of LTU-rib after forming at room temperature.

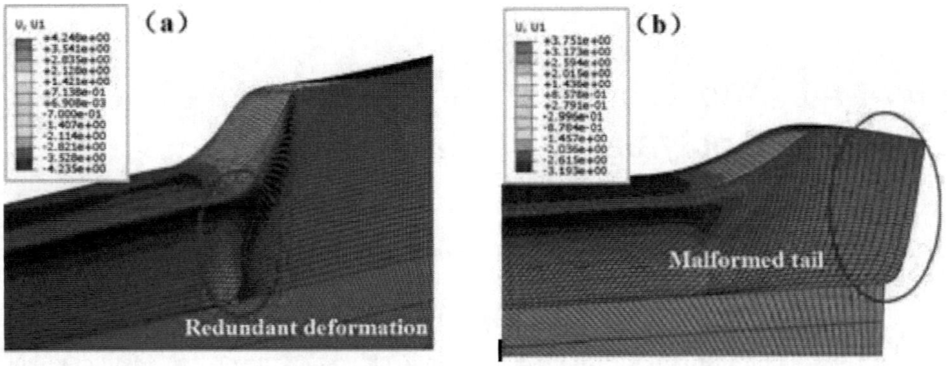

Figure 10. Redundant deformation of LTU-rib at 1000°C.

4. Conclusion

In the simulation process of LTU-rib, two kinds of passdesign schemes are given. The two schemes are compared from metal flow law, section size and roll force characteristics. In first scheme, there is slight instability, and the precision of corner is poor, but rolling torque is small at second pass. Although the ideal corner can be obtained in second scheme, the deformation is concentrated at the upper end and the deformation area is insufficient.

The forming temperature has a significant effect on the rolling force of the rollers. The maximum rolling force is only 110.9kN when the roll-forming temperature is 1000°C. And the rolling force decreases by nearly 61% compared with the value at room temperature. However, redundant deformation occurs if the forming temperature is too high.

Overall, the simulation results confirmed hot roll-forming process is more suitable for LTU-rib of bridge and provided a reference for the parameters design of actual hot roll-forming process.

5. References

[1] XanthakosP P 1994*Theory and Design of Bridges* (New York)
[2] Battista R C, Pfeil M S andCarvalho E M L 2008*J. Constr. Steel Res.***64** 134
[3] Oh C, Hong K andBae D 2011*Int. J. Steel Struc.***11**227
[4] Lv PM andLi DT 2013*J. Zhengzhou Uni. (Eng. Sci.)***34**89
[5] Ren ZR 2014 a new type U-rib *Pat.* CN203977292 U
[6] Tian XW, Zhang YSh and Wang Zh 2011*Adv. Mat. Res.***421**147
[7] Tian XW, Zhang YSh andLi J 2012 *Tri. Let.***45** 3
[8] Zhang XZH andLiu JSh 2013*For. Sta. Tech.***04** 148
[9] Xu YB, Wang XG andFan M 2009 *Mec. Eng. Aut.***05** 84
[10] Ma B, Peng Y and, Liu YF 2010 *T. Mater. Heat Treat.***04** 141
[11] Chen JL, Li ZhX andShu WY2015 *J. Sou. Uni. (E. E.)***06** 1145

Effects of the off-axis layer on the tensile failure of carbon fiber reinforced polymer $[0/\theta]_{ns}$ laminates

Xi Deng[1], Wen-Xue Wang[2] and Terutake Matsubara[2]

[1]Department of Aeronautics and Astronautics, Engineering School, Kyushu University, Motooka Nishi-ku Fukuoka 819-0395, Japan

[2]Research Institute for Applied Mechanics, Kyushu University, Kasuga, Fukuoka 816-8580, Japan

Abstract. Off-axis plies have a great influence on the tensile failure stress of carbon fiber reinforced polymer (CFRP) multidirectional laminate. Matrix cracks in the off-axis plies will cause negative effect on the layers, leading to fatal failure of the laminate. A microscale cohesive zone model is proposed to explore the influence of matrix cracks in the off-axis plies on the tensile failure of $[0/\theta]_{ns}$ laminates. Simulations of tensile failure of $[0/\theta]_s$ laminate were performed by varying the off-axis ply angle θ ranging from 15° to 90° and the cohesive properties to investigate the effects of off-axis plies to the failure of the laminates. Tensile tests of [0/θ]2s laminates with various off-axis ply angles were conducted to validate the simulations of microscale model. The simulation results with matrix cracking are in high agreement with the experimental results, illustrating the influence of off-axis plies on the tensile failure of $[0/\theta]_{ns}$ laminates.

1. Introduction

Carbon fiber reinforced polymer (CFRP) multi-directional laminates are widely used in recent years. The off-axis layers play an important role in the CFRP multidirectional laminates to improve the in-plane shear properties although the final failure of a CFRP multidirectional laminate under tension is always controlled by the failure of 0° layers which are parallel to the load direction. However, the matrix crack which easily occurs in the off-axis layers and propagates parallel to the fibers is the most commonly observed damage in the relatively low tensile load [1-2]. This damage usually causes negative effects on the laminate properties, and even leads to laminate failure indirectly. Therefore, the investigation of the effects of the off-axis layer with matrix cracking on the laminate failure is helpful to analyze the strength of the whole laminate.

According to previous studies, matrix cracks are largely observed in off-axial layers. The most commonly observed cracks always take place in 90° plies normal to uniaxial loading. To explain the failure mechanics, various failure models have been proposed to illustrate the matrix cracking generating on the total failure. Tohgo [3] used a 2-D simulation to investigate ply-cracking damage for cross-ply laminate. The results showed that transverse cracking damage in 90° plies was first propagated. Li et al. [4] proposed a damage representation for cracked laminates with matrix cracking in the direction parallel to the fibers, in which a damage growth law was also combined.

In this study, aiming at investigating the effects of the off-axis layer on the tensile strength and failure behavior of $[0/\theta]_{ns}$ laminates, numerical simulation of $[0/\theta]_s$ (n=1) using finite element method and tensile tests of $[0/\theta]_{2s}$ (n=2) were conducted. A three-dimensional microscale model is proposed, in which the transverse matrix cracking is modeled by cohesive zone model by using cohesive interface elements. Different cohesive energy release rate (G_c) values are used to reveal the effect of the transverse cracking resistance of the matrix on the failure of the laminate. Moreover, tensile tests of $[0/\theta]_{2s}$ laminates with various off-axis layer angles were performed. The failure processes of laminates with various off-axis layer angles were investigated in situ by using a digital microscope (Keyence VHX-2000) and a long distance high performance zoom lens (Keyence VH-Z50L).

2. Numerical simulation

In the numerical simulation, for the sake of simplicity, a $[0/\theta]_s$ laminate was modeled. A common geometry of a $[0/\theta]_s$ laminate is depicted in Figure 1. As is shown in Figure 1, the matrix cracks always occurred in the off-axis plies and extended along the fiber direction.

According to the symmetry of the stacking sequence of the $[0/\theta]_s$ laminate, only half of the laminate is needed to be simulated. Meanwhile, based on the periodicity of the microscale structures of the $[0/\theta]_s$ laminate, the whole laminate can be simulated by a representative unit cell (RUC) at a microscale level [6-8], as seen in Figure 1. This microscale RUC is generated using a rhombohedral unit cell, and it is located at the internal region of the laminate. The initiation and extension of a matrix crack is modeled at the center of the off-axis layer in the RUC by using the cohesive interface element. Since the matrix crack plane is parallel to the fiber direction in the off-axis layer, two boundary surfaces of the RUC that carry tensile load are also taken to be parallel to the matrix crack plane so that periodic boundary conditions are easily applied to the RUC. In this way, this RUC is useful for investigating the influence of off-axis layer.

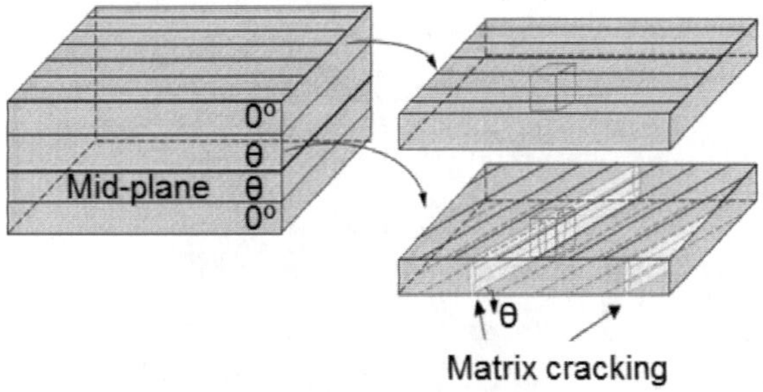

Figure 1. Sketch of $[0/\theta]_s$ laminate with matrix cracking.

The periodic boundary conditions must enforce the stress and displacement continuity between adjacent RUCs, and create the desired stress state in the RUC. In this way, the microscale model with the periodic boundaries is under same load condition as the whole laminate. For this reason, the periodic boundary conditions request the displacements of pair nodes in the corresponding surface should satisfy the relations of the displacements as follows [5]:

$$u_i^+ - u_i^- = \delta_i^* \quad i = x, y, z \tag{1}$$

where u_i^+, u_i^- denote the displacement components of pair nodes in the i-direction coordinate axis, δ_i^* means the stretch in the i-direction between the pair nodes on the corresponding surfaces. Applying the periodic boundary conditions to the RUC, the components of macroscopic tensile stress versus the tensile strain in this RUC can be calculated as follows:

$$\overline{\sigma_{ij}} = \frac{1}{V}\int \sigma_{ij}\, dV \tag{2}$$

where V is the volume of the microscale model.

The present finite element simulation was conducted by using MSC.Marc 2011. A three dimensional microscale model is shown in Figure 2. A single layer is 0.125 mm in thickness. The length of the microscale model varies with the off-axis angle θ, ranging from 0.618 mm to 0.4 mm for θ = 15° ~ 90°

to ensure the periodicity of the microscale model which includes a whole matrix crack. The width of the microscale model is 0.16 mm. In this microscale model, matrix, fibers and the cohesive elements are simulated separately and are described in different colors, respectively. One layer of the laminate includes three arrays of carbon fiber bundle, which is utilized to approximately simulate a single layer with fiber volume fraction of 60% according to the prepreg used in the fabrication of the laminates for the tensile tests. High order 10-nodes tetrahedral solid elements are adopted to make it better for simulating the plasticity of matrix and the failure of fibers. 6-nodes pentahedral cohesive elements are chosen to mimic the matrix crack initiation and propagation. In order to strike a balance between computational cost and simulation accuracy, the fibers in the laminate are assumed to be arranged in a square array and that the matrix cracks are uniformly spaced.

A general estimation of the total number of elements in such a microscale model is approximately 110000. In this microscale RUC model, the mesh of the microscale model is carefully constructed that the nodes on the pair corresponding surface are arranged with the same distribution. This mesh is used to ensure the application of periodic boundary conditions. Furthermore, the fiber and matrix nodes are also constructed to be coincident at the fiber-matrix interface, since this manner can greatly reduce numerical iteration cost in contact analysis along the surface. The possible debonding between fibers and matrix was not considered in the present simulation. Hence, the matrix cracking in the off-axis layer and fiber fracture are the main damages which cause the laminate failure.

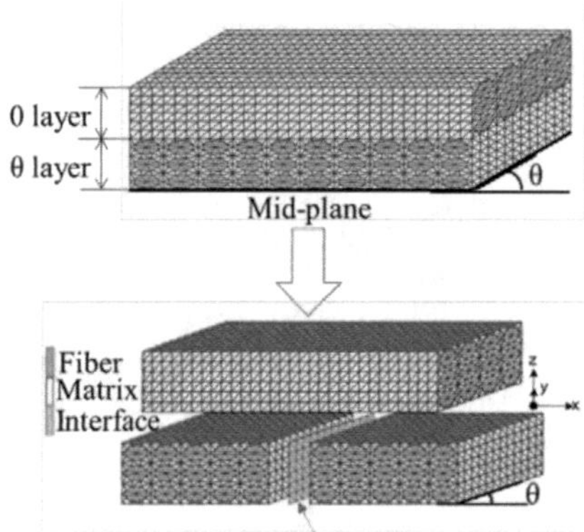

Figure 2. Description of the $[0/\theta]_s$ microscale model (in this figure, $\theta = 45°$).

In the MSC.Marc, one of the ways to implement the periodic boundary conditions is to use the option "servo-link". This option allows to prescribe multi-point boundary conditions for nodal displacements expressed as a linear function with constant coefficients. The relation is described as follows [9]:

$$\mathrm{u} = a_1 \mathrm{u}_1 + a_2 \mathrm{u}_2 + \cdots \quad (3)$$

where u sis a degree of freedom to be constrained; u_i $(i = 1, 2, \cdots)$ are the other retained degrees of freedom in this structure; a_i $(i = 1, 2, \cdots)$ are constants provided in this option. With this option, the periodic boundary condition is easily applied to the microscale model.

The cohesive interface elements are inserted into the matrix region where the matrix crack may initiate. In principle, the matrix crack can be positioned anywhere within the matrix region in the microscale model. In this simulation, the crack position is located at the center of the microscale model, as shown in Figure 2. These cohesive interface elements simulate the onset and propagation of the matrix crack based on the cohesive zone model (CZM). The cohesive zone elements do not need to represent any physical material so that it can be modeled with an initial zero thickness, just as Figure 2 depicted. The properties of cohesive elements are defined by a bilinear curve, as is described in following Figure 3. The relations among the maximum critical traction T_c, the energy release rate G_c, the maximum opening displacement δ_m, the critical opening displacementδ_c, and the initial stiffness K satisfy the following equations [9-10]:

$$\delta_m = 2G_c/\delta_c \tag{4}$$

$$\delta_c = T_c/K \tag{5}$$

Through the simulation of the cohesive elements, the matrix crack initiates if the cohesive energy reaches the critical energy release rate. After then the cohesive traction reduces, and as the traction reduced to zero, the cohesive zone disappears and the matrix crack occurs.

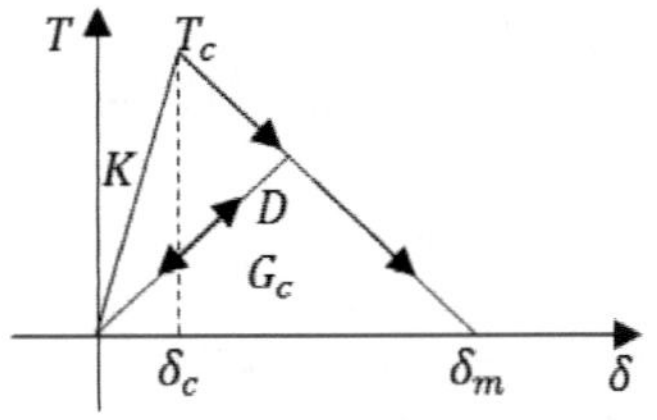

Figure 3. Bilinear curve of traction-displacement relation.

The material properties of the microscale model are defined separately. The matrix is considered as an elastic-plastic isotropic material, carbon fiber is defined as a transverse isotropic material, and the interface element is of zero-thickness, respectively. The material constants of matrix and fibers are listed in Table 1 and the constants of the cohesive interface element are listed in Table 2. To estimate the failure of the carbon fiber, the maximum stress criteria of the carbon fiber material are also listed in Table 1. The elastic-plastic property of the matrix is prescribed through a power law as follows [11]:

$$\sigma_m = A\left(\varepsilon_m^p\right)^r + \sigma_y \tag{6}$$

where A = 256, r = 0.259, σ_y = 30 (MPa) are employed in this simulation to approximately describe the relation between the equivalent plastic strain and the stress.

Table 1. Constants of materials

Constants	Carbon fiber	Matrix
E_1 (GPa)	235	3.3
E_2, E_3 (GPa)	13	
μ_{12}, μ_{13}	0.2	0.38
μ_{23}	0.3	
G_{12}, G_{13} (GPa)	15	1.2
G_{23} (GPa)	5	
σ_{t11} (MPa)	4936	
σ_{c11} (MPa)	2820	
τ_{12} (MPa)	700	

Table 2. Constants of cohesive elements

Constants	Value
Cohesive energy G_c (N/mm)	0.001~0.4
Critical opening displacement δ_c(mm)	$8*10^{-7}$
Maximum opening displacement δ_m (mm)	0.01
Stiffness of cohesive element (N/mm^3)	10^8

In order to investigate the effect of the off-axis layers on the failure of the laminate, microscale models with off-axis angle θ ranging from 15°~90° are simulated, respectively. The effect of off-axis layers on the global stress can be clarified easily by analyzing the failure stress with the variation of off-axis angle θ. Furthermore, more simulations were conducted with various cohesive energy release rate G_c to investigate the effect of fracture toughness of the matrix on the failure strength of the laminate.

Several cases of initially complete crack, crack-free, and those between these two ideal situations by altering cohesive energy release rate G_c of cohesive elements are simulated.

3. Tensile tests

Tensile tests were carried out to validate the availability of the proposed microscale model. However, the $[0/\theta]_s$ laminate is too thin that it is hard to handle during the tensile test because a small scratch may lead to large decrease on the failure strength. Therefore, $[0/\theta]_{2s}$ laminates are fabricated for the tensile tests. The off-axis angle of θ were chosen as 15°, 30°, 45°, 60°, 75° and 90°, respectively. T700S/2500 carbon fiber/epoxy prepreg (ply thickness $t_0 = 0.125$ mm) was used to fabricate the laminates. The carbon fiber volume fraction of the prepreg was 60%. Each laminate was cured at 135°C for 2h by autoclave. The size of laminates was 250mm × 250mm × 1 mm to ensure reliability of the experimental results, 9 specimens were fabricated for each $[0/\theta]_{2s}$ laminate. The specimen is 240 mm in length and 25 mm in width, respectively, as described in Figure 4. The end tabs have a thickness of 2 mm, length of 55 mm, and a slope of 5 mm.

In order to observe damage events such as matrix crack initiation, propagation and fiber breaking at the free edge of the specimen, one side section of each specimen is firstly grinded through 600#, 1500#, 2000#, 2500#, and 4000# sandpaper, and is secondly polished by 1 μm and then 0.06 μm Al_2O_3 powder. The surface failure processes of each specimen were monitored in situ using a digital microscope (Keyence VHX-2000) and a long distance high performance zoom lens (Keyence VH-Z50L) during the tensile test. In particular, at the three load levels of 30%, 50% and 80% of the average tensile strength of the laminate, the load state was held on and the detail surface damage modes were recorded by enlarged photographs. Matrix crack density was investigated based on the observation data. Tensile tests were conducted by using a MTS 810 materials testing system with 1 mm/ min. crosshead rate.

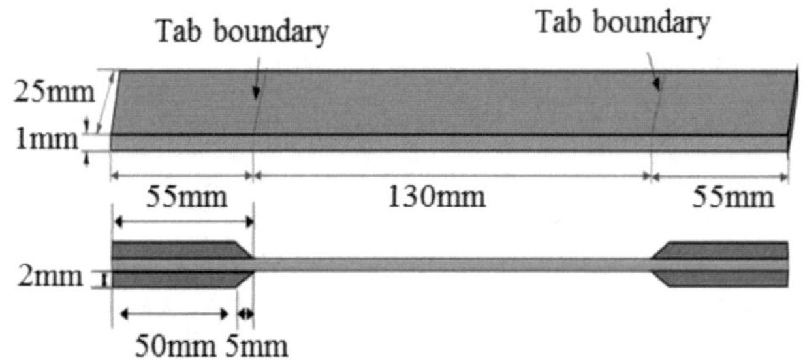

Figure 4. The size of the specimen in tensile tests.

4. Results and discussions

4.1. Numerical results

The variation of tensile failure strength of the laminate with the off-axis layer angle is presented in Figure 5 with the energy release rate of the matrix as parameter. The crack-free case indicates the microscale model without crack. In this case, no cohesive interface elements were inserted in the microscale model. With the off-axis angle increasing from 15° ~ 90°, the tensile failure stress decreases rapidly when the off-axis angle θ is smaller than 45°. After that, the tensile failure stress has relative small variation with the increase of the off-axis layer angle. A minimum value seems to exist in the range of θ = 60° ~ 75°. A little increase trend is observed conversely as the θ increases to 90°. In the cases that adding the cohesive interface elements to simulate the matrix cracking, the variation of the tensile failure stress versus θ demonstrates similar tendency, except that the values of tensile failure stress of the laminate with matrix cracking is always smaller than that without matrix cracking. These results indicate that the off-axis angle has a significant effect on the failure of the $[0/\theta]_s$ laminate. The case that $G_c = 0.001\, N/\mathrm{mm}$ approximately simulate that the matrix crack exists initially in the microscale model.

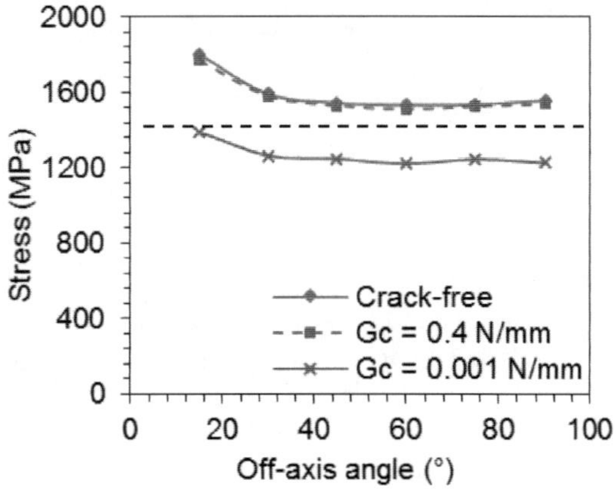

Figure 5. Relations of failure stress to off-axis angle.

The horizontal dash line plotted at the stress 1390 MPa represents the failure stress carried by only the 0° plies if the off-axis plies are assumed to be failed and cannot carry any load. Comparing this reference failure stress with those of failure stress of laminates with or without matrix cracking, it is observed that the off-axis layer always plays positive role if there is no crack in the off-axis layers. The tensile strength of the laminate with off-axis layer angle smaller than 45° is many higher than the reference failure stress. The tensile strength of the laminate with off-axis layer angle in the range of 45° ~ 90° is slightly higher than the reference failure stress. In the cases of small off-axis layer angle and high energy release rate, the off-axis layer still plays positive role and the tensile failure stress is still higher than the reference failure stress. However, in the cases of large off-axis layer angle and low energy release rate, the off-axis layer usually plays negative role and the tensile strength of the laminate is many lower than the reference failure stress. From these results, it is realized that off-axis layer angle and fracture toughness of the matrix have significant influence on the tensile strength of the multidirectional laminates.

4.2. Experimental results

The experimental results are plotted in Figure 6 ~ 8. The variation of averaged tensile strength of the laminate with the off-axis layer angle is presented in Figure 6. The tensile failure stress illustrated similar tendency with the numerical simulation results. The region above the horizon line means that the off-axis layer in laminate has positive effect on the tensile failure strength of the laminate. However, the off-axis layer show negative influence on the tensile strength of the laminate when the off-axis angle θ is in the range of 60° ~ 75°. Comparing the experimental curve with those simulation curves shown in Figure 5, the experimental curve agreed well with the case that $G_c = 0.4\, N/\mathrm{mm}$, as is shown in Figure 6. The difference between two methods were due to the inexact assumptions and parameters in simulation. Therefore, to accurately predict the tensile strength of a laminate, correct energy release rate value of the matrix is necessary.

The typical in situ observation results of damage images on the free edge of each type laminate at the load level of 80% are described in Figure 7. Fiber breakage and matrix cracks were observed by the microscope. For the tensile load level of 30% in the tensile process, there was no damage found on the observing free edge of all the $[0/\theta]_{2s}$ laminates. As tensile load increased continuously to the tensile load level of 50%, fiber breakage initiated first in the 0° layers. The dashed open arrows plotted in Figure 7 referred to the fiber breakages in the laminates. On the other hand, there was still no matrix crack initiated in off-axis layers in this period.

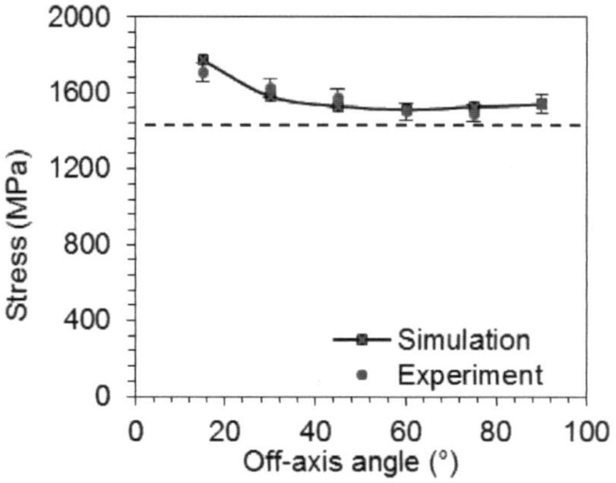

Figure 6. Experimental results of failure stress to off-axis angle. Dashed line indicates the assumed stress that off-axis layer cannot carry any load.

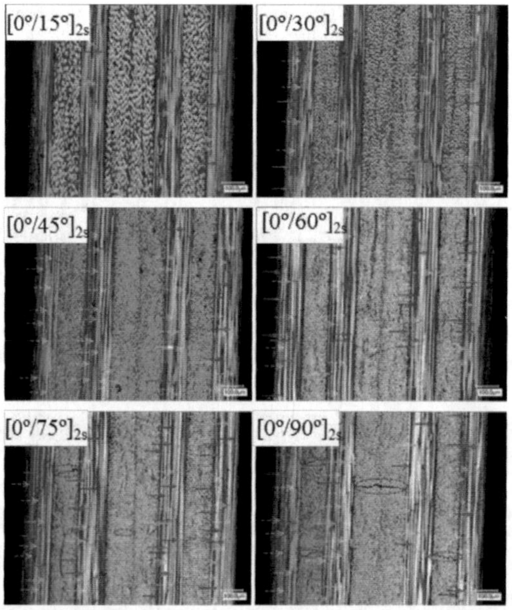

Figure 7. Failure modes of laminates under tension. Long dashed open arrows refer to the fiber breakages before 50% load, the solid arrows refer to new fiber breakages under 80% load, and the dashed circles marked the matrix cracks.

When the tensile load reached the level of 80%, new fiber breakages initiated intensively at the positions among the existing damages, accompanied with the propagation of previous existing fiber breakages. Solid arrows portrayed in Figure 7 pointed to the new fiber breakages developed at the load level of 80%. The number of the fiber breakage in 0° layers was almost saturated at this tensile load level. Meanwhile, matrix cracks were observed in two types of layups barely, which are the $[0/75°]_{2s}$ and $[0/90°]_{2s}$ laminates. The matrix cracks are signed by dashed circles in Figure 7. As this result, it indicated that the matrix cracks merely occurred in the laminate that the off-axis angle was larger enough.

According to the experimental results of all the $[0/\theta]_{2s}$ laminates, for the laminates with off-axis angle θ lower than 75°, the matrix crack did not occur in the specimens until the final tensile failure. For the cases that the off-axis angle θ in $[0/\theta]_{2s}$ laminate reached 75°, as the off-axis angle increased, the matrix crack initiated under a smaller strain. Meanwhile, the matrix crack initiated and propagated more intensive in the $[0/\theta]_{2s}$ laminate with larger off-axis angle. The relations of matrix crack density in central off-axis layers to tensile strain are depicted in Figure 8. In addition, there is no interfacial delamination in all the specimens during the tensile tests, hence the effect of delamination is negligible.

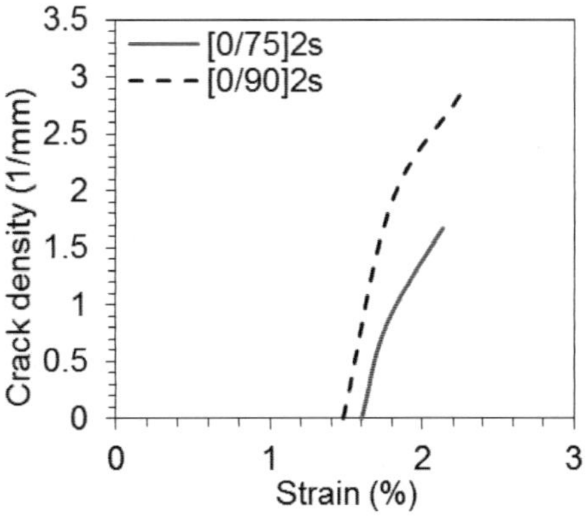

Figure 8. Relations of matrix crack density to failure strain.

5. Conclusions

The effect of off-axis layer on the failure of CFRP $[0/\theta]_{ns}$ laminates is investigated by finite element analysis and tensile tests of $[0/\theta]_{2s}$ laminates. Based on the present numerical and experimental results, the following conclusions are obtained.

1. The variation of off-axis angle leads to a reduction on the failure stress for most conditions. As the off-axial angle increases to 75°, the failure stress keeps decreasing directly. When θ>75°, A little increase is observed conversely as the θ increases to 90°.
2. Lower G_C of the matrix leads to low tensile failure stress. Comparing the numerical results to the reference stress, revealing the effect of the off-axis layer on the failure of the laminates intuitively.
3. Series of $[0/\theta]_{2s}$ specimens(15°≤θ≤90°) were tested to validate the simulations. Matrix cracks and fiber breakages were observed to analyze the failure modes of laminates. Matrix cracks initiated merely in the laminate with large off-axis angle. The matrix crack density was higher in the laminate with large off-axis angle as well.

6. References

[1] Nairn J A 2001 *Polym. Matrix Compos.* **2** pp 1-33.
[2] Abrate S 1991 *Compos Eng.* **1** pp 337–53
[3] Tohgo K, Sugiyama Y and Akizuki K 2001 *JSME Int. J. Ser. A.* **44** pp 282-90.
[4] Li S, Reid S R, Soden P D 1998 *Phil. Trans. R. Soc. Lond. A.* **356** pp 2379-412.
[5] Drach A, Drach B and Tsukrov I 2014 *Adv. Eng. Softw.* **72** pp 18-27.
[6] Romanowicz M 2016 *Compos Part B.* **90** pp 45-57.
[7] Tran T, Kelly D, Prusty B G, Pearce G and Gosse J 2013 *Compos. Struct.* **103** pp 1-8.
[8] Ellyina F, Zhang YF and Xia ZH 2011 *Procedia Eng.* **10** pp 63-8.
[9] MSC Software. *MSC Marc 2011 User Documentation, Volume A : Theory and User Information* (2011).
[10] Vandellos T, Huchette C and Carrère N 2013 *Compos. Struct.* **105** pp 199-206.
[11] Sun C T and Chen J L 1991 *Compos. Sci. Technol.* **40** pp 115-29.

Energy Output Efficiency of Shielded Mild Detonation Cord in Cylinder Structures

Lina Feng[1], **Dong Li** [2], **Jiandong Tian**[1], **Zhenxing Hu**[1] **and Jing Sun**[1]

[1] Beijing Institute of Astronautical Systems Engineering, 100076 Beijing, China
[2] China Academy of Launch Vehicle Technology, 100076 Beijing, China

Abstract. Expanding tube separation device is a commonly used linear separation device, whose energy resource is a shielded mild detonation cord (SMDC). In this research, the SMDC is a mild detonation cord sheathed in flat polyethylene. Two experiments were achieved to discuss the energy output efficiency of the SMDC. In the first experiment, the SMDC was inserted into 1060 aluminum cylinder directly. While in the other experiment, the SMDC was insert into aluminum cylinder after contained within a flat steel tube. The velocity and displacement of the characteristic points on the outer surface of the aluminum cylinder during explosion were measured by photonic Doppler velocimetry (PDV). The simulation results with Ls-dyna fluid-solid coupling algorithm correspond well with the experimental results. According to the simulation analysis, the distribution of total energy of explosives were obtained. The plastic strain energy of the aluminum cylinder corresponds to the ability of the SMDC to do external work. The result shows that, 39.3% of the total energy in the explosive is output to aluminum cylinder with SMDC acting on it directly, while the energy output efficiency turns to 20.8% after passing through the flat tube.

1. Introduction

Expanding tube separation device is a typical linear separation device. The function of the separating device is to supply reliably connection of the structures before separated. Moreover, when the separation is required, the specified part of the separating device is destroyed to achieve the separation of connecting structures. Therefore, it is a significant part for the whole system [1, 2]. The shielded mild detonation cord (SMDC) in this research is a mild detonation cord coated with polyethylene in flat ribbon shape, that is the energy resource of the device. During the working process, high energy of the SMDC is fast released. The energy of the explosive is transmitted to the flat tube through the polyethylene, then transmitted to external structure. To study the mechanism of this structure for a better guidance in actual structural design, it is necessary to analyse the energy output efficiency of this kind of SMDC.

However, the quantity of explosive energy cannot be measured directly. It is usually converted to other energy forms. Traditional explosive energy testing methods including lead-cast, blasting funnel, ballistic mortar, underwater explosion and detonation cylinder test [3-9], which convert the explosive energy to kinetic energy, plastic strain energy, or bubble energy. For theoretical analysis, classical Gurney formula is used to describe energy distribution of explosive acting on regular symmetrical structures, such as cylindrical specimens and plates [10]. However, for expanding tube assemblies, the flat structure changes the energy distribution of the explosive in the circumferential direction, which is more complicated. Until recently, various researches are presented on the transfer and distribution of explosive energy. Rao studied the explosive energy output characteristics of aluminum-containing explosives in water, confined space and air [11]. Feng studied the output characteristics of explosive energy in rocks, the energy of explosion shock waves accounts for about 40%, and the energy generated by explosive gases accounts for about 60% [12]. Zhao studied the characteristics and energy distribution of blast waves in concrete by experiments [13]. Cao used cylinder experiment to measure the work ability of the mild detonating fuse [14]. In most of these studies, the explosives are applied to the object directly. For SMDC, the energy of the explosive energy is transmitted outside through polyethylene, and the energy transmission efficiency is unclear.

In this paper, deformation cylinder test is used to measure the energy output efficiency of the SMDC in cylinder structures. The experiment is designed in the next section. Then simulation method is introduced in Section 3. Experimental and simulation results are compared and analysed in Section 4. The paper is then concluded in Section 5.

2. Experiment design

Considering 1060 aluminum has well plastic properties, it is selected as the material of the outer cylinder to absorb the output energy of the SMDC in this paper. The inner hole matches with the SMDC and the flat tube, respectively. The expansion velocity and displacement of the four feature points on the middle section of the cylinder were measured by a photonic Doppler velocimetry (PDV). The experimental device was composed of a bottom plate, a top cover and four screws. Four PDV probe brackets were mounted on the bottom plate. Adjusting the brackets, laser outputs from the fiber optic probes can be aligned with the measured points. The specimens were fixed in the middle of the device by clamps. The combination of experimental specimens includes two types. In condition 1, the SMDC is directly penetrated into the aluminum cylinder. In condition 2, the SMDC is penetrated into the aluminum cylinder after being presented in a flat steel tube. Experiment device and specimens are showed in Figure 1.

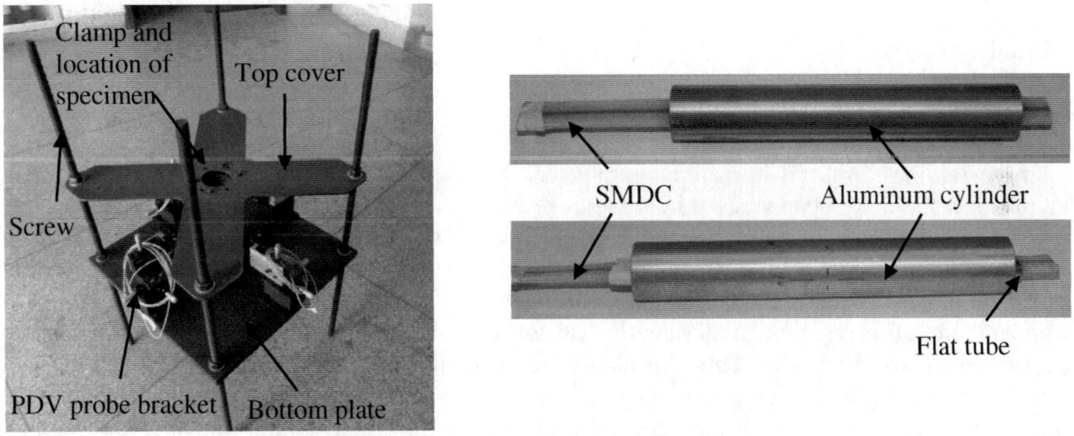

Figure 1. Experiment device and specimens.

3. Simulation method

3.1. Geometry and finite element models

The structure was modelled by Hypermesh. The 10 mm in the middle of the structure were selected as separate parts to avoid the influence from the boundaries in analysing energy output efficiency. The aluminum cylinder and the flat tube were meshed as lagrange elements, while the air, polyethylene, lead and explosive were euler elements. In the Ls-dyna fluid-solid coupling analysis, the lagrange and euler meshes can overlap each other, therefore, the air mesh and the aluminum cylinder mesh were overlapped. Two kinds of finite element mesh models are showed in Figure 2.

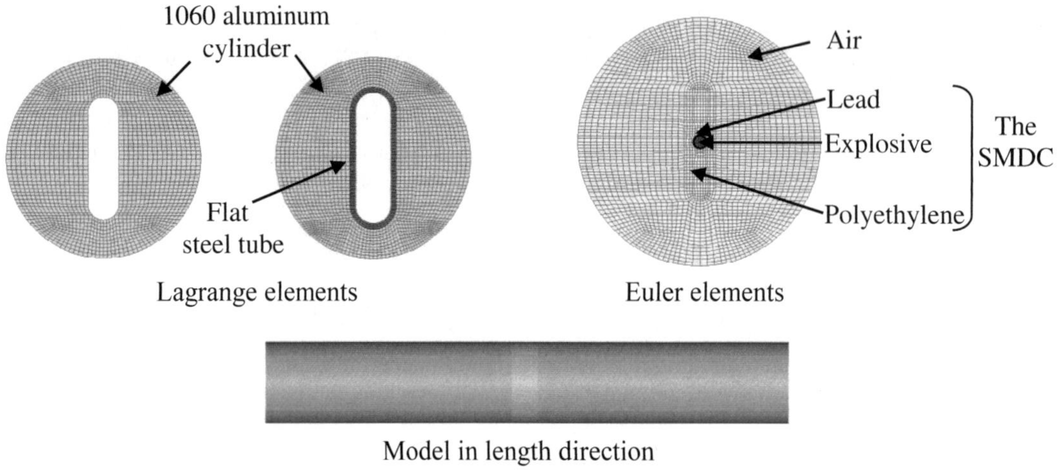

Figure 2. Models for the simulations.

3.2. Material models

The relevant material parameters of the simulation were referred to the existing studies [15-18]. The material model of the explosive (RDX) was Mat high explosive burn and the JWL equation of state was set, wherein the explosive velocity of the explosive was 7344 m/s to the average value provided by the explosive's manufacturer. The air and lead were set to the Mat null model. The lead used the Gruneisen equation of state, while the air was the gamma criterion equation. The polyethylene used the Mat piecewise linear plasticity material model, the flat steel tube and 1060 aluminum cylinder used the Mat power law plasticity material model as an elastoplastic model considering strain hardening.

4. Results analysis

4.1. Comparison of simulation and experimental results

In the simulations with condition 1 at first, the simulation results were somewhat different from the experimental ones. In terms of material parameters, parameters for explosive, lead, polyethylene and flat tube selected have been verified in previous researches, therefore, they are suitable for simulation analysis with explosive load [15-17]. Reference [18] indicates that 1060 aluminum is strain hardening and sensitive to strain rate and temperature, but the Mat power law plasticity material model for 1060 aluminum in the keywords of Ls-dyna only considered the strain hardening and strain rate effects. A large amount of heat released during the explosion loading process, which would change the temperature of the 1060 aluminum. Meanwhile, considering the different loading method or other uncertain factors from the reference which may cause deviation in material parameters. In the later simulation, a coefficient η was introduced to amend the constitutive equation of 1060 aluminum material. In the material parameter setting of the keywords, η was introduced by multiplying to the strength coefficient. A large number of simulation analysis and comparisons showed that when η is 0.686, the simulation results agree well with the experimental results. There were three experiments for each condition. Point 1 and 3 are symmetric points, so as point 2 and 4. The symmetric points results were consistent.

The experimental and simulation results are showed in Figure 3. The velocity of point 1 and 3 on the aluminum cylinder drops faster after reaching the extreme value, and increases in the reverse direction. While expand time of point 2 and 4 is much longer. After the velocity of point 2 and 4 drops to zero, it gradually stabilized after a small fluctuation. In the corresponding displacement curve, when the expansion displacement of point 1 and 3 reach the extreme value, the displacement of point 2 and 4 is still increasing. Due to the expansion traction of point 2 and 4, point 1 and 3 are contracted inward, which indicates that the SMDC can do more work in the short axis. After about 80 μs, the velocity of the four points began to slowly approach to the final stable value.

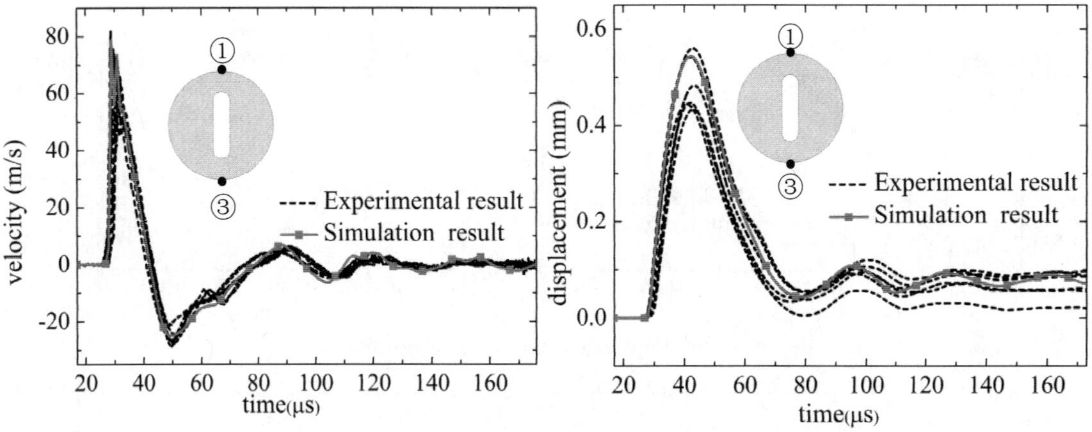

Velocity and displacement curves of point 1 and 3

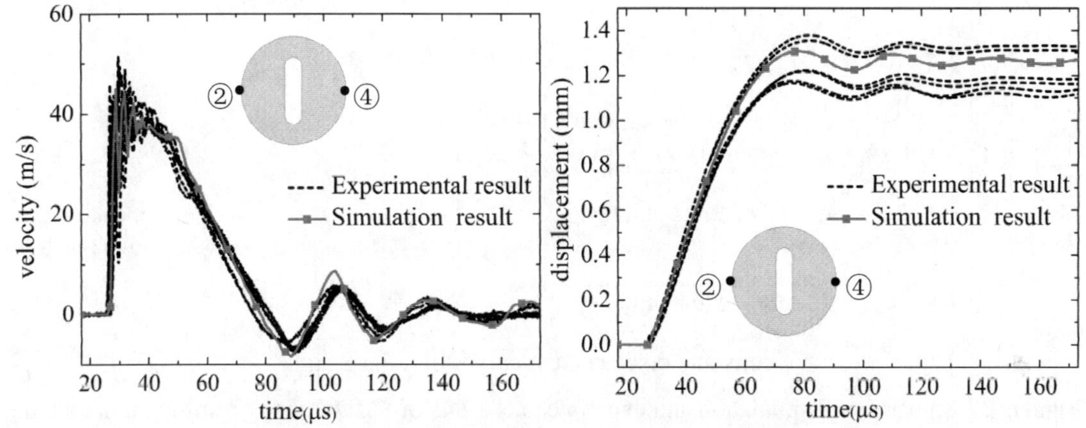

Velocity and displacement curves of point 2 and 4

Figure. 3 Comparison of simulation and experimental results of the SMDC acting on the aluminum cylinder directly.

Keeping all parameters unchanged, a model of the flat tube was added to obtain another simulation result. The experimental and simulation results are showed in Figure 4. Due to the addition of the flat tube, the amplitudes of velocity and displacement are reduced. Compared with the final expanding of point 1 and 3 in condition 1, they are contracted in condition 2. After the restriction of the flat tube, the energy was output more concentrated on the direction of short axis of the SMDC.

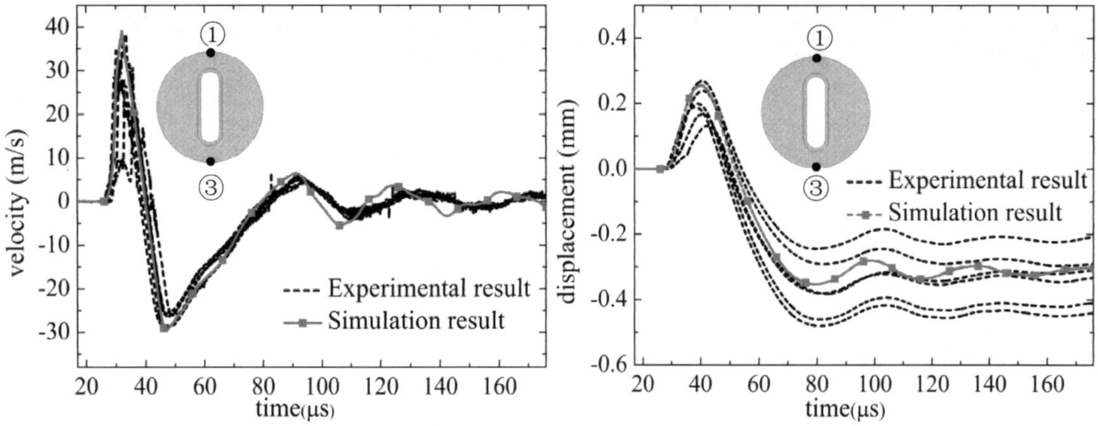

Velocity and displacement curves of point 1 and 3

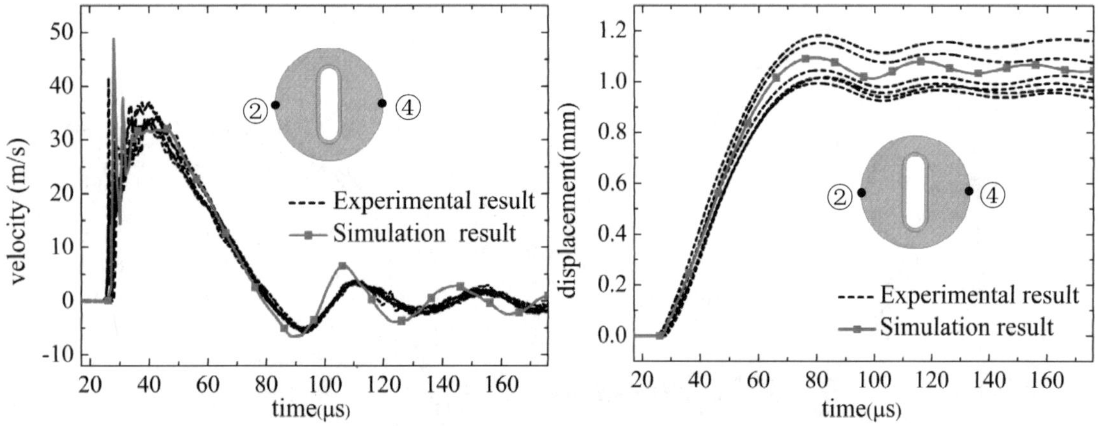

Velocity and displacement curves of point 2 and 4

Figure 4. Comparison of simulation and experimental results of SMDC acting on aluminum cylinder after passing through a flat tube.

The aluminum cylinders were cut at the middle section of the longitudinal direction after the experiment. The final states of the experiment and simulation results are showed in Figure 5. The inner holes are obviously enlarged, and a large local plastic deformation occurs. After the flat tube was applied, the amount of expansion was reduced.

(a) (b)

Figure 5. Sectional view of aluminium cylinders under two working conditions:

(a) condition 1, (b) condition 2.

4.2. Energy output results

The energy distribution process was obtained by simulation analysis. The energy history of each structure in the middle parts of the specimen were divided by the total energy of the part of the explosive to get the energy output efficiency, as shown in Figure 6.

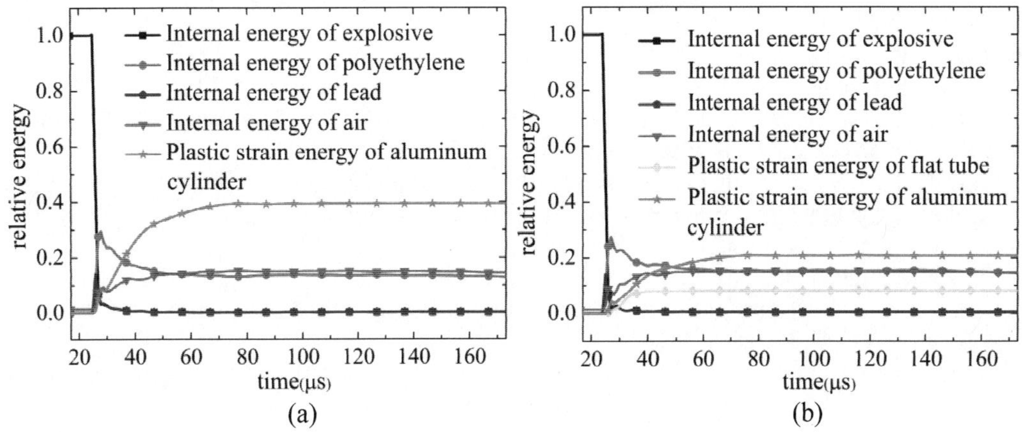

Figure 6. Energy histories of each structure under two conditions: (a) condition 1, (b) condition 2.

When the SMDC acts on the aluminium cylinder directly, the plastic stain energy of the aluminum cylinder accounts for 39.3% of the total energy of the explosive. In addition, the polyethylene and air accounts for 12.9% and 14.5% of the total energy, respectively. When the SMDC acts on the aluminum cylinder through the flat tube, the plastic stain energy of aluminum cylinder is 20.8%, the plastic stain energy of the flat tube is 8.1%, the polyethylene is 14.6%, and the air is 14.7%. For both conditions, the final energy of lead approach to zero. For condition 2, after the flat tube added, more interfaces were introduced, the explosion shock wave was attenuated. Besides, the flat tube consumed part of the energy, so the energy finally transmitted to the outer aluminum cylinder was reduced.

5. Conclusions

In this paper, the energy output efficiency of the SMDC is studied through experiments and simulation analysis. The velocity and displacement of the characteristic points on the outer surface of the aluminum cylinder during explosion were measured by PDV. And energy distribution was researched by simulations. The result shows that, 39.3% of the total energy in the explosive is output to aluminum cylinder with SMDC acting on it directly, while the energy output efficiency turns to 20.8% after passing through the flat tube. It can be seen from the deformation form of the aluminum cylinder that the SMDC does work in directionally way. That is, the energy is output more concentrated on the direction of short axis of the SMDC. This paper supplies a guidance for separation device design.

6. References

[1] Tao C, Longfei G, Zhen Z, et al 2017 Dynamic Model and Experimental Study of Low-shock Separation Device *Astronautical Systems Engineering Technology.* **1** 44-53
[2] Zhusheng L, Xiaojun Wang, Xuechang Z, et al 2012 Aerospace Firework Device *China Aerospace Publishing House.* 64-67
[3] Scilly N F 1995 Measurement of the Explosive Performance of High Explosives *J. Loss Prev. Process Ind.* **8** 265-73
[4] Matyas R and Selesovsky J 2009 Power of TATP Based Explosives *J. Hazard. Mater.* 165, 95-99
[5] Chunhai D 2003 Discussion of lead-cast Method to Test the Capability of Industry Explosive's Making Strength *Blasting.* **20** 120-21

[6] Wen L and Gang W 2016 Study on work ability of RDX/HMX-based aluminized explosive by cylinder test *China Science Paper*. **11** 551-53
[7] Mengju Z and Tongchang Y 1990 Property and Testing Technology of Explosives *Ordnance Industry Press*.20
[8] Siyou Z and Yikun Z 2009 Summary on Power Test Method of Industrial Explosives *Coal Mine Blasting*. **2** 22-26
[9] Zhaozhong W and Guozhong L 2009 Discussion on Standard Methods for Work Ability Testing of Industrial Explosives *National Defense Technology Foundation*. **8** 16-20
[10] Gurney R W 1943 The Initial Velocities of Fragments from Bombs, Shells and Grenades. Ballistic Research Laboratory *Report* (Aberdeen, Maryland) *No* 405
[11] Guoning R 2007 Study on the output characteristics of explosive energy and the role of blast wave and target. *Nanjing University of Science and Technology* (Nanjing)
[12] Cai F and Zegong L 2014 Energy distribution of columnar explosive blasting in rock *Electronic Journal of Geotechnical Engineering*. **19** 4221-31
[13] Jianping Z 2009 Experimental Study on Characteristics and Energy Distribution of Explosion Wave *BLASTING*. **26** 1-5
[14] Lei C, Yuliang L, Fangyun L, et al 2016 Study on energy output efficiency of mild detonating fuse in cylinder tube structure *Materials & Design*. **92** 690-00
[15] Jing S and Zhiguang Y 2006 Simulation and Optimization of Explosive Separation Process for Flat Tube Separator. *Proc. of the 9th National Conf. on Explosion and Safety* (Shenyang) 140-43
[16] Zhiguo M, Rong C, Jing S, et al 2017 Separation Impact Characteristics of Expansion Tube Separation Device. *Journal of Projectiles, Rockets, Missiles and Guidance*. **2** 99 -03
[17] Lina F, Dong L, Jiandong T, et al 2018 Simulation and Tests for Expansion Performance of a flat tube. *Journal of Vibration and Shock* **37** 149-56
[18] Yongguo G and Dehui W 2015 Investigation on Dynamic Mechanical properties of 1060 pure Aluminium *Tool Engineering* **49** 48-51

IOP Conf. Series: Materials Science and Engineering **576** (2019) 012026 doi:10.1088/1757-899X/576/1/012026

A revisit of circular hole problem by the equivalent inclusion method

R Zhang[1], Y H An[1], P Li[1], X N Zhang[1], and X Q Jin [1,2]

[1]State Key laboratory of Mechanical Transmissions, Chongqing University, Chongqing 400030, China

[2]College of Aerospace Engineering, Chongqing University, Chongqing, 400030, China

Corresponding Author's E-mail: jinxq@cqu.edu.cn

Abstract. Stress and displacement analysis of an infinite extended isotropic elastic plane with a circular cavity is a classical elasticity problem. To solve this kind of problem, finite element method (FEM) and complex variable function are often used which have their pros and cons to some degree. In classical textbooks of elasticity, the closed-form solutions are usually given for the *stresses*, but the displacement solutions are seldom listed. In this paper, we revisit the circular hole problem via the equivalent inclusion method (EIM), and attempt to derive both the stress and displacement solutions simultaneously under the combined tensile and shear loading. With the knowledge of Eshelby's inclusion in two dimensional and EIM, the solutions to be sought may be constructed in a straightforward and elegant manner. Several examples are examined to validate the present solutions, demonstrating that the EIM may provide a more concise and convenient way to solve the circular cavity problem.

1. Introduction

Structures with circular holes are very common in practical engineering, covering a lot of fields such as architectural engineering, water supply and drainage, aerospace materials, complex mechanical construction and biomedical materials. The corresponding elasticity solutions of the stress and displacement are essential to their engineering design. It is widely known that the cavities or pores in structure may cause crack or local stress concentration under external load, leading to severe deterioration of the mechanical properties of the engineering materials. The problem is also of theoretical interest, because studies on stress and displacement solutions of pore structure might be a corner-stone for other complex elasticity problems, such as notched circular openings in rocks, an infinite wedge with a circular cavity, elastic half plane and composite plate, micromechanical research of ceramic strength [1-4].

Generally, the stress and displacement in the vicinity of the cavities may exhibit strong effect of concentration, which are usually solved by either numerical or analytical method. The finite element method (FEM) is most widely used in the numerical method. FEM is powerful and cost-efficient; however, it has to solve the problem case by case, and tends to be difficult to achieve a systematic and exhaustive parametric study for design guidelines. Moreover, accurate determination of the concentration effects usually requires extremely fine meshes in the vicinity of the geometric discontinuities, such as pores or cracks. Sometimes may even cause inaccurate simulation of real situation without initial knowledge. The complex variable function method is often used in analytical solution. Although this method can formulate the stresses or displacements effectively, it often requires profound knowledge of complex variable function, complicated calculations and well mathematical basis. This could be a challenge for general engineering and technical person.

An infinite plate containing a circular hole is a classical elasticity problem, and has been well-documented in many textbooks. However, most of the articles only concern about the stress solution, while the displacement results are less noted. This paper attempts to solve the circular hole problem by the equivalent inclusion method (EIM). A merit of the current work is that both the stresses and displacements are derived simultaneously in closed-form, facilitating a convenient and comprehensive reference for many engineering designs in practice.

2. Formulation

The equivalent inclusion method is a well-known method and has a significant influence on the development of micromechanics [5, 6]. In 1957, Eshelby [7] first proposed that perturbed field due to the presence of inhomogeneity can be simulated by the disturbance field of the appropriately distributed equivalent eigenstrain in an inclusion. This constitutes the essence of the well-known Eshelby's equivalent inclusion method. When applying EIM to 2D circular cavity problem, the corresponding formulation may be greatly simplified since the elastic moduli of the inhomogeneity becomes zero, i.e. $C^*_{ijkl}=0$ (Figure. 1).

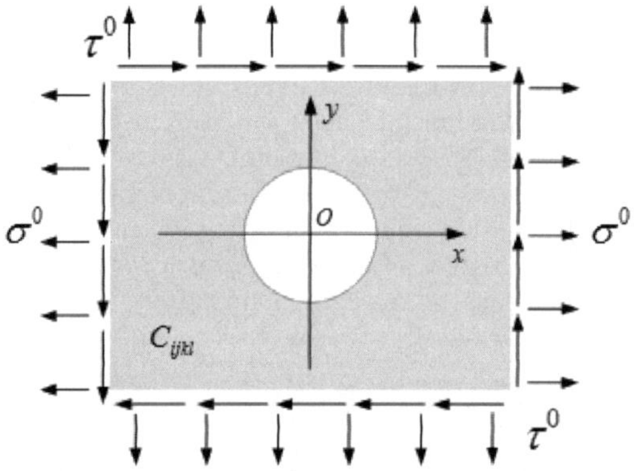

Figure 1. The inhomogeneity problem

Through the construction of stress equivalence, the homogeneous part σ^0_{ij} Figure. 2(a) plus the inclusion part Figure. 2(b) in form of stress Eshelby tensor i.e. $T_{ijkl}\varepsilon^*_{kl}$ based on our previous work [8]. Accordingly, the EIM of the circular cavity may be represented as:

$$\sigma^0_{ij}+T_{ijkl}\varepsilon^*_{kl}=0 \tag{1}$$

where σ^0_{ij} is the uniform load applied at remote, also representing the homogeneous material solution in the absence of the cavity [9]. The influence coefficients T_{ijkl} are termed as the Eshelby tensor for interior stresses [10]. It is noted that the present EIM formulation are established for *plane stress*; however, the final results may be used to interpret the *plane strain* case as well [10].

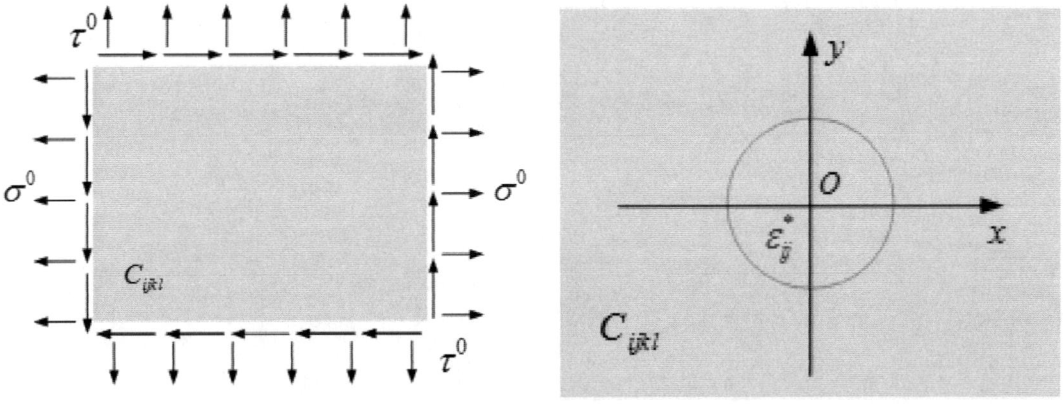

(a) (b)

Figure 2. (a) The homogeneous part of EIM; (b) The inclusion part of EIM

2.1. Plane circular inclusion solution

For the current circular inclusion solution, both stresses and displacements can be derived from a degenerate case of a two-dimensional elliptical inclusion, by letting $a=b$, (cf. [8]). The detailed formulae for the stress and displacement results are given as follows in sequel.

2.1.1. Interior stress field solution. Two independent elastic constants, i.e. shear modulus, μ, and the Kolosov constant κ, are chosen with $\mu = E/2(1+\nu)$, and for *plane stress*

$$\kappa = \frac{3-\nu}{1+\nu} \tag{2}$$

For any points (x, y) inside the circular inclusion, i.e., $x^2+y^2<a^2$, where a is the radius of the circular hole, the eigenstress components are

$$[\boldsymbol{\sigma}^*] = \begin{bmatrix} \sigma_x^* \\ \sigma_y^* \\ \tau_{xy}^* \end{bmatrix} = [\mathbf{T}][\boldsymbol{\varepsilon}^*] = \frac{-\mu}{(\kappa+1)} \begin{bmatrix} 3 & 1 & 0 \\ 1 & 3 & 0 \\ 0 & 0 & 1 \end{bmatrix} \begin{bmatrix} \varepsilon_x^* \\ \varepsilon_y^* \\ \gamma_{xy}^* \end{bmatrix} \tag{3}$$

2.1.2. Exterior stress field solution. Although the exterior solution is more complex, it can also be expressed in explicit closed-form and the results are concisely listed here in detail. First of all, for any point $p(x, y)$ located outside the circle, i.e., $x^2+y^2>a^2$, the stress components of the exterior field are listed as follows in which $r=\left(x^2+y^2\right)^{1/2}$,

$$[\boldsymbol{\sigma}] = \begin{bmatrix} \sigma_x \\ \sigma_y \\ \tau_{xy} \end{bmatrix} = \frac{4\mu a^2}{(\kappa+1)r^2} \begin{bmatrix} H_{11} & H_{12} & H_{14} \\ H_{21} & H_{22} & H_{24} \\ H_{41} & H_{42} & H_{44} \end{bmatrix} \begin{bmatrix} \varepsilon_x^* \\ \varepsilon_y^* \\ \gamma_{xy}^* \end{bmatrix} \tag{4}$$

where,

$$
\begin{aligned}
&H_{14}=H_{41}=\sin\theta\cos\theta\left[1-\frac{3a^2}{r^2}+\left(\frac{6a^2}{r^2}-4\right)\cos^2\theta\right]\\
&H_{24}=H_{42}=\sin\theta\cos\theta\left[1-\frac{3a^2}{r^2}+\left(\frac{6a^2}{r^2}-4\right)\sin^2\theta\right]\\
&H_{11}=\frac{2+3\rho^2}{4}+\cos^2\theta\left[2-\frac{6a^2}{r^2}+\left(\frac{6a^2}{r^2}-4\right)\cos^2\theta\right]\\
&H_{22}=\frac{2+3\rho^2}{4}+\sin^2\theta\left[2-\frac{6a^2}{r^2}+\left(\frac{6a^2}{r^2}-4\right)\sin^2\theta\right]\\
&H_{12}=H_{21}=H_{44}=\frac{1}{2}-\frac{3a^2}{4r^2}+\left(\frac{6a^2}{r^2}-4\right)\sin^2\theta\cos^2\theta
\end{aligned}
\tag{5}
$$

Noting that all the H_{ij}, $(i,j=1,2,3,4)$, in Eq. (5) are independent of the elastic constants and the $[\mathbf{H}]$ matrices are symmetric in Eq.(4). The normal and the shear components are decoupled in the interior field in contrast to the exterior field. The strain solution in exterior field may be obtained directly from the stress field through Hooke's law, and is therefore omitted here, but the interested readers are referred to Ref. [9] for detailed expressions.

2.2. The Displacement Eshelby Tensor

2.2.1. Exterior displacement field solution. The Voigt notation [10] is applied to represent the matrix form tensorial equations, e.g., $[\boldsymbol{u}]=[u_1\ \ u_2\ \ u_3]^T$, $[\boldsymbol{\varepsilon}^*]=[\varepsilon_{11}^*\ \ \varepsilon_{22}^*\ \ \varepsilon_{33}^*\ \ 2\varepsilon_{12}^*]^T$, where the superscript T indicates the transpose.

$$
\begin{bmatrix}u_1\\u_2\end{bmatrix}=\begin{bmatrix}W_{111} & W_{122} & W_{133} & W_{112}\\ W_{211} & W_{222} & W_{233} & W_{212}\end{bmatrix}\begin{bmatrix}\varepsilon_{11}^*\\ \varepsilon_{22}^*\\ \varepsilon_{33}^*\\ 2\varepsilon_{12}^*\end{bmatrix}
\tag{6}
$$

in which,

$$
\begin{aligned}
&W_{111}(\mathbf{x})=\frac{2x}{\kappa+1}\left[(\kappa-1)\frac{a^2}{4r^2}+\frac{3a^2}{4r^2}+\frac{a^2}{r^2}\left(1-\frac{a^2}{r^2}\right)\cos^2\theta\right]\\
&W_{112}(\mathbf{x})=\frac{2}{\kappa+1}\left\{\left[(\kappa-1)\frac{a^2}{4r^2}+\frac{a^4}{4r^4}\right]y+\left[\frac{a^2}{r^2}\left(1-\frac{a^2}{r^2}\right)\sin\theta\cos\theta\right]x\right\}\\
&W_{122}(\mathbf{x})=\frac{2x}{\kappa+1}\left[(1-\kappa)\frac{a^2}{4r^2}+\frac{a^4}{4r^4}+\frac{a^2}{r^2}\left(1-\frac{a^2}{r^2}\right)\sin^2\theta\right]\\
&W_{133}(\mathbf{x})=\frac{(3-\kappa)}{2(\kappa+1)}\frac{a^2}{r^2}x
\end{aligned}
\tag{7}
$$

2.2.2. Interior displacement field solution. A benefit of the displacement solution is that the interior field solution may be immediately derived from the exterior one in a way of letting $\lambda=0$ and dropping the trigonometric terms. To be specific, the displacement Eshelby tensor inside the inclusion can be expressed as

$$\begin{bmatrix} u_1 \\ u_2 \end{bmatrix} = \begin{bmatrix} W_{111} & W_{122} & W_{133} & W_{112} \\ W_{211} & W_{222} & W_{233} & W_{212} \end{bmatrix} \begin{bmatrix} \varepsilon_{11}^* \\ \varepsilon_{22}^* \\ \varepsilon_{33}^* \\ 2\varepsilon_{12}^* \end{bmatrix} \tag{8}$$

in which,

$$\begin{aligned} &w_{111}(\mathbf{x}) = \frac{2+\kappa}{2(\kappa+1)}x, \quad w_{222}(\mathbf{x}) = \frac{2+\kappa}{2(\kappa+1)}y, \quad w_{122}(\mathbf{x}) = \frac{2-\kappa}{2(\kappa+1)}x, \quad w_{211}(\mathbf{x}) = \frac{2-\kappa}{2(\kappa+1)}y \\ &w_{133}(\mathbf{x}) = \frac{3-\kappa}{2(\kappa+1)}x, \quad w_{233}(\mathbf{x}) = \frac{3-\kappa}{2(\kappa+1)}y, \quad w_{112}(\mathbf{x}) = \frac{\kappa}{2(\kappa+1)}y, \quad w_{212}(\mathbf{x}) = \frac{\kappa}{2(\kappa+1)}x \end{aligned} \tag{9}$$

3. EIM in circular cavity

3.1. Implementation and stress solution

For an infinite matrix in a state of uniaxial tensile loading together with a shear loading, the remote stresses are recorded in matrix form as $\left[\sigma_x^0, 0, \tau_{xy}^0\right]$. From Eqs. (1) and (3), the EIM is established as

$$\begin{bmatrix} \sigma^0 \\ 0 \\ \tau^0 \end{bmatrix} - \frac{\mu_1}{\kappa_1+1} \begin{bmatrix} 3 & 1 & 0 \\ 1 & 3 & 0 \\ 0 & 0 & 1 \end{bmatrix} \begin{bmatrix} \varepsilon_x^* \\ \varepsilon_y^* \\ \gamma_{xy}^* \end{bmatrix} = 0 \tag{10}$$

Note that the normal and shear eigenstrain components are decoupled. Consequently, the solution of the unknown equivalent eigenstrains in Eq. (10) is obtained after some algebraic simplification:

$$\begin{aligned} \begin{bmatrix} \varepsilon_x^* \\ \varepsilon_y^* \end{bmatrix} &= \frac{\kappa_1+1}{8\mu_1} \begin{bmatrix} 3 & -1 \\ -1 & 3 \end{bmatrix} \begin{bmatrix} \sigma_x^0 \\ 0 \end{bmatrix} \\ \gamma_{xy}^* &= \frac{\kappa_1+1}{\mu_1}\tau_{xy}^0 \end{aligned} \tag{11}$$

The stress solution outside the circular hole may be expressed as

$$[\boldsymbol{\sigma}] = \left[\boldsymbol{\sigma}^0\right] + [\mathbf{H}]\left[\boldsymbol{\varepsilon}^*\right] \tag{12}$$

Substituting Eq. (4) into (12) and after transforming it into the polar coordinates, one obtains the stress components as follows.

$$\begin{aligned} \sigma_r &= \frac{\sigma_x^0}{2}\left(1-\frac{a^2}{r^2}\right)\left[1+\cos 2\theta\left(1-\frac{3a^2}{r^2}\right)\right] + \tau_{xy}^0\left(\frac{3a^2}{r^2}-1\right)\left(\frac{a^2}{r^2}-1\right)\sin 2\theta \\ \sigma_\theta &= \frac{\sigma_x^0}{2}\left[1+\frac{a^2}{r^2}-\left(\frac{3a^4}{r^4}+1\right)\cos 2\theta\right] - \tau_{xy}^0\left(\frac{3a^4}{r^4}+1\right)\sin 2\theta \\ \tau_{r\theta} &= \frac{3a^4-2a^2r^2-r^4}{2r^4}\left(\sigma_x^0 \sin 2\theta - 2\tau_{xy}^0 \cos 2\theta\right) \end{aligned} \tag{13}$$

3.2. Displacement solution

The homogeneous strain field may be obtained by the given stress $\left[\sigma_x^0, 0, \tau_{xy}^0\right]$ through Hooke's law directly

$$\varepsilon_x^0 = \frac{\kappa_1+1}{8\mu_1}\sigma^0, \quad \varepsilon_y^0 = \frac{\kappa_1-3}{8\mu_1}\sigma^0, \quad \gamma_{xy}^0 = \frac{1}{\mu_1}\tau^0 \tag{14}$$

Hence, the displacement state of this part is

$$\begin{aligned} u_x^0 &= \sigma_x^0\frac{\kappa_1+1}{8\mu_1}x + \tau_{xy}^0\frac{1}{2\mu_1}y \\ u_y^0 &= \sigma_x^0\frac{\kappa_1-3}{8\mu_1}y + \tau_{xy}^0\frac{1}{2\mu_1}x \end{aligned} \tag{15}$$

The inclusion solution of the displacement may be obtained through Eqs. (6) and (11)

$$\begin{aligned} u_x^* &= -\frac{\rho^2}{4\mu_1}\sigma^0 x\left[2\cos 2\theta\left(\rho^2-1\right)-\rho^2-\kappa_1\right]+4\tau^0 x\sin 2\theta\left(\rho^2-1\right)+2\tau^0 y\left(1-\rho^2-\kappa_1\right) \\ u_y^* &= -\frac{\rho^2}{4\mu_1}\sigma^0 y\left[2\cos 2\theta\left(\rho^2-1\right)+\rho^2+\kappa_1-2\right]+4\tau^0 y\sin 2\theta\left(\rho^2-1\right)+2\tau^0 x\left(1-\rho^2-\kappa_1\right) \end{aligned} \tag{16}$$

On the grounds of EIM, the resultant displacement outside the circular hole may be derived by superposing the inclusion solution (16) with the homogeneous material solution Eq. (15). After converting the final result into polar coordinates, one obtains

$$\begin{aligned} u_r &= \frac{1}{8\mu_1 r}\sigma_x^0\left[\left(\kappa_1-1\right)r^2+2a^2\right]+\frac{1}{4\mu_1 r^3}\left(\sigma_x^0\cos 2\theta+2\tau_{xy}^0\sin 2\theta\right)\left[r^4-a^4+a^2r^2\left(\kappa_1+1\right)\right] \\ u_\theta &= \frac{1}{4\mu_1 r^3}\left(-\sigma_x^0\sin 2\theta+2\tau_{xy}^0\cos 2\theta\right)\left[r^4+a^4+a^2r^2\left(\kappa_1-1\right)\right] \end{aligned} \tag{17}$$

4. Results and discussions

Although the textbook [11] does not explicitly list the final solutions of the present elasticity problem, the exact solution may be derived from the tables given in [11], and the details are omitted here. On the basis of the closed-form solution in this paper, a double precision code is programmed by FORTRAN language to examine the variation trend along some target lines and verify the validness of the present work. In the benchmark study, $\nu = 0.3$ and all the three target lines are chosen along the radial direction, with θ equals to $0, \pi/4$, and $\pi/2$, respectively (Figure. 3).

The center of the circular hole is placed at the origin of the coordinate system, as shown in Figure. 3 Contrast between normalized stress and displacement along target lines are demonstrated in Figures. 4 and 5 correspondently. It is worth noting that the solutions of this paper are in perfect consistency with the exact solution. As can be seen from Figure. 4, stress concentration occurs at the boundary of the circular hole, when the points are far away from the boundary of the hole, the magnitude of stress components all decrease gradually to the applied stresses at remote. For displacement contrast in Figure. 5, we notice that the variations of the displacement along different radial directions show somewhat distinct patterns.

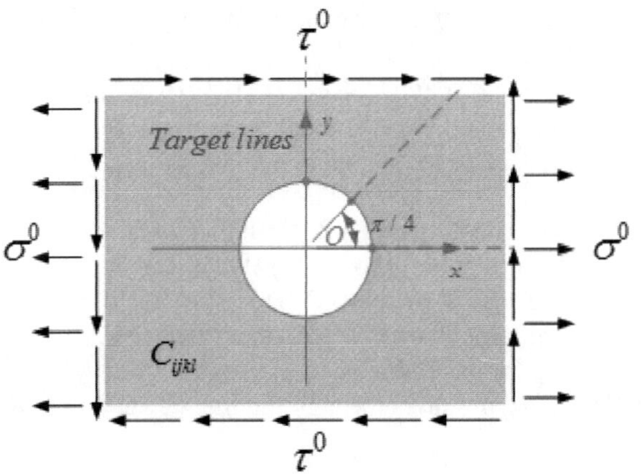

Figure 3. The comparison between EIM and exact displacement solutions

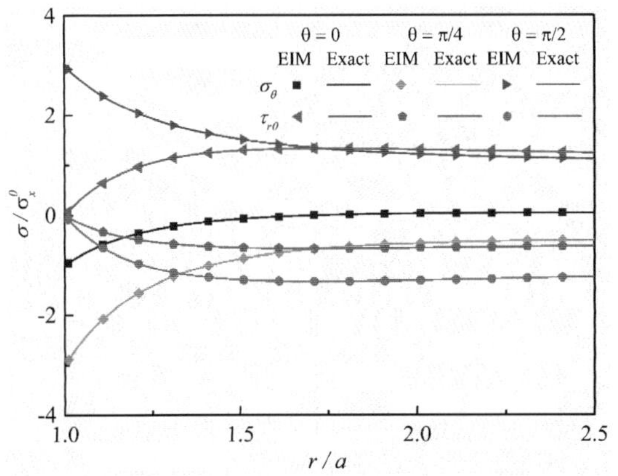

Figure 4. The comparison between EIM and exact stress solutions

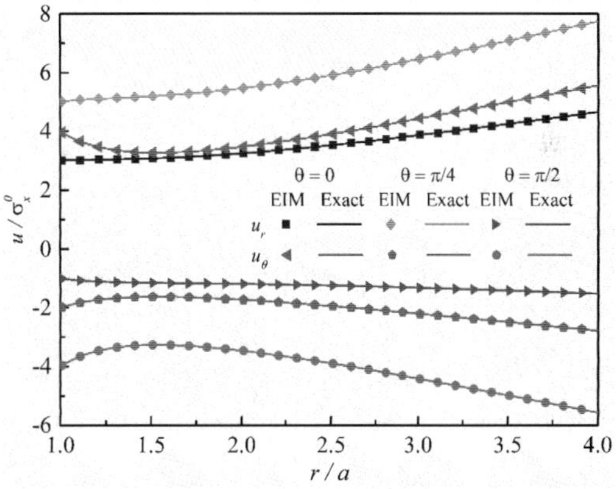

Figure 5. The comparison between EIM and exact displacement solutions

5. Conclusion

The present study revisits the circular cavity problem in an infinite elastic plane, and the equivalent inclusion method (EIM) is employed to obtain the elastic field solution under the combined tensile and shear loading. The disturbance solution due to the plane circular inclusion may be derived from a degenerate 2D case of an elliptical inclusion. With the help of the equivalent inclusion method, both the displacement and stress solutions are derived in closed-form. Benchmark examples are presented to validate the current solutions. The analytical solutions in this work may provide a convenient reference for engineering design.

6. References

[1] Exadaktylos G E, Liolios P A, and Stavropoulou M C 2003 *Int. J. Solids Struct* **40** 1165-1187.
[2] Liu G, Ji B, Chen H, Liu D 2009 *J. Appl. Mech* **76** 061008-1-9.
[3] Lu A, Zeng X, Xu Z 2016 *Int. J. Rock Mech* **89** 34-42.
[4] Toubal L, Karama M, Lorrain B 2005 *Compos. Struc* **68** 31-36.
[5] Jarali C S, Madhusudan M, Vidyashankar S, Raja S 2018 *Compos. Part B-Eng* **152** 17-30.
[6] Lee J, Nozaki H, Lee H 2018 *Compo. Interfaces* **25** 221-249.
[7] Eshelby J D, 1957 *Proc. R. Soc* **241** 376-396.
[8] Jin X, Wang Z, Zhou Q, Keer L M, Wang Q, 2014 *J. Elast* **114** 1-18.

[9] Mura T 1982 *Micromechanics of defects in solids. 2nd ed* (Dordrecht: Kluwer Academic Publishers).
[10] Jin X, Lyu D, Zhang X, Zhou Q, Wang Q, Keer L M, 2016 *J. Appl Mech* **83** 121010-1-12.
[11] Barber J R 1992 *Elasticity* (Netherlands: Springer Dordrecht) p 749.

Acknowledgments
This work is supported by Fundamental Research Funds for the Central Universities (2018CDYJSY0055). The authors are grateful to the Graduate Research and Innovation Foundation of Chongqing, China (Grant Nos. CYB18020 and CYB17025). X.J. would acknowledge the support from Fundamental Research Funds for the Central Universities (No. 106112017CDJQJ328839), the National Natural Science Foundation of China (Grant Nos. 51475057 and 51875059), and the State Key Laboratory of Mechanical Transmissions through funding (SKLMT-ZZKT-2017M15).

IOP Conf. Series: Materials Science and Engineering **576** (2019) 012027 doi:10.1088/1757-899X/576/1/012027

Natural frequencies of a Timoshenko beam subjected to axial forces by the differential transform method

Omar Bensabri[1] and Jingjun Yu[1]

[1]School of Mechanical Engineering and Automation, Beihang University, Beijing 100191, China

E-mail: bensabri-omar@hotmail.fr

Abstract. Rotating beams are extensively used in different mechanical and aeronautical installations. In this paper, a systematic approach is presented in order to solve the eigenvalues problem through the Timoshenko beam theory. The equations of motion are deduced by using the Hamiltonian approach. These equations are then solved by the differential transform method (DTM). The obtained numerical results using DTM are compared with the exact solution. Natural frequencies are determined, and the effects of the rotational speed and axial force on the natural frequencies are investigated. Results show high accuracy and efficiency of the differential transform method.

1. Introduction

Rotating structures can be found in turning machinery systems such as motors, engines, and turbines. The bending vibrations analysis of the beams aroused considerable interest for the engineers. The natural frequencies and mode shapes of such systems are indispensable in the design of structures. Zu and Han [1] analytically solved the free flexural vibration of a spinning Timoshenko beam with classical boundary conditions. Zhang [2] studied the free vibration of axially loaded shear beam column and obtained a very simple frequency equation to utilize for axially loaded beam as well as to obtain the buckling loads by setting the natural frequencies to disappear.

Farchaly and Shebl [3] determined two sets of exact general frequency and mode shape equations to study the vibration and stability of a Timoshenko beam that carries an end masses of finite length. Lee [4] enclosed the constant axial force and found that it had a considerable effect on the magnitude of the dynamic response. Ouyang [5] established a dynamic model for a rotating Timoshenko beam subjected to three force components acting on the surface. The deflection of the beam examined and found it had proportional increases with respect to the deflection and the frequency components when the axial force component is included.

The effect of such parameters such as moving velocity, the skew force angle, and the rotating speed on the system dynamic response is investigated by utilizing the global assumes mode method by considering boundary conditions [6]. The dynamic green function is used to introduce the free vibration of elastically supported Timoshenko beam on a partly Winkler foundation [7]. The finite element method is used the investigate the behaviour of the natural frequencies and to determine the influence of the rotating speed profile on the vibration of the cantilevered beam based on the dynamic modelling method by using the stretch deformation [8].

Among different types of numerical techniques, finite element method, the finite difference approach, the Galerkin method, etc. In the present study, the differential transform method is used. The proposition of the concept of the differential transform method was presented for the first time by Zhou [9] in 1986 when the problems of linearity and no-linearity that involved the electrical circuit problems had been solved.

2. Governing equation of motion

Figure 1 shows a uniform beam of the circular cross-section A, in an inertial coordinate system $oxyz$ subjected to axial force P along the x-axis

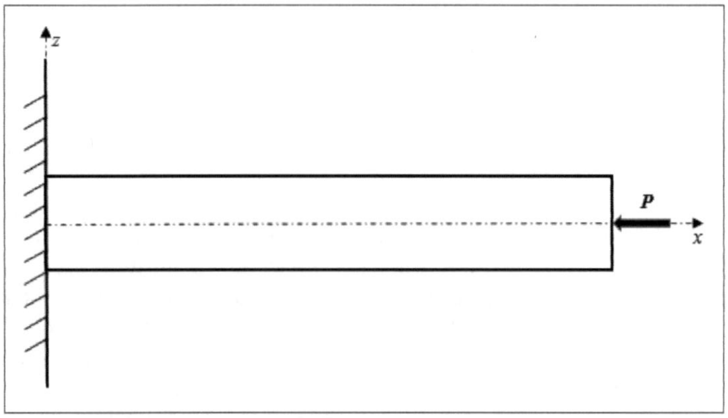

Figure 1. Cantilevered beam geometry.

The governing differential equations of motion as well as the boundary conditions can be derived through Hamilton's principle which can be stated as follows:

$$\delta I = \delta \int_{t_1}^{t_2} L dt = 0 \tag{1}$$

$$L = T - U_1 - U_2 \tag{2}$$

Where δ is the variational operator and L is the Lagrangian of the model. T and U_1 are the kinetic energy and the potential energy respectively of the Timoshenko beam adapted from reference [10].

$$T = \frac{\rho}{2}\int_0^l \left\{ A\left[\left(\frac{dv}{dt}\right)^2 + \left(\frac{dw}{dt}\right)^2\right] + I\left[\dot{\phi}^2 + \dot{\psi}^2 - 2\Omega\left(\dot{\phi}\psi - \phi\dot{\psi}\right) + 2\Omega^2\right]\right\} dx \tag{3}$$

$$U_1 = \frac{1}{2}\int_0^l \left\{ EI\left[\left(\frac{d\phi}{dt}\right)^2 + \left(\frac{d\psi}{dt}\right)^2\right] + \kappa GA\left(\gamma_{xz}^2 + \gamma_{yz}^2\right)\right\} dx \tag{4}$$

U_2 is the potential energy caused by the work of the axial load P according to the Engesser approach [11] given by the following equation:

$$U_2 = -\frac{1}{2}\int_0^l \left[\left(\frac{dv}{dx}\right)^2 + \left(\frac{dw}{dx}\right)^2\right] dx \tag{5}$$

Where v, w are the deflections of the rotating beam in the y, z directions respectively, and the angular rotations around y and w axes by ϕ and ψ . A is the cross-section area of the beam, and I is the moment of the area.

$$I = \frac{\pi r^4}{4}$$

Where r is the radius of the beam ρ is the mass density of the beam material. E and G are Young's modulus and shear modulus respectively, and κ is the shear coefficient. γ_{xy} and γ_{xz} denote the shear strain expressed in (6) and (7).

$$\gamma_{xy} = \frac{\partial v}{\partial x} - \phi \tag{6}$$

$$\gamma_{xz} = \frac{\partial w}{\partial x} - \psi \tag{7}$$

By introducing the following non-dimensional variables and parameters:

$$\mu^{*2} = \frac{\rho A L^4}{EI}\omega\Omega\ ,\ \mu^2 = \frac{\rho A L^4}{EI}\omega^2$$

$$r^2 = \frac{I}{AL^2}\ ,\ s^2 = \frac{EI}{\kappa GAL^2}\ ,\ \overline{P} = \frac{P(x)L^2}{EI} \tag{8}$$

And presuming that the beam incurs harmonic motion:

$$\overline{Z}(x,t) = \Gamma(x)e^{-j\omega t}\ ,\ \alpha(x,t) = \Phi(x)e^{-j\omega t} \tag{9}$$

The differential equations of motion are obtained after integration, and its dimensionless form are as follows:

$$\left(\frac{1}{s^2} - \overline{P}\right)\frac{\partial^2\Gamma}{\partial x^2} + \mu^2\Gamma + \frac{1}{s^2}\frac{\partial\Phi}{\partial x} = 0 \tag{10}$$

$$\frac{\partial^2\Phi}{\partial x^2} + \frac{1}{s^2}\left(\Phi - \frac{\partial\Gamma}{\partial x}\right) - 2\omega\mu^2 r^2\Phi = 0 \tag{11}$$

As well as the associated boundary conditions:

$$\left[\left(\frac{1}{s^2} + \overline{P}\right)\frac{\partial\Gamma}{\partial x} - \frac{1}{s^2}\Phi\right]\delta\Gamma\Big|_0^l = 0 \tag{12}$$

$$\left(\frac{\partial\Phi}{\partial x}\right)\delta\Phi\Big|_0^l = 0 \tag{13}$$

3. Differential transform method

The differential transform method (DTM) is a semi analytical-numerical method that is suitable for solving initial and/or boundary value problems and can provide highly accurate results with small computational effort. [12, 13] The DTM has gained the attention of many researchers recently.

Let $f(x)$ be an analytic function in a domain D and let $x = x_0$ represent any point in domain D; therefore, the differential transform of $f(x)$ is given by [14]:

$$F(k) = \frac{1}{k!}\left[\frac{d^k f(x)}{dx^k}\right]_{x=x_0} \tag{14}$$

Where $F(k)$ is the transformed function in the transformation domain, $f(x)$ is the original function and k is the transformation parameter. The inverse transformation is defined as:

$$f(x) = \sum_{k=0}^{\infty}(x - x_0)^k F(k) \tag{15}$$

Combining equations (13) and (14), we obtain:

$$f(x) = \sum_{k=0}^{\infty}\frac{(x - x_0)^k}{k!}\left[\frac{d^k f(x)}{dx^k}\right]_{x=x_0} \tag{16}$$

In the real case, the number of series is limited to N and equation (16) can be written as follows:

$$f(x)=\sum_{k=0}^{N}\frac{(x-x_0)^k}{k!}\left[\frac{d^k f(x)}{dx^k}\right]_{x=x_0} \tag{17}$$

Where the number N is defined according to the convergence criterion. The basic theorems of the differential transform are listed in Table 1 [14].

Table 1. Formatting sections, subsections and subsubsections.

Original function	Transformed function
$f(x)=h(x)\pm g(x)$	$F(k)=H(k)\pm G(k)$
$f(x)=\beta g(x)$	$F(k)=\beta G(x)$
$f(x)=h(x).g(x)$	$F(k)=H(k)\sum_{k=0}^{l}(k-l)G(l)$
$f(x)=\dfrac{dg(x)^n}{d(x)^n}$	$F(k)=\left[(k+n)!\right]G(k+n)$
$f(x)=x^n$	$F(k)=\delta(k-n)=\begin{cases}0 si k\neq n\\ 1 si k=n\end{cases}$

4. Application of the DTM

The following recursive expressions are determined by applying DTM to equations (10) and (11) using theorems listed in Table 2.

$$F(k+2)=\frac{-s^2}{\overline{P}}H(k+2)-\frac{s^2\mu^2\omega^2}{\overline{P}(k+1)(k+2)}F(k) \tag{18}$$

$$H(k+2)=\frac{1}{s^2(k+1)(k+2)}H(k)+\frac{1}{s^2(k+2)}F(k+1)+\frac{r^2\omega\left(2j\mu^{*2}+\omega\mu^2\right)}{(k+1)(k+2)}H(k) \tag{19}$$

Applying DTM to equations (12) and (13), the boundary conditions are as follows:

At $x=0$:

$$\begin{cases}F(0)=0\\ H(0)=0\end{cases} \tag{20}$$

At $x=l$:

$$\sum_{k=0}^{N}kH(k)=0 \tag{21}$$

$$\sum_{k=0}^{N}H(k)=\sum_{k=0}^{N}kF(k) \tag{22}$$

$$(22)\Rightarrow\sum_{k=0}^{N}\left[kF(k)-H(k)\right]=0 \tag{23}$$

For any $F(k)$ and $H(k)$ with $k\geq 2$ are expressed regarding the first two terms in function of ω, c_1, and c_2. By inserting the terms of $F(k)$ and $H(k)$ into the boundary conditions in (21) and (23) the system of equations obtained will be as follows:

$$A_{i1}^{N}(\omega)c_1+A_{i2}^{N}(\omega)c_2=0\ ,\ i=1,2,3...N \tag{24}$$

Where $A_{i1}^{N}(\omega)c_1, A_{i2}^{N}(\omega)c_2$ are polynomials of ω corresponding to n^{th} term. The equation (24) can be rewritten in the matrix form as follows:

$$\begin{bmatrix} A_{11}^N(\omega) & A_{12}^N(\omega) \\ A_{21}^N(\omega) & A_{22}^N(\omega) \end{bmatrix} \begin{Bmatrix} c_1 \\ c_2 \end{Bmatrix} = \begin{Bmatrix} 0 \\ 0 \end{Bmatrix} \tag{25}$$

The natural frequencies are obtained by equating the characteristic equation of the determinant of the equation (25).

Table 2. DTM theorems for boundary conditions

$x=0$		$x=l$	
Original BC	Transformed BC	Original BC	Transformed BC
$f(x)=0$	$F(0)=0$	$f(1)=0$	$\sum_{k=0}^{N} F(k)=0$
$\frac{\partial f}{\partial x}(0)=0$	$F(1)=0$	$\frac{\partial f}{\partial x}(1)=0$	$\sum_{k=0}^{N} kF(k)=0$
$\frac{\partial^2 f}{\partial x^2}(0)=0$	$F(2)=0$	$\frac{\partial^2 f}{\partial x^2}(1)=0$	$\sum_{k=0}^{N} k(k-1)F(k)=0$

5. Numerical results and discussion

For the determination of the natural frequencies a for a cantilevered beam, a computer package Matlab is used. In order to validate the computed results, an associate illustrative example taken from [3] is solved and also the results are compared with those within the same reference paper.

The results for cantilevered beam are tabulated in Table 3 for different values of *P*. The results demonstrate the effect of the parameter of the axial force on the natural frequency. A good agreement up to the third digit in the first mode and up to the first digit in the second mode are shown between the present study and that mentioned in reference [3]. The found results show a decrease of the natural non-dimensional frequencies parameter by the increasing of the non-dimensional of the axial force parameter P .

Table 3. Variation of the natural non-dimensional frequencies for a cantilever beam for various values of *P* concerning ($r^2=0.01$ and $s^2=3r^2$).

	First mode		Second mode	
P	Present	Ref.[3]	Present	Ref.[3]
0	1.799	1.799	3.819	3.820
1	1.579	1.579	3.708	3.715
10	-	-	2.333	2.199

The results presented in Table 4 show the effect of the rotational speed $\Omega(rad/s)$ on the fundamental natural frequencies of the beam concerning the material and geometric properties of the beam are: $\rho=8480kg/m^3, L=0.4m, I=1.886\times10^{-5}m^4, E=2\times10^7\,pa$ By raising the rotational speed gives rise to all natural frequencies ω which occurs as a result of the increase of the centrifugal tension force.

Table 4. Variation of the fundamental natural frequency of a rotating Timoshenko beam for various of rotational speed Ω.

P	1st mode	2nd mode	3rd mode	4th mode
0	6.121	12.998	15.89	17.012
50	6.159	13.073	15.898	17.173
100	6.197	13.147	15.91	17.330
150	6.235	13.221	15.925	17.486
200	6.272	13.295	15.942	17.639
250	6.309	13.369	15.894	17.792
300	6.346	13.443	15.979	17.944

Figure 2 gives the variation of the natural frequencies as a function of the variation of the speed of rotation. The influence of the speed of rotation on the natural frequencies of the beam is illustrated as can be seen from the Figure 2. We notice that for each speed of rotation, there are two frequencies, there are increasing values according to the speed of rotation called modes in direct precession, and other decreasing ones called retrograde modes. The gyroscopic effect causes this phenomenon. Indeed, we also notice that there is a proportional variation between the speed of rotation and the natural frequencies caused by the centrifugal stiffening

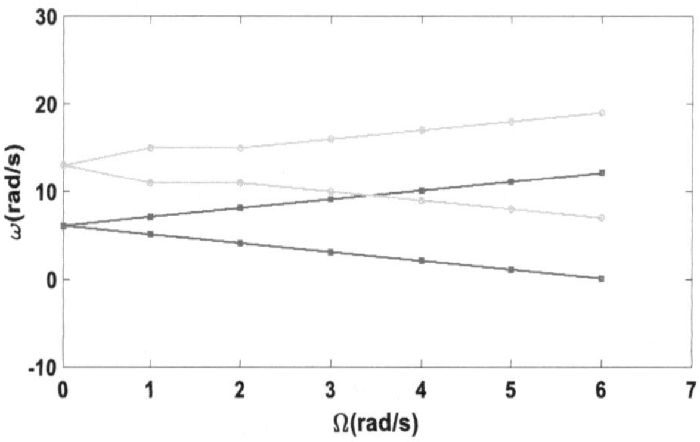

Figure 2. Whirl speed map of the rotating beam.

6. Conclusion.

In this paper, the natural frequencies of a non-rotating and rotating cantilevered Timoshenko beam are studied by using the DTM. The effects of the axial force and rotational speed on the natural frequencies are investigated. A good agreement is observed through a comparison of the DTM results and exact solutions from the reference [3] mentioned above.

7. References

[1] Zu, J.W.-Z. and R.P. Han, *Natural frequencies and normal modes of a spinning Timoshenko beam with general boundary conditions.* J Appl Mech, 1992. **59**(2S): p. S197-S204.

[2] Zhang, H., Y. Kang, and X. Li, *Stability and vibration analysis of axially-loaded shear beam-columns carrying elastically restrained mass.* Appl Math Model, 2013. **37**(16-17): p. 8237-8250.

[3] Farchaly, S. and M. Shebl, *Exact frequency and mode shape formulae for studying vibration and stability of Timoshenko beam system.* J sound vib, 1995. **180**(2): p. 205-227.

[4] Lee, H., *Dynamic response of a rotating Timoshenko shaft subject to axial forces and moving loads.* J sound vib, 1995. **181**(1): p. 169-177.

[5] Ouyang, H. and M. Wang, *A dynamic model for a rotating beam subjected to axially moving forces.* J sound vib, 2007. **308**(3-5): p. 674-682.
[6] Shiau, T., et al., *Dynamic response of a spinning Timoshenko beam with general boundary conditions under a moving skew force using global assumed mode method.* JSME International Journal Series C Mechanical Systems, Machine Elements and Manufacturing, 2006. **49**(2): p. 401-410.
[7] Ghannadiasl, A. and M. Mofid, *An analytical solution for free vibration of elastically restrained Timoshenko beam on an arbitrary variable Winkler foundation and under axial load.* Latin American Journal of Solids and Structures, 2015. **12**(13): p. 2417-2438.
[8] Chung, J. and H.H. Yoo, *Dynamic analysis of a rotating cantilever beam by using the finite element method.* J sound vib, 2002. **249**(1): p. 147-164.
[9] Zhou, J., *Differential transformation and its applications for electrical circuits*. 1986, Huazhong University Press, Wuhan, China.
[10] Chen, L.-W. and D.-M. Ku, *Dynamic stability of a cantilever shaft-disk system.* J Vib Acoust, 1992. **114**(3): p. 326-329.
[11] Yim, K.B. and J. Yim, *Dynamic behavior of overhung rotors subjected to axial forces.* Int J Precis Eng Man, 2012. **13**(9): p. 1575-1580.
[12] Ayaz, F., *Solutions of the system of differential equations by differential transform method.* Appl Math Comput, 2004. **147**(2): p. 547-567.
[13] Karakoç, F. and H. Bereketoğlu, *Solutions of delay differential equations by using differential transform method.* Int J Comput Math, 2009. **86**(5): p. 914-923.
[14] Bervillier, C., *Status of the differential transformation method.* Appl Math Comput, 2012. **218**(20): p. 10158-10170.

The Irrigation Uniformity Analysis Based on Translation Variable-Rate Sprinkler

Yan Run[1, 2]

[1] Jiangsu Vocational College of Agriculture and Forestry, China
[2] Jiangsu University, Jurong, China

Corresponding author: paperiset@163.com

Abstract. Research and compare various modeling of three-dimensional model, the paper chooses CATIA software to model the translation irrigation machine, and analyzes its working principle, structure, mode of operation and obtains to various parts of the structure size parameters, organize and remove redundant parameters, to facilitate the establishment of the model. We proposed here two indexes which were spray shape and water distribution uniformity to evaluate hydraulic performance of variable-rate sprinkler. And analyzed the influence of the throw radius caused by working pressure, nozzle area and trajectory angle. By calculating instantaneous application rate, it is indicated that the irrigation uniformity will go bad when change single factor to accomplish variable-rate irrigation.

1. Introduction

Variable irrigation is the key technology in the precision irrigation, and variable-rate sprinkler is one of the ways to accomplish variable irrigation. There has been lots of researches variable-rate sprinklers both from home and abroad, but according to the existing products, the common variable-rate sprinklers have the disadvantage of complicated structure, unreliable performance and so on. The design of the existing variable-rate sprinkler is to add a variable-rate device to the strike sprinkler. The variable-rate device adjusts the structural parameter to accomplish the variable-rate spray [1]. The undesirable irrigation uniformity was a common problem for those variable-rate sprinklers. In order to improve the irrigation uniformity of the variable-rate sprinkler, a new kind of variable-rate sprinkler with a pressure and flow rate adjusting device and a nozzle-changeable device was put forward in this dissertation. The pressure and flow rate adjusting device was utilized to adjust the working pressure of the sprinkler and the nozzle-changeable device was used to change the discharge area. By regulation the two factors synchronized, the water distribution uniformity of the variable-rate sprinkler will be improved. Meanwhile, this dissertation also studied the water distribution of the variable-rate sprinkler applied in slope land with small angle. The results showed that this kind of sprinkler can dove the water distribution problem caused by the factor of the terrain.

Our freshwater resources are currently available not only effective, but utilization is also low, polluted and warming caused by the greenhouse effect, resulting in below normal rainfall combination of factors, seriously hamper the development of China's economy, the largest of which agricultural water used, conservation and control are very essential therefore can be performed on agricultural water [2]. In this paper, with reference to the market-related parameters of translated sprinkling machine, the design of a large span translated irrigation machine, according to research purposes, either to reduce the occupied area of farmland irrigation machine translation, to maximize the use of land, but also to ensure the uniformity of crop irrigation, increase production, and save water resources. This paper will use CATIA software to design three dimensional modeling which a sprinkling machine structural parameters of structural stability analysis to ensure a structure design to meet the design requirements. By ANSYS FLOTRAN software to structure of sprinkling machine will be simplified with the mesh, and using its own fluid simulation software ANSYS FLOTRAN pipeline fluid simulation analysis, flow rate at the inlet certain circumstances, each outlet relations between the velocity distribution. Experimental prototype in the field of five test groups to testing, the resulting distribution of water, irrigation water depth data to analysis are calculated based on the calculated results to determine the design of translation sprinkler irrigation uniformity Meets "Irrigation engineering design specifications" requirement.

2. Structure of paper

The status of freshwater resources and the significance of the development of agriculture sprinkler irrigation, and domestic and foreign research and test methods for translation sprinkler machine irrigation uniformity of structure and other aspects of the analysis.

The use of CATIA software, to finite element analysis, unit setup, mesh, other loads and constraints imposed on the translation irrigation machine which across the whole body and structural stability was analyzed and the corresponding results obtained proposed structure optimization tips.

Refer to the relevant literature and standards, and summarized translation impact of sprinkler irrigation uniformity factor by calculation and analysis, fluid analysis to determine the relevant parameters, including the inlet pressure, flow rate, emitter selection, emitter spacing parameter.

Using ANSYS FLOTRAN software to simplifies structures of translation irrigation machine. Meshing pipes were simulations to determine the translation irrigation machine which the relationship between of job stress, emitter operating pressure, speed, and flow, etc.

The experimental prototype testing in the field trials, collected and organized testing data to calculate the irrigation uniformity coefficient span translation irrigation, and then with ANSYS FLOTRAN simulation data are compared to determine the large-span sprinkler irrigation uniformity performance meet the design requirements under the corresponding emitter spacing, sprinkler operating pressure, inlet flow rate and other parameters.

3. Influence factors of sprinkler irrigation uniformity and evaluation of hydraulic performance

Based on equal nozzle area theory, non-circle nozzles with four different shapes were designed and their hydraulic characteristics were compared through experiments. The result indicated that for the star shaped nozzle, the range rarely decrease and the tendency of the water distribution under different pressure were of the same [3]. The hydraulic characteristics were compared between the star shaped nozzle and the circle nozzle, as shown in Figure 1. The result showed that water distribution of the star-shaped nozzle was better than that of the circle nozzle.

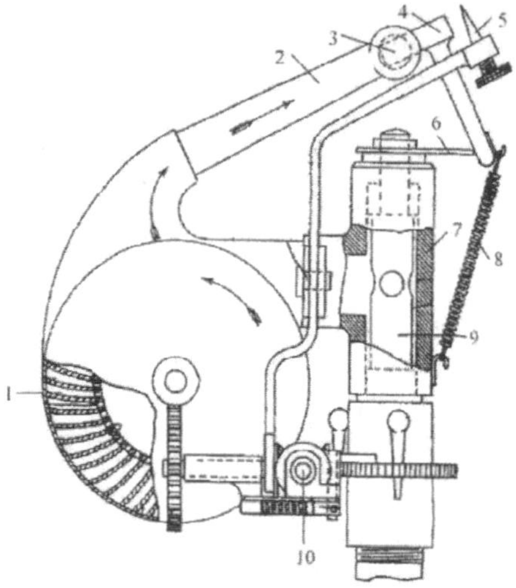

Figure 1. Donald developed the nozzle

Variable-rate sprinkler BPY series with nozzle-regulator were designed. This kind of sprinkler uses a pressure and flow rate adjusting device to adjust the working pressure, and a nozzle-changeable device to change the discharge area [4]. By regulation of these two factors synchronized, the water distribution uniformity of the variable-flow rate sprinkler will be improved. Through feasibility analysis on this sprinkler, the relation among nozzle area, the range and the working pressure was obtained. The result indicated that, when the spray shape is square, the range of the working pressure for BPY20 is 0.54 to 1, and for BPY30 is 0.57 to 1, for BYP with nozzle-regulator is 0.46 to 1.When the spray shape is triangle, the range of the working pressure for BPY30 is 0.33 to 1, and for BPY20 is 0.21 to 0.8, for BYP with nozzle-changeable device is 0.21 to 1.The variable-rate sprinkler with nozzle- changeable device could accomplish given spray area under a smaller range of pressure.

$$Q_p = 3600\mu A\sqrt{2gH} \tag{1}$$

Three kinds of nozzle-changeable device with different orifice areas for the sprinklers BPY20 and BPY30 were designed respectively. Impact of the nozzle-regulators to the stress introduced on the spring, the discharge coefficient of the sprinkler, the shape factor, and the throw radius were tested. It was obtained that: the premium radius of the orifice for the nozzle-regulator in BPY20 is D=6mm, for BPY30, it was D=7mm. The hydraulic performance of the variable-rate sprinkler with nozzle-regulator was measured. By compared with that of the star-shaped variable-rate sprinkler, the variable-rate sprinkler with nozzle-regulator provided better anticipated areas so that the water distribution uniformity was improved.

$$R = \frac{2v^2}{g}\sin 2\alpha = 2\varphi^2 H \sin 2\alpha \tag{2}$$

Based on the kinetic equations of the rain drops, a mathematic model for the transformation from the water distribution pattern on a flat ground to that on a sloping land was derived. Through the comparison between the field test and calculation data, it was found that: among 50200, obtains from the transformation model were in good consistence with that from test. Based on the transformation from the water distribution data on the flat ground, a large amount of data on the sloping land were acquired to provide an experimental basis for the study of irrigation on the sloping land.

$$\bar{\rho} = \frac{\sum_{i=1}^{n} S_i p_i}{\sum_{i=1}^{n} S_i} \tag{3}$$

4. Experiments

According to the water distribution model, the impact of slopes on the hydraulic performance was investigated. For the sprinkler on the slopes of 50-200, the variable-rate sprinkler can help to solve the problem of non-uniform distribution caused by the terrain conditions on an upslope or an down slope. The variable-rate sprinkler suitable to the slope of 100 was designed and the hydraulic performance in field operation was tested. The test indicated that: the variable-rate sprinkler will solve the problem of long range, high sprinkler intensity on the upslope and short range, low sprinkler intensity on the down slope.

$$\bar{\rho} = \sum_{i=1}^{n} \frac{2i}{n(n+1)} \cdot \rho_i \tag{4}$$

The low energy consumption has become one of the focuses in the development of sprinkler irrigation systems under the growing water shortage and energy consumption problem worldwide. In China, the terrain conditions, the crops and the rural economic levels are various across the land. Hence, the diversity in the configuration of Small- Scale Sprinkler Irrigation Systems is a trend. The increasingly shrinking farm labor as a result of urbanization, combined with the large- scale operations of farmland have challenged the present sprinkler irrigation machines with one-line pipe layout which is accompanied with inconvenience in transferring the pipeline to the next location. Therefore, the flexibility in the combination of components and the possibility in the application of larger areas within limited power need to be improved, and the energy consumption needs to be reduced so as to provide some basis for the further promotion of Small-Scale Sprinkler Irrigation Systems. The theoretical simulation and experimental validation were used in the optimization design and experimental study of Small-Scale Sprinkler Irrigation Systems with Low Energy Consumption and Multi Purposes. Spray water distribution curve was shown in Figure 2.

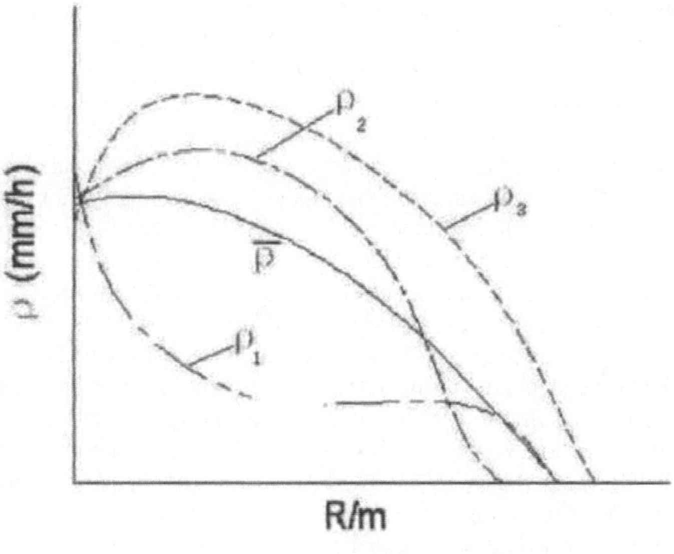

Figure 2. Spray water distribution curve

A Fixed-Movable Multi-Purpose Sprinkler Irrigation System was proposed to improve the flexibility of the small-scale sprinkler irrigation system and the adaptability to the crops. The key parts in the system were designed and produced including a quick-connect coupling for connection of the riser and the pipe tee, and a fixture for the sprinkler and the pipe in the system.

A Multiple-Sprinkler Irrigation System with Pairs of Laterals was developed. Theoretical analyses show that it has the minimal operation time, the minimal total cost in the life span, the highest irrigation uniformity coefficient and proper energy consumption compared to the traditional systems with one pipeline or two parallel lines.

Eleven kinds of pipe layouts were summarized for different areas according to the combination of different lateral layouts used in the two new irrigation systems, and the mobility of components of the system. These will provide various alternatives of pipe networks for different users and applications. The stress figure of Cross-body pipes was shown in Figure 3.

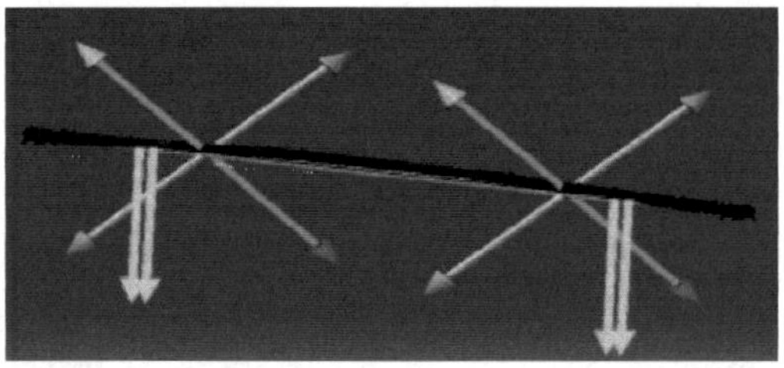

Figure 3. The stress figure of Cross-body pipes

A mufti- criteria evaluation model of small-scale irrigation system considering technical, economic, environmental and social aspects with ten sub-criteria and processed with Grey Relational Analysis (GRA} was developed. The Analytical Hierarchy Process and entropy method was employed in the evaluation of irrigation systems to take into account the subjective and objective aspects. Analyses indicate that the evaluation model can be adjusted easily and will provide an efficient method for the mufti- criteria and mufti- objective evaluation of irrigation systems.

And the impact factors of the evaluation indicators of irrigation system were explored. For the technical aspect, the impact of working pressure and sprinkler spacing on the overlapping irrigation uniformity were investigated to provide a basis for the optimization of the system. For the economic aspect, the components of Life Cycle Cost of irrigation systems were analyzed. Results show that: the energy consumption accounts for the major part of LCC of the systems, followed by the initial cost and the labor cost. Thus, the configuration optimization of irrigation system is very important to the reduction of cost and energy consumption. For the environmental aspect, the orthogonal design of four factors on three levels considering the interaction between factors was applied in the study of impacts of configuration parameters on the energy consumption. The results demonstrate that the impact of working pressure at the end sprinkler comes first, followed by that of pipe diameter, the impact of number of sprinklers is also obvious, and that of sprinkler spacing is manifested in the interactions with the other factors. Therefore, the multi-objective optimization of irrigation systems is needed.

The traditional hydraulic model was improved. And three models were developed according to the co-operation characteristics of pump-pipeline-sprinkler in the Small-Scale Irrigation Systems for there were generally no pressure regulators in the systems. They are the back step method, and the forward method considering the constraint of discharge of pump for the system with one line. And combination of both with the Golden Section Search was first used for the complicated networks. Results show that: the forward method will reflect the characteristics of irrigation pipelines better compared with the back step method. In a sprinkler irrigation system, the outlets are far less and the flow rate for each is far higher compared to a micro-irrigation lateral, thus there is an abrupt pressure drop at some length along the pipeline, not as continuous as that calculated by the back step method. The newly-designed forward-backward method will better meet the requirement of minimal pressure at the end sprinkler on each lateral in a pipe network compared to previous researches. The Von*Mises stress at maximum force diagram was shown in Figure 4.

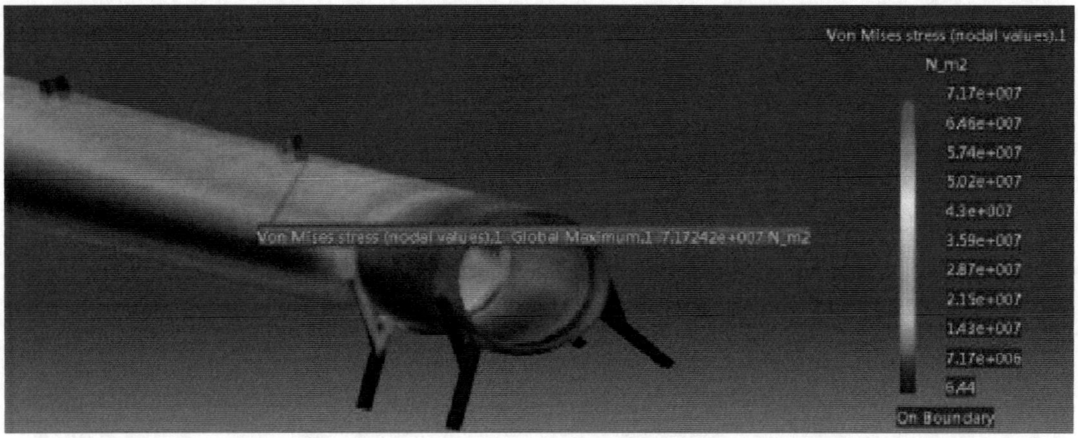

Figure 4. Von*Mises stress at maximum force diagram

Ant Colony Optimization (ACO) was originally introduced in the optimization of sprinkler irrigation systems for the lowest specific energy consumption and compared with Genetic Algorithm (GA) and the field tests. Results show that the optimal number of sprinklers and pressure profiles along the pipeline calculated with ACO are closer to the experimental results compared to that by GA. The multi-objective optimization model of small-scale irrigation systems was then built with Linear Weight Sum Method based on the analyses of impact factors of indicators.

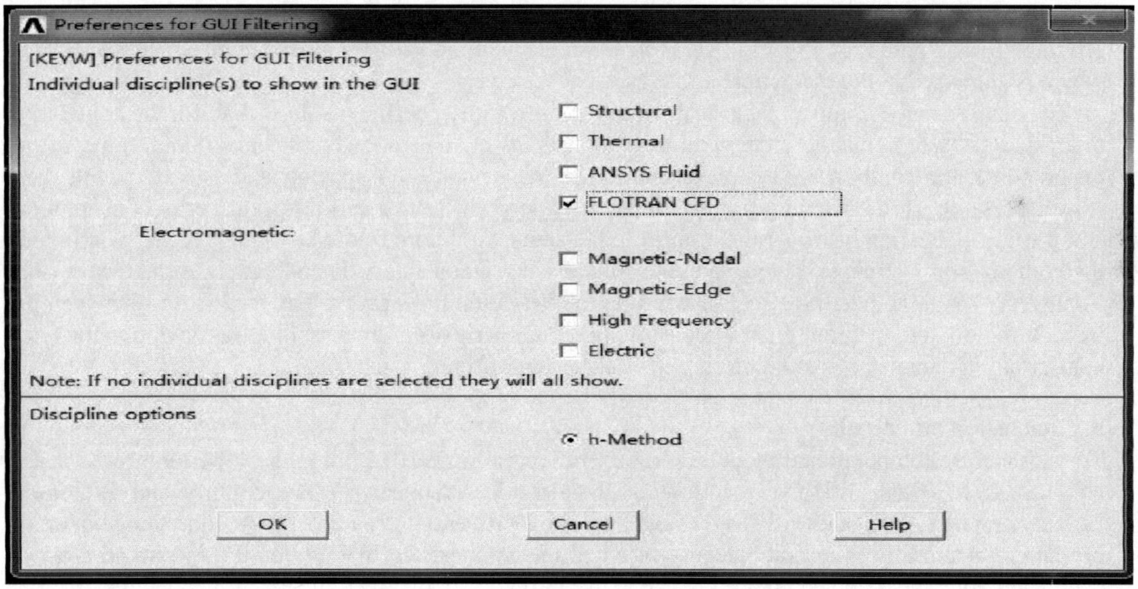

Figure 5. Setting the analysis options dialog

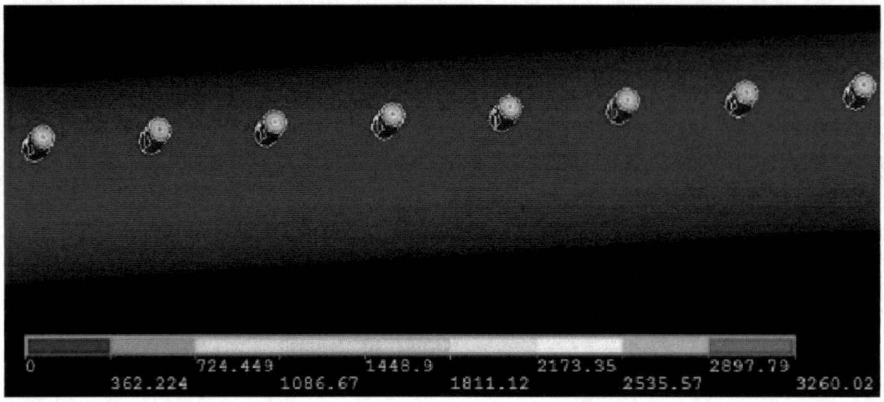

Figure 6. Outlet velocity contours

The specific energy consumption and the irrigation uniformity are the major criteria in the environmental and technical aspects, and will reflect the operation condition and the irrigation quality of the system, respectively. Moreover, they are more sensitive to the change of configuration parameters. Hence, the field experiment was carried out on the system typed PC45-4.4 equipped with sprinklers 15PY to investigate the impact of configuration parameters of system on the energy consumption and irrigation uniformity, to validate the practicability of model and optimization method, and to derive the optimal configuration in the test condition. The analyses on the impact factors of energy consumption demonstrate that: the theoretical analysis method is generally practical. But in the field test the energy consumption the pipe diameter is increased for the pressure head of sprinklers provided by the pump are higher. This indicates the interaction between the pump-pipeline-sprinkler, which is neglected in the theoretical analyses.

The analyses on the impact factors of irrigation uniformity coefficient show that: for the application of irrigation systems in the areas with wind speed lower than 2.7m1s (the maximum value in the experiment), the combination of rectangle layout, paper sprinkler spacing and lateral spacing, and Lower pressure can be used to ensure a high irrigation uniformity and a lower energy consumption. But for the application of irrigation systems in the areas with more frequent wind or higher wind speed, the combination of triangle layout, closer sprinkler and lateral intervals and higher sprinkler pressure is suggested to minimize the effect of wind on the irrigation uniformity. The interactions between the impact factors of specific energy consumption and irrigation uniformity are obvious, thus the configuration parameters should be chosen considering different aspects.

5. Conclusion and results

To validate the computational model, field experiments were carried out under low stable wind speed conditions (less than 1.5 ms 1) following the American Society of Agricultural and Biological Engineering (ASABE) standard. Two categories of experiments were carried out at the same operating pressure concurrently. The first category of experiments consisted of four sprinklers with an array of catch cans equally spaced and placed between the sprinklers constituted the experimental area.

Table 1. Irrigation Uniformity under tables test pressure

No.	Pressure Kpa	High mm	Index %
1	200	12.72	91.01
2	250	14.50	95.21
3	300	19.35	96.27
4	350	25.80	96.63
5	400	35.89	96.11

obtained by performing uniaxial tensile tests on an aluminium alloy. Zener-Hollomon parameter can also be introduced into this standard model to take the effect of temperature into account [4]. It is generally assumed that the model parameters are not dependent of strain. Otherwise, as proposed by Li et al. [5], to describe the behavior of a 7050 aluminium alloy, the strain compensation can be used. Parameters of the flow rule are thus expressed as polynomial functions of strain. However, these models do not describe the decrease in stress when exceeding a given strain due to damage development in the specimen. However, these models do not describe the decrease in stress when exceeding a given strain due to damage development in the specimen.

Damage in superplastic deformation is of ductile fracture type, it is mainly linked to cavities germination and growth [6]. The knowledge of its evolution during deformation allows predicting the forming limit of the material. Different approaches can be considered to take this phenomenon into account in the material model.

The material fracture can be predicted by coupling the ductile fracture criterion with the constitutive equation of the material. Ragab et al. [7] have indeed shown that the association of flow rules with some laws governing void growth such as Gurson's, Green's or Shyma-Oyane's models leads to good prediction of experimental data. Another approach consists in introducing some state variables in the flow rule. Costa Mattos et al. [8] have for instance modeled the superplastic behavior of anMg alloy by multiplying the flow rule by the term $1-D$ where D is a damage variable. When D is equal to 0, no damage is present; while D reaches 1, failure occurs.Furthermore, as proposed by Kim et al. [9], by tracking the evolution of work, the failure is predicted to occur when the fracture work reaches a critical value.

The use of an appropriate pressure control algorithm during Finite Element Modeling (FEM) simulation is also essential to predict the formability of a given part. Pressure control strategies in FEM simulation are generally defined in terms of pressure versus time and are able to generate an optimal pressure cycle which permits to deform a part at the optimal strain rate. Yang has summarized some often used methods [10]. Bonet et al. [11] used an analytical solution to define the pressure cycle for free bulging tests. The results are quite precise but they do not allow simulating industrial problems. Ollivier et al. [12], Rama et al. [13] and Carrino et al. [14] have proposed proportional rules between the equivalent stress and the applied pressure to generate the pressure law by coupling with a logarithmic correction. Bellet [15] and Robert [16] have proposed pressure controls by strain rate or stress ranges, which are more preferred by the industry due to their simplicity and sufficiency.

In the present work, the superplastic behavior of a 7475 aluminium alloy has been investigated by performing uniaxial tensile tests at different temperatures and strain rates. A rheological model using Zener-Hollomon parameter and strain compensation has been identified. Numerical simulations based on this material model have been realized on ABAQUS© to study the ability of the studied material to form complex parts. Failure is predicted by comparing the predicted fracture work to a critical value and a specific pressure control algorithm has been developed. These simulations have finally been checked by comparing numerical results to experimental results obtained by performing some bulging experiments.

2. Material and experimental methods

2.1. Material

The material used was a rolled 7475 aluminium alloy plate with a thickness of 2mm.The composition is given in Table 1. The initial microstructure was composed of elongated grains in the rolling plane with an average grain size of 12μm calculated by linear intercept method.

Table 1. Composition (in wt%) of the 7475 aluminium alloy [1].

Al	Si	Fe	Cu	Mn	Mg	Cr	Ni	Zn	Ti	Zr	Pb
balance	0.03	0.06	1.64	87 *	2.25	0.23	15 *	5.79	0.02	24 *	16 *
En %, * en ppm											

2.2. Uniaxial tensile tests

The experimental device used for the tensile tests is shown in Figure 1. The force was measured by a 1kN load cell. The samples shape was chosen in order to concentrate the deformation in their central part. They were taken from the plate with their tensile direction parallel to the rolling direction. A previous study [17] has shown that the anisotropy for this alloy is negligible. The crosshead displacement was piloted in real-time to ensure a constant strain rate within the central part of the specimens. The temperature was measured by a K-Type thermocouple welded in the central part of the sample. It was monitored in real-time to get as close as possible to the desired thermal cycle and allowed an oscillation of 2°C maximum at the desired temperature. An argon flow in the furnace allows preventing the corrosion of the specimen during the experiment. Tensile tests were conducted until fracture at constant strain rates of 5.10^{-5}, 2.10^{-4} and 5.10^{-4} s^{-1}. Different test temperatures were studied (477, 497 and 517°C).

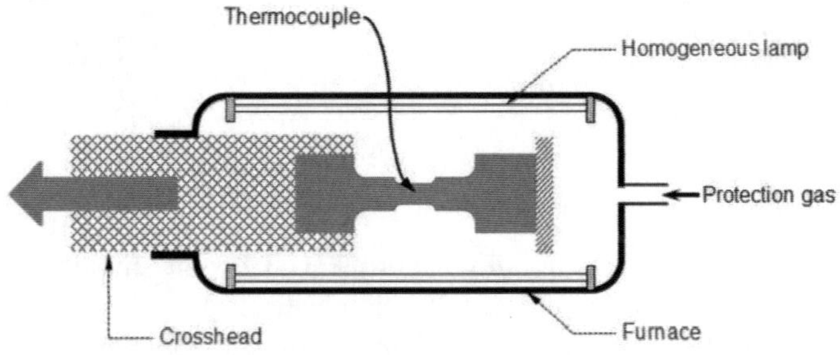

Figure 1. Sketch of the experimental device used for uniaxial tensile tests.

2.3. Bulging experiments

The experimental device used to perform bulging experiments is presented in Figure 2. This demonstrator has the ability to perform the experiments at a temperature within the range of [300-1100°C] and the accuracy of the displacement sensor is about 0.3%. The forming pressure is applied by using an inert gas (argon) which is introduced in the upper part of the enclosure. It is piloted by either the displacement of the sheet or the forming time. A back pressure can also be used during the forming process to limit the development of the cavitation encountered for some materials. It is thus introduced by an inert gas in the downer part of the enclosure. The die shape used during the experiments allows forming a complex revolution part with a height of 65mm and a maximum radius of 145mm.

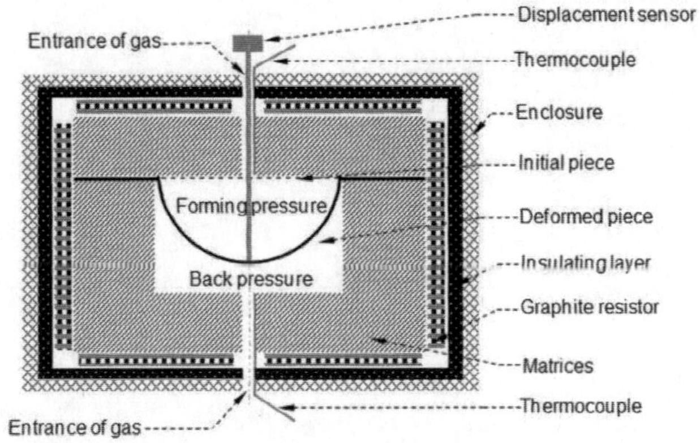

Figure 2. Experimentalsketch used for bulging experiments.

3. Experimental results

3.1. Rheological behavior of the 7475 aluminium alloy under hot tension

Uniaxial tensile experiments at various temperatures and strain rates have been performed. The true stress-strain curves can be found in [17]. Figure 3 shows a 3-D representation of these results for a strain rate of 2.10^{-4} s^{-1} and different temperatures (Figure 3a) and for a temperature of 517°C and different strain rates (Figure 3b).

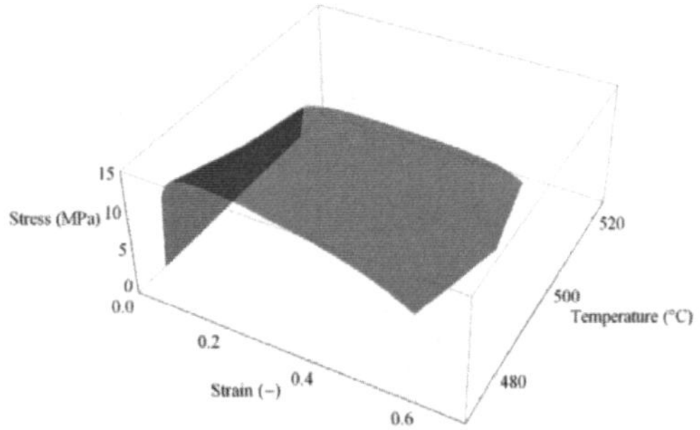

(a) strain rate of 2.10^{-4} s^{-1} and different temperatures

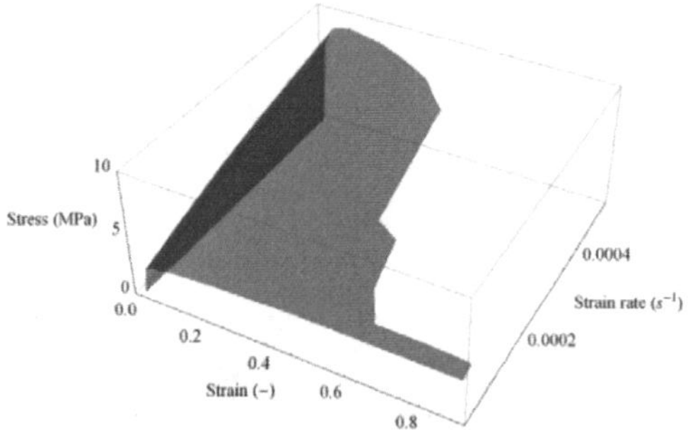

(b) temperature of 517°C and different strain rates

Figure 3. Experimentalsketch used for bulging experiments.3-Drepresentation of the stress-strain curves obtained during tensile tests.

As shown in Figure 3, stress increases to a plateau where it remains relatively constant, thus showing negligible strain hardening during the deformation. Moreover, stress increases obviously with increasing strain rate and decreasing temperature, which highlights the viscoplastic behavior of the alloy. Damage becomes predominant from a critical strain which depends on strain rate and temperature, which leads to a decrease in stress. Strain rate exhibits a significant effect on the strain to failure. The higher the strain rate is, the earlier the damage occurs. The optimal thermo-mechanical conditions for superplastic forming of 7475 alloy can thus be defined as the conditions for which the highest elongation to failure is obtained, i.e. 517°C and 5.10^{-5} s^{-1}.

3.2. Rheological behavior of the 7475 aluminium alloy under hot tension

The experimental fracture work is evaluated from the true stress-strain curves by integrating flow stress with respect to the strain until fracture. Its evolution as a function of strain rate at different temperatures is given in Figure 4. It can be noticed that fracture work tends to increase with increasing strain rate at a given temperature. However, it remains stable at 517°C which is the highest testing temperature.

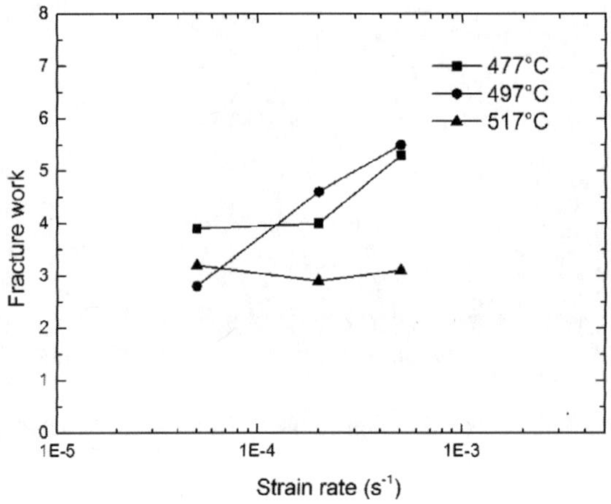

Figure 4. Fracture work as a function of strain rate at different temperatures.

4. Discussion

A model of the rheological behaviour of the material including strain compensation will be discussed first. Then a failure criterion based on fracture work evolution will be proposed. Finally, these models will be introduced in a numerical simulation to determine the pressure law to apply to form a complex part. The pressure law will then be used to perform bulging experiments with which the numerical simulations results are compared to validate the proposed protocol.

4.1. Modeling of the rheological behavior

During superplastic deformation of 7475 aluminium alloy, the steady state flow stress can be related to the strain rate and the temperature in the form of an Arrhenius kinetic rate equation [18]:

$$\dot{\bar{\varepsilon}} = A\left[\sinh\left(\alpha\bar{\sigma}\right)\right]^{n}\exp\left(-\frac{Q}{RT}\right) \tag{1}$$

with $\dot{\bar{\varepsilon}}$ is the strain rate, A and α are constants, $\bar{\sigma}$ is the steady flow stress, n is the stress exponent, Q is the activation energy, R is the gas constant and T is the absolute testing temperature.

The effect of strain rate and temperature on the hot deformation behavior of a material can be described by the Zener-Hollomon parameter [19]:

$$Z = \dot{\bar{\varepsilon}}\exp\left(\frac{Q}{RT}\right) \tag{2}$$

Then, the flow stress can be rewritten as a function of Z parameter:

$$\bar{\sigma} = \frac{1}{\alpha}\ln\left\{\left(\frac{Z}{A}\right)^{1/n} + \left[\left(\frac{Z}{A}\right)^{2/n} + 1\right]^{1/2}\right\} \tag{3}$$

After calculating the material constants (α, n, Q and A) in above equation, it can be shown that these constants are strain-dependent. Values of material constants calculated at four different strains (0.05, 0.15, 0.25 and 0.35) are given in Table 2. They increase when the strain increases. Strain compensation has thus to be taken into account to accurately predict the flow stress. It has been assumed that the material constants are polynomial functions of the strain [5, 20]. The best fitting correlation is obtained with a second order polynomial. The material constants expressions are summarized in Table 3. The identified parameters can be checked by comparing experimental and predicted data, as shown in Figure 5. It can be seen that the predicted data can efficiently represent the flow behavior of the alloy under the entire studied experimental conditions.

Table 2. Material constants for an Arrhenius-type equation calculated at different strain levels.

Parameter	Strain Level				
	0.05	0.15	0.25	0.35	Average
α	0.1325	0.1285	0.1369	0.1543	0.1381
n	1.608	1.799	2.041	2.411	1.964
Q	232.2	247.5	263.9	302.0	261.4
$\ln A$	27.19	29.51	31.97	37.71	31.60

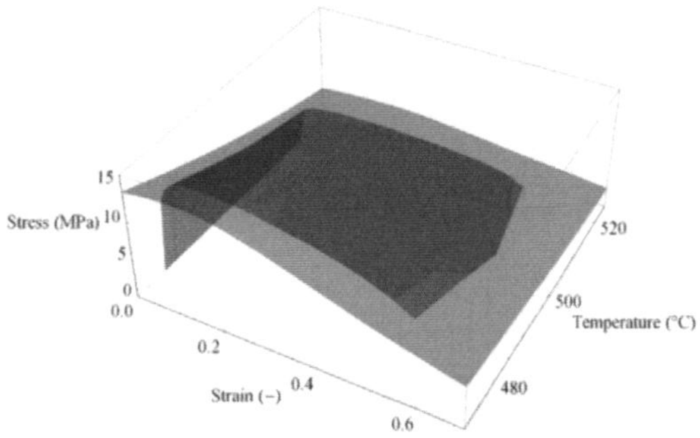

(a) at different temperatures

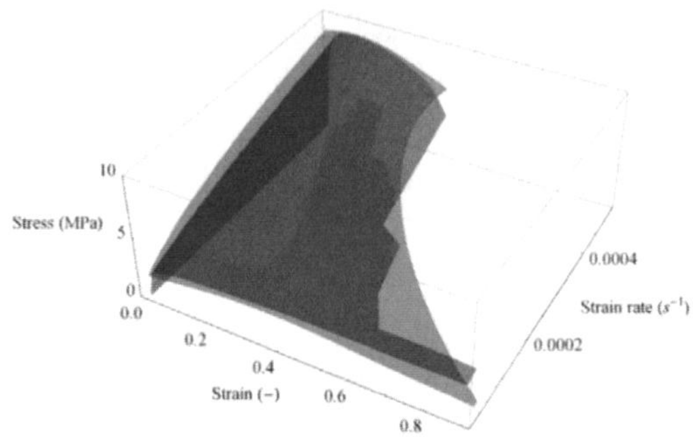

(b) at different strain rates

Figure 5. Comparison between experimental (gray) and predicted flow stress data by the constitutive model.

Table 3. Polynomial expressions of material constants for an Arrhenius-type equation.

Parameter	Expression
α	$0.5369\varepsilon^2 - 0.1402\varepsilon + 0.1378$
n	$4.077\varepsilon^2 + 0.937\varepsilon + 1.559$
Q	$338.4\varepsilon^2 + 61.49\varepsilon + 228.9$
$\ln A$	$50.60\varepsilon^2 + 9.483\varepsilon + 26.67$

4.2. Formulation of a failure criterion

The approach proposed by Kim et al. [9] for hot forming of a magnesium alloy is adopted in this study. A damage variable Ψ for ductile fracture is used and compared with the fracture damage value Ψ_f which is obtained at the effective fracture strain $\bar{\varepsilon}_f$:

$$F = \frac{\Psi}{\Psi_f} = \int_0^{\bar{\varepsilon}} \left[\frac{f(\sigma_{ij})}{\Psi_f S(\dot{\bar{\varepsilon}}, T)} \right] d\bar{\varepsilon} \tag{4}$$

where F is the normalized damage value, $f(\sigma_{ij})$ is a stress function, $\bar{\varepsilon}$ is the effective strain, and $S(\dot{\bar{\varepsilon}}, T)$ is a damage strength parameter which depends on the effective strain rate $\dot{\bar{\varepsilon}}$ and the temperature T. Material fracture occurs when F reaches 1. The expression of $f(\sigma_{ij})$ is a function of the damage criterion chosen (Freudhental [21] or Cockroft-Latham [22]). In the case of Freudhental criterion, the stress function is the effective stress $\bar{\sigma}$, which gives the normalized damage value as:

$$F = \frac{\Psi}{\Psi_f} = \int_0^{\bar{\varepsilon}} \left[\frac{\bar{\sigma}}{\Psi_f S(\dot{\bar{\varepsilon}}, T)} \right] d\bar{\varepsilon} \tag{5}$$

The temperature and strain rate dependence of S can be represented by using the Zener-Hollomonparameter Z. The denominator in equation (5) is thus replaced with a function $H(Z)$:

$$F = \frac{\Psi}{\Psi_f} = \int_0^{\bar{\varepsilon}} \left[\frac{\bar{\sigma}}{H(Z)} \right] d\bar{\varepsilon} \tag{6}$$

Since $F_{\bar{\varepsilon}=\bar{\varepsilon}_f} = 1$, equation (6) gives:

$$F = \frac{\Psi}{\Psi_f} = \int_0^{\bar{\varepsilon}} \left[\frac{\bar{\sigma}}{H(Z)} \right] d\bar{\varepsilon} \tag{7}$$

where W_f is the fracture work. The expression of $H(Z)$ can thus be found by using the tensile data. Kim et al. [9] have shown that $H(Z)$ can be expressed as a polynomial function of $\ln Z$. The fracture work obtained for the 7475 alloy as a function of $\ln Z$ is shown in Figure 6.

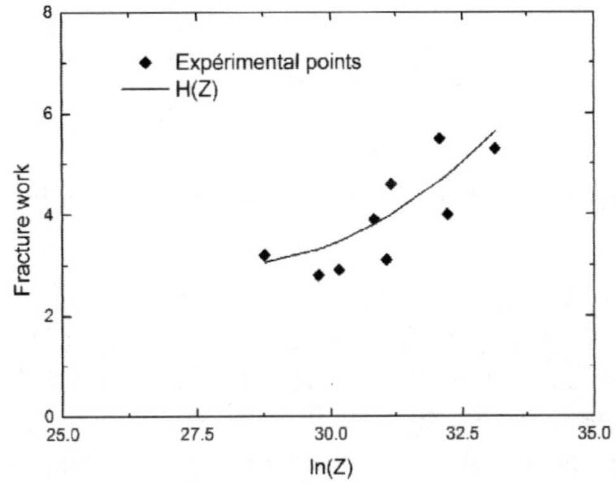

Figure 6. Fracture work as a function of ln(Z) determined from the stress-strain curves.

To calculate Z, an average value of Q is used i.e. $Q = 261.4\,\text{kJ/mol}$. $H(Z)$ can be expressed as follows:

$$H(Z) = 0.1[\ln(Z)]^2 - 5.6\ln(Z) + 81.4 \tag{8}$$

4.3. Application to the forming of a complex part

The pressure control algorithm proposed by Bellet [15] is implemented by default in ABAQUS©/Standard. This algorithm generates pressure law by taking a referenced strain rate value into account. This method gives globally good results and is commonly used in industrial applications.However,it needs to decrease the time increment to compensate the instabilities in the generated pressure law. Moreover, this method is limited by the use of implicit solver and the creep law. In this study this algorithm has been modified based on the form proposed by Robert [16] in order to improve the pressure law stability and to reduce the CPU time. The modifications have been implemented in ABAQUS©/Standard by FORTRAN subroutine, which allows to use a wide range of material behavior by programming a UMAT subroutine and to control more efficiently the pressure evolution. The major correction proposed by Yang [23] besides the form proposed by Robert [16] consists in introducing a viscous factor μ to pilot the pressure evolution and a strain corrector curve of $\dot{\varepsilon}_{ref}$ to adjust the maximal strain $\varepsilon_{\max}$:

$$\frac{P_{n+1}}{P_n} = \left[-5(r_{\max} - 1)^3 - 0.5(r_{\max} - 1) + 1 \right] \exp(-\mu r_{\max}) \tag{9}$$

P_{n+1} and P_n are the pressures at the increments $n+1$ and n, respectively; $r_{\max} = \dot{\varepsilon}_{\max} / \dot{\varepsilon}_{ref}$ is the strain rate ratio between the computed maximum value $\dot{\varepsilon}_{\max}$ and the referenced optimum value $\dot{\varepsilon}_{ref}$. Figure 7a compares a pressure law corrected with a μ value of 0.4 and a non-corrected one. It can be noticed that this correction acts like a viscous pressure which generates a much more continuous pressure evolution. The Figure 7b shows the corrector curve of $\dot{\varepsilon}_{ref} = f(\varepsilon_{\max})$.

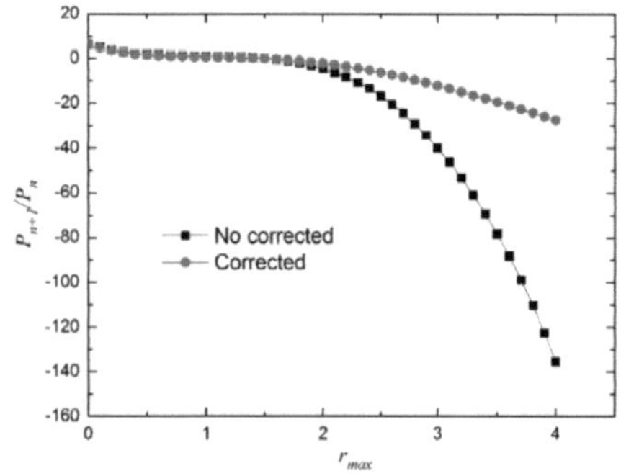

(a) comparison between non-corrected and corrected pressure law

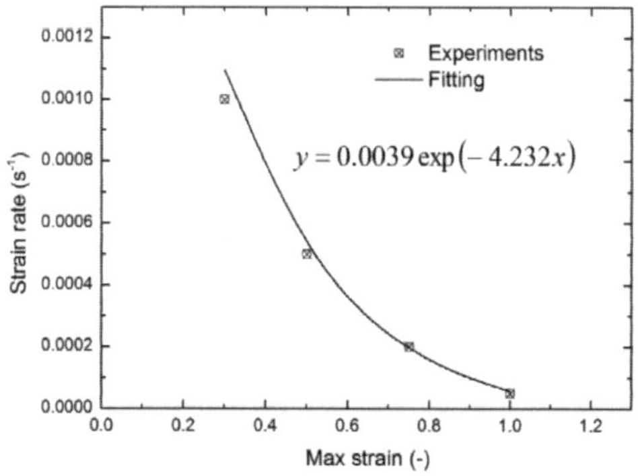

(b) strain rate corrector for the optimal strain rate

Figure 7. Major corrections proposed.

This algorithm has been used to predict the ability to form a complex revolution part. The ABAQUS© simulation is built using a quarter of the revolution part with symmetry conditions on the straight edges. The sheet is meshed with thin shell elements. The die is modeled by a discrete rigid surface. The implicit solver is chosen and the contact model uses classical "hard contact" options and Coulomb friction with a coefficient of 0.2. The material behavior and failure criterion detailed above are implemented by user subroutines. The reference strain rate is chosen within the range [5.10^{-5}–$5.10^{-4}s^{-1}$]. Figure 8 presents the model and the thickness evolution at 6000s.

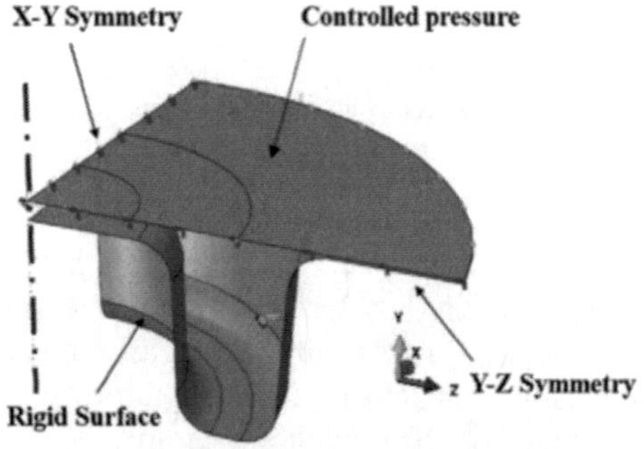

(a) numerical model

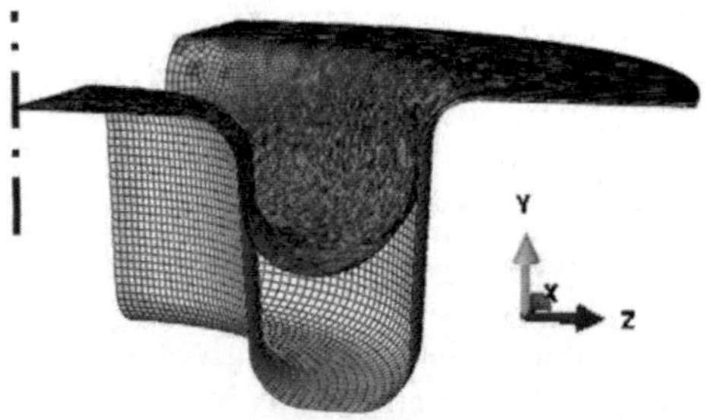

(b) simulation results

Figure 8. Numerical simulation of the superplastic forming of a complex revolution part.

The generated pressure law allows forming the component with the new algorithm. The optimal strain rate is corrected by the law presented in Figure 7. The generated law is used to perform bulging experiments. The thickness evolution in the formed component is measured. Figure 9 shows the location of the thickness measurements on the component and a comparison between numerical and experimental results. It can be noted that numerical results are in good agreement with experimental results. Moreover, as shown in Figure 10, the damage criterion is maximal in the zone near points 5 and 6 (on Figure 9) which have already been identified as a critical zone in previous study [24].

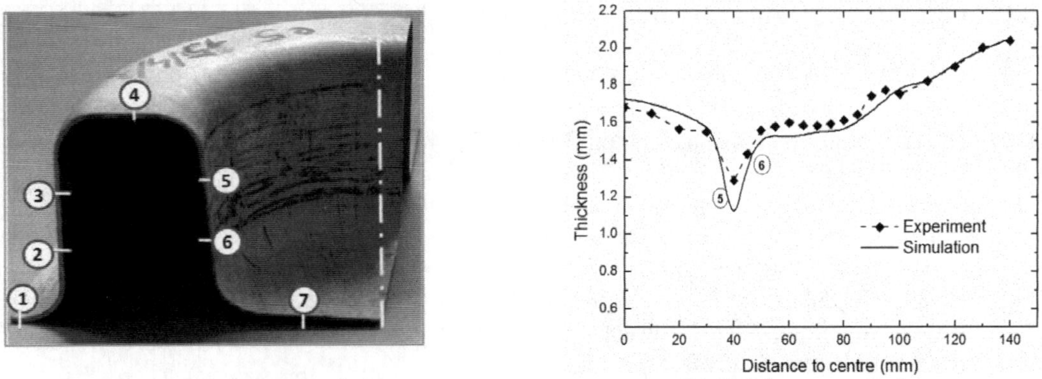

(a) Piece with measured points (b) Thickness evolution

Figure 9. Comparison of thickness evolution between the part obtained by numerical simulation and that by bulging experiment.

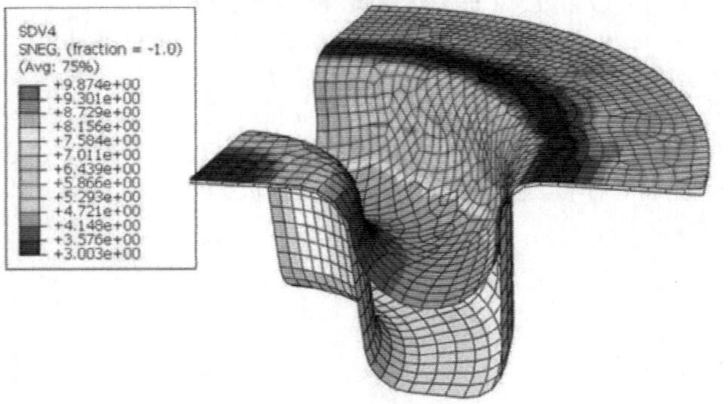

Figure 10. Damage-risk zones predicted by numerical simulation.

5. Conclusions

A material model allowing the description of the superplastic behavior of a 7475 aluminium alloy has been identified by performing isothermal hot tension tests within the following temperature and strain rate ranges: [477 - 517°C] and [5.10^{-5} - 5.10^{-4} s^{-1}]. After the optimization of the pressure control algorithm and the development of user subroutine, numerical simulations of the superplastic forming of a complex revolution part have been realized by ABAQUS. The conclusions are as follows:

(1) An Arrhenius-type constitutive equation including strain compensation allows predicting the mechanical response of a 7475 aluminium alloy under the thermo-mechanical conditions studied.

(2) A failure criteria based on the evolution of fracture work is sufficient to predict the damage-risk zones during the superplastic forming process.

(3) The use of a viscous factor in the pressure control algorithm allows improving the stability of the pressure law.

6. References

[1] Nieh T, Wadsworth J and Sherby O 1997 *Superplasticity in metals and ceramics* (Cambridge:Cambridge University Press)

[2] Barnes A 2007 *JMEPEG***16** 440-454

[3] Yang H and Mukherjee A 1992 *Int J MechSci* **34** 283-297

[4] Zhang C, Li X, Li D, Jin C and Xiao J 2012 *Trans Nonferrous Met Soc China* **22** s457-s464

[5] Li J, Li F, Cai J, Wang R, Yuan Z and Xue F 2012 *Mater Des* **42** 369-377

[6] LemaitreJ, Chaboche J, BenallalA and Desmorat R 2009 *Mécanique des matériaux solides* (Paris : Dunod)

[7] RagabA and Saleh C 1999 *Int J Plast* **15** 1041-1065

[8] Costa Mattos H, Minak G, Gioacchino F andSolda A 2009 *Mater Des* **30** 1674-1679
[9] Kim H and Kim W 2010 *Mech mater* **42** 293-303
[10] Yang J 2018 *Civ. Air. Des. & Res* **129** 52-58
[11] Bonet J and Wood R 1994 *J Mater Process Technol* **42** 147-165
[12] Ollivier D, Aoura Y, Ambari A and Boude S 2004 *MecIndust* **5** 511-517
[13] Rama S and Chandra N 1991*Int J NonLinMech* **26** 711-725
[14] Carrino L, Giuliano G and Napolitano G 2003 *Finite Elem Anal Des* **39** 1083-1093
[15] Bellet M 1988 *Modélisation numérique du formage superplastique de tôles* (Ecole Nationale Supérieur des Mines de Paris)
[16] Robert C 2009 *Contribution à la simulation numérique des procédés de mise en forme - Application au formage incrémental et au formage superplastique* (Ecole Nationale Supérieure d'Arts et Métiers, Angers)
[17] Yang J, Boude S, Giraud E and Dal Santo P 2013 *AIP ConfProc* **1532**, 623-629
[18] Lin Y and Chen X 2011 *Mater Des* **32** 1733–1759
[19] Zener C andHollomon J 1944 *J ApplPhys* **15** 22-27
[20] Peng W, Zeng W, Wang Q and Yu H 2013 *Mater Des* **51** 95-104
[21] Freudenthal A 1950*The inelastic behavior of solids* (New York: Wiley)
[22] Cockroft M and Latham D 1968 *J Inst Met* **96** 33-39
[23] Yang J 2014 *Étude expérimentale et modélisation du formage superplastiqued'un alliage d'aluminium Al7475* (Ecole Nationale Supérieure d'Arts et Métiers, Angers)
[24] Bellet M, Massoni E and Boude S 2004 *AIP ConfProc* **712** 1142-1147

Effect of Nanometer Antimony Trioxide on Properties of Flame Retardant Polypropylene Matrix Composites

J L Xu[1,2], Y G Lin[1,2], W L Yang[1,2], L Niu[1,2] and C H Kang[1,2]

[1] State Key Laboratory of Advanced Processing and Recycling of Non-ferrous Metals, Lanzhou University of Technology, Lanzhou 730050, China
[2] Baiyin Research Institute of Novel Materials, Lanzhou University of Technology, Baiyin 730900, China

Corresponding author's e-mail: ggdjlxu@sina.com

Abstract. Using antimony trioxide nanoparticles (nano-Sb_2O_3), montmorillonite (MMT) and brominated polystyrene (BPS) as additive and using polypropylene (PP) as matrix material, a novel nano-Sb_2O_3/MMT/BPS-PP flame retardant composite was synthesized via mechanical ball milling dispersion and melt blending method and characterized by X-ray diffraction, scanning electron microscopy, differential scanning calorimeter and tensile properties testing. The flame retardant performance of the composite was investigated by limiting oxygen index and vertical burning test (UL-94). The results show that the mechanical properties and flame retardant properties of the composites are improved compared to that of pure PP when PP matrix retardant composites are added with different mass fraction of nano-Sb_2O_3. When the mass fraction of nano-Sb_2O_3 is 3wt%, the PP matrix composite exhibits excellent tensile strength and crystallization properties. When the mass fraction of nano-Sb_2O_3 is more than 3wt%, the PP matrix composite exhibits excellent flame retardancy.

1. Introduction

With the rapid development of science and technology, polymer materials occupy an increasingly essential position in life and production. Polypropylene (PP) materials and their products are widely used in many fields due to their excellent properties of insulation, corrosion resistance and plasticity. However, PP materials and their products have some shortcomings such as poor flame retardancy, lower mechanical properties and low temperature, which limit the further development of PP matrix composites in engineering and life. Therefore, improving the flame retardant properties and related mechanical properties of PP materials and making PP more widely used in various fields have become an inevitable trend[1]. The antimony trioxide nanoparticles (nano-Sb_2O_3) have an essential position in flame retardant materials because nano-Sb_2O_3 has a good synergistic flame retardant effect with halogen flame retardants and organic flame retardants [2,3]. Since nano-Sb_2O_3 particles have high surface activity, large specific surface area and high strength bonding with polymers, adding appropriate amount of nano-Sb_2O_3 into the matrix material can not only maintain good flame retardancy, but also have little effect on the mechanical properties of the matrix material. Brominated polystyrene (BPS) is widely used in flame retardant engineering plastics as bromine organic flame retardant due to its high flame retardancy, thermal stability and light stability. BPS is made from polystyrene (PS) by bromination. BPS is compatible with engineering plastics and has little effect on mechanical properties of materials[4]. Montmorillonite (MMT) has also received extensive attention in the field of flame retardancy of polymers. Adding an appropriate amount of MMT to the composite material can improve the mechanical properties, flame retardancy and thermal stability of the material, which is becoming a hot spot of the flame retardant material with non-toxic and low-smoke[5]. On the one hand, the synergistic flame retardant effect of nano-Sb_2O_3 and BPS will produce the effect of suffocating flame retardant and endothermic flame retardant[6]; On the other hand, MMT added into the matrix can improve to form a carbon layer on the surface of the substrate, which can owe forming a flame retardant effect to isolate the oxygen and heat[7]. Therefore, nano-Sb_2O_3/MMT/BPS-PP flame retardant composite material samples will be prepared in this work. The influences of different mass fraction of nano-Sb_2O_3 on flame retardancy and mechanical properties of composites were researched.

2. Experimental

2.1. Materials

MMT with 15 μm of particle size was provided by Guangzhou Yifeng Chemical Technology Co., Ltd. of China. Nano-Sb_2O_3 were prepared and modified by ball milling, as reported in our study[8]. The particle size was 50-100nm. BPS with 1600 of average molecular weight was provided by Xingsheng Chemical Co., Ltd. of China. PP with 1 g/min of melt index was produced by China Petroleum & Chemical Corporation Beijing Yanshan Branch.

2.2. Sample Preparation

Firstly, PP powder, BPS powder, MMT powder and nano-Sb_2O_3 particles were dispersed by QM-3SP04 high-energy ball milling (Nanjing Laibu Science and Technology Industry Co. Ltd. of China) with 400 r/min of ball grinding speed for about 6 hours. Secondly, PP matrix composites were prepared by melted blending basing on a twin screw extruder (SJZS-10A, Wuhan ruiming Plastic machinery Co. of China). Finally, the test specimens were moulded by an injection mould machine (SZS-20, Wuhan Ruiming Machinery Manufacture Co. Ltd. of China). The compositions of experimental materials are listed in Table 1.

Table 1. Compositions of experimental materials (mass fraction)

Samples No.	Sample name	Component (wt%)			
		MMT	BPS	nano-Sb_2O_3	PP
1	PP	0	0	0	100
2	1.5%nano-Sb_2O_3/3%MMT/8%BPS-PP	3	8	1.5	87.5
3	2.0% nano-Sb_2O_3/3%MMT/8%BPS-PP	3	8	2.0	87.0
4	3.0% nano-Sb_2O_3/3%MMT/8%BPS-PP	3	8	3.0	86.0
5	3.5% nano-Sb_2O_3/3%MMT/8%BPS-PP	3	8	3.5	85.5

2.3. Characterizations

XRD patterns of samples were recorded at room temperature by Ultima IV XRD (Rigaku Corporation of Japan) with diffraction angle 2θ range from 5° to 35° at 2°/min of scanning speed.

The crystallization and melting behaviours of samples were investigated by QT-DSC-500C differential scanning calorimeter (DSC, Shanghai Qiantong Instruments Co. Ltd. of China). Samples of about 15 mg were encapsulated in aluminium pans and heated from room temperature to 220 °C, and kept thermal for 5 minutes to erase the thermal history, then cooled rapidly to room temperature and subsequently heated again to 200 °C. Heating rate is 10°C/min as well as cooling rate.

The tensile testing of samples were carried out by tensile tester (HS-100KN, Yangzhou Huahui testing instrument Co. Ltd. of China) with a crosshead speed of 20 mm/min. The results were the average values of at least five specimens. The tensile fractures were observed by SEM (JSM-6700F, JEOL).

The limiting oxygen index (LOI) was analyzed by PX-01-005 instruments (Suzhou Phinix quality inspection instrument Co. Ltd., of China) according to GB/T2406.2-2009. The UL-94 grade was performed by STD-94 vertical burning tester (Shanghai Pan Standard Textile Testing Technology Co. Ltd. of China) according to GB/T2408-2008.

3. Results and Discussion

3.1. Crystallization properties

Figure 1 shows the XRD spectra of pure PP and experimental composites with different content of nano-Sb_2O_3. PP is a polymer with higher crystallinity and contains five different crystal forms of α, β, γ, δ and pseudo hexagon in its crystal structure, in which α crystal is the most common form[9]. As can be seen from figure 1 that 2θ angles were 13.9°, 16.7°, 18.4°, and 21.7° corresponding to the (110), (040), (130), (-131) crystal planes of the PP-αcrystal form, respectively. A diffraction peak of MMT appears at 2θ were 16.2° and 31.2°, and the diffraction peak of Sb_2O_3 at 2θ were 32.1°. It can be seen from the figure 1 that the addition of MMT and nano-Sb_2O_3 did not change the diffraction position of the crystal plane of PP, indicating no alteration of the crystal structure of PP. With increasing of the mass fraction of nano-Sb_2O_3, the diffraction intensity of the PP matrix composites first increases and then weakens. The heterogeneous nucleation of the matrix material caused by the nanoparticles played a leading role resulting in increasing the crystallinity of the PP matrix composites when the mass fraction of Sb_2O_3 was less (<3wt%); While the mass fraction of Sb_2O_3 was higher (>3wt%), the nanoparticles hindered the movement of the PP molecular chain during the cooling and crystallization process of the PP matrix composites, which caused the matrix material insufficiently crystallized[10,11].

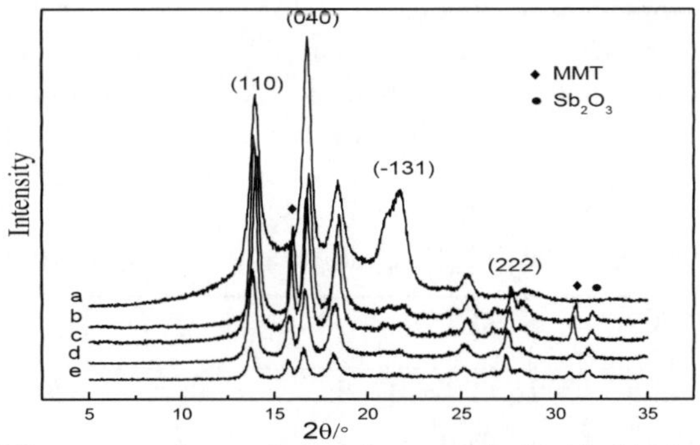

Figure 1. XRD spectra of experimental materials (a) 1#; (b)2#; (c)3#; (d)4#; (e)5#

3.2. Non-isothermal crystallization behavior

The DSC crystallization curve and melting curve of PP and PP matrix composites are shown in figure 2. It can be comprehended from the figure that the crystallization curve of the PP matrix composites was similar to that of pure PP, demonstrating that the addition of MMT and nano-Sb_2O_3 did not change the crystal form of the PP matrix. The crystallization peak position of the PP matrix composites was obviously shifted to the right, but the change of the melting peak position was not obvious. It revealed that the addition of MMT and nano-Sb_2O_3 played a role of heterogeneous nucleation effect, which increased the initial crystallization temperature and crystallization temperature of PP material, but the effect on melting temperature was not obvious.

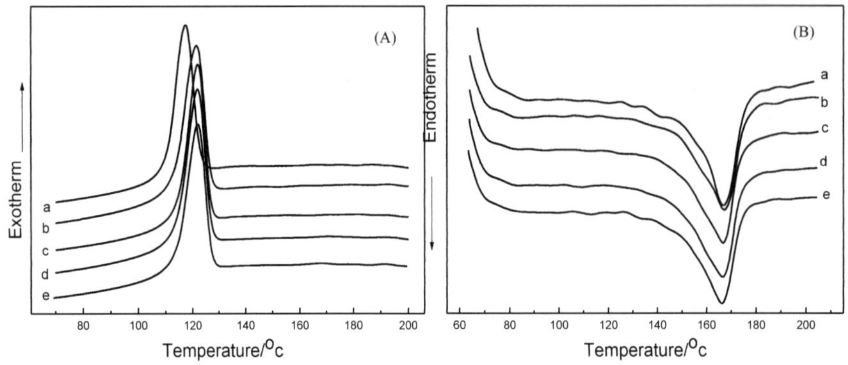

Figure 2. DSC melting curve (A); DSC crystallization curve (B) (a) 1#; (b)2#; (c)3#; (d)4#; (e)5#

Table 2 shows the non-isothermal crystallization parameters of PP and PP matrix composites. T_p is the peak temperature of crystallization; T_m is the peak temperature of melting peak; ΔT is the degree of supercooling, and the value is the difference between T_m and T_p, reflecting the crystallization rate. The smaller the value of ΔT, the greater the crystallization rate of the composite[12]; ΔH is crystallization enthalpy; X(t) is the degree of crystallinity. The crystallinity of the sample can be calculated by the following formula:

$$X(t) = \frac{\Delta H_m - \Delta H}{\Delta H_m^o} \times 100\% \quad (1)$$

where ΔH_m is a sample melting enthalpy, and ΔH is a sample crystallization enthalpy. ΔH_m^0 is the melting enthalpy when the PP crystallinity is 100%, and its value is 207.1 J/g[13,14].

The results of Table 2 show that the crystallization rate and crystallinity of PP matrix composites are significantly higher than that of pure PP. While the mass fraction of nano-Sb_2O_3 increased, the crystallization rate of PP matrix composites became getting faster and the crystallinity showed a tendency to rise first and then decrease. Pure PP had homogeneous nucleation and heterogeneous nucleation in the crystallization process, while PP matrix composites were mainly heterogeneous nucleation. As we all known homogeneous nucleus has longer nucleation time than heterogeneous nucleus. With increasing of the mass fraction of nano-Sb_2O_3, MMT and nano-Sb_2O_3 particles obstructed the movement of macromolecular chains in the process of cooling and crystallization of matrix material, which caused the molecular chain movement insufficient leading to the crystallinity increases first and then decreases[15]. This is consistent with the results of the above XRD analysis.

Table 2. Non-isothermal crystallization parameters of experimental materials

Sample	T_p/℃	T_m/℃	ΔT	ΔH/(Jg-1)	X(t)/%
1	117.35	168.50	51.15	77.23	52.99
2	121.48	167.11	45.63	84.56	54.53
3	121.97	166.59	44.62	85.87	67.65
4	121.87	166.47	44.40	87.76	69.88
5	121.92	166.17	44.25	85.49	67.59

3.3. Tensile properties

The tensile strength and Young's modulus of PP and PP matrix composites are shown in figure 3. It can be known from the figure that the tensile strength of the PP matrix composites was significantly higher than that of the pure PP because the addition of the MMT and the nano-Sb_2O_3 particles

IOP Conf. Series: Materials Science and Engineering **576** (2019) 012031 doi:10.1088/1757-899X/576/1/012031

Bond characteristics of Al/Mg clad materials produced by vacuum roll bonding

J F Shu[1] and T K Yamaguchi[1]

[1]Department of Materials Science, Graduate School of Engineering, Kyushu Institute of Technology, Kitakyushu 804-8550, Japan

E-mail: yamaguchi.tomiko374@mail.kyutech.jp

Abstract. The A1050/AZ31 clad materials were produced by vacuum roll bonding. The effects of rolling temperature and reduction rate on the interface of A1050/AZ31 clad materials were investigated, and the fracture behaviors of A1050/AZ31 interface were studied. A very thin intermetallic compound layer was formed at the A1050/AZ31 interface at rolling temperature of 350°C and reduction rate of 25%. Therefore the A1050/AZ31 interface mainly was ductile fracture, and the average tensile strength of A1050/AZ31 interface reached the maximum (43MPa). When the rolling temperature was 450°C and the reduction rate was 25%, the fracture occurred on the brittle intermetallic compound Al_3Mg_2 layer (thickness 7.68μm) and $Al_{12}Mg_{17}$ layer (thickness 1.95μm), mainly on Al_3Mg_2 layer, and the average tensile strength of A1050/AZ31 interface reached the minimum (0.61MPa).

1. Introduction

Mg alloys as the lightest structural material and green engineering material are attracting much attention in the application of engineering [1, 2]. However, the applications of Mg alloys are still limited due to the poor corrosion resistant performance and weak deformability [3]. Therefore, it is particularly important to improve the disadvantage of Mg alloys.

In recent year, clad materials are being used in various fields since they exhibit excellent comprehensive performance [4-6]. Al is also a very light structural material, which is widely used in industry [7]. It can overcome demerits of Mg alloys by cladding an Al layer with corrosion resistance and good deformability on Mg alloy surfaces [8, 9]. At present, there are many processes can be used to produce Al/Mg clad materials. Roll bonding is an important technique to produce clad materials in industry [4, 10]. However, in the process of warm roll bonding, the oxide film is easy to produce and affect the performance of the interface [11]. The deterioration of oxide film can be effectively reduced under the vacuum condition. In this paper, Al/Mg clad materials were prepared by vacuum roll bonding.

2. Experimental

2.1. Materials and equipment

In this study, the pure industrial aluminium A1050 and Mg-Al-Zn alloy AZ31 plates were used. The dimensions of A1050 and AZ31 plates were 3mm×80mm×300mm, 1.5mm×65mm×200mm, respectively. The chemical compositions of A1050 and AZ31 are given in Table 1. A1050 and AZ31 plates were annealed at 350°C and 400°C for 30min respectively. Before vacuum roll bonding, the AZ31 plate used #2000 sandpaper, and the A1050 plate used steel brush with the diameter of 3mm to remove the oxide films of the surface. The schematic diagram of the vacuum rolling equipment is shown in Figure 1. The materials were heated to rolling temperature through the high-frequency induction heating device and passed through the bonding roll. After rolling, the clad materials were cooled in vacuum.

Table 1. Chemical compositions of raw materials (mass %).

	Si	Cu	Mn	Zn	Ti	Fe	Mg	Al
A1050	0.25	0.05	0.05	0.05	0.03	0.4	0.05	Bal.
AZ31	0.07	0.05	0.27	0.69	-	0.003	Bal.	2.26

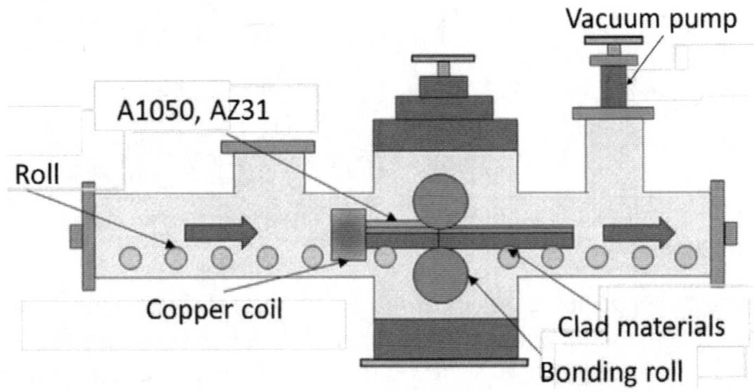

Figure 1. The schematic diagram of the vacuum rolling equipment.

2.2. Conditions

Keeping the vacuum degree (6mPa) and rolling speed (7mm/s) constant, the A1050/AZ31 clad materials were produced at various rolling temperature (350°C, 400°C, 450°C) and reduction rate (20%, 25%) by vacuum roll bonding. The reduction rates referred to in this paper are the total reduction rate of A1050/AZ31 clad materials. The bonding interfaces of the clad materials were observed and analyzed by scanning electron microscope (SEM) and electron probe X-ray microanalyzer (EPMA). To examine the interfacial properties, the tensile tests were performed at room temperature, and energy dispersive X-ray spectrometry (EDS) and X-ray diffraction (XRD) analyses were performed on the fracture surfaces.

3. Results and discussion

The reduction rate was an important parameter in the roll bonding process, affecting the bonding strength of the clad materials. Figure 2 shows the relationship between the reduction rate of A1050 and AZ31 and total reduction rate under different rolling temperatures. The abscissa is the total reduction rate of clad materials, and the longitudinal coordinate is the reduction rate of A1050 and AZ31. It can be seen from the graph that the reduction rates of A1050 are higher than the total reduction rate, and the reduction rates of AZ31 are lower than the total reduction rate. And the higher the rolling temperature at the same total reduction rate, the greater the difference between the reduction rate of A1050 and AZ31. This is because A1050 is FCC structure, and AZ31 is HCP structure. A1050 with more slip systems than AZ31 was easier to produce plastic deformation, and with the increase of rolling temperature plastic deformation capacity was greater.

3.1. Microstructures of the bonding interface

Figure 3 shows the interfacial microstructures of A1050/AZ31 clad materials at different rolling temperatures and reduction rates. It can be found from Figure 3(a) that no obvious diffusion layer was observed at the A1050/AZ31 interface when the rolling temperature was 350ºC. It can be seen from Figure 3(b)(c)(d) that a certain thickness of diffusion layers at the A1050/AZ31 interfaces were formed at rolling temperature of 400ºC and 450ºC. And at the same temperature, the greater the reduction rate, the thicker the diffusion layer; at the same reduction rate, the higher the temperature, the thicker the diffusion layer. Therefore, when the rolling temperature was 450ºC and the reduction rate was 25%, the thickness of the diffusion layer was the largest.

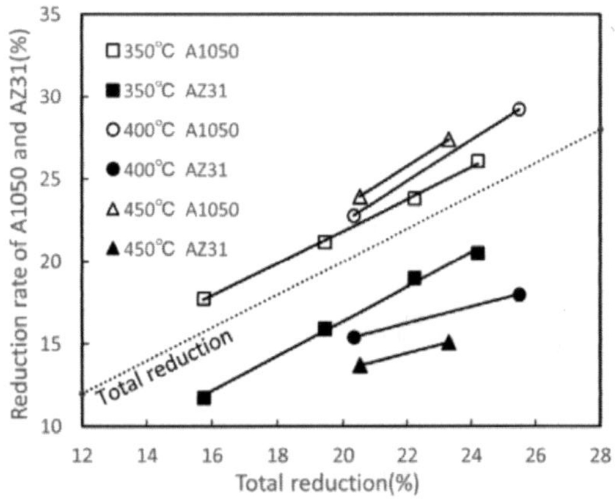

Figure 2. The relationship between the reduction rate of A1050 and AZ31 and total reduction rate under different rolling temperature.

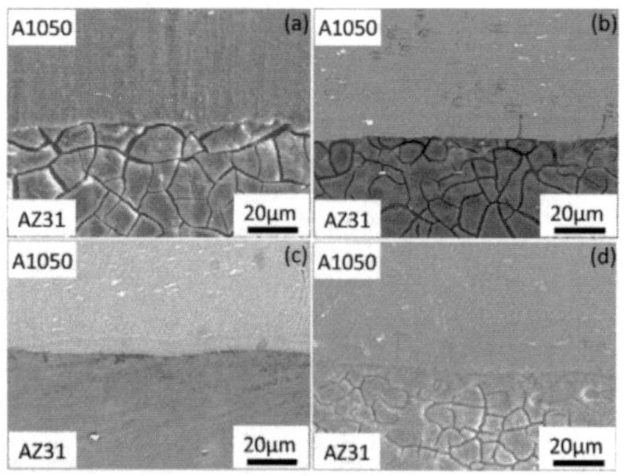

Figure 3. Interfacial microstructures of A1050/AZ31 clad materials at different rolling temperatures and reduction rates: (a) 350ºC 25%, (b) 400ºC 25%, (c) 400ºC 20%, (d) 450ºC 25%.

3.2. EPMA analysis
Figures 4 shows the results of EPMA area analysis of the A1050/AZ31 interface at rolling temperature of 350ºC and reduction rate of 25%. As seen in Figure 4, a very thin intermetallic compound layer was formed at the A1050/AZ31 interface. Figures 5 shows the results of EPMA area analysis of the A1050/AZ31 interface at rolling temperature of 450ºC and reduction rate of 25%. The results show that a certain thickness of the intermetallic compound layer was formed at the A1050/AZ31 interface. And the intermetallic compound of the A1050/AZ31 interface was formed on the A1050 side. This is because the diffusion activation energy of Mg atom in Al is much smaller than the diffusion activation energy of Al atom in Mg. Table 2 shows the results of EPMA point analysis of the diffusion layers. The atomic percentages of Mg in diffusion layer I and II were 40.57% and 58.23%, respectively. According to the Al-Mg alloy phase diagram, the chemical compositions in the diffusion layer I and II were intermetallic compounds Al_3Mg_2 (thickness 7.68μm) and $Al_{12}Mg_{17}$ (thickness 1.95μm), respectively.

3.3. Tensile strength of the bonding interface
The effects of rolling temperature and reduction rate on the tensile strength of A1050/AZ31 bonding interface are shown in Figure 6. From Figure 6, the tensile strength of A1050/AZ31 interface decreased with increasing rolling temperature. This was due to the formation of intermetallic compound of certain thickness during high temperature rolling, resulting in the decrease of tensile strength. In addition, when the rolling temperature was 350ºC, the greater the reduction rate, the greater the tensile strength. However, low reduction rate got higher tensile strength at rolling temperature of 400ºC and 450ºC. The main reason is that the higher reduction rate led to thicker intermetallic compound layer at high rolling temperature, as shown in Figure 3.

The average tensile strength of A1050/AZ31 interface reached maximum (43MPa) at rolling temperature of 350ºC and reduction rate of 25%. It was indicated that even though intermetallic compound was formed at the A1050/AZ31 interface, the thickness of the intermetallic compound layer was too thin to play a leading role in the fracture. When the rolling temperature was 450°C and the reduction rate was 25%, the average tensile strength of A1050/AZ31 interface reached minimum (0.61MPa). The main reason is that the thickness of intermetallic compound reached the maximum (Al_3Mg_2 7.68μm and $Al_{12}Mg_{17}$ 1.95μm) at this time. These intermetallic compounds are brittle and likely to deteriorate the bonding strength of A1050/AZ31 interface.

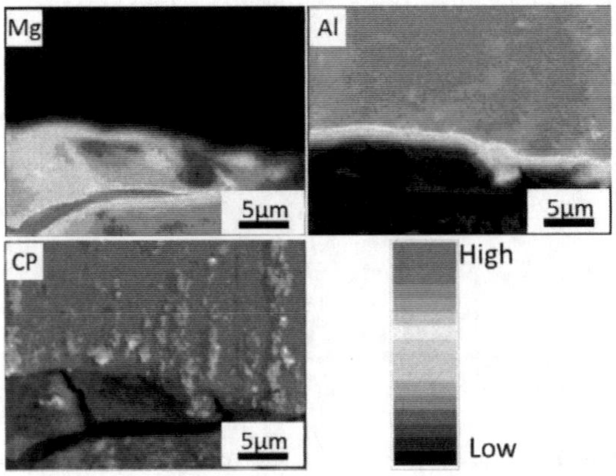

Figure 4. EPMA area analysis of A1050/AZ31 interface at rolling temperature of 350°C and reduction rate of 25%.

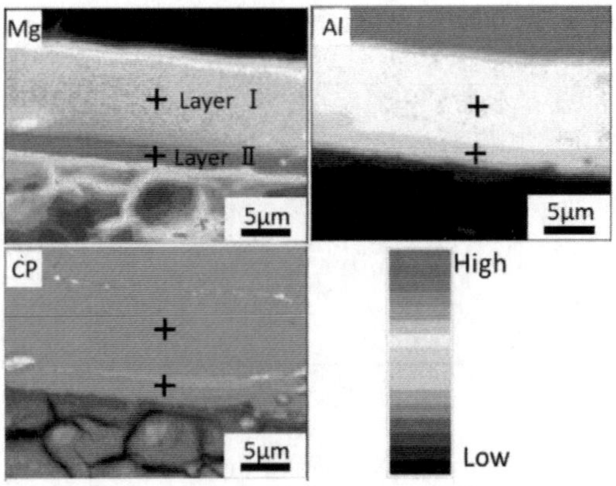

Figure 5. EPMA area analysis of A1050/AZ31 interface at rolling temperature of 450°C and reduction rate of 25%.

Table 2. EPMA point analysis of the diffusion layers.

Diffusion layer	Mg/(Mg+Al) (at%)	Chemical composition
I	40.57%	Al_3Mg_2
II	58.23%	$Al_{12}Mg_{17}$

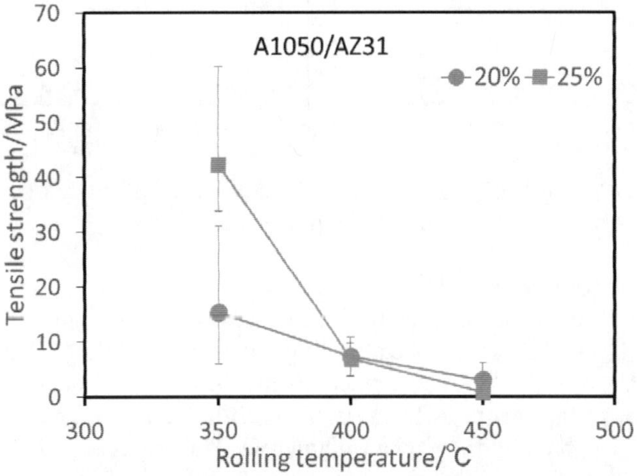

Figure 6. Effects of rolling temperature and reduction rate on the tensile strength of A1050/AZ31 interface.

3.4. Phase analysis of the fracture surface

The fracture surfaces of the tensile samples were analysed by XRD, and the results of A1050/AZ31 fracture surfaces are shown in Figure 7 and Figure 8. The results from figure 7 show that when the rolling temperature was 350°C and the reduction rate was 25%, a small amount of the diffraction peaks of intermetallic compound Al_3Mg_2 were found on both sides of A1050 and AZ31, and Mg diffraction peaks were found on the A1050 side, while Al diffraction peaks were not detected on the AZ31 side. It is concluded that the fracture occurred mainly near the Mg side.

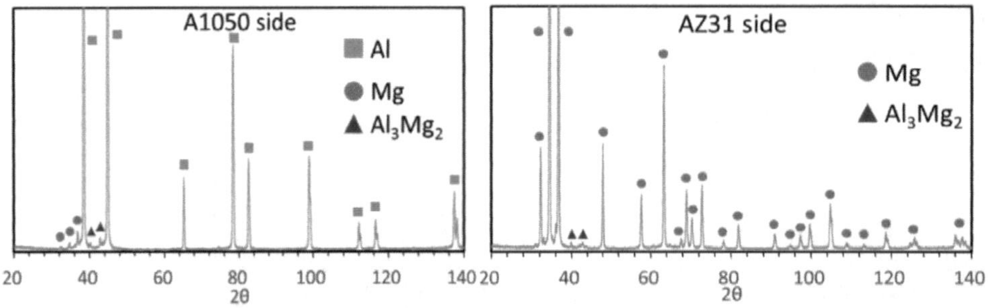

Figure 7. XRDPs of the A1050/AZ31 fracture surface at rolling temperature of 350°C and reduction rate of 25%.

It can be seen from Figure 8 that in addition to the diffraction peaks of the main elements, the diffraction peaks of intermetallic compounds Al_3Mg_2 and $Al_{12}Mg_{17}$ appeared on both sides of A1050 and AZ31 at rolling temperature of 450°C and reduction rate of 25%. Accordingly, combined with the analyses of EPMA, it is obvious that the fracture occurred mainly on intermetallic compound layers I (Al_3Mg_2) and II ($Al_{12}Mg_{17}$) in Figure 5. Furthermore, the amount of diffraction peaks of $Al_{12}Mg_{17}$ was very small. It can be further judged that most of the fracture occurred on layer I (Al_3Mg_2).

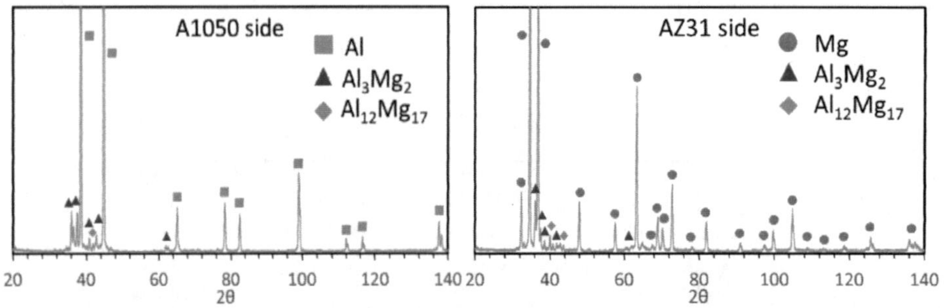

Figure 8. XRDPs of the A1050/AZ31 fracture surface at rolling temperature of 450°C and reduction rate of 25%.

3.5. Morphology analysis of fracture surface

The fracture surfaces of the tensile samples were examined by SEM. Figure 9 shows the morphology of fracture surface of A1050/AZ31 interface at rolling temperature of 350°C and reduction rate of 25%. According to Figure 9, the A1050/AZ31 interface mainly was ductile fracture, and there were a lot of dimples on the fracture surface of the AZ31 side. The brittle fracture also occurred simultaneously at the A1050/AZ31 interface, but it didn't play a leading role. This is because the intermetallic compound layer formed at the A1050/AZ31 interface was very thin.

Figure10 shows the morphology of fracture surface of A1050/AZ31 interface at rolling temperature of 450°C and reduction rate of 25%. It can be seen that the A1050/AZ31 interface was brittle fracture. The EDS analysis of AZ31 side fracture surface is shown in Figure11. As can be seen from Figure 11, the distribution of Mg and Al elements on AZ31 side fracture surface was not very uniform, and the content of Mg elements was obviously less. The content of Mg elements in area 1, 2, 3 was 38.05%, 58.01% and 63.38% respectively, as indicated in Table 3. Combined with the analyses of EPMA, The fracture of area 1 occurred on intermetallic compound layer I in Figure 5, and area 2, 3 occurred on intermetallic compound layer II in Figure 5. Furthermore, the fracture occurred mainly in the layer I (Al_3Mg_2). This result is the same as that of XRD.

Figure 9. Morphology of fracture surface of A1050/AZ31 interface at rolling temperature of 350°C and reduction rate of 25%.

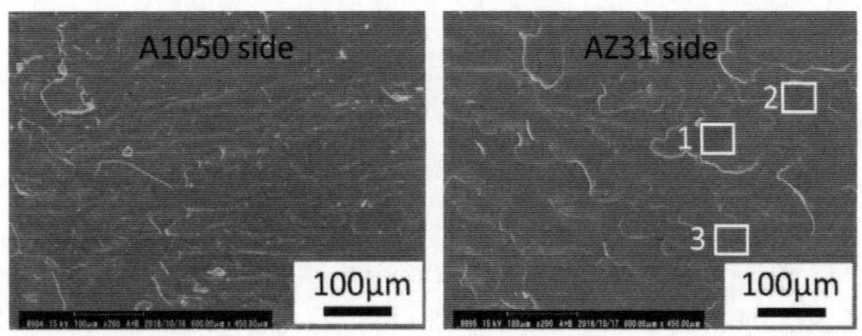

Figure 10. Morphology of fracture surface of A1050/AZ31 interface at rolling temperature of 450°C and reduction rate of 25%.

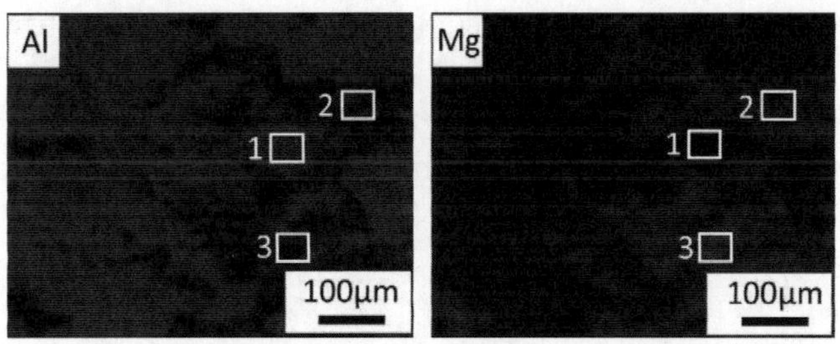

Figure 11. EDS analysis of AZ31 side fracture surface at rolling temperature of 450°C and reduction rate of 25%.

Table 3. Element content in different regions of the AZ31 side.

Area	Mg(at%)	Al(at%)	Zn(at%)
1	38.05	60.87	0.8
2	58.01	41.02	0.66
3	63.38	35.29	1.04

4. Conclusions

In this paper, the A1050/AZ31 clad materials were produced by vacuum roll bonding, and bond characteristics of Al/Mg clad materials were studied. The conclusions are as follows:

(1) A very thin intermetallic compound layer was formed at the A1050/AZ31 interface at rolling temperature of 350°C. A certain thickness of intermetallic compound layers at the A1050/AZ31 interfaces were formed at rolling temperature of 400°C and 450°C, and the greater the reduction rate, the greater the thickness.

(2) The tensile strength of A1050/AZ31 interface decreased with increasing rolling temperature. The average tensile strength of A1050/AZ31 interface reached the maximum (43MPa) at rolling temperature of 350°C and reduction rate of 25%. Low reduction rate got higher tensile strength at rolling temperature of 400°C and 450°C, so the average tensile strength of A1050/AZ31 interface reached minimum (0.61MPa) at rolling temperature of 450°C and reduction rate of 25%.

(3) The influences of intermetallic compound on the fracture behaviors of A1050/AZ31 interface also were discussed. The intermetallic compound layer formed at the A1050/AZ31 interface was very thin at rolling temperature of 350°C and reduction rate of 25%, therefore the A1050/AZ31 interface mainly was the ductile fracture. The intermetallic compound layers of Al_3Mg_2 (thickness 7.68μm) and $Al_{12}Mg_{17}$ (thickness 1.95μm) were formed in the diffusion zone at rolling temperature of 450°C and reduction rate of 25%. Accordingly, the fracture occurred on Al_3Mg_2 layer and $Al_{12}Mg_{17}$ layer, and mainly on Al_3Mg_2 layer.

5. References

[1] Zhang J H, Liu S J, Wu R Z, Hou L G and Zhang M L 2018 *J. Magnes. Alloy* **6** 1-15
[2] Wei G B, Peng X D, Hu F P, Hadadzadeh A, Yang Y, Xie W D and Wells M A 2016 *Trans. Nonferrous Met. Soc. China* **26** 508-518
[3] Feng J W, Li H N, Deng K K, Fernandez C, Zhang Q R and Peng Q M 2018 *Corros. Sci.* **143** 229-239
[4] Zhang L, Meng L, Zhou S P and Yang F T 2004 *Mater. Sci. Eng. , A* **371** 65-71
[5] Yu H L, Lu C, Godbole A, Su L H, Sun Y, Liu M, Tang D L and Kong C 2013 *Sci. Rep.* **3** 1-9
[6] Nie H H, Liang W L, Chen H S, Zheng L W, Chi C Z and Li X R 2018 *Mater. Sci. Eng. , A* **732** 6-13
[7] Abbass M K, Hassan K S and Alwan A S 2015 *IJMMM* **3** 31-35
[8] Tokunaga T, Ohno M K and Matsuura K T 2018 *J. Mater. Sci. Technol.* **34** 1119-1126
[9] Mraied H, Cai W J and Sagüés A A 2016 *Thin Solid Films* **615** 391-401
[10] Wang Q, Leng X S, Yan J H, Guo W B, Fu Y and Luan T M 2013 *J. Mater. Csi. Technol.* **29** 948-954
[11] Xu Y, Hu L N and Sun Y 2013 *J. Alloy. Compd.* **580** 262-269

IOP Conf. Series: Materials Science and Engineering **576** (2019) 012032 doi:10.1088/1757-899X/576/1/012032

Analysis of the precipitated phase changes for Super304H heat resistant steel during shot time aging at 800°C

Song Tao[1] and Wang Zhiwu[2]

[1]Zhixing college of Hubei University, 437200 Hubei Wuhan, China
[2]Wuhan University, 437200 Hubei Wuhan, China

Corresponding Author' E-mail: songtao185880@163.com

Abstract. For analyzing the change trend of Super304H heat resistant steel steel precipitated phases, this article adopts the method of high temperature aging to simulate the material conditions during actual working environment. Super304H steel aged under 800 °C for short time 10 h, 20 and 30 h. And analyzed SEM morphology, precipitate phase statistics, composition changes of the original sample and the samples after aging. The results showed that the number of precipitated phase increased first, then decreased and then increased with the prolong of aging time, while the average size and area fraction of precipitated phase increased all the time. In the precipitated phase, Cr continued to increase while Ni decreased.

1. Introduction

In order to improve the efficiency of power generation and reduce pollution emission of coal-fired power plants, it is an effective development direction to improve the parameters of power plant boilers. However, the development of high-parameter power plant boilers has put forward higher requirements for materials, which are required to have good high-temperature strength, oxidation resistance, high-temperature creep strength and welding performance[1-2]. Super304H steel is a kind of fine austenitic heat resistant steel developed by sumitomo metal corporation and 1nalysed1i heavy industries[3-4]. By adding Cu, Nb and N, the material can precipitate Cu rich phase, $M_{23}C_6$, MX and Z phase to produce precipitation strengthening effect[5-6]. In order to study the variation trend of precipitated phase of Super304H heat resistant steel, the samples with different aging time were simulated by high temperature and short time aging method, and the number, size and composition of the precipitated phase were 1nalysed by SEM and EDS, in order to understand the change of precipitated phase during the aging process of Super304H steel and guide the practical application of the material.

2. Test materials and procedures

Super304H heat-resistant steel used in this test is the steel tube used for the boiler superheater of thermal power plant. The size of the steel tube is up to φ45 * 9mm, and the chemical composition is shown in table 1. Sample line bar samples into 5 * 5 * 20 mm, after surface treatment clean, the samples were aged at 800 °C in the high temperature furnace respectively for 10h, 20h and 30h. SEM was used to observe the changes of precipitated phases, and EDS was used to measure the changes of chemical composition of precipitated phases.

It can be seen from table 1 that the composition of Super304H steel used in the test meets the ASME standard, and the harmful elements such as S and P of the original sample are well controlled.

Therefore, it can be inferred that the mechanical properties of the original steel sample of Super304H used in the test can reach the engineering application.

Table 1. Composition of Super304H (mass fraction %)

Element	C	Cr	Ni	Cu	Nb	N	Si	Mn	S	P	Fe
Content	0.09	18.20	9.62	2.77	0.41	0.10	0.25	0.42	0.03	0.04	Bal

3. Test results and analysis

3.1. Analysis of SEM morphology of the samples

The SEM morphology observation results of the original sample and Super304H heat-resistant steel samples after high temperature aging are shown in Figure. 1.

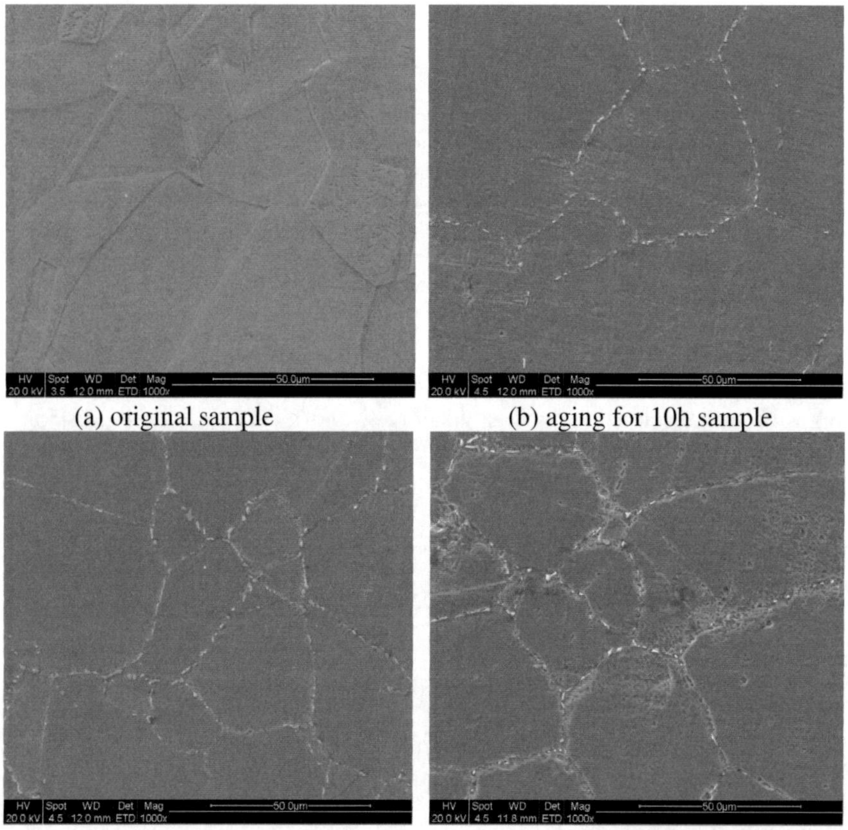

(a) original sample (b) aging for 10h sample

(c) aging for 20h sample (d) aging for 30h sample

Figure 1. SEM morphology test results of the samples

It is generally believed that Super304H steel benefits from fine grain strengthening and precipitation strengthening. The fine grain metal at room temperature has higher strength, hardness, plasticity and toughness than the coarse grain metal. This is because the plastic deformation of fine grains can be dispersed in more grains due to external forces, and the plastic deformation is more uniform and the stress concentration is less. In addition, the smaller the grain size, the larger the grain boundary area, the more zigzag the grain boundary, the more unfavorable crack expansion. The mechanism of precipitation strengthening is that the second phase particles in the metal materials are precipitated out from the supersaturated solid solution, which causes the strain, and thus leads to the

IOP Conf. Series: Materials Science and Engineering **576** (2019) 012032 doi:10.1088/1757-899X/576/1/012032

reinforcement of the metal lattice. The atoms that form the second phase tend to pile up in groups, and they remain coherent to the parent phase.[7-8]

According to the results in Figure. 1, the grain size of the original Super304H sample was the smallest, and the size of the precipitated phase was very small and a large quantity of the precipitates phases, which could provide the effect of fine grain strengthening and precipitated phase strengthening, and guarantee the excellent mechanical properties of the original material and the strength of creep fracture resistance. After 10h of aging, the grain size of the sample was larger than that of the original sample. After aging for 20h, the precipitated phase grew further, and the precipitated phase connected growth bars appeared in some places. At the time of aging for 30h, the integrated mass of the grains was precipitated and shed, and the phase distribution inside the grains was also precipitated, showing serious aging of the material.

3.2. *Statistical results of precipitated phase of the sample*

For further analysis of precipitation changes of Super304H steel ageing at 800 °Calong with aging time, using ImageJ software to analysis the quantity of precipitates, average size of the precipitated phase per unit area, precipitation phase area percentage (total characterization of precipitated phase), and the result is shown in figure 2.

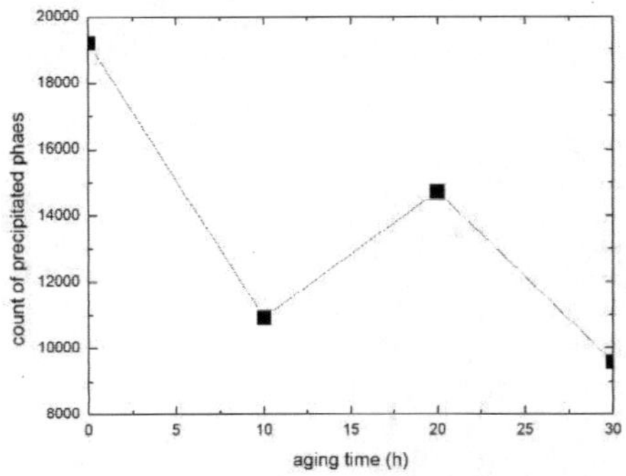

(a) number of precipitated phases

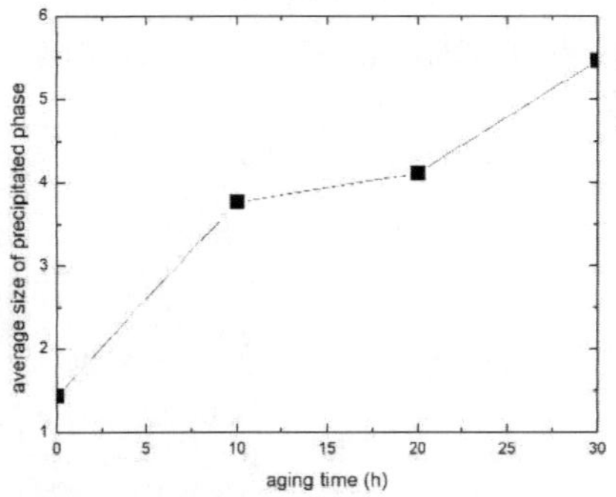

(b) average size of precipitated phases

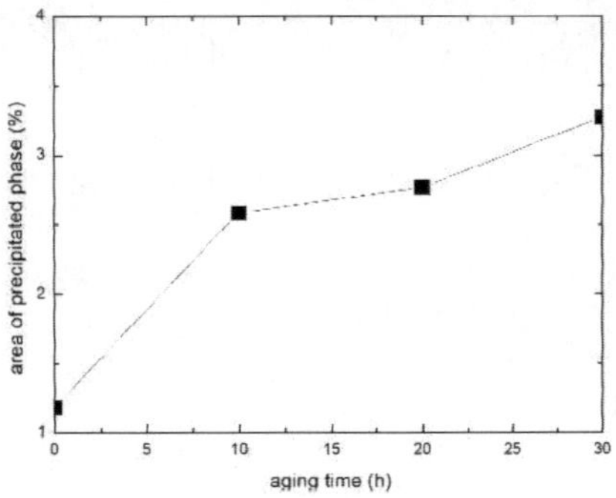

c) area fraction of precipitated phase

Figure 2. Results of statistical analysis of precipitated phases

As can be seen from the results in Figure. 2, the number of precipitated phase per unit area shows a trend of decrease followed by increase and then decrease, while the average size and area fraction of the precipitated phase show a trend of continuous increase. Therefore, it can be speculated that with the increase of aging time, the original sample will show the optimal mechanical performance, while the mechanical performance of the sample will not change much after 10h and 20h of aging, and the mechanical performance will decline sharply after 30h of aging.

3.3. Chemical component analysis of precipitated phases

The original sample and the precipitated phase composition of the sample after high temperature aging were analyzed by EDS attached with SEM. The EDS analysis pictures are shown in Figure.3. The results are shown in Figure. 4.

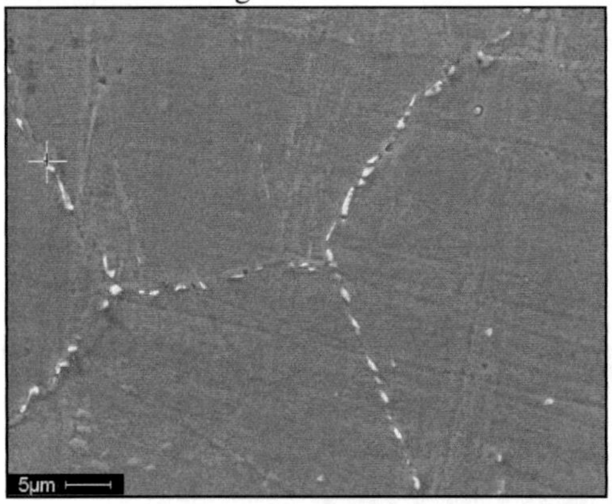

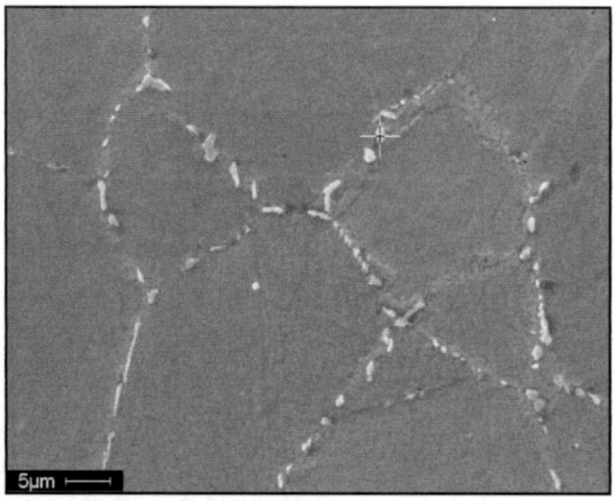

(a) aging for 10h (b) aging for 20h

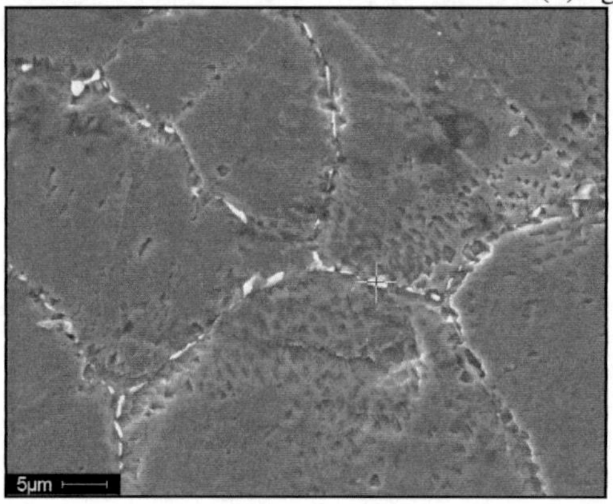

(c) aging for 30h

Figure 3. Pictures for EDS analysis

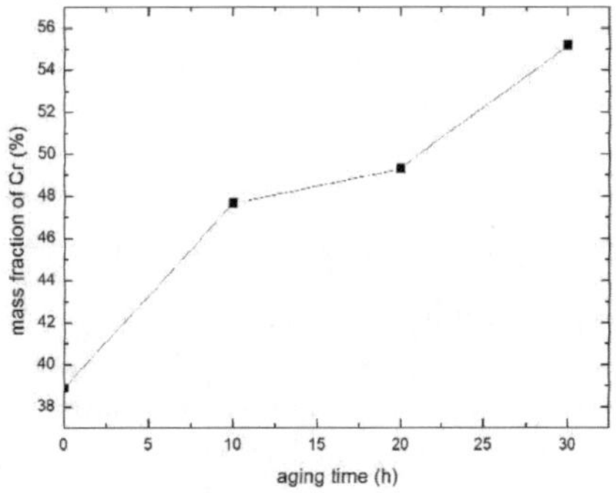

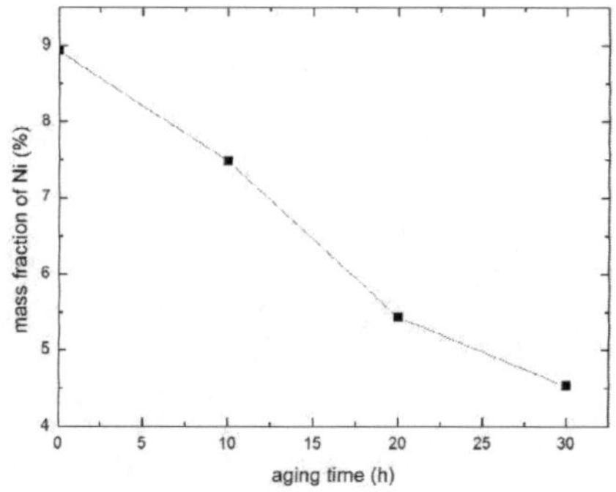

(a) mass fraction of Cr (b) mass fraction of Ni

Figure 4. Analysis results of the precipitated phase composition of the sample

Chromium in high performance and heat resistant steel can significantly improve strength, hardness and wear resistance, but at the same time reduce plasticity and toughness. Chromium also improves the oxidation resistance and corrosion resistance of steel. Nickel increases the strength of steel while maintaining good plasticity and toughness. Nickel has a higher corrosion resistance to acid and alkali, rust prevention and heat resistance at high temperature.[9-10]

As shown in Figure. 4, Cr in the precipitated phase showed a trend of continuous growth with the extension of aging time, while Ni showed an opposite trend. This indicates that with the increase of aging time, the precipitated phase is mainly transformed into $M_{23}C_6$ type precipitated phase and precipitated along the grain boundary, which will lead to decreased creep fracture strength of the material and intergranular corrosion.

4. Results

1) Super304H steel at 800 °C high temperature aging after 10h, 20h and 30h, the precipitated phase quantity reduction first, then increases, and then decreases. And precipitated phase size and area fraction increased all the time.

2) Super304H steel after 800 °C high temperature aging, the Cr in precipitation phases increased while Ni show the opposite trend.

5. References

[1] Ou P, Xing H, Wang X, Sun J, Cui Z and Yang C 2015 *Metall. Mater. Trans. A* **46** 1-8
[2] Qi Y, Wu Z, Zhang X and Ma C 2018 *Sci. Rep.* **8** 2618.
[3] Huang X, Zhou Q, Wang W, Li W and Gao Y 2017 *Mater. High Temp.* **35** 1-13.
[4] Wang W, Huang X, Li W and Gao Y 2017 *Eng. Fail. Anal.* **80** 79-86
[5] Prabha B, Sundaramoorthy P, Suresh S, Manimozhi S and Ravishankar B. 2009 *J. Mater. Eng. Perform.* **18** 1294-1296
[6] Li X, Zhou Y, Zhang Z and Zou Z. 2010 J. Mater. Sci. Technol. **18** 256-261
[7] Bai J, Liu P, Zhu Y, Li X, Chi C, Yu H, Xie X and Zhan Q 2013 *Mater. Sci. Eng. A* **584** 57-62
[8] Li D, Chen X and Dai Q 2011 *A. M. M.* **84-85** 337-341
[9] Jiao J, Fan Z, Zhong F, He C and Wu B 2016 *Res. Nondestruct. Eval.* **27** 193-203
[10] Shim J, Kozeschnik E, Jung W, Lee S, Kim D, Suh J, Lee Y and Cho Y 2010 *Calphad: Comput. Coupling Phase Diagrams Thermochem.* **34** 105-112.

IOP Conf. Series: Materials Science and Engineering **576** (2019) 012033 doi:10.1088/1757-899X/576/1/012033

Synthesis of N-doped TiO_2 Nanoparticle by Solvothermal Method for Dye Treatment

S Samangsri[1, 2], S Chiarakorn[3] and T Areerob[4]

[1]The Joint Graduate School of Energy and Environment, King Mongkut's University of Technology Thonburi, Bangkok, Thailand
[2]Center of Excellence on Energy Technology and Environment, PERDO, Bangkok, Thailand
[3]Division of Environmental Technology, School of Energy, Environment and Materials, King Mongkut's University of Technology Thonburi, Bangkok 10140, Thailand
[4]Faculty of Technology and Environment, Prince of Songkla University Phuket Campus, Phuket 83120, Thailand

Abstract. Nitrogen-doped titanium dioxide (NT) was synthesized by solvothermal method. Titanium (IV) butoxide (TBOT) was used as a titanium source and diethyleneamine (DEA) was used as a nitrogen source. The DEA concentration was fixed at 5% mol. The effect of the synthesized temperatures was studied (150 °C to 200 °C). The characterization results showed that the XRD patterns of NT synthesized by solvothermal method exhibited only anatase structure of TiO_2 at 25.56° (101), 37.76° (004), 48.04° (200), 48.18° (105), 54.06° (211), 55.24° (204), 62.76° (116), 69.04° (220), 70.32° (215) and 75.30° (303). The crystallite sizes of NT (13-14 nm) were smaller than that of TiO_2 (22.39 nm) while the surface area of NT (31-47 m^2g^{-1}) was larger than that of TiO_2 (6.42 m^2g^{-1}). This indicated that doping N on TiO_2 had a significant effect on crystallite size and surface area of TiO_2. The photocatalytic activities of NT were evaluated by decolorization of 10^{-5} M methylene blue (MB) under LED visible light irradiation. The result showed that the photocatalytic performance of TiO_2 was significantly improved by N dopant with the efficiency of 90%. However, the synthesized temperatures had no significant effect on the physical properties and the photocatalytic reactivity of NT.

1. Introduction

TiO_2 is commonly used as a catalyst for the photocatalytic degradation of organic pollutants. It has interesting physical and chemical properties such as good stability, low cost and non-toxicity [1]. Therefore, the TiO_2 is mostly used for environmental treatment such as air purification [2], water treatment [3], self-cleaning [4] and degradation of organic pollutants [5,6]. However, TiO_2 must be activated only under ultraviolet light for photocatalytic activities because it has wide band gap energy about 3.2 eV. Several researchers studied the enhancement of TiO_2 efficiency under visible light by doping non-metal and metal, which are carbon (C), phosphorus (P), bromine (B) and nitrogen (N) for non-metal dopants and silver (Ag), iron (Fe) and copper (Cu) for metal dopants.

Lee, et al. (2005) studied the photocatalytic reactivity of Ag doped TiO_2 by sol-gel method. The photocatalytic efficiency of the doped photocatalyst was examined by p-nitrophenol treatment under UV light irradiation. It was reported that 1 mmol of $AgNO_3$ provided the highest photocatalytic efficiency compared to the bare TiO_2 [7]. Tongon, et al. (2014) prepared Ag doped TiO_2 and

IOP Conf. Series: Materials Science and Engineering **576** (2019) 012033 doi:10.1088/1757-899X/576/1/012033

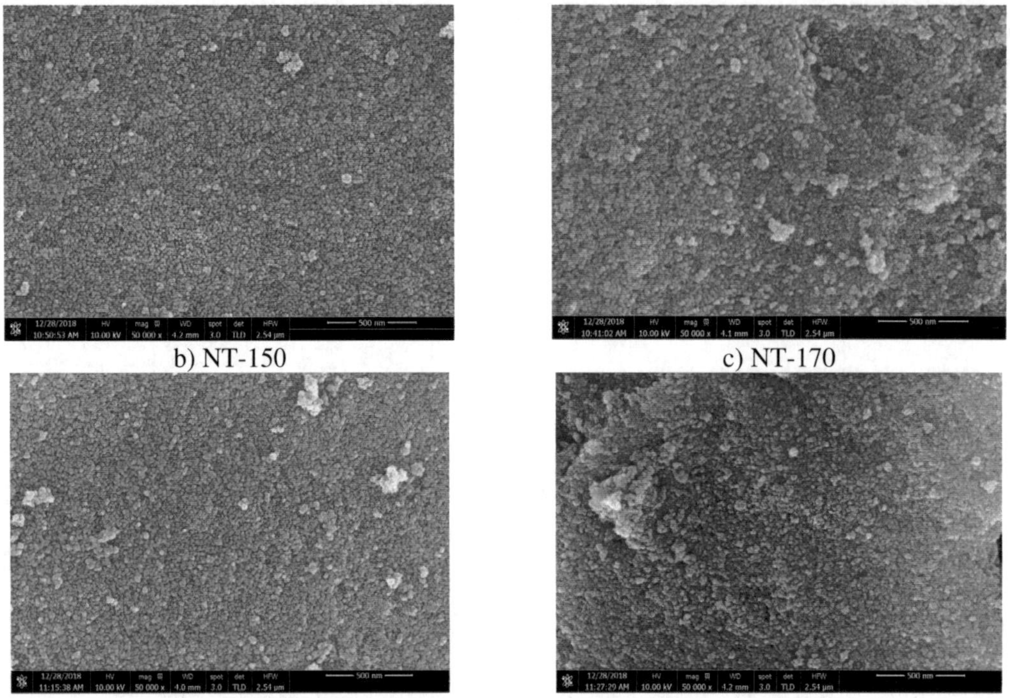

b) NT-150 c) NT-170

d) NT-190 e) NT-200

Figure 2. SEM images N-doped TiO_2 and undoped TiO_2 prepared by solvothermal with different synthesized temperatures.

3.3. Porosity

The specific surface area, total pore volume and average pore diameter of N-doped TiO_2 prepared by different temperatures were analyzed by BET technique as shown in Table 2. The specific surface area and total pore volume of the N-doped TiO_2 samples were significantly increased when compared with the undoped TiO_2. This is caused by the incorporation of nitrogen in anatase TiO_2 structure. Nitrogen atoms could interfere the growth of anatase TiO_2 crystals leading to the decrease of crystallite size. The NT-190 and NT-200 had larger specific surface area than the NT-150 and NT-170. Figure 3 presents the N_2 adsorption-desorption isotherms of N-doped TiO_2 which are identified as type IV isotherms according to the IUPAC classification [16]. It was indicated that N-doped TiO_2 was classified as a mesoporous material (diameter in the range of 2-50 nm) with uniform pore size. This result is agreement with the calculated crystallite sizes as discussed previously.

Table 2. Porosity N-doped TiO_2 and undoped TiO_2 prepared by solvothermal with different synthesized temperatures.

Sample	Specific surface area (m^2 g^{-1})	Total pore volume (cm^3 g^{-1})	Average pore diameter (nm)
TiO_2-150	6.42	0.02	12.94
NT-150	38.16	0.15	15.29
NT-170	31.96	0.14	16.94
NT-190	47.08	0.15	12.65
NT-200	45.06	0.14	12.49

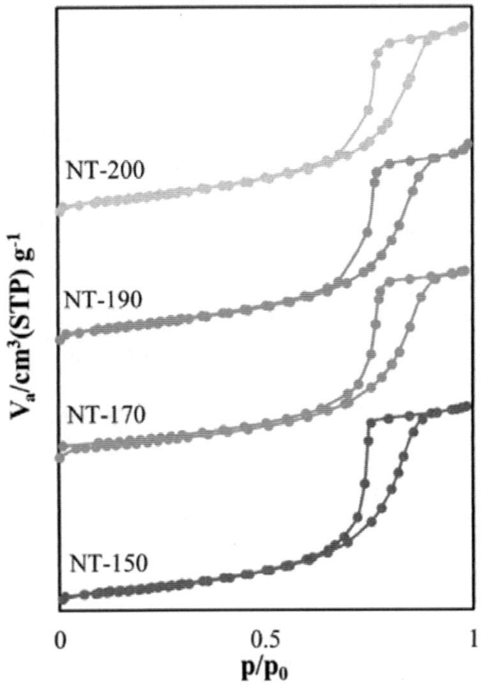

Figure 3. N_2 adsorption-desorption isotherm of N-doped TiO_2 prepared by solvothermal with different synthesized temperatures.

3.4. Absorption edge and band gap energy

Figure 4 shows the UV-visible adsorption spectra of N-doped TiO_2 and the undoped TiO_2 prepared by different temperatures. The adsorption edges of the doped samples were not shifted to higher wavelength, resulting to not decrease of band gap energy. The band gap energy of N-doped TiO_2 and the undoped TiO_2 were 3.07 eV. However, the change of synthesis temperatures did not have the significant effect on the band gap energy.

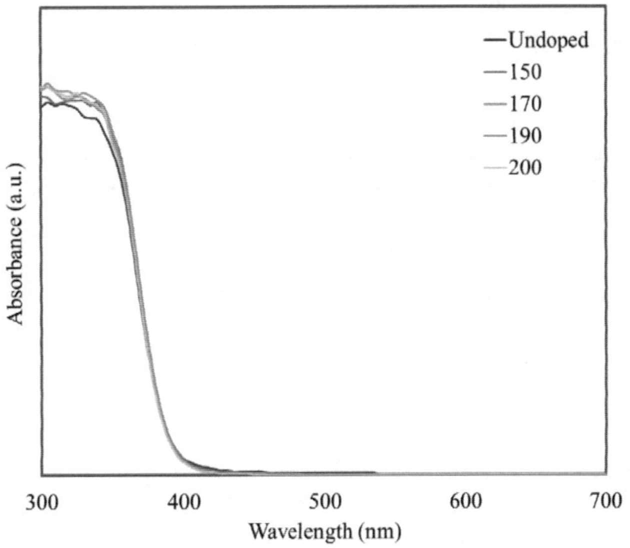

Figure 4. UV-vis absorption spectra N-doped TiO_2 and undoped TiO_2 prepared by solvothermal with different synthesized temperatures.

3.5. Photocatalytic activity

The photocatalytic reactivity of doped and undoped TiO_2 was tested by decolorization of methylene blue (MB) under LED visible light irradiation. The results were shown in Figure 5 and 6. The decolorization efficiency of undoped TiO_2 was 23% within 300 min. While that of NT-150, NT-170, NT-190 and NT-200 were 89, 85, 84 and 92%, respectively within 500 min. It was found that N dopant could enhance the photocatalytic reactivity of TiO_2. Even though the NT-200 had the highest MB decolorization efficiency, its efficiency was not significantly different from the other doped samples. Thus, this paper recommended NT-150 as the promising photocatalyst. The photocatalytic mechanism of N-doped TiO_2 was shown in Figure 7. The role of N atom was creating mid energy band gap above valence band of TiO_2. As a result, the band gap energy was reduced, and the electrons could excited under visible light irradiation. The excited electrons in valence band move to conduction band. This moving creates electron-hole and leaves a hole in valence band. When electron-hole and hole react with oxygen and water (H_2O) in atmosphere to produce super dioxide (O_2^-) and hydroxyl radical (·OH). The super dioxide and hydroxyl radical can oxidize the molecules of MB. Finally, carbon dioxide (CO_2), water and intermediate product are formed.

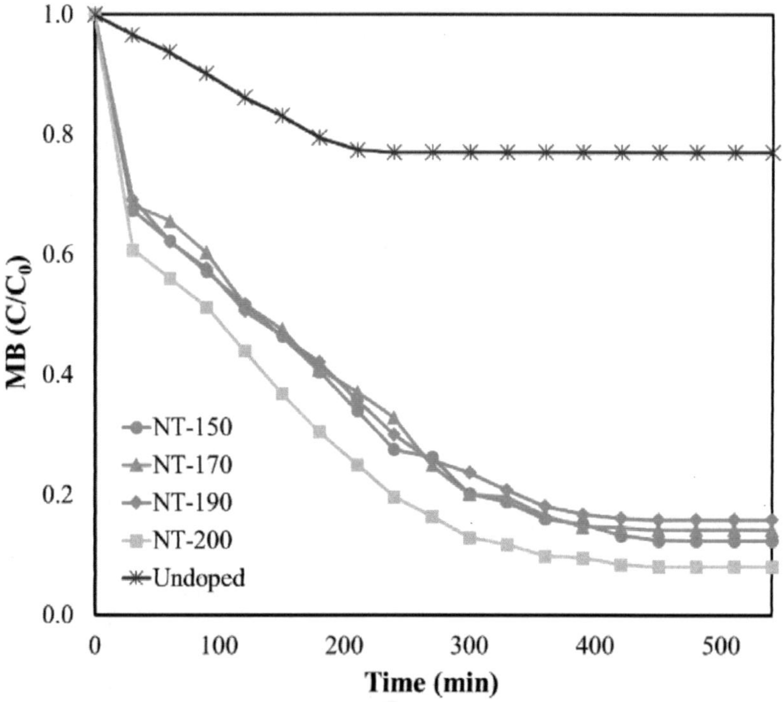

Figure 5. MB decolorization under visible light irradiation by N-doped TiO_2 and undoped TiO_2 prepared by solvothermal with different synthesized temperatures.

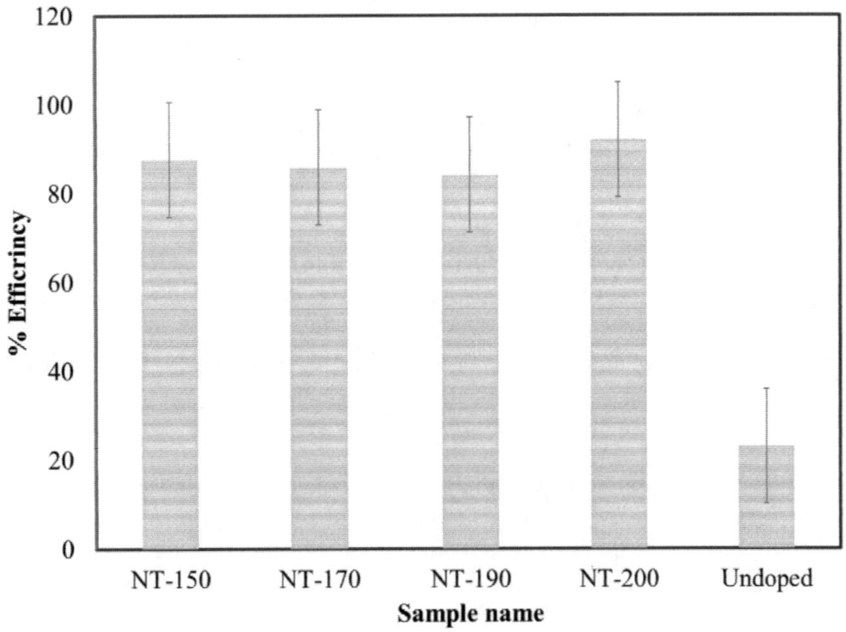

Figure 6. MB decolorization efficiency of N-doped TiO_2 and undoped TiO_2 prepared by solvothermal with different synthesized temperatures under visible light.

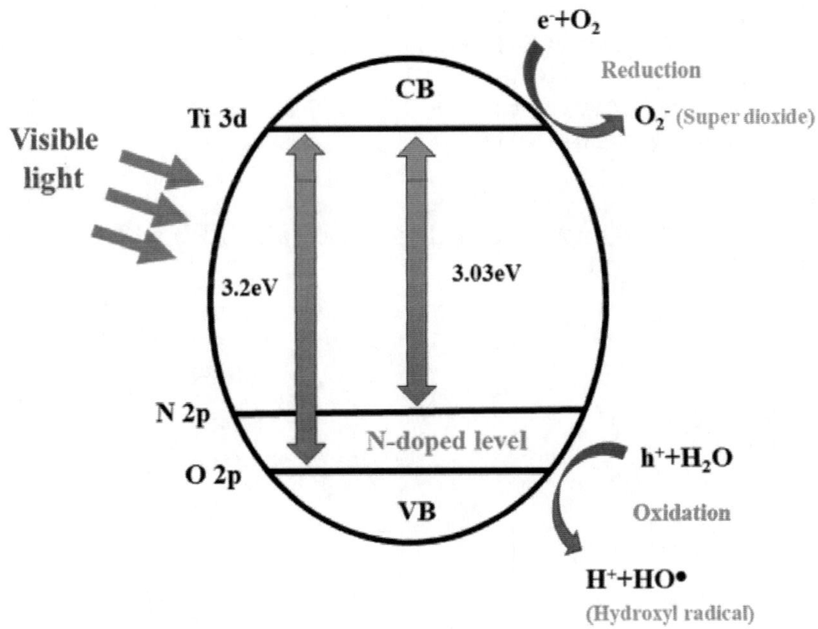

Figure 7. The mechanism of photocatalytic reaction of N-doped TiO_2.

IOP Conf. Series: Materials Science and Engineering **576** (2019) 012033 doi:10.1088/1757-899X/576/1/012033

4. Conclusion

The photocatalytic reactivity of TiO_2 was successfully enhanced by N dopant via solvothermal method. The synthesized temperature at 150°C was recommended as the optimal condition to prepare N-doped TiO_2. N dopant could reduce the band gap energy, decrease crystallite size and increase surface area of TiO_2.

5. References

[1] Kalantari K, Kalbasi M, Sohrabi M and Royaee S J 2016 *Ceram. Int.* 42 14834-14842
[2] Nischk M, Mazierski P, Gazda M and Zalesk A 2014 *Appl. Catal. B-Environ.* 144 674-685
[3] González E I C, Riccò R, Montorsi M, Montorsi M, Falcaro P and Siligardi C 2014 *Build. Environ.* 71 7-14
[4] Rtimi S, Baghriche O, Sanjines R, Pulgarin C, Bensimon M and Kiwi J 2013 *J. Photoch. Photobio. A.* 256 52-63
[5] Chen J, Li G, He Z and An T 2011 *J. Hazard. Mater.* 190 416-423
[6] Thevenet F, Guillard C and Rousseau A 2014 *Chem. Eng. J.* 244 50-58
[7] Lee M S, Hong S S and Mohseni M 2005 *J. Mol. Catal. A-Chem.* 242 135-140
[8] Tongon W, Chawengkijwanich C and Chiarakorn S 2014 *Superlattice. Microst.* 69 481-489
[9] Choi W, Termin A and Hoffmann M R 1994 *J. Phys. Chem.* 98 13669-13679
[10] Nah Y C, Paramasivam I and Schmuki P 2010 *Chem. Phys. Chem.* 11 2698-2713
[11] Daghrir R, Drogui P and Robert D 2013 *Ind. Eng. Chem. Res.* 52 3581-3599
[12] Li Y Z, Hwang D S, Lee N H and Kim S J 2005 *Chem. Phys. Lett.* 404 25-29
[13] Zou Z G, Zhao J C, Fu X Z and Zhang P Y, Chen J, Zhu H M and Ye J H 2004 *J. Funct. Mater.* 35 83-88
[14] He H Y 2010 *Mater. Res. Innov.* 14 165-168
[15] Yang G, Jiang Z, Shi H, Xiao T and Yan Z 2010 *J. Mater. Chem.* 20 5301-5309
[16] Toe E D, Kurniawan W, Mariquit E G and Hinode H 2018 *J. Environ. Chem. Eng.* 6 5125-5134
[17] Demazeau G and Naturforsch Z 2010 65 999-1000
[18] Kolen'ko Y V, Maximov V D, Garshev A V, Meskin P E, Oleynikov N N and Churagulov B R 2004 *Chem. Phys. Lett.* 388 411-415

Acknowledgments

This study is supported by Research and Researcher for Industries (RRI) scholarship from Thailand Research Fund (TRF).

IOP Conf. Series: Materials Science and Engineering **576** (2019) 012034 doi:10.1088/1757-899X/576/1/012034

Fiber Metal Laminate Structure, a good replacement for monolithic and composite materials

Ehsan Sherkatghanad[1,*], Lihui Lang[1], Hamza Blala[1], Lei Li[1] and Sergei Alexandrov[1]

[1]School of Mechanical Engineering and Automation, Beihang University, 100191, Beijing, China

E-mail: ehsan_shgh@yahoo.com

Abstract. Lightweight materials such as aluminium alloys, magnesium alloys, and composite material present a significant possibility for applications in automobile and aerospace structures. In order to consider the feasibility of selecting and replacing such materials in the form of hybrid laminated materials, calculations and simulations are investigated with different lightweight materials such as single layer aluminium, fiber-reinforced composite materials, and the aluminium composite laminates, the carbon and glass cloth as the middle layer, under the constant force condition. The stiffness and deflections of these materials are discussed and compared in details. According to calculations, FE and experimental results validate that the deflections and the stiffness of different materials can be predicted well. In order to finalize the material selection, especially for aerospace-based components, the weight ratios are very determinative. Actually, a material with the superior specific stiffness and strength will be the best one considering some other limitations such as corrosion. Calculating the mass ratio of the suggested lightweight materials were done to find an appropriate one. In fact, these kinds of materials have the advantages of both metallic alloys and composites. In addition, in cases where corrosion resistance, environmental protection, fatigue resistance, and impact resistance are needed, FMLs can be the best choice.

1. Introduction

In order to decrease the fuel consumption and also to reduce pollutions in transportation, research on advanced lightweight materials having equal stiffness can be a popular research issue [1, 2]. After procurement of advanced laminated hybrid materials like fiber metal laminates (FMLs), the tendency to produce modern aerospace or automobile parts with composite materials is also growing manifolds [3, 4, 5]. In one hand, some composite materials such as carbon-reinforced composites, consisted of fabric and epoxy, are widely utilized for components production of aileron, landing gear doors, flaps, and other artefacts used in aeronautical industry [6, 7]. In addition, the composite materials can be applied in other applications like home construction, military, automobile and sport industries [8, 9]. Carbon/Epoxy composites show mechanical properties similar or superior than the conventional monolithic and metallic materials such as Aluminium. In most cases an advanced polymeric composite offers high specific strength and specific stiffness ratios [7, 10, 11]. FMLs are hybrid materials consisting of alternating thin layers of metal sheets and composites (fiber-reinforced epoxy prepreg). Aluminium is the most conventionally used metal, and Kevlar, glass or carbon can be used as fibers. GLARE is the trade name of FML using glass fibers in the middle layers, ARAL is the trade name of

FML using Kevlar fibers in the middle layers and CARAL is the trade name of FML using carbon fibers in the middle layers. They have been evaluated for potential usages in aerospace structures. Glare contains thin aluminium layers stuck together with unidirectional/biaxial reinforced prepregs of high-strength glass fibers. This material offers a pioneer combination of characteristics such as distinguished fatigue resistance, superior specific static properties, significant impact resistance, flame resistance and corrosion properties. In addition, for normal and easy-shaped parts, manufacture and repair are not challenging. GLARE laminate was selected for the upper fuselage skin structures of Airbus A380, as shown in figure 1. This is the first structural application of GLARE laminate in a commercial airline. This structure resulted in providing a new group of the hybrid laminate system which is able to prevent and stop crack growth caused by dynamically cyclic loading, with excellent impact and damage tolerance properties while having low density and high specific strength and stiffness [12, 24].

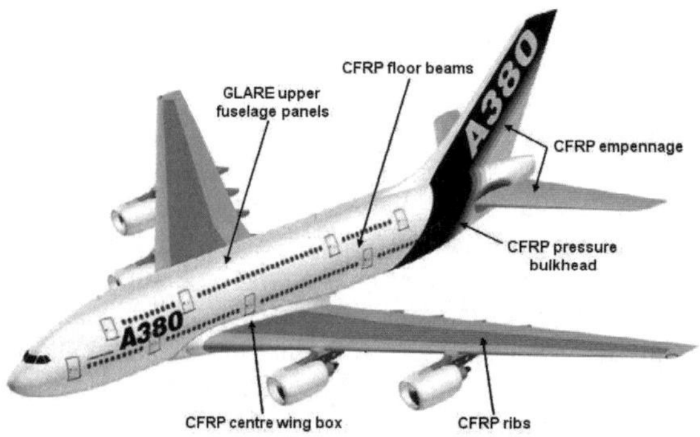

Figure 1. Glare application in Airbus A-380 [24].

As already mentioned, FMLs and conventional composite materials have been demonstrating prosperity in their applications especially in low volume aerospace components and defence applications because of their superior fatigue and impact characteristic [13]. While, there has been restricted use of these kinds of hybrid materials in the high volume productions such as automotive, construction, and consumer goods due to long production times. Also, for complex-shaped parts forming using these materials there are still big challenges. In addition, design for this kind of materials is not very simple and in many cases, former calculations for metals and isotropic materials cannot be useful. In fact, the experiences and findings obtained over the years about the metallic materials' behaviour do not apply to hybrid and composite structures. Composites and Hybrid Materials offer significant variety in design and manufacturing, but still there are some difficulties in these two fields.

Some investigations have been occurred to solve the manufacturing process of these materials. Zafar R. et al. modified an innovative way named as "3A method". Their methodology is based on the generating of any number of metallic blanks in the desired shape simultaneously utilizing hydroforming process [14]. Also, Sherkatghanad E. et al. investigated an innovative approach using hydroforming method and applying semi-cured Glare blanks instead of a solidified blank to form complex shaped FML components [15].

Another problem is how to satisfy the design requirements for a component while a FML is desired to use and how it can be replaced by these materials. Actually, in some cases monolithic materials and composite materials are not appropriate and they must be replacing by FML to enjoy their advantageous properties. In one hand, in the condition of replacing a FML or composite material, usually, there is not enough information to design the component. In the other hand, in most applications of lightweight materials especially in structures, the mainly loading mode is bending. Generally, bending stiffness and bending strength are the limiting parameters in replacing parts. In fact, according to the "materials performance index" concept, in the condition of bending load for a component, the minimum feasible thickness (t), can be calculated when the mass (m) is minimized [12,

16]. Also, mechanical properties predictions for textile composites have been widely surveyed with most investigations focusing on plain woven composites. Primary models for the woven lamina analysis can be traced back to the 1970s when Halpin et al. investigated the stiffness of 2D and 3D composites [17]. After that, the theory was improved by Chou and Ishikawa [18, 19]. Great reviews were expressed for modelling, design, and calculations in this field and most of our main formulas were derived which are mentioned in References [20, 21].

In fact, the result of this study, help us to understand the behaviour of FMLs behaviour.

2. Materials and method

A material index evaluates the superiority of a material in a given demand and application. Material indices are applied to rank materials that satisfy the requirements imposed by a design. Depending on the condition, deflections can be calculated considering the principle static equations. Generally, in the material replacement of a component, the part geometry is kept constant excluding the thickness of the part. In fact, in most cases thickness changing is possible and the mass is the limiting factor. So as it is obvious, a material selection with a high specific stiffness and strength is needed. It means that mass, stiffness, and strength must be considered simultaneously to find the most appropriate material with the minimum mass. In fact, to boost the lightweight design popularity in its different applications in the industries, this research is carried out in the area of stiffness limited design. There are some calculations for understanding and measuring the stiffness of monolithic material as well as multilayer materials. In this research Al 2024-T3, epoxy prepreg glass fiber (50%-50%), epoxy prepreg medium strength carbon fiber (50%-50%), Glare, and Caral are compared to each other to find the most suitable material in the condition of Bending Stiffness. The standard properties of each material are depicted in table 1 and the whole calculations and simulations are according to these properties. Al layer thicknesses are 0.3 and 0.8 mm and each fibers epoxy laminate is 0.2 mm. In fact, it is tried to find the stiffness of a specimen with different material to show which of them is the best one related to requirements of the design. After that, some simulations are carried out to validate the calculations as well as the experiments. Finally, the most important parameters are discussed in details.

Table 1. Nominal Properties of suggested materials.

	Elasticity Modules (Gpa)	**Density (gr/cc)**
Aluminium 2024-O	73	2.78
Carbon Epoxy (50%-50%)	75	1.6
Glass Epoxy (50%-50%)	31	1.8
GLARE	60	2.5
CARAL	72	2.4

3. Calculations

In this part, some calculations and material performance indices are presented. The bending stiffness for the one-layer material can be calculated as follows:

$$D_{Bending} = \frac{Et^3}{12(1-v^2)} \tag{1}$$

Where E is Elastic Modulus, t is the thickness of material, v is the Poisson ratio, and D is the stiffness in the bending condition. In addition, the main formula for multi-layer material, to calculate the bending stiffness, is as follow [21, 23]:

$$D = \frac{1}{3}\sum_{k=1}^{n} \overline{D}(z_{k+1}^3 - z_k^3) \tag{2}$$

So, by substituting the parameters for three-layer (2+1) FMLs condition, the bending stiffness can be calculated as follow [21]:

$$D_{Bending_3} = \frac{1}{3}\left\{\frac{2E_f}{(1-v_f^2)}\left[\frac{3}{2}h_c t_f^2 + \frac{3}{4}h_c^2 t_f + t_f^3\right] + \frac{E_C}{(1-v_c^2)}\frac{h_c^3}{4}\right\} \quad (3)$$

Where E_f is Elastic Modulus of face, v_f is the Poisson ratio of the face, t_c is the thickness of the core, t_f is the thickness of the faces and v_c is the Poisson ratio of the core. The parameters are shown in figure 2.

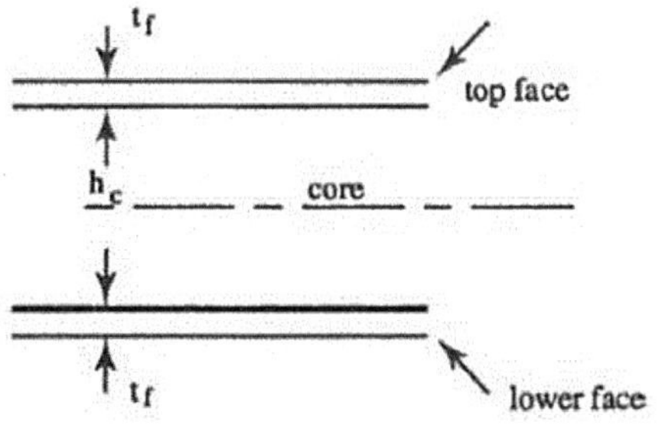

Figure 2. The definitions of the 2+1 FMLs parameters, used in the formulas.

In addition, the bending stiffness for five-layer (3+2) FMLs can be calculated as follow:

$$D_{Bending_5} = \frac{1}{3}\left\{\frac{2E_f}{(1-v_f^2)}\left[15h_c t_f^2 + \frac{15}{2}h_c^2 t_f + 8t_f^3\right] + \frac{E_C}{(1-v_c^2)}\left[\frac{27}{4}h_c^3 + 12h_c^3 t_f + 6h_c t_f^2\right]\right\} \quad (4)$$

As a case study, it is considered that displacement, force, width, and length are 5 mm, 5 N, 30 mm and 150 mm respectively. In addition, according to the use of woven fabrics, our materials are quasi-isotropic. So it can be stated that:

$$\delta = \frac{FL^3}{48EI} \quad (5)$$

And

$$I = \frac{bh^3}{12} \quad (6)$$

So,

$$\delta = \frac{12FL^3}{48Ebh^3} \quad (7)$$

Finally, the thickness equation is obtained as follows:

$$t = \left(\frac{12\left(\frac{F}{\delta}\right)L^3}{48bE}\right)^{1/3} \quad (8)$$

Where δ is the displacement in the middle, F is the applied force, L is the length, b is the width, and h is the thickness of the sheet as shown in figure 3.

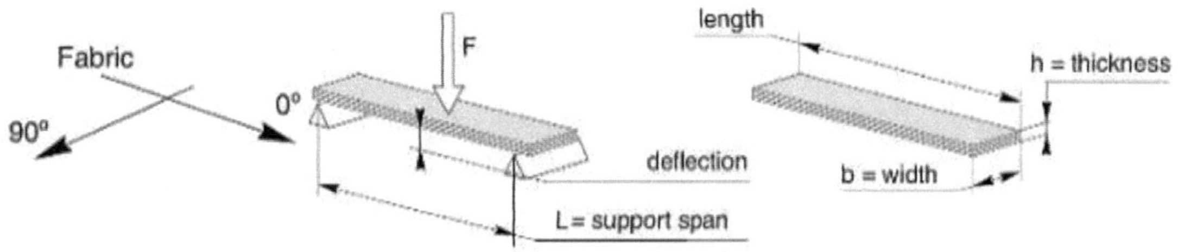

Figure 3. The parameters in 3-point Bending test

The first step is to calculate the bending stiffness for the above-mentioned lightweight materials in table 1. Bending stiffness has been obtained by their own Formula according to material structures. The calculated results are shown in table 2.

Table 2. Calculated Stiffness for Suggested Materials.

Material	Bending Stiffness (N·m2)
Aluminium 2024-O (t=0.8 mm)	4.34
Carbon/Epoxy (three layers, total thickness=0.8 mm)	3.98
Glass /Epoxy (three layers, total thickness=0.8 mm)	1.6
GLARE 2+1 (Al, t=0.3 – Glass, t=0.2 – Al, t=0.3)	4.26
CARAL 2+1 (Al, t=0.3 – Carbon, t=0.2 – Al, t=0.3)	4.33

In the following diagram, figure 4, the stiffness ratios of different materials vs. Aluminium, based on calculations, is depicted. As it is depicted the lowest ratio belongs to Glass/epoxy composite and it has a big space in comparison to the others, whereas Caral has the highest one with a minimum space and CARAL and GLARE stiffness ratios are very close together.

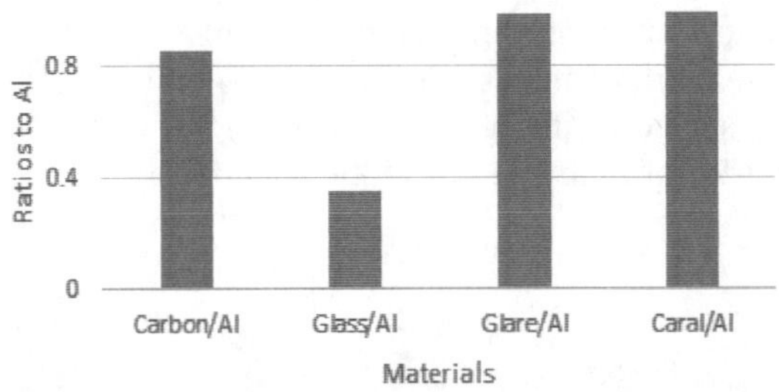

Figure 4. Calculated Stiffness Ratio for Carbon/Epoxy, Glass/Epoxy, Glare and Caral vs. Aluminium

The calculated thicknesses in the case of equal stiffness are as mentioned in table 3 which means, as a cases study, if having the same stiffness in the condition of force 10 Newton is desired, the thicknesses of different suggested materials must be as table data. In this case as well as stiffness ratios, it is shown that Carbon Epoxy composite, Caral and Glare have really close results and it may be found as a good replacement for each other.

Table 3. Calculated thickness in the condition of equal stiffness while a force of 10 Newton applied

Material	Calculated thickness in the condition of equal stiffness (mm)
Carbon Epoxy	0.721
Glass Epoxy	0.968
Al	0.727
GLARE 3	0.776
CARAL 3	0.731

In other word, in the case of the material replacement, if the whole parameters and the geometry, excluding the elasticity modulus and thickness, are considered as a constant parameter, so Eh^3 must be constant.

According to the above-mentioned formula 5, 6, and 7, in the condition of comparing two materials for the same stiffness, the effective parameters can be calculated as follows:

$$\frac{h_1}{h_2} = \sqrt[3]{\frac{E_2}{E_1}} \quad (9)$$

Equation No. 9 can be calculated by using Formula 5, 6, and 7 and comparing two kinds of materials as well as using material performance index table in the references of material selection [16]. Also, mass is calculated as $m = \rho bhl$. So, it can be obtained that:

$$m = \rho bl \left(\frac{12\left(\frac{F}{\delta}\right)L^3}{48bE} \right)^{1/3} \tag{10}$$

Hence,

$$m = \left(\frac{\rho}{E^{1/3}}\right)\left[bl\left(\frac{(F/\delta)L^3}{4b}\right)^{1/3}\right] = \left(\frac{\rho}{E^{1/3}}\right) * cte \tag{11}$$

So it is obvious that Material Performance Index to maximize for minimum mass is $\left(\frac{E^{1/3}}{\rho}\right)$. In order to compare two or more different materials, it can be defined as:

$$\frac{m_1}{m_2} = \frac{\rho_1}{\rho_2}\left(\frac{E_2}{E_1}\right)^{1/3} \tag{12}$$

In fact, for selection and replacement of materials, some other parameters such as thickness ratio, mass ratio, and even cost ratio must be considered. It should be noted that in order to select the superior materials, the total mass ratio and the thickness ratio must be minimized. In this paper, after the calculation of bending stiffness, in order to have an accurate selection of the material, a comprehensive survey for thickness ratio and mass ratio are done.

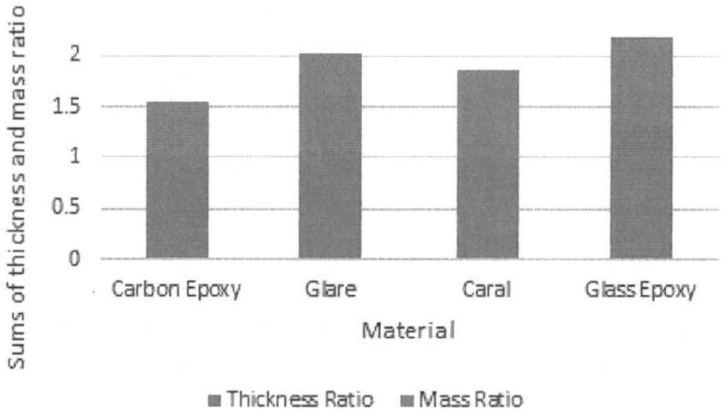

Figure 5. Calculated Sums of thickness and mass ratios for Carbon/Epoxy, Glass/Epoxy, Glare and Caral vs. Aluminium

The calculate sums of thickness and mass ratio for different material are depicted in figure 5. Considering this factor, carbon epoxy can the best choice but in many cases this kind of material is susceptible to expose to the environment as well as corrosion. So, depending on the ambient condition of the component, an appropriate material should be selected.

4. Simulations and Experiments

4.1. Simulations

In this section, simulation results are discussed. In fact, in this research, hired software to simulate the process of tests is ABAQUS. The supports are completely constrained in the process where the specimen is able to move and the force applied in the Z-axis direction in the middle of the specimen to simulate the 3-point bending test. In fact, the same force (5 newton) is applied for the whole combinations of materials and the related displacements were measured in both simulations and experiments. The resultant data are depicted in table 4. The whole experiments and simulations were done in the condition of the same thickness. The simulation results as typical ones for stress and strain distributions in the specimen are depicted in figure 6. As mentioned in the previous section, simulations and experiments including the investigations into the mechanical properties of different kinds of fiber metal laminates, the Aluminium 2024-T3, glass fiber epoxy composite (EWR200-100 prepreg), carbon fiber epoxy composite (T700-12K prepreg) are considered. In another word, specimens of Aluminium alloy, Caral, Glare, Carbon epoxy, and glass epoxy were cut into rectangular shaped specimens with mentioned size according to the test standards. In addition, in order to avoid the debonding phenomenon between the metal layer and glass fiber layer, the phosphoric acid anodizing of the aluminium alloy utilized before making the FML for both methods.

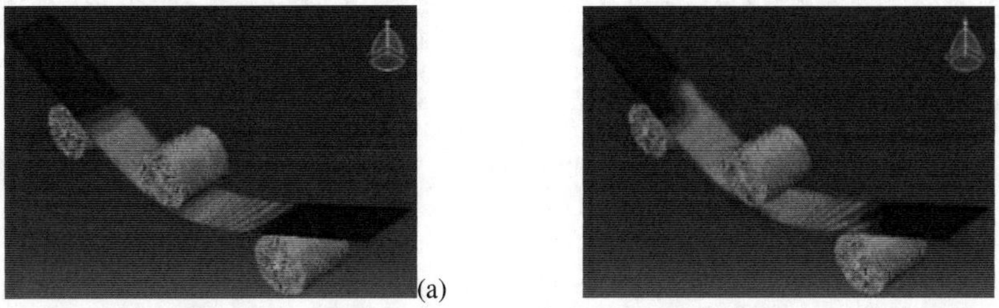

(a) (b)

Figure 6. Typical simulation result of (a) stress and (b) strain distributions in 3-point Bending Test

4.2. Experimental results

Different thicknesses of this Al alloy and prepregs were obtained by 0.8, 0.3, 0.2 and 0.1 mm of thicknesses, to see the cumulative effect of the different combinations of materials by simulations and experiments, which are depicted in Table 4. Furthermore, the differences between simulations and experiments value can be understood as a quality parameter of the FML manufacturing. In fact, there are many effective factors on the manufacturing of an FML such as anodizing the Al, laying up the prepreg, the quality of the prepreg fibers, optimizing the curing curve and temperature. The method of 3-point bending test is shown in figures 7.

Table 4. Resultant Displacement in the condition of the same force of 5 newton in the conditions of both numerical calculation and experiment

No.		Material definition	Layers Orientations	Layers Th. (mm)	Displacement (mm)	Experimental Results (mm)
1	A	*Glass Composite*	*0/0/0/0*	4*0.2	0.718	0.90
	B	*Glass Composite*	*0/90/0/90*	4*0.2	1.150	1.3
	C	*Glass Composite*	*Woven*	4*0.2	0.945	1.1
2	D	*Carbon Composite*	*0/0/0/0*	4*0.2	0.247	0.30
	E	*Carbon Composite*	*0/90/0/90*	4*0.2	0.45	0.52
	F	*Carbon Composite*	*Woven*	4*0.2	0.427	0.50
3	G	*Glare (Al/Glass/Glass/Al)*	*Al/0/0/Al*	0.3/2*0.1/0.3	0.427	0.47
	H	*Glare (Al/Glass/Glass/Al)*	*Al/0/90/Al*	0.3/2*0.1/0.3	0.429	0.50
	I	*Glare (Al/Woven Glass/Al)*	*Al/0,90/Al*	0.3/0.2/0.3	0.427	0.48
4	J	*Caral (Al/Carbon/Carbon/Al)*	*Al/0/0/Al*	0.3/2*0.1/0.3	0.419	051
	K	*Caral (Al/Carbon/Carbon/Al)*	*Al/0/90/Al*	0.3/2*0.1/0.3	0.424	0.49
	L	*Caral (Al/Woven carbon/Al)*	*Al/0,90/Al*	0.3/0.2/0.3	0.423	0.50
5	M	*Al 2024 – T3*	-	0.8	0.423	0.47

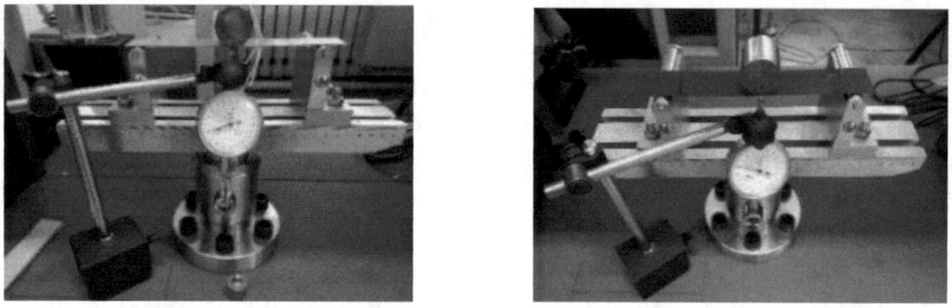

Figure 7. The applied method for 3-point Bending test and measuring the displacement

5. Conclusions

It is found that the behaviour of the specimens in the experimental results are in accordance with 3-points bending simulations and calculation. Although, the values of the experiments are not completely the same as the simulations and calculations but the trend of the results are in an agreement with each other and prove that the material behaviour can be predicted in order to our method. According to close results for the stiffness, it is validated that Fiber Metal Laminates with composite fibers as the middle layer are able to be an appropriate substitution for lightweight metal alloys and composites materials, especially where the equal-stiffness is the limiting factor. In fact, material design can be done innovatively and conditionally. Changing the middle layer thickness, number of layer and their orientation are other flexible parameters which can be investigated.

References

[1] Mustafa KK, Magnesium and its alloys applications in automotive industry *Int J Adv Manuf Technol*, vol. **39**, 2008, pp. 851–65.
[2] Fredrich H and Schumann S, Research for a new age of magnesium in the automotive industry *J Mater Process Technol*, vol. **117**, 2001, pp. 276–81
[3] Reyes G, and Kang H, Mechanical behavior of lightweight thermoplastic fiber–metal laminates *J Mater Process Technol*, vol. **186**, 2007, pp. 284–90
[4] Sokolova O, Carrado A and Palkowski H, Metal–polymer–metal sandwiches with local metal reinforcements: A study on formability by deep drawing and bending *Composite Structures*, vol. **94**, 2011, pp. 1–7
[5] Pilato LA and Michno MJ, Advanced Composite Materials, Berlin, Germany. *Springer Verlag Berlin Heidelberg*, 1994
[6] Baker AA, Callus PJ, Georgiadis S, Falzon PJ, Dutton SE and Leong KH, An affordable methodology for replacing metallic aircraft panels with advanced composites *Composites: Part A*. 2002, Vol **33**, issue 5, pp 687-96
[7] Park SY, Choi WJ and Choi HS, A comparative study on the properties of GLARE laminates cured by autoclave and autoclave consolidation followed by oven postcuring *Int J Adv Manuf Technol*, vol. **49**, 2010, pp. 605-13
[8] Cândido GM, Almeida SFM and Rezende MC, Processamento de laminados de compósitos poliméricos avançados com bordas moldadas *Polímeros: Ciência e Tecnologia*. 2000, Vol. **10**, issue 1, pp. 31-41
[9] Schwartz MM, Composite Materials: Properties, Nondestructive Testing, and Repair. v. **1**. New Jersey, USA. *PrenticeHall Inc*, 1997
[10] Buehler FU and Seferis JC, Effect of reinforcement and solvent content on moisture absorption in epoxy composite materials *Composites Part A*, 2000, Vol **31**, issue 7, 741-48
[11] Hayes BS, Gilbert EN and Seferis JC, Scaling complications of dual temperature cure resin prepreg systems in airplane part manufacture *Composites Part A; Applied Science and Manufacturing*, 2000, Vol **31**, issue 7, pp. 717-25
[12] Luo AA, Magnesium, Current and Potential Automotive Applications *JOM*, 2002, Vol. **54**,

TEM with the focused ion beam (FIB; FEI Nova Nanolab 200) was used to observe the cross-section of the thin film to determine the thickness of thin film. X-ray photoelectron spectroscopy (XPS; ESCALAB 250Xi spectrometer) was employed to analyse the chemical composition. The I-V curves were measured by Autolab 302N electrochemical workstation connected to Nova software.

3. Results and Discussion

3.1. Phase and Chemical composition

The XRD patterns shown in Figure 2a demonstrated that all diffracted peaks can be indexed as the face-centered-cubic pure phase with space group $Fm\bar{3}m$ of cerium oxide (JCPDS 34-0394). No impurity peaks were detected which indicates that no Pr-related oxide was formed during synthesis. To verify the absence of dopant Pr, an ICP measurement was examined, shown in Figure 2(b). It can be seen the actual dopant concentration is close to the nominal dopant concentration. Combining with the XRD result, it's suggested the Pr element is successfully doped into the cerium oxide lattice.

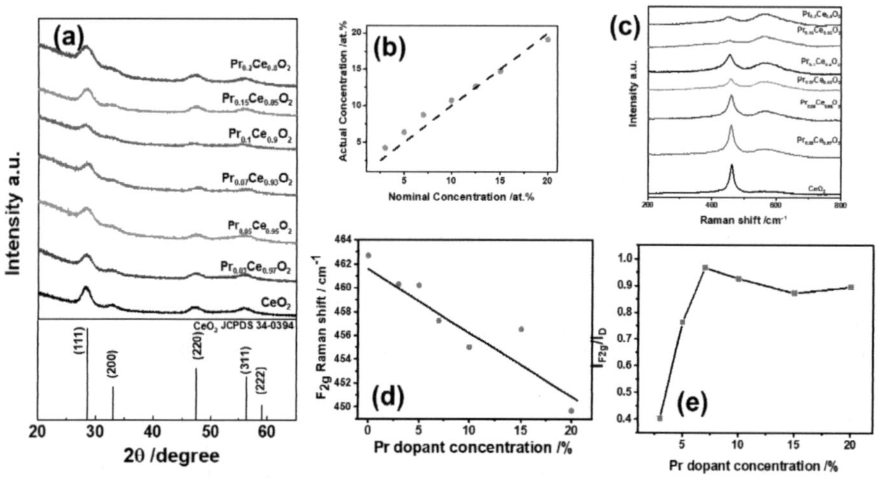

Figure 2. (a)XRD patterns; (b)dopant concentration; (c) Raman spectra; (d) Raman peak shifts of F2g peak; (e)Intensity ratio of F2g and D Raman peaks of as-prepared samples.

The chemical bonding of as-prepared samples was investigated by Raman spectroscopy, as shown in Figure 2(c). A dominant peak of all samples is located at ~461 cm^{-1}, which refers to active F2g mode originating from a symmetrical stretch of the eight surrounding O^{2-} ions with respect to the central stationary Ce ion in the fluorite structure[9].

The substitution of Pr^{4+}/Pr^{3+} contributes to a local disorder in the sublattice since the radius of Ce^{3+} (0.103 nm), Ce^{4+} (0.970 nm) and Pr^{3+} (0.113 nm), Pr^{4+}(0.960 nm) are different. In addition, the introduction of Pr^{3+} resulted in oxygen defects to maintain the electronic natural state. In addition, the peak shifts of Ce-O bond can be attributed to the replacement of Ce^{4+} ions with Pr^{4+}/Pr^{3+} after doping. As the dopant concentration increased, the degree of shifts increased because more Ce-oxygen bond was replaced by Pr-O bond. Additional broad peaks at around ~597 cm^{-1} can be assigned to the second-order optical mode and to the presence of oxygen vacancies[15]. These oxygen vacancies may originate from the presence of Ce^{3+} and Pr^{3+} in the lattice. The ratio of the intensity of F2g peak and defect peak can be used to determine the defects level. As shown in Figure 2(d), as the dopant concentration increased, more defects were formed which indicates that the high concentration results in high degree of oxygen vacancies.

3.2. Microstructure observation

The TEM images of the Pr doped CeO_2 nanocrystals (Figure 3(a)-(d)) exhibits well-dispersed nanocubes forming a monolayer of the self-assembly structure on a copper grid. The nanoparticles in all samples have a consistent size at around 10 nm and uniform morphology which ensures a long-range ordered structure via a simple drop-coating method. On the substrate, a dense and tightly packed self-assembly thin film can be produced owing to the distinct structure of nanocubes. By repeating dropping the solution on the substrate, multiple layers of nanocube can form into a 3D architect. The single-crystalline nature of an individual 3% Pr doped CeO_2 nanocube was determined by a high-resolution TEM (HRTEM) image. The lattice spacing was

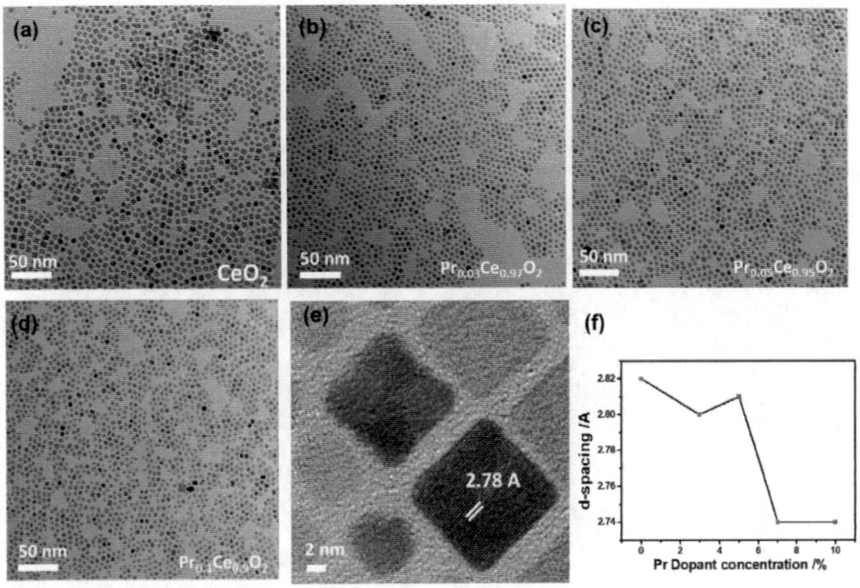

Figure 3. TEM images of (a)CeO2; (b)$Pr_{0.01}Ce_{0.9}O_2$; (c) $Pr_{0.1}Ce_{0.9}O_2$; (d) $Pr_{0.1}Ce_{0.9}O_2$;(e) $Pr_{0.1}Ce_{0.9}O_2$ (f) The relationship between the d-spacing and dopant concentration

measured to be 0.278 nm, corresponding to the (100) planes of CeO_2, showing that the as-prepared doped CeO_2 nanocubes are exposed with (100) facets. The d-spacing of all as-prepared samples were measured and shown in Figure 3(f). The distance of each crystal plane is decreased in a subtle degree from 0.282 nm to 0.274 nm which indicates the difference of radius of Pr and Ce elements would not hinder the doping process.

3.3. Oxygen vacancies tailoring

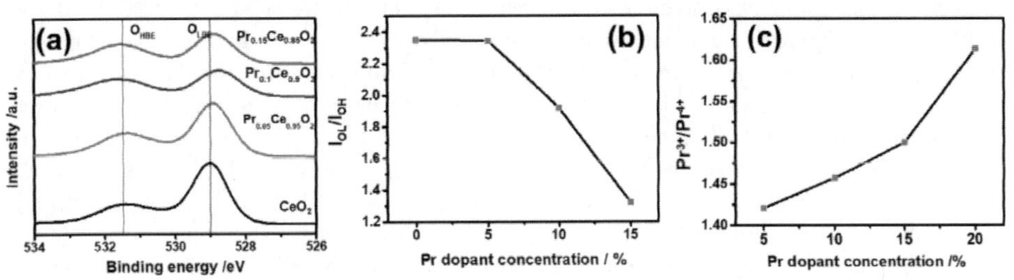

Figure 4. (a) XPS spectra of O1s peaks; (b) peak intensity ratio of O_l and O_h peaks; (c) concentration ratios of Pr^{3+} and $Pr^{4+}$of different as-prepared samples.

To investigate the chemical states of as-prepared samples, the XPS was employed. The O1s curve of all samples was shown in Figure 4(a), which can be fitted into two peaks. One peak located at the higher binding energy level at ~531.5 eV can be attributed to the oxygen ions in the oxygen-deficient region, while the other one at the lower binding energy level at ~529 eV refers to lattice oxygen bonding to cerium. The ratio of the intensity of peaks of two O1s can be viewed as an indicator of oxygen defects level as shown in Figure 4 (b). As dopant concentration increased, the ratio decreased which suggests high doping level results in a high concentration of oxygen vacancies in accordance with Raman results. To verify the valance state of Praseodymium ions, the curves of Pr 3d orbit were investigated closely. One peak located at ~933 eV attributed to Pr^{3+} while the other at ~ 929 eV may originate from Pr^{4+}[16]. The intensity ratio of Pr^{3+} and Pr^{4+} was calculated and could found in Figure 4(c). It is obvious that the as the doping level increased the 3+ valence Pr accounts the larger portion of all Pr elements.

3.4. Current-voltage (I-V) response of the device

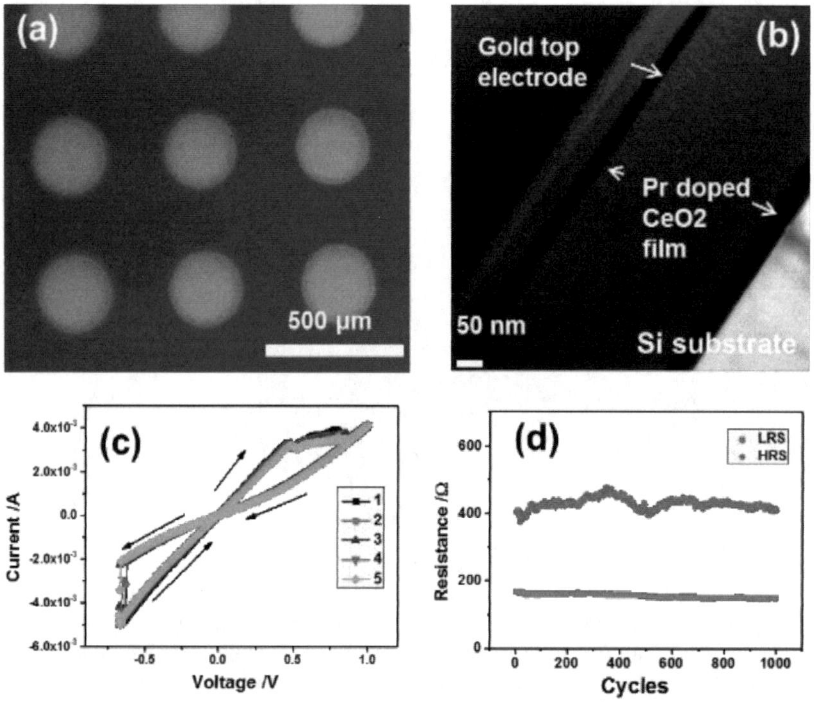

Figure 5. The (a)top view (b)the cross-section of the 15% Pr-doped CeO_2 thin film with gold top electrodes; (c)the I-V curve response under sweeping voltage; (d) The endurance curve of device

The I-V response of the sandwich-structured device with gold top electrode was tested. The display of top electrodes was shown in Figure 5(a) and the cross-section image was shown in Figure 5(b). From the top view, the drop-coating thin film is crack-free and dense. The cross-section shows that the thickness of the Pr doped CeO_2 film is around 300 nm and no crack can be found which indicates a higher dimensional structure could be achieved through layer-by-layer fabricating the 1D nanocube. Figure 5(c) shows the I-V characteristics of the Au/Pr doped CeO_2/FTO structure under voltage sweeping mode, at a speed of 1 V s^{-1}, in the following sequence 0 → 1 → 0 → −0.6 → 0 V. The resistance state of the device is transformed from ON (low-resistance state (LRS)) to OFF (high-resistance state (HRS)) on the first sweep from 0 V to 1 V. During the subsequent voltage sweep from 1 V to -0.6 V, the cell remained in the HRS until the next coming sweep. When switching the voltage from -0.6 V to 0 V, the current dramatically jumped to LRS level around 0.5 V. And such resistance switching behaviour is reproducible for 5 repeating IV sweepings. In order to evaluate the potential for applications in non-volatile memory devices, a cycling switching test (endurance) was implemented. The electrical pulse-induced resistance change effect in the memory device is shown in Figure

5(d). A positive and a negative pulse voltage were programmed and applied to the device to activate the resistance switching process, respectively. The cell was repeatedly switched between the HRS and the LRS. The endurance test indicates that the memory cell exhibits negligible degradation during 1000 ON/OFF switching cycles. This endurance result indicates that Pr doped CeO_2 is a potential candidate material for RRAM application.

As it is acknowledged that the oxygen migration model could be one of the frequently reported for the origin of the bipolar resistance switching effect. The previous study reported that the resistive switching behaviour in the CeO_2 films mainly attributes to switching in oxygen concentration which results in different resistance switching[9]. Without any external stimuli, the oxygen vacancies would gather on the surface of nanomaterials. Under the trigger of the electric field, the positively charged oxygen vacancies would align with the electric force forming a conductive filament towards the anode. These localized filaments give rise a current flow through the oxide layer from the metal electrode (Au) to the conductive substrate (FTO) which leads to LRS and turn ON the memory cell. While a reverse voltage was applied, the oxygen vacancies would reorder results in breaking the oxygen vacancies' interactions hence rupturing the conductive filaments, which switch the memory cell from ON state to OFF. The doping of Pr element into the cerium oxide lattice as an effective way to induce the oxygen vacancies defects greatly, therefore, a prominent resistive switching property can be exhibited.

4. Conclusion

In summary, a unique Pr doped CeO_2 monodispersed nanocubes were successfully synthesized. The characterization suggested that the oxygen defects level can be tuned easily by adjusting the dopant concentration during a facile solvothermal method. Doping Pr with Ce atom introduces the oxygen vacancies which give rise to resistive switching behaviours. The thin film was prepared by drop-coating the nanoparticle suspension solution and further fabricated to sandwich structure devices (Au/Pr doped CeO_2/FTO) which exhibit resistance switching characteristics after 1000 cycles. The distinct monodispersed nanocubes and self-assembly film make the materials a promising candidate to fabric 3D memory device as nano building blocks which has potential in filed such as data-depended fields such as AI system and machine learning.

5. References

[1] Wang Y, Liu Q, Long S, Wang W, Wang Q, Zhang M, Zhang S, Li Y, Zuo Q, Yang J and Liu M 2010 Investigation of resistive switching in Cu-doped HfO_2 thin film for multilevel non-volatile memory applications *Nanotechnology* **21** 045202

[2] Kim D C, Seo S, Ahn S E, Suh D-S, Lee M J, Park B-H, Yoo I K, Baek I G, Kim H-J, Yim E K, Lee J E, Park S O, Kim H S, Chung U-I, Moon J T and Ryu B I 2006 Electrical observations of filamentary conductions for the resistive memory switching in NiO films *Appl. Phys. Lett.* **88** 202102

[3] Nili H, Walia S, Kandjani A E, Ramanathan R, Gutruf P, Ahmed T, Balendhran S, Bansal V, Strukov D B, Kavehei O, Bhaskaran M and Sriram S 2015 Donor-Induced Performance Tuning of Amorphous $SrTiO_3$ Memristive Nanodevices: Multistate Resistive Switching and Mechanical Tunability *Adv. Funct. Mater.* **25** 3172–82

[4] Li Y, Sun X-Y, Xu C-Y, Cao J, Sun Z-Y and Zhen L 2018 Ferroelectric resistive switching behavior in two-dimensional materials/$BiFeO_3$ hetero-junctions *Nanoscale* **10** 23080–6

[5] Waser R, Dittmann R, Staikov G and Szot K 2009 Redox-Based Resistive Switching Memories - Nanoionic Mechanisms, Prospects, and Challenges *Adv. Mater.* **21** 2632–63

[6] Cho S, Yun C, Tappertzhofen S, Kursumovic A, Lee S, Lu P, Jia Q, Fan M, Jian J, Wang H, Hofmann S and MacManus-Driscoll J L 2016 Self-assembled oxide films with tailored nanoscale ionic and electronic channels for controlled resistive switching *Nat. Commun.* **7** 12373

[7] Younis A, Chu D, Lin X, Yi J, Dang F and Li S 2013 High-Performance Nanocomposite Based Memristor with Controlled Quantum Dots as Charge Traps *ACS Appl. Mater. Interfaces* **5** 2249–54

[8] Imagawa H, Suda A, Yamamura K and Sun S 2011 Monodisperse CeO_2 Nanoparticles and Their Oxygen Storage and Release Properties *J. Phys. Chem. C* **115** 1740–5

[9] Younis A, Chu D and Li S 2012 Oxygen level: the dominant of resistive switching characteristics in cerium oxide thin films *J. Phys. D. Appl. Phys.* **45** 355101

[10] Tok A I Y, Boey F Y C, Dong Z and Sun X L 2007 Hydrothermal synthesis of CeO2 nano-particles *J. Mater. Process. Technol.* **190** 217–22
[11] Sun C, Li H, Zhang H, Wang Z and Chen L 2005 Controlled synthesis of CeO_2 nanorods by a solvothermal method *Nanotechnology* **16** 1454–63
[12] Huang M, Yan Y, Feng W, Weng S, Zheng Z, Fu X and Liu P Controllable Tuning Various Ratios of ZnO Polar Facets by Crystal Seed-Assisted Growth and Their Photocatalytic Activity
[13] and S Y and Gao* L 2006 Controlled Synthesis and Self-Assembly of CeO2 Nanocubes
[14] Qian L, Zhu J, Du W and Qian X 2009 Solvothermal synthesis, electrochemical and photocatalytic properties of monodispersed CeO2 nanocubes *Mater. Chem. Phys.* **115** 835–40
[15] Yang Y, Mao Z, Huang W, Liu L, Li J, Li J and Wu Q 2016 Redox enzyme-mimicking activities of CeO2 nanostructures: Intrinsic influence of exposed facets *Sci. Rep.* **6** 35344
[16] Chen N, Younis A, Chu D and Li S 2018 Controlled Fabrication of Pr(OH)3 Nanowires for Enhanced Photocatalytic Activities *J. Rare Earths*

IOP Conf. Series: Materials Science and Engineering **576** (2019) 012036 doi:10.1088/1757-899X/576/1/012036

Red Shift in Optical Properties of Type-I $Al_{0.45}Ga_{0.55}As/GaAs_{0.84}P_{0.16}/Al_{0.45}Ga_{0.55}As$ Nano-heterostructure under External Strain

Amit Rathi[1], Amit Kumar Singh[1], Md. Riyaj[1], S.Dalela[2], P. A. Alvi[3]

[1] Department of Electronics and Communication Engineering, Manipal University Jaipur, Rajasthan, India
[2] Department of Pure and Applied Physics, University of Kota, Kota, Rajasthan, India
[3] Department of Physics, Banasthali University, Rajasthan, India

Corresponding author: amitrathi1978@gmail.com

Abstract. The application of external pressure on a heterostructure produces changes in the lattice parameter and symmetry of the material. These in turn produce significant changes in the electronic bandstructure. Due to the application of external pressure energy gaps are altered. This paper reports the energy bandstructure, wavefunctions and optical gain in type-I n-$Al_{0.45}$ $Ga_{0.55}$ $As/GaAs_{0.84}P_{0.16}$ / p-$Al_{0.45}$ $Ga_{0.55}$ As nano-heterostructures under uniaxial strain (-1 to +1 GPa) along [110] directions in the red visible region. Numerical calculations of the photonic energy and optical gain of TE and TM modes in GaAsP/AlGaAs laser diode structure have been carried out for various uniaxial strain at temperature 300 K. The band structure was calculated using 6×6 Luttinger-Kohn Hamiltonian to determine sub-band dispersion and corresponding wavefunctions. The Heterostructure is observed to operate in the energy range 1.62 to 1.78 eV (696 to 765 nm).

1. Introduction

Specification of the lasing wavelength and substrate material availability usually influence the active region material selection. In a quantum well heterostructure design for the active region, there are a number of parameters that need to be determined, including the number of quantum wells, the well thickness, the well band gap energy and the strain on the well (i.e., the well material composition), the barrier band gap energy and the strain on the barrier. Strain modifies the bandstructure and crystal symmetry of the semiconductor. The k.p method was used to specify conduction and valance band shift. In semiconductors, lattice mismatch, applied external stress leads to strain. For Optical gain, Hongping Zhao et. al demonstrated compensated InGaN-AlGaN quantum well active regions for laser emitting at 420-500 nm. A calculation for reduction of threshold carrier density and current density for strain compensated InGaN-AlGaN quantum well active region has been presented. It was found that biaxial strain would enhance the optical gain and minimum peak gain is obtained under a small tensile strain [1]. E V Bogdanov et al. have studied p-AlGaAs/GaAsP/n-AlGaAs heterostructures and polarization mode switching has been reported under the application of compressive stress [2]. Mixing of wavefunction character between contiguous sub bands can be reduced by strained structure. [3]. In heterostructures optical transitions can be specified in terms of momentum matrix elements or dipole matrix elements [4].

The effect of Uniaxial strain and compression on optical properties of heterostructures have been investigated in several studies recently [5, 6, 7, 8, 9]. To know the bandstructure and the wavefunctions of bound and unbound states in both lattice matched and strained quantum well k.p method was used.

2. Structure & Theory

The active region consists of a single quantum well (5 nm) of $GaAs_{0.84}P_{0.16}$ material and the barrier layers consists of ternary alloys $Al_{0.45}Ga_{0.55}As$ and $Al_{0.7}Ga_{0.3}As$ (16 nm) The subband dispersion and corresponding wavefunctions were calculated using Luttinger-Kohn 6x6 model hamiltonion. Equilbrium and injected electrons/holes concentration were 1.6×10^{12} /cm^2 and 3.5×10^{12}/cm^2 respectively. The entire heterostructure was grown on the substrate of binary compound GaAs. For the calculations, both external and internal strains have been considered. Internal strain appears because of lattice mismatch of the layers. For the lattice mismatched strain, We consider

$$\varepsilon_{xx} = \varepsilon_{yy} = \frac{a_0 - a}{a}; \qquad \varepsilon_{zz} = \frac{2C_{12}}{C_{11}}\varepsilon_{xx}$$

Where a and a_0 are the lattice constants of the layer and substrate. C_{12} and C_{11} are the stiffness constants. The energy band structure is shown in Fig. 1 with various energy bands. The type-I QW structure is based on GaAs substrate and has a of length 21 nm.

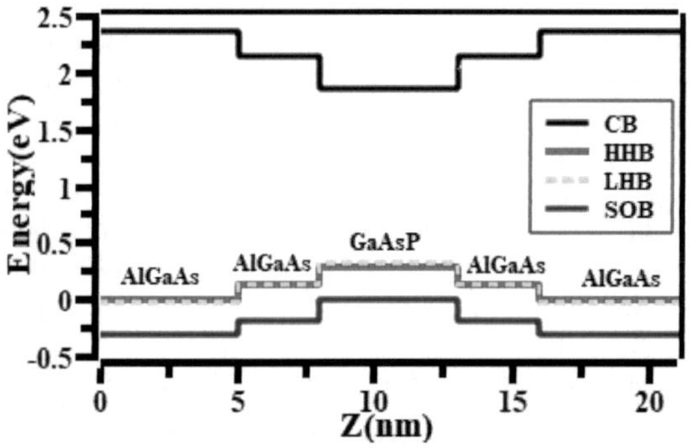

Figure. 1. Energy banddiagram of $Al_{0.45}Ga_{0.55}As/GaAs_{0.84}P_{0.16}$ heterostructure.

The fundamental theory for the study of motion of holes and electron in perturbed fields is extensively presented [10,11,22,13,14]. The E-k dispersion curve of the structure calculation is the basis of succeeding computations. The block diagonalised system Hamiltonian is applied for the numerical computation of electronic energy bands and wavefunctions in the QW heterostructure.

For calculation of the optical gain coefficient of the nano-scale heterostructure, Fermi's golden rule is applied [15,16,17,18,19]. The valance and conduction band profiles and wavefunctions of electrons and holes for the type-I n-$Al_{0.45}$ $Ga_{0.55}$ $As/GaAs_{0.84}P_{0.16}$ / p-$Al_{0.45}$ $Ga_{0.55}$ have been numerically calculated for various values of external pressure along [110] directions. Numerical calculations were performed with the help of HD studio software. Wavefunctions, Optical matrix element of electron photon interaction and momentum matrix element as well as optical gain in TM and TE mode for the type-I n-$Al_{0.45}Ga_{0.55}As/GaAs_{0.84}P_{0.16}$/p-$Al_{0.45}Ga_{0.55}$ were performed under external pressure up to 1 GPa along [110] directions at 300 K. Fig. 2, Fig.3 and Fig. 4 show the bandstructure and the wavefunctions of type I $Al_{0.45}Ga_{0.55}As/GaAs_{0.84}P_{0.16}$ heterostructure [20,21,22].

In the following section the results from simulations are discussed.

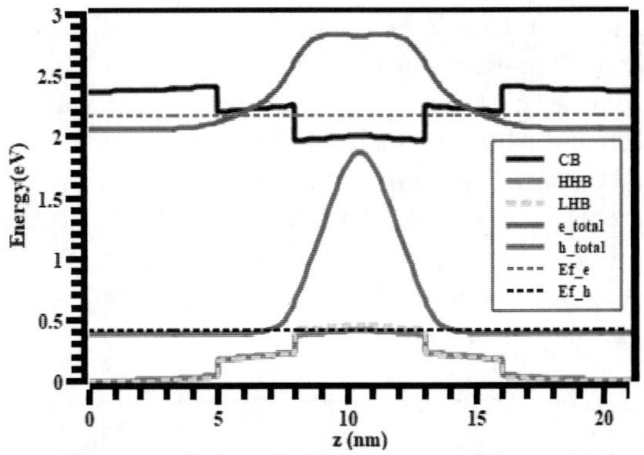

Figure 2. e-h (total) wavefunctions in $Al_{0.45}Ga_{0.55}As$/ $GaAs_{0.84}P_{0.16}$ heterostructure

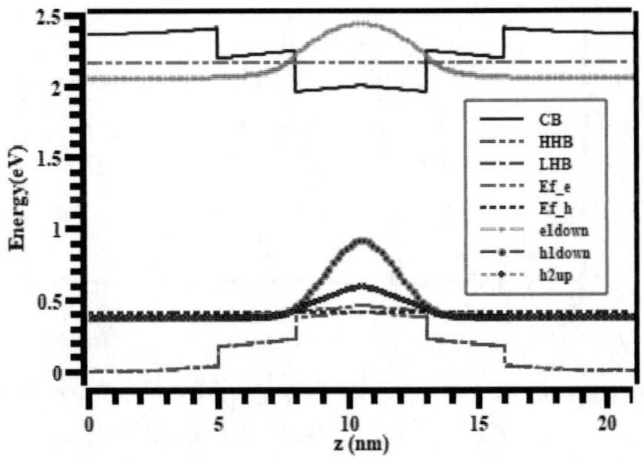

Figure 3. e_1-h_1(down) wavefunctions in $Al_{0.45}Ga_{0.55}As$/ $GaAs_{0.84}P_{0.16}$ heterostructure

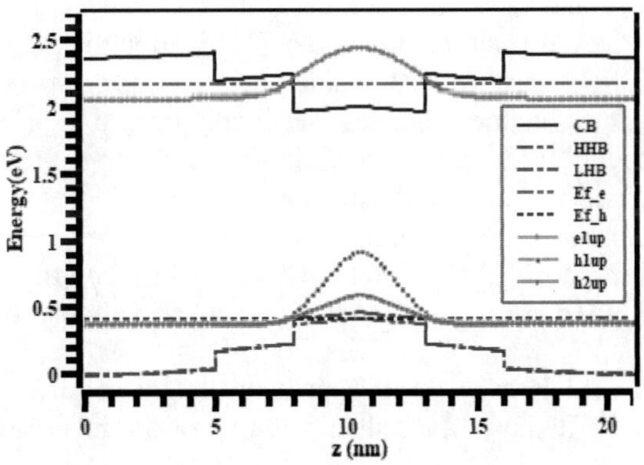

Figure 4. e_1-h_1 (up) wavefunctions in $Al_{0.45}Ga_{0.55}As$/ $GaAs_{0.84}P_{0.16}$ heterostructure

3. Results & Discussion

This paper investigates the effect of external pressure on optical gain associated with the optical matrix element for a 5nm n-$Al_{0.45}$ $Ga_{0.55}$ As/$GaAs_{0.84}P_{0.16}$/p-$Al_{0.45}$ $Ga_{0.55}$ quantum well along with [110] direction at temperature 300 K. To investigate the optical gain in n-$Al_{0.45}Ga_{0.55}As$/$GaAs_{0.84}P_{0.16}$/p-$Al_{0.45}Ga_{0.55}$ type-I nanoscale heterostructure, initially the Schrodinger equation was solved for finding the wavefunctions related to conduction band where as six band hamiltonian was solved to find out the wavefunctions related to valance subbands. After wavefunction determination, energy band dispersion for conduction band electrons and valance band holes were calculated. The 6×6 Luttinger-Kohn model and self-consistent calculations is used for computation of wavefunctions of the structure.For this e_1-h_1 and e_1-h_2 transitions are selected for computation in this present case.

The application of external pressure on a heterostructure produces changes in the lattice parameter and symmetry of the material. These in turn produce significant changes in the electronic band structure. Due to the application of the external pressure energy gaps was altered. Effective masses are affected by change in the interband matrix elements and variations in the energy gaps.

The optical gain in z polarization as well width variations is shown in Fig. 5. Optical gain is computed on various strains (-0.5, -1, 0, 1, 0.5 GPa) at 300K. The optical gain of the heterostructure under z polarization is 4177/cm under no pressure condition. A right shift is also observed in the optical gain spectrum for increasing compression. A left shift along with a drop is observed in the optical gain spectrum for increasing tensile strains.

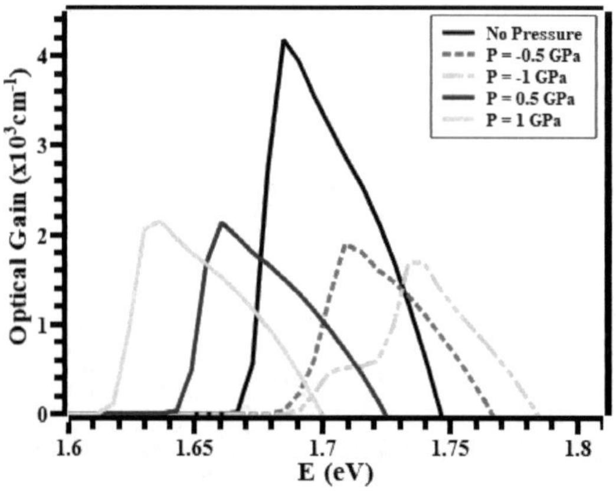

Figure 5. Optical Gain spectra of $Al_{0.45}Ga_{0.55}As$/ $GaAs_{0.84}P_{0.16}$ heterostructure as strain variations along [110] direction at T=300 K

4. Conclusion

This paper reports the analysis of optical matrix element and optical gain in $Al_{0.45}Ga_{0.55}As/GaAs_{0.84}P_{0.16}$ heterostructure along the [110] directions. Initially, the wavefunctions associated with the conduction and valance band, energy band dispersion for conduction band electrons and valance band holes, matrix elements and finally optical gain have been computed under different strain conditions. Equilbrium and injected eletron/hole charge carrier concentration were 1.6 & 3.5 ($\times 10^{12}$ cm^{-2}). The band structure was calculated using 6x6 Luttinger-Kohn Hamiltonian to determine sub-band dispersion and corresponding wavefunctions. This heterostructure is observed to operate in the energy range 1.62 to 1.78 eV (696 to 765 nm). Based on the outcomes it is realised that the optical gain of the InGaAs/GaAsP heterostructure can be suitably red shift tuned as per the requirements.

Overall comparision between Energy (eV), gain, wavelength at different strain along the [110] directions are shown in table 1. It shows the red shift tuning of the InGaAs/GaAsP heterostructure on different strain.

Table 1. Comparison of Energy (eV), gain, wavelength at different strain along the [110] directions

Strain [110]	E (eV))	Gain (max)	λ (nm)
-1	1.739	1688	712.96
-0.5	1.7	1903	729.3
0	1.684	4177	736.24
0.5	1.66	2141	746.89
1	1.636	2157	757.84

5. References

[1] Hongping Zhao, Ronald A. Arif "Optical gain analysis of strain-compensated InGaN-AlGaN quantum well active regions for laser emitting at 420-500nm. Opt Quant Electron (2008) 40:301-306

[2] E V Bogdanov, K I Kolokolov, N V Melnikova, N Ya Minina and G V Tikhomirova, "Polarization mode switching in p-AlGaAs/GaAsP/n-AlGaAs diodes in presence of compressive stress", IOP Conf. Series: Journal of Physics: Conf. Series 950 (2017) 042047

[3] Wood, A.C.G., "Strain effects in Semiconductor quantum wells", PhD thesis, Durham University, Sept 1990.

[4] B.G.U, N.H Kwong, "Relation between Interband Dipole and momentum matrix elements in Semiconductor" physical review. B, Condensed matter March 2013

[5] H. K. Nirmal, S. G. Anjum, Pyare Lal, Amit Rathi, S. Dalela, M. J. Siddiqui, P. A. Alvi, "Field effective band alignment and optical gain in type-I $Al_{0.4s}Ga_{0.55}As/GaAs_{0.84}P_{0.16}$ nano-heterostructures", International Journal of light and electron optics: Optik 127,7274-7282 (2016).

[6] Md. Riyaj, A.K. Singh, Amit Rathi, Sandhya Kattayat, Shalendra Kumar, S Dalela, P.A. Alvi, "High Pressure affects on Optical Characteristics of AlGaAs/GaAsP/AlGaAs nano-heterostructure", Optik, Volume 181, March 2019, Pages 389-397, Elsevier, Dec. 2018.

[7] A. K. Singh, Amit Rathi, Md. Riyaj, P.A. Alvi, "Optical gain tuning within IR region in type-II $In_{0.5}Ga_{0.5}As_{0.8}P_{0.2}/GaAs_{0.5}Sb_{0.5}$ nano-scale heterostructure under external uniaxial strain", Superlattices and Microstructures 98 (2016) , Elsevier, June 2017.

[8] A. K. Singh, Md. Riyaj, S.G. Anjum, Nisha Yadav, Amit Rathi, M.J. Siddiqui, P.A. Alvi, "Anisotropy and optical gain improvement in type-II $In_{0.3}Ga_{0.7}As/GaAs_{0.4}Sb_{0.6}$ nano-scale heterostructure under external uniaxial strain", Superlattices and Microstructures 98 (2016), Elsevier, June 2016, Volume 98, October 2016, Pages 406–415.

[9] Vibha Kumari, Ashish, Swati Jha, Amit Rathi, H. K. Nirmal, P. A. Alvi, "Optical Gain of InGaAlAs Quantum well with Different Barriers, Claddings and Substrates", Journal of Optoelectronics Engineering, Science and Education Publishing, 2014, Vol. 2, No. 2, pp. 42-45

[10] J.M. Luttinger, W. Kohn, "Motion of electrons and holes in perturbed periodic fields", Physical Review 97 (4) (1955) 869.

[11] Wallace C. H. Choy, Tailoring Light and Heavy Holes of GaAsPAIGaAs Quantum Wells by Using Interdiffusion for Polarization - Independent Amplifier Applications, IEEE Journal of Quantum Electronics, Vol. 36, No. 2, February 2000.

[12] J.M. Luttinger, Quantum theory of cyclotron resonance in semiconductors: general theory, Physical Review 102 (4) (1956) 1030.

[13] Paul Harrison, Quantum Wells, Wires and Dots: Theoretical and Computational Physics of Semiconductor Nanostructures, John Wiley & Sons, 2005.

[14] Shun Lien Chuang, Physics of Optoelectronic Devices, John Wiley & Sons, 1995.

[15] Chih-Sheng Chang, Shun Lien Chuang, "Modeling of strained quantum well lasers with spin-orbit coupling", IEEE Journal of Selected Topics in Quantum Electronics I (2) (1995) 218-229.

[16] Amit Rathi , Amit Kumar Singh, "Optical Response Computations in Type-II Doped AlSb/InAs Nano-Heterostructure under External uniaxial strain in SWIR Range" , IEEE 3rd International Conference on Engineering, Technologies and Social Sciences, (2017), ISBN No.-978-1-5386-1611-6.

[17] Jiirgen Sebastian, Gert Beister, Frank Bugge, E. Buhrandt, Goetz Erbert, H. G. Hansel, R. Hulsewede et al., High-power 810-nm GaAsP-AIGaAs diode lasers with narrow beam divergence, IEEE Journal of Selected Topics in Quantum Electronics 7, no. 2 (2001): 334-339.

[18] Md. Riyaj, A. K. Singh, Sandhya K. , Amit Rathi, P. A. Alvi, Optical properties of Type-I GaAsPI AIGaAs nano-heterostructure under external uniaxial strain, In AIP Conference

Proceedings, vol. 1832, no. I, p. 120022. AIP Publishing, 2017.
[19] Hidenao Tanaka, Jun-ichi Shimada, and Yoshio Suzuki, Highly efficient TE/TM mode switching of GaAsPI AIGaAs strained quantum-well laser diodes, Applied physics letters 64, no. 2 (1994): 158-160.
[20] E.V. Bogdanov, H. Kissel, K.1. Kolokolov, N. Ya Minina, TM/TE polarization tuning and switching in tensile strained pAIGaAs/ GaAsP/n-AIGaAs heterostructures by uniaxial compression, Semiconductor Science and Technology 31 (3) (2016) 035008.
[21] E.V. Bogdanov, N. Ya Minina, J.W. Tomm, H. Kissel, Effect of uniaxial stress on electroluminescence, valence band modification, optical gain, and polarization modes in tensile strained p-AIGaAs/GaAsP/n-AIGaAs laser diode structures: numerical calculations and experimental results,Journal of Applied Physics 112 (9) (2012) 093113.
[22] E.V. Bogdanov, N. Ya Minina, S.S. Shirokov, A.E. Yunovich, H. Kissel, Electroluminescence in laser diode nanostructures p - AlxGa1_xAs/ GaAs1_yPy/n - AlxGal_xAs under uniaxial compression, in: Physics, Chemistry and Application of Nanostructures: Reviews and Short Notes, 2009, pp. 609-612.

Acknowledgments
Authors would like to thank Dr. Konstantin I. Kolokolov (Faculty of Physics, M V Lomonosov Moscow State University, Moscow 119991, Russia) for supporting the research work. Authors are thankful to Manipal University Jaipur, Jaipur 303007, Rajasthan, India for the financial support under the project seed grant: MUJ/REGR/1467/13.

Parametric Study for Calculating Surface Deflection Using the Method of Images

Y An[1], R Zhang[1], P Li[1], D Luo[1], and X Jin[1,2]

[1]State Key laboratory of Mechanical Transmissions, Chongqing University, Chongqing 400030, China

[2]College of Aerospace Engineering, Chongqing University, Chongqing, 400030, China

Corresponding Author's E-mail: jinxq@cqu.edu.cn

Abstract. The solution of the elastic field for inclusions in an elastic half space has been applied widely in many contact analyses and engineering designs. A popular method to solve the half-space inclusion problem resorts to the method of images, where the solution is decomposed into 3 components consisting of the full space inclusion, mirrored inclusion, and the surface traction cancellation. In the process of cancelling the redundant surface tractions determined from a full space inclusion problem, the computation domain is supposed to be limited in a finite size and there is inevitably truncation error. It has not been quantitatively investigated how the truncation error will influence the accuracy of the numerical computations based on the method of images. This work studies the deflection of the boundary surface of a half-space containing an Eshelby inclusion. Errors due to mesh refinement and domain truncation are quantitatively analyzed. Parametric studies are performed for a systematic examination of surface redundant tractions and their influences.

1. Introduction

The inclusion problem provides a unified treatment for a variety of subjects and has been regarded as being of fundamental importance to many research branches in material science. The solutions for the half-space inclusions usually tend to be more intricate and complex, and the derivation for this problem may be laborious and time-consuming. The general problem of arbitrarily shaped inclusion in half space usually solved by the method of images which is straightforward and effective.

Eshelby [1] first solved the problem of the interior elastic field of an ellipsoidal inclusion subjected to uniform eigenstrain in an infinite space in 1957. However, the exterior elastic field is much more complicated. The explicit complete solutions of an ellipsoidal inclusion in full space was analytically proposed by Jin et al [2]. Chiu [3] solved the displacement gradient due to cuboidal inclusion with uniform eigenstrains in a full elastic space, and further [4] proposed a strategy for solving the inclusion in a semi-infinite space via the method of images. In practical engineering design, the inclusions might be any arbitrary shape and distributed in a random manner. Zhou et al. [5] proposed a fast numerical method to solve 3D arbitrarily shaped inclusions in half-space. Liu et al [6] derived a set of explicit formulas in terms of the Galerkin vectors for the displacements and stresses in half space and ingeniously proposed the Fast Fourier Transform (FFT) techniques combining convolution and correlations for numerical computations of the half-space inclusion problem. Their half-space solutions to the cuboidal inclusion subjected in uniform eigenstrains are exact and represented in closed-form, and therefore could be used as the elementary solution for solving arbitrarily shaped inclusion by breaking up the inclusion domain into multiple cuboids. Further, Zhou et al [7] proposed a mesh differential refinement scheme for solving elastic fields of half-space inclusion problems.

This paper intends to examine the numerical error issues in determining the surface deflection caused by distribution of eigenstrains (or inclusion) in a half space using the method of images [4]. The method of images is often adopted for solving a half-space inclusion problem, which is constructed by cancelling the redundant surface tractions determined from a full space inclusion problem. However, truncation errors would inevitably appear in the numerical implementation, particularly when an inclusion is located close to the surface of the half space. Several benchmark examples are provided to demonstrate the effects on the surface displacement for a variety of relevant parameters.

2. Formulation

2.1. Method of images

Consider an isotropic and homogeneous arbitrarily shaped inclusion embedded in an elastic semi-infinite space which is defined by $x_3 \geq 0$ Chiu [4] proposed that, the solution of inclusion in a semi-infinite space may be obtained by utilizing the method of images. According to the superposition principle, an isotropic inclusion in a semi-infinite space may be decomposed into three components (figure 1): (1) and (2) are the original and the mirror inclusions with specified eigenstrains in the full space respectively, and (3) is the solution of normal stress distributed on the free surface of semi-infinite space.

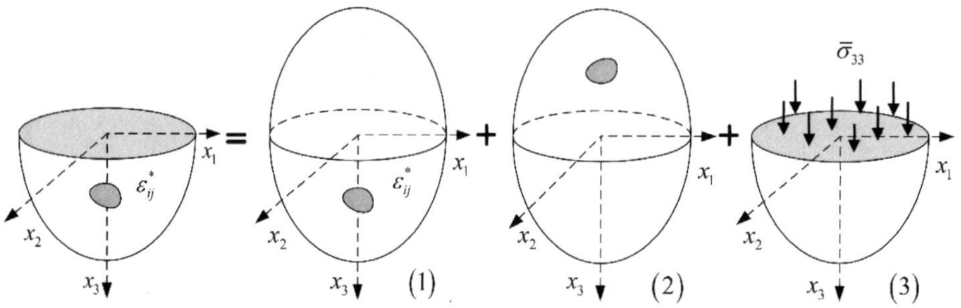

Figure 1. Decomposition of the half-space solution into three components.

By the Voigt notation [8], the prescribed eigenstrain uniformly distributed in the inclusion of domain (1) may be written as

$$\varepsilon^*_{ij} = \left[\varepsilon^*_{11},\ \varepsilon^*_{22},\ \varepsilon^*_{33},\ 2\varepsilon^*_{23},\ 2\varepsilon^*_{13},\ 2\varepsilon^*_{12}\right]^T \tag{1}$$

The resulting displacement at an arbitrary field point in the inclusion of domain (1) may be represented in a matrix form as follows

$$\begin{bmatrix} u_1 \\ u_2 \\ u_3 \end{bmatrix} = \begin{bmatrix} w_{111} & w_{122} & w_{133} & w_{123} & w_{113} & w_{112} \\ w_{211} & w_{222} & w_{233} & w_{223} & w_{213} & w_{212} \\ w_{311} & w_{322} & w_{333} & w_{323} & w_{313} & w_{312} \end{bmatrix} \begin{bmatrix} \varepsilon^*_{11} \\ \varepsilon^*_{22} \\ \varepsilon^*_{33} \\ 2\varepsilon^*_{23} \\ 2\varepsilon^*_{13} \\ 2\varepsilon^*_{12} \end{bmatrix} \tag{2}$$

The second, category consisted of isolated sprinkler experiments with catch cans placed along eight radii at equal distance from each other within the radius of throw of the sprinkler forming an orthogonal pattern as shown above in Table 1. Same operating pressure conditions were ensured with the aid of pressure gages and valves. This ensured that the distribution pattern of the isolated sprinkler experiments were as much as possible similar to each of the four overlapping sprinklers. This is a necessary requirement for comparison with the simulated overlaps with the measured pattern for similarity of the contributing sprinklers.

The highest absolute error registered was 2.9% for the 16x1.8 rectangular layouts which also recorded the least measured coefficient of uniformity. This observation may be attributed to the relatively larger spacing. The lowest absolute deviation from the measured was 0.736% for the 17x15 rectangular layouts. The relatively small overall mean absolute error is an indication of the efficiency of the model to overlap the single leg distributions to mimic field situation.

The 14x14 rectangular spacing gave the highest coefficient of uniformity values of 85% and 83.45% for both measured and simulated overlaps and the rectangular layout of 1bx18 registered the lowest simulated coefficient of uniformity value of 79%. And the lowest uniformity coefficient for the measured data was registered by the 18x18 triangular layout. In general the coefficient of uniformity increased as the spacing between sprinklers reduced for both observed and computed patterns (Figure 5-6).

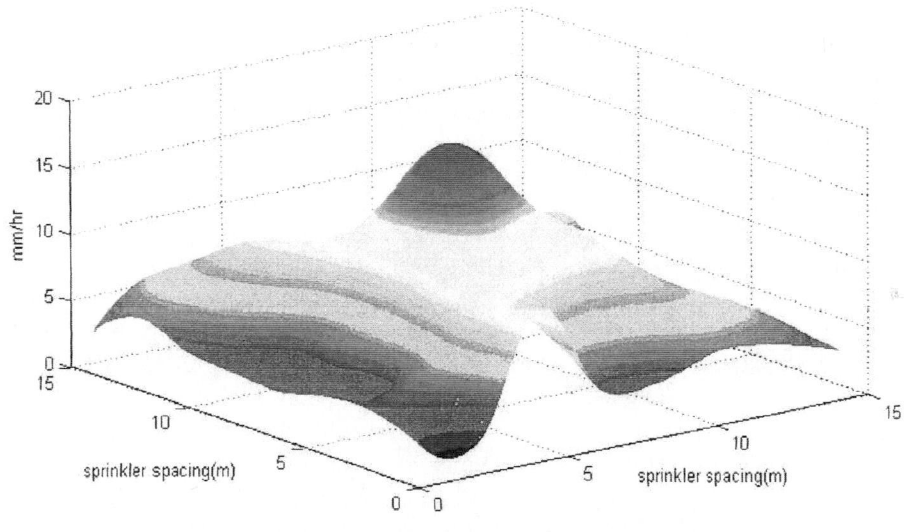

Figure 7. Measured overlapped square layout

From the generated overlap data other comparative performance indices such as distribution uniformity, Heerman and Hein's uniformity coefficient, standard deviation and statistical uniformity could be evaluated either from original expressions or from established empirical relationships to support decision making. Simulated and observed 3D plots for four overlapping sprinklers are shown in Figure 7, for square and triangular layouts. The vertices of the plots are the locations of the sprinkler positions. The plots show the effect of sprinkler spacing at the operating conditions.

For the single sprinkler field experiment, catch cans were placed in the form of a grid with the sprinkler head mounted on a riser of height 1.4 m from the ground, placed in the middle of the grid. Cans were placed 2 m apart from each other. In the solid-set experiments, layouts of 16x16, 16x18, 16x20 and 16x22 were performed to determine the relationship between spacing and uniformity of application while varying the operating pressure. Catch cans were placed in the center of the four overlapping sprinklers. Water application rates were determined for both single sprinkler and solid set experiments, concurrently, at pressures of 200, 250, 300 and 350 kPa respectively by means of graduated measuring cylinders. For each operating pressure configuration, the sprinkler flow rate was measured by connecting flexible rubber hose to the sprinkler nozzle and the time required to collect a known volume of water into a bucket was measured.

The maximum, minimum and average wind speed (m/s), air temperature and relative humidity (%) that prevailed during the field test were as shown in Table 1. They were typical of the weather conditions at the test site at the onset of summer. The measured sprinkler flow rates ranged from 3.45 to 4.15 m/hr, with an average of 3.98 m3/h within operating pressure range of 200 to 350 kPa. CU values were computed according for each experiment. In general the field data exhibited an inherent variability depending on the wind speed category, operating pressure and spacing.

The first step was to set the reference conditions and numerically compute the droplet drift for the reference conditions; droplet diameter of 2.5 mm, wind speed of 2.Srn/s, initial droplet velocity of 23.79 m/s and nozzle height of 1.2 m. Subsequently, one parameter was varied from the reference value, while the remaining independent variables were held constant. The droplet drift was then computed. The sensitivity of the droplet drift to each of the influencing factors was similarly, assessed.

6. Conclusions and results

By reference to market-related entity machine parameters to determine the structure of sprinkling machine design parameters and three-dimensional modeling, using of CATIA software prototype assembly and using of their finite element analysis for structural stability analysis, in order to verify the design reliability of the translation irrigation machine, provide a theoretical basis for the future market to product.

Research on the design of translation sprinkler irrigation uniformity in case emitter spacing of 2.9m, fluid simulation analysis and field testing to determine the impact of different design working pressure of the inlet of the irrigation uniformity, the translation irrigation machine from further research and development is important.

7. References

[1] Chen, Chao, et al. Irrigation uniformity of nozzle-changeable variable-rate sprinkler. *Journal of Drainage & Irrigation Machinery Engineering*, 2011, pp. 25-38.

[2] Chen, Chao, G. U. Zhe, and Y. Tang. Evaluation of Variable-rate Irrigation Sprinkler Based on Open-loop Control Strategy. *China Rural Water & Hydropower*, 2015, pp. 654-667.

[3] Fukui, Y., K. Nakanishi, and S. Okamura. Computer evaluation of sprinkler irrigation uniformity. *Irrigation Science 2.1*, 2008, pp.23-32.

[4] Ayars, J. E, et al. Influence of cotton canopy on sprinkler irrigation uniformity. *Transactions of the Asae,* 2011, Vol. 34 No. 3, pp. 0890-0896.

IOP Conf. Series: Materials Science and Engineering **576** (2019) 012029 doi:10.1088/1757-899X/576/1/012029

Prediction of superplastic forming of 7475 aluminium alloy sheets

J YANG[1], E GIRAUD[2], P DAL SANTO[2], S BOUDE[2] and J-L LEBRUN[3]

[1]Shanghai Aircraft Design & Research Institute - COMAC, 5188 Jinke Road, 201210 Shanghai, China

[2]Arts et Métiers ParisTech - LAMPA, 2 Bd du Ronceray - BP 93525, 49035 Angers Cedex 01, France

[3]ESTP Paris, 28 avenue du Président Wilson, 94234 Cachan Cedex, France

Abstract. Superplastic forming is mainly used in the aeronautic and astronautic domains and to fabricate the complex shape parts with the materials hard to process. The numerical simulation is used to predict the superplastic forming process due to the difficulty to predict a forming pressure by a traditional trial and error method. In the article, the superplastic behaviour of a 7475 aluminium alloy has been investigated. Uniaxial tensile tests over a wide range of temperatures and strain rates have been performed. An Arrhenius-type model using strain compensation has been identified. A failure criterion based on the evolution of fracture work is also proposed. This material model has been implemented by user subroutine in numerical simulations (ABAQUS) to predict the ability of the material to form a complex revolution part by superplastic forming. A pressure control algorithm has been developed to minimize the pressure instability and decrease the simulation CPU time. The numerical results, in terms of thickness evolution in the formed component and location of damage-risk zones, have finally been compared to experimental results after the realization of some bulging experiments using the generated pressure cycle.

1. Introduction

Superplastic forming is commonly used to elaborate deep drawn surfaces from thin metallic sheets by bulging processes. It consists in deforming a flange at high temperature by applying a time-dependent pressure [1]. This time-dependent pressure (pressure law) is chosen in order to deform the plate at a constant and low strain rate (usually called the optimal strain rate which is allowed to obtain the best deformation performance) while preventing the formation of defects. The ability to form a given part is generally studied by numerical simulations. They allow generating an optimized pressure law to pilot the strain rate evolution during the process. And the quality of the prediction depends on the material model and the stability of the pressure control.

A good knowledge of the mechanical behavior of the material under superplastic conditions, in terms of mechanical response or damage evolution, is necessary to obtain realistic predictions by numerical simulations.

Superplastic rheological behavior has to be identified by applying representatively thermo-mechanical conditions of superplastic forming, i.e. high temperature, low strain rate and biaxial stress state [2].

Empirical models are generally sufficient to describe the mechanical response of superplastic material. Yang et al. [3] have used a simple Norton-Hoff law to fit experimental strain-stress curves increased the tensile strength of the matrix material. First and foremost, the specific diameter of MMT was larger, which played a role in fiber reinforcement during the stretching process of the matrix material. Last but not least, the nano-Sb_2O_3 particles were separated from the matrix material during the stretching process to produce voids, which were effective in preventing the diffusion of cracks and could absorb the tensile stress of the matrix material. This played a leading role in increasing the tensile strength of the matrix material. It can't ignored in figure 3 that the tensile strength of the PP matrix composites was different with different of the mass fraction of nano-Sb_2O_3. When the mass fraction of nano-Sb_2O_3 was 3wt%, the tensile strength of 3.0%nano-Sb_2O_3/3%MMT/8%BPS-PP composites had a maximum value, which was 22.6% higher than that of pure PP. When the mass fraction of nano-Sb_2O_3 was 3.5wt%, the tensile strength of 3.5%nano-Sb_2O_3/3%MMT/8%BPS-PP composites was slightly lower than that of the former. Similar to tensile strength, the Young's modulus value of the PP matrix composite material was also significantly higher than that of the pure PP, and the Young's modulus value of PP matrix composites reached a maximum value when the mass fraction of nano-Sb_2O_3 was 3wt%.

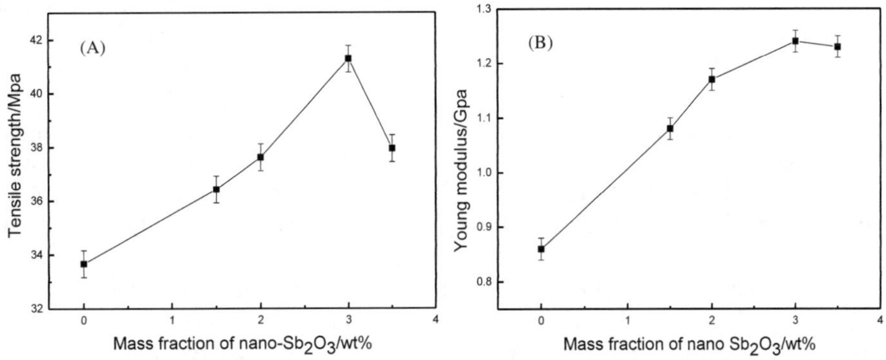

Figure 3. Tensile strength (A) and Young's modulus (B) of PP matrix composites with different mass fraction of nano-Sb_2O_3

Figure 4 is tensile fracture SEM images of PP matrix composites containing nano-Sb_2O_3 with 2.0wt% and 3.5wt% of mass fraction. As can be seen from the figure, a peeling phenomenon occurred between the nano-Sb_2O_3 particles and the substrate under the action of the tensile force when the mass fraction of nano-Sb_2O_3 was 2.0wt%. The appearance of voids could effectively alleviate the tensile stress on the matrix material, and could effectively prevent the crack from spreading. At this time, the tensile strength of the PP matrix composites was obviously increased. When the mass fraction of nano-Sb_2O_3 was 3.5wt%, the agglomeration took place due the higher content of nano-Sb_2O_3 with the higher surface activity of the nanoparticles. During the tensile process, the agglomeration of the nano-Sb_2O_3 caused stress concentration and produces cracks in matrix material, resulting in decreasing of the tensile strength of the PP matrix composites[8,16].

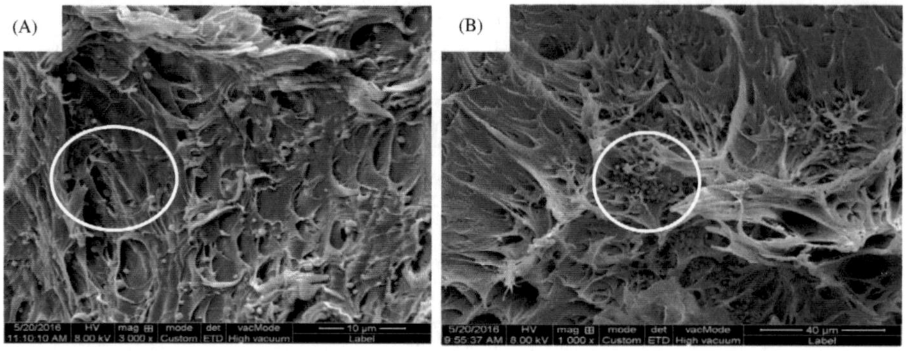

Figure 4. Tensile fracture SEM images of PP matrix composites (A)3#; (B)5#

3.4. Flame retardant performance

In the vertical burning test, the UL-94 grade of PP matrix composites containing MMT and nano-Sb_2O_3 was significantly better than that of pure PP. For pure PP, it was flammable accompany with serious low drop phenomenon and had a large fire risk, which its UL-94 grade reached V-2 degree. When the mass fraction of the nano-Sb_2O_3 was 1.5wt%, although the 1.5% nano-Sb_2O_3/3%MMT/8%BPS-PP composite had droplets in the vertical combustion, the burning time of the composite was reduced and its UL-94 grade reached V-2 degree. When the mass fraction of nano-Sb_2O_3 was 2.0wt%, MMT and nano-Sb_2O_3 could effectively suppress the droplets generation of the 2.0%nano-Sb_2O_3/3%MMT/ 8%BPS-PP, which the UL-94 grade reached V-1 degree. When the mass fraction of nano-Sb_2O_3 was higher than 3.0wt%, the PP matrix composites was extinguished within 10s of the fire source, which their UL-94 grade reached V-0 degree without droplets.

The LOI values and UL-94 grades of PP and PP matrix composites are shown in Table 3. It can be seen from the table that the LOI value of the PP matrix composite material was significantly higher than that of pure PP. The LOI value of the PP matrix composites increased with the increase of the mass fraction of nano-Sb_2O_3. The reason is that the noncombustible gas generated by the BPS and nano-Sb_2O_3 in the PP matrix composites had a suffocating flame retardant effect on the combustion. At the same time, nano-Sb_2O_3 could catalyze BPS during combustion, which generated Br ions to react with active free radicals H^+ and OH^- ions in the combustion process, and absorbed heat to play a flame retardant role[17,18]. Moreover, MMT has also improved the flame retardancy of PP matrix composites. MMT formed a high temperature resistant layer during combustion, and gradually moved to the surface layer of the combustion body to form a dense carbon layer as the gas and the solution flow generated during the combustion process. The formation of the carbon layer conduces to hinder the contact of the combustion body with the air and heat transfer, thus providing a good flame retardant effect[19,20]. Therefore, MMT and nano-Sb_2O_3/BPS show a good synergistic flame retardant effect.

Table 3. The LOI values and UL-94 grades of experimental materials

Samples	1	2	3	4	5
LOI (%)	16.9	23.8	25.1	27.3	28.8
UL-94	V-2	V-2	V-1	V-0	V-0

4. Conclusion

A novel polypropylene (PP) matrix flame retardant composite containing antimony trioxide nanoparticles (nano-Sb_2O_3) and brominated polystyrene (BPS) and montmorillonite (MMT) flame retardants was synthesized and characterized and the following conclusions were drawn from the study:

(1) The incorporation of nano-Sb_2O_3 could improve the crystallization temperature and crystallization of the PP matrix. When the mass fraction of nano-Sb_2O_3 was 3.0wt%, the crystallinity of the PP matrix composite was highest with a value of 69.88%.

(2) Due to the addition of MMT and nano-Sb_2O_3, the tensile properties of the PP matrix composites were significantly increased. When the mass fraction of nano-Sb_2O_3 was 3wt%, the tensile strength of the PP matrix composites reached the maximum, which was 22.6% higher than that of pure PP.

(3) The flame retardancy of the PP matrix composites was remarkably improved due to the incorporation of nano-Sb_2O_3 and MMT. As the mass fraction of nano-Sb_2O_3 reached 3.5wt%, the LOI value of the PP matrix composites was 28.8 %, and its UL-94 grade reached the V-0 level.

5. References

[1] Dong J H 2012 *Plastics Sci. Technol.* **40** 100
[2] Si M M, Hao J W, Xu L S and Du J X 2013 *China Plastics.* **27** 1
[3] Porter D, Metcalfe E and Thomas M J K 2000 *Fire Mater.* **24** 45
[4] Zhang X L, Xing Q and Zhu S L 2010 *Thermosetting Resin.* **25** 32
[5] Gholamian F, Nabiyouni G, Ghanbari D, Jalajerdi R and Aminifazl A 2013 *High Temp. Mat. Pr-Isr.* **32** 125
[6] Xie Y L, Zhang M and Zhou Y H 2012 *Chem. Ind. & Eng. Pro.* **31** 844
[7] Bai M J, Liu Y, Xiao F Y and Zhang J 2012 *Acta Polym. Sin.* **43** 539
[8] Xu J L, Yang W L, Niu L, Dan X M, Wang C C and Wen C 2016 *Inorg. Nano-Met. Chem.* **47** 697
[9] Lei H, Li Y J, Wu B and Xu T 2013 *Plastics.* **42** 27
[10] Zhang L, Hu B and Li C Z 2011 *Acta Polym. Sin.* **42** 1374
[11] Liu J, Jiang Z W, Yu H O and Tang T 2011 *Polym. Degrad. Stabil.* **96** 1711
[12] Yang J, Liu W J, Chen G X and Liu J J 2001 *Acta Polym. Sin.* **32** 383
[13] Chen M, Tang M, Ma Y H, Chen X L, Qin J, He W D and Zhang Z B 2015 *Polym. Eng. Sci.* **55** 1355
[14] Rafiee F, Otadi M, Goodarzi V, Khonakdar H A, Jafari S H, Mardani E and Reuter U 2016 *Compos. Part B-Eng.* **103** 122
[15] Zuiderduin W C J, Westzaan C, Huétink J and Gaymans R J 2003 *Polymer*, **44**, 261
[16] Xu J L, Zhang L, Zhou S G and Feng C 2014 *Rare Metal Mat. Eng.* **43** 3003
[17] Li H, Wang X H, Liu P, Li B, Sun G H and Yang C B 2017 Engineering Plastic Application. **45** 51
[18] Lu H, Lu G H, Xie X L and Wang L J 2016 *Materials Review B.* **30** 71
[19] Li B, Liu L, Luo H X, Luo Y F and Jia D M 2007 *Acta Polym. Sin.* **38** 456
[20] Zuo L Z, Fan W, Zhang Y F, Zhang L S, Gao W, Huang Y P and Liu T X 2006 *Compos. Sci. Technol.***139** 57

Acknowledgments
The work was supported by the National Natural Science Foundation of China (No. 51761025) and the Science and Technology Program Project Funds of Gansu Province of China (No. 17CX2JD075).

incorporated into as-synthesized MCM-41 by microwave assisted sol-gel technique. Then, the nanocomposite was coated on the glass substrate by dip coating. The performance of the thin film was tested by MB dye reduction under UV and visible light irradiation. The optimal Ag/Ti/Si molar ratio was 0.1/1/2, which gave the highest photocatalytic efficiency about 81% under UV and 30% under visible light irradiation [8].

Although metal doping can reduce the band gap of TiO_2, metal dopant can act as a recombination site which reduces the photocatalytic reactivity [9-13]. Non-metal doping is another attractive method. He, et al. (2016) prepared N-doped TiO_2 by solvothermal method using tetrabutyl titanate and titanium sulphate as silica sources and trimethylamine, triethanolamine and diethylene-triamine as nitrogen sources. The photocatalysis efficiency was evaluated by methyl orange (MO) reduction under visible light irradiation. The result indicated that N-doped TiO_2 prepared from titanium sulphate and diethylenetriamine gave high photocatalytic degradation efficiency of MO about 90% [14]

Yang, et al. (2010) prepared N-doped anatase TiO_2 by solvothermal method and used titanium butoxide and ethylenediamine as a titanium source and a nitrogen source, respectively. The photocatalysis efficiency of N-doped TiO_2 was tested by methylene blue (MB) and methyl orange (MO) decolorization under visible light irradiation. It was found that N-doped TiO_2 (N:Ti molar ratio of 1:1) gave the highest decolorization rate about 92% [15].

Toe, et al. (2018) modified mesoporous TiO_2 by N-doping. The studied parameters were molar ratio of nitrogen and titanium (N:Ti), synthesized temperatures and calcined temperatures. The optimal condition of N-doped TiO_2 synthesis was N:Ti molar ratio of 1:1, synthesized temperatures at 175 °C to 200 °C and calcined temperatures at 550 °C. The performance of N-doped TiO_2 was tested by MB and MO decolorization under visible light irradiation. N-doped TiO_2 gave the highest decolorization rate of MB and MO about 98% and 90%, respectively. According to the previous studies, nitrogen doping was widely used to enhance the photocatalytic properties of TiO_2 [16].

In the recent years, solvothermal method become more attractive for the synthesis of N-doped TiO_2. Because this method uses a solvent (aqueous and non-aqueous solution) and a temperature higher than that of the boiling point of a solvent in a closed system to synthesize highly reactive TiO_2 [17]. The advantages of this method are controllable crystallite size and tailor-made morphology by varying synthesis parameters such as temperature, pressure, process duration and the pH value of the solution [18].

Therefore, this study focused on the improvement of the photocatalytic reactivity of TiO_2 by doping nitrogen via solvothermal method. The effect of synthesis temperatures on the photocatalytic reactivity was studies. Additionally, the photocatalytic decolorization of methylene blue by N-doped TiO_2 under LED visible light irradiation was examined.

2. Experimental

2.1. Material

All chemicals used in this research were analytical grade. Titanium (IV) butoxide (TBOT, 97%, Sigma-Aldrich), and diethyleneamine (DEA, Merck) were used as sources of titanium and nitrogen, respectively. Ethanol (Merck) was used as a solvent.

2.2. Synthesis of N-doped TiO_2 and TiO_2

N doped TiO_2 were prepared by solvothermal method. N-doped TiO_2 was prepared by mixing TBOT and ethanol, then the solution was stirred for 30 min. Deionize water and nitric acid were added into the TBOT solution with the molar ratio of $1Ti(OC_4H_9)_4$:$18C_2H_5OH$: $2H_2O$:$0.2HNO_3$. After that, the mixture was added with 5% mol of DEA. Then, the solution was stirred for 30 min at room temperature. After that, the mixture was poured into a Teflon-lined stainless-steel autoclave and heated at 150 °C (NT-150), 170 °C (NT-170), 190 °C (NT-190) and 200 °C (NT-200) °for 3 h. The suspended solid was centrifuged and washed by ethanol and deionized water. The powder of N-doped TiO_2 was dried by oven at 110 °C overnight. Finally, the powder was calcined at 450 °C for 3 h.

TiO_2 was prepared by solvothermal method (TiO_2-150). TBOT was used for titanium source. TBOT and ethanol were mixed and then added deionize water and nitric acid. Then, the solution was stirred for 2 h at room temperature. After that, the mixture was poured into a Teflon-lined stainless-steel autoclave and heated at 150ºC for 3 h. The suspended solid was centrifuged and washed by ethanol and deionized water. The TiO_2 powder was dried by oven at 110 ºC overnight. Finally, the powder was calcined at 450 ºC for 3 h.

2.3. Characterization of N-doped TiO_2

The crystalline structure of the synthesized N-doped TiO_2 was analyzed by X-ray diffractometer (XRD: Rigaku TTRAX III). The porosity was analysed by Brunauer-Emmett-Teller surface analyzer (BET:Belsorp mini II, Osaka) and the morphology was examined by scanning electron microscope (SEM:JSM-5420LV, JOEL). Finally, the absorption edge and band gap energy were analyzed by UV-vis NIR spectrophotometer (Perkin Elmer, Lambda 950).

2.4. Photocatalytic decolorization of methylene blue (MB)

The photocatalytic reactivity of N-doped TiO_2 was evaluated by the decolorization of MB under visible light irradiation (>400 nm). A LED lamp of 15 W was used as a light source for testing. 1g of the N-doped TiO_2 powder was added into 100 ml of 10^{-5} MB solution. Before open LED light, the suspension was put in the reactor and stirred until dye adsorption was completed. After that, the LED light was turned on. The dye solution was evaluated by UV-vis spectrophotometry at wavelength of 664 nm (λ_{max}) every 30 min until the concentration of MB was stable

3. Results and discussion

3.1. Crystalline structure

Figure 1 shows the XRD patterns of N-doped TiO_2 prepared by different temperatures. The results show that all N-doped TiO_2 samples exhibit similar diffraction peaks appeared at 25.56º, 37.76º, 48.04º, 48.18º, 54.06º, 55.24º, 62.76º, 69.04º, 70.32º and 75.30º. These peaks represent to (101), (004), (200), (105), (211), (204), (116), (220), (215) and (303) planes, respectively, which is anatase structure of TiO_2 as followed Joint Committee on Powder Diffraction Standard (JCPDS no. 84-1286). All peaks of N-doped TiO_2 are slightly border than the peaks of bare TiO_2 (undoped) due to incorporating N atom into the anatase TiO_2 structure. The average crystallite size of the samples was calculated by Debye-Scherrer equation based on the strong peak (101). It was found that the crystallite sizes of N-doped TiO_2 were smaller than that of undoped TiO_2 (see Table 1). This indicated that N doping into TiO_2 anatase structure had significant effect on crystallites size [16]. However, these results can be concluded that the synthesized temperatures had no effect on crystallites size.

3.2. Morphology

SEM images illustrate the morphology of N-doped TiO_2 prepared by different temperatures compared with the undoped TiO_2 as shown in Figure 2. All samples exhibit the agglomerated spherical particles. The particle size of N-doped TiO_2 is approximately 0.02 μm while that of undoped TiO_2 is approximately 0.06 μm. The particle size of N-doped TiO_2 is smaller than that of undoped TiO_2. Among all doped samples, NT-200 had very small and well-dispersed particles, compared with the others.

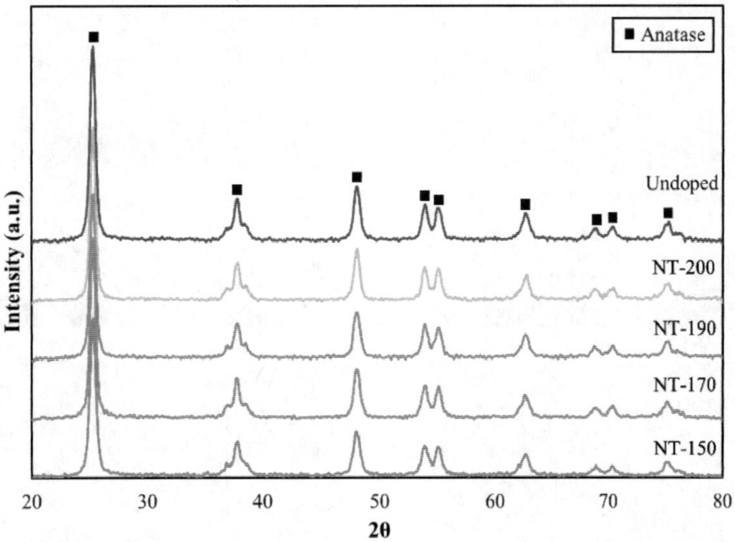

Figure 1. The XRD patterns of N-doped TiO_2 and undoped TiO_2 prepared by solvothermal with different synthesized temperatures

Table 1. The crystallite sizes of N-doped TiO_2 and undoped TiO_2 prepared by solvothermal with different synthesized temperatures.

Sample	Crystallite size (nm)
TiO_2-150	22.39
NT-150	14.67
NT-170	13.72
NT-190	13.72
NT-200	14.18

a) TiO_2-150

issue 2, pp. 42-48

[13] Vlot A, Vogelesang LB and de Vries TJ, Fibre metal laminates for high capacity aircraft *30th international SAMPE technical conference*, San Antonio, October 1998. p. 456–70

[14] Zafar R, Lang L and Zhang R, Experimental and numerical evaluation of multilayer sheet forming process parameters for light weight structures using innovative methodology *International Journal of Material Forming,* 2016, Vol. **9**, Issue 1, pp. 35-47

[15] Sherkatghanad E, Lang L, Liu S and Wang Y, Innovative approach to mass production of fiber metal laminate sheets *Materials and Manufacturing Processes*, Vol. **33**, issue 5, pp. 552-63

[16] ASM: *Metals Handbook*, Vol. **20** (Materials Park, OH: ASM Int., 1998).

[17] Halpin J, Jerine K and Whitney J, The laminate analogy for 2 and 3 dimensional composite material *Journal of Composite Materials*, Vol. **5**, pp. 36–49

[18] Ishikawa T and Chou TW, Stiffness and strength behaviour of woven fabric composites *Journal of Materials Science*, 1982, Vol **17**, pp. 3211–20

[19] Ishikawa T and TW Chou, Elastic behavior of woven hybrid composites *Journal of Composite Materials*, 1982, Vol **16**, pp. 2–19

[20] Aliabadi MH, *Woven Composite*; Volume **6** of Computational and experimental methods in structures , *Imperial College Press*, 2015

[21] Mallik PK, FIBER-REINFORCED COMPOSITES; Materials, Manufacturing, and Design *CRC Press; Taylor & Francis Group*, 2007

[22] Wu G, Yang J, The mechanical behavior of GLARE laminates for aircraft structures *JOM*, Vol **57**, issue 1, pp. 72-29

[23] Ashby M, Shercliff H and Cebon D, Materials: Engineering, Science, Processing and Design *Materials and Manufacturing Processes*, 2008, Vol. **23**, issue 2, pp. 222-23

[24] Wanhill RJH, Aerospace applications of aluminum-lithium alloys *Aluminum-Lithium Alloys, Processing, Properties and Applications*, Eds. Prasad, N.E., Gokhale AA and Wanhill RJH (*Butterworth-Heinemann, Elsevier Inc*., Oxford, UK), 2013, pp.503-535.

Synthesis of oxygen-deficient and monodispersed Pr doped CeO_2 nanocubes with enhanced resistive switching properties

Nan Chen* **and Sean Li**

School of Materials Science and Engineering, University of New South Wales, Sydney Australia

Abstract Monodispersed praseodymium (Pr) doped cerium oxide (CeO2) nanocubes (~ 10 nm) were successfully synthesized via a facile two-phase solvothermal method and further assembled into a dense film through a drop-coating procedure for resistive switching applications. The XRD and Raman results showed that Pr element was successfully doped into the CeO2 lattice structure while maintaining the highly uniform nano-cubic structure. In addition, the concentration of oxygen vacancy in CeO_2 nanocubes could be effectively modulated by varying dopant concentration evidenced by XPS and Raman analysis. The two-terminal structured device of Au/Pr doped CeO2/fluorine-doped tin oxide (FTO) exhibits stable resistive switching behaviour and ON/OFF endurance for more than 1000 cycles at a small operational voltage ranging from -0.6 to 1V. The present study may extend the potential of using praseodymium (Pr) doped cerium oxide nanocube in nanoscale building block for novelty 3D architectural memory and logic design in data-dependent applications such as artificial intelligence (AI) system and machine learning.

1. Introduction

Facing the extensive growth of demand for data storage in a big-data era, resistive random-access memories (RRAMs) have attracted huge attention because of its high speed, long retention time, and high density. In general, the building block for RRAM is the memory cell based on a two-terminal metal-insulator-metal (MIM) sandwich structure, where a resistive switching layer is the layer between two electrodes A wide range of metal oxides in various nanostructures have been studied for potential RRAM application, for example, simple binary oxides HfO[1] and NiO[2]; complex oxides such as perovskites $SrTiO_3$[3] and $BiFeO_3$[4]. Many theory models such as filamentary conduction, charge trapping defects states, trap-controlled space-charge-limited current and a change of a Schottky-like barrier have been proposed according to previous reports[5]. Among all the various models, the migration of oxygen vacancies and electrons under certain exerted electric voltage has been extensively studied in metal oxides based switching behaviour[6].

It is believed that the resistive switching phenomena are associated with the migration of oxygen vacancy and the accompanied oxidation state change of the cations under the external stimuli[7] which indicates that the oxygen vacancies engineering is a promising way to tune the performance of memory devices.

Among all the candidates, cerium oxide (CeO_2) has been extensively investigated[7] owing to its unique properties such as high dielectric constant, and excellent redox properties with various valence states of cerium atom. Specifically, the reversible transition of Ce^{3+} to Ce^{4+} give rise to the formation and elimination of oxygen vacancies[8]. Furthermore, it is widely accepted that doping can introduce more surface defects such as oxygen vacancies in CeO_2 lattice[9]. These oxygen vacancies lead to form conductive filaments which may play a vital role in tuning the resistive switching properties. Therefore, it is of great interest to investigate the resistive switching behavior in relation to the dopant level. In the meantime, the self-assembly of monodispersed nanocrystals is currently attractive for the large-scale fabrication of metal oxide thin film devices as with the advancing in the synthesis process. The highly uniform nanostructure can be employed as building blocks to assemble a higher dimensional structure. Moreover, the morphology of nanocrystal, size effect, ionic defect tuning as well as the interfacial carrier transportation can be well studied through this chemical route. The nanocube can be arranged into a highly coordinated square array and then layer-by-layer assembled into a 3D architecture. Many other approaches have been applied to synthesize doped cerium oxides such as hydrothermal[10], solvothermal[11], and electrochemical deposition[12]. A facile two-phase solvothermal method was chosen[13] for this study because of its unique superiority of forming dense switching film by preparing the monodispersed nanocube dispersion. With the help of a capping agent such as oleic acid, the as-prepared nanocube can be easily assembled into 2D or 3D superstructure[14]. The report on resistive switching behaviours for Pr doped cerium oxide nanocube is still very limited. As rare earth elements, both cerium and praseodymium have unique 4f electronic structure. Pr possesses different valances 3+ and 4+ which leads to more surface defects such as oxygen vacancies in the oxide. Moreover, the similar ion radius makes doping feasible.

In this work, oxygen-deficient Praseodymium (Pr) doped cerium oxide ($Pr_xCe_{1-x}O_{2-\delta}$, x=0-0.2) nanocubes with the size of around 10 nm were synthesized via a facile two-phase solvothermal method and then self-assembled to a dense film on fluorine-doped tin oxide (FTO) coated glass substrate simply by drop-coating. 10 at. % Pr doped CeO_2 with adequate oxygen vacancy concentration exhibited a high resistance difference and optimal stability in the voltage ranging from -0.6 to 1V. The underlining mechanism for resistive switching process is claimed to be the result of the migration of oxygen vacancies inside the nanoparticles.

2. Experimental section

All chemicals were purchased from Sigma without further purification. Pr doped CeO_2 nanocubes were prepared by using a solvothermal process as shown in Figure 1(a). In a typical reaction, a certain amount of cerium (III) nitrate and praseodymium (III) nitrate (0,5,10,15,20 at. % of cerium) was dissolved in15 mL of DI water then transferred into a 50 mL autoclave. Meanwhile, a miscible solution of 0.56ml oleic acid (OLA) and 14ml of toluene was prepared separately until they are fully mixed. The ratio of OLA to the metal ions is fixed to be 8:1. Afterward, 0.15 ml of tert-butylamine was added into the mixture and then the organic solution was carefully transferred into the previous autoclave upon the aqueous solution to maintain the two-phase state. The sealed autoclave was heated at 180 °C for 24 h and then cooled to room temperature. The upper organic solution was centrifuged to separate the nanocrystals. The separated CeO_2 nanocrystals were redispersed into 4 mL of toluene. Subsequently, the suspension solution was drop-coated on an FTO coated glass substrate to obtain a thin film with self-assembled Pr doped CeO_2 nanocube. To assemble a device, the round gold top electrode (diameter 250 μm) was sputtered on the surface of the thin film. The schematic illustration of the obtained device was shown in Figure 1(b).

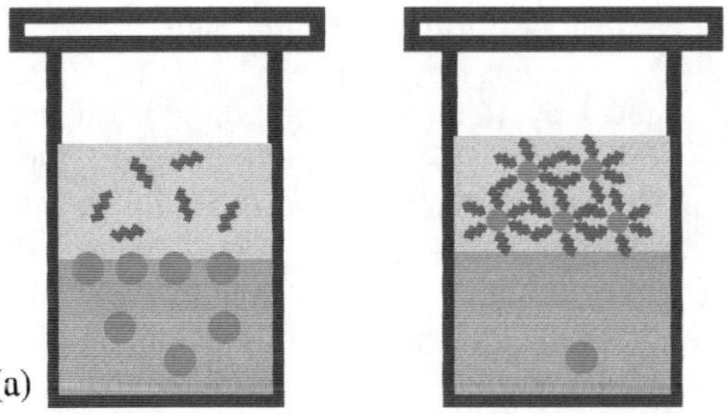
(a)

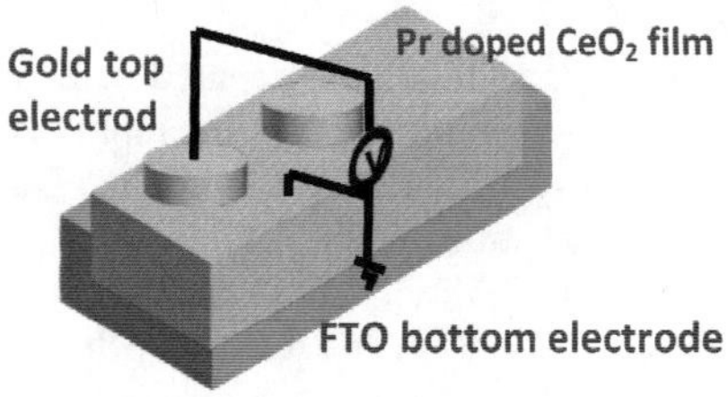

(b)

Figure 1. (a)Schematic illustration of solvothermal synthesis and (b)

Structural analysis of the as-prepared Pr doped CeO_2 film was carried out using an X-ray diffractometer (XRD; PANalytical Empyrean 1) with Cu Kα radiation. The accurate chemical composition was studied by inductively coupled plasma mass spectrometry (ICPMS; PerkinElmer Nexion). Raman Microscope (Renishaw inVia) was employed to determine the chemical bond of the as-prepared sample with a 514 nm laser beam. The microstructures of the as-prepared materials were studied by transmission electron microscopy (TEM; Phillips CM200) and field-emission scanning electron microscope (FE-SEM; FEI Nova NanoSEM 450). The where w_{ijk} denote the influence coefficients of the original inclusion in full space. Moreover, the symmetric inclusion of domain (2) is subjected to a mirror eigenstrainIt is worthwhile to note that the eigenstrain

$$\varepsilon_{ij}^{\prime *} = \left[\varepsilon_{11}^{*},\ \varepsilon_{22}^{*},\ \varepsilon_{33}^{*},\ -2\varepsilon_{23}^{*},\ -2\varepsilon_{13}^{*},\ 2\varepsilon_{12}^{*}\right]^{T} \tag{3}$$

After the summation of the first two solutions, the normal stress on the symmetry plane $x_3 = 0$ is twice and the symmetry plane is free of shear tractions. Besides, the normal displacement along the surface cancels. According to the method of Images [4], the traction-free condition at the boundary surface is obtained by superimposing the solution for a half space under normal plane stress σ_{33} which is opposite to the resultant stresses after superposing problems (1) and (2) in figure 1.

In view of the superposition of three components, the displacement of an arbitrary shaped inclusion in half space may be expressed as

$$u_i(\mathbf{x}) = \int_{\Omega_1} w_{ikl}^{(1)}(\mathbf{x}-\mathbf{x}')\varepsilon_{kl}^{*} d\Omega_1 + \int_{\Omega_2} w_{ikl}^{(2)}(\mathbf{x}-\mathbf{x}')\varepsilon_{kl}^{\prime *} d\Omega_2 + \Sigma\Sigma\chi_{i3}(\mathbf{x}-\mathbf{x}')p_{30} \tag{4}$$

where $\chi_{i3}(\mathbf{x}-\mathbf{x}')$ is the response function, also termed as the influence coefficient [8], of elemental normal contact force uniformly distributed on the rectangular patch, and p_{30} is a normal stress whose magnitude is evaluated at the centre of the rectangle.

2.2. Love's rectangular contact solution

In abovementioned equation, Solutions (1) and (2) (figure 1) have been determined in closed-form, whose details can be found in [6]. The solution (3) is solved numerically by discretizing the computational domain into a number of rectangular patches at the symmetry plane. The elastic field produced by a uniform rectangular patch loading is solved by employing Love's rectangular contact solution, which has been comprehensively discussed and recorded by an effective notation in conjunction with the FFT algorithm [9].

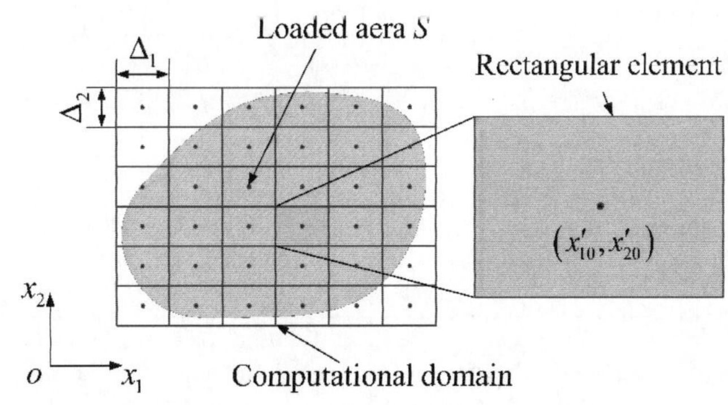

Figure 2. Discretization of the computation domain using a uniform rectangular mesh.

Consider a loaded area S of the symmetry surface as a computational domain which is discretized by $N_1 \times N_2$ uniform rectangular meshes with each sides of length Δ_1 and Δ_2 (figure 2). Each grid point labelled by $[i, j]$ is placed at the centre of the elements and assumed to be subjected uniform stress whose magnitude is evaluated at the centre of the rectangle. The indices i and j refer to the columns and rows of each grid, respectively. All of the grid points are denoted by I_g

$$I_g = \{[i,j]:\ 1 \le i \le N_1,\ \ 1 \le j \le N_2\} \tag{5}$$

The surface displacement of single element could be written in matrix form as follows

$$\begin{bmatrix} u_1 \\ u_2 \\ u_3 \end{bmatrix} = \begin{bmatrix} \chi_{11} & \chi_{12} & \chi_{13} \\ \chi_{21} & \chi_{22} & \chi_{23} \\ \chi_{31} & \chi_{32} & \chi_{33} \end{bmatrix} \begin{bmatrix} p_{10} \\ p_{20} \\ p_{30} \end{bmatrix} \tag{6}$$

where p_{j0}, $(j=1,2,3)$ is the magnitude of the uniform stress acting on the centre of the rectangle along the x_j axis; and χ_{ij} is termed as the elementary influence coefficient which relate the stress to the surface displacement at the surface point.

The computational domain in problem (3) (cf. figure 1) is numerically discretized by rectangular patches, each of which is assumed to be subjected uniform normal stress p_{30} .The surface displacement could be solved by superposing contributions of each element. For example, the surface deflection at grid point [i, j] may be written as

$$\bar{u}_3^{[i,j]} = \sum_{k=1}^{N_1}\sum_{l=1}^{N_2} \chi_{33}^{i-k,j-l} p_{30}^{[k,l]}, \quad [i,j] \in I_g \tag{7}$$

As a result, the solution (3) can be solved by discretizing the computational domain into a number of rectangular patches and superposing contributions of each element.

3. Results and discussions

A benchmark example of a cuboidal inclusion is given in this section to explore the effects on the surface redundant traction and the surface displacement for different depth, shape and meshes. The related parameters for the baseline case in the computation are listed in table 1. The different results can be compared to each other by considering that the parameters for the baseline case are common between all examples of simulations.

Table 1. Parameters of matrix and the cuboidal inclusion for the baseline case.

Parameters	Values
Young's modulus, E (GPa)	210
Poisson's ratio, v	0.3
Eigenstrain components	$[\varepsilon^*]=10^{-3}\times[1\ 1\ 1\ 0\ 0\ 0]^T$
side length	$a=1,\ b=1,\ c=1$
Distance to the Surface	$s=1$

The variation of relative surface redundant traction and surface displacement are calculated along the x_1 axis for different depth in dimensionless form, as depicted in figures. 3 and 4 respectively. It clearly shows that the effect of the inclusion on the stress field is decreasing when the cube is located deeper below the surface. It can be noted that when the cube is located at $s=2a$, the stress field tends to be uniform and the redundant traction and the displacement on the surface is not distinct. This means that once the inclusion is far away from the surface, the value of its depth play a negligible role in determining the redundant surface traction and surface deflection. Moreover, when the value of s/a is close to zero, the redundant traction on the surface above the cube is almost a constant except at the boundary edges.

The comparison of the surface redundant traction and the surface displacement along the x_1 axis for different value of b/a and c/a is shown in figures. 5 and 6 respectively. It can be observed that

IOP Conf. Series: Materials Science and Engineering **576** (2019) 012037 doi:10.1088/1757-899X/576/1/012037

both the redundant traction and the displacement on the surface are reduced with the decreasing of the cuboidal size. Moreover, comparing the figures of the surface displacement for different shapes, it can be concluded that the effect of c on the surface displacement is concentrated inside, and the influence of b is scattered.

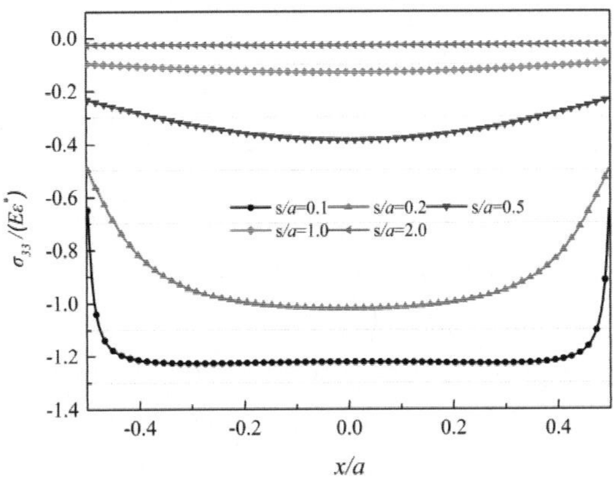

Figure 3. Variation of dimensionless surface redundant traction along the x_1 axis for different depth.

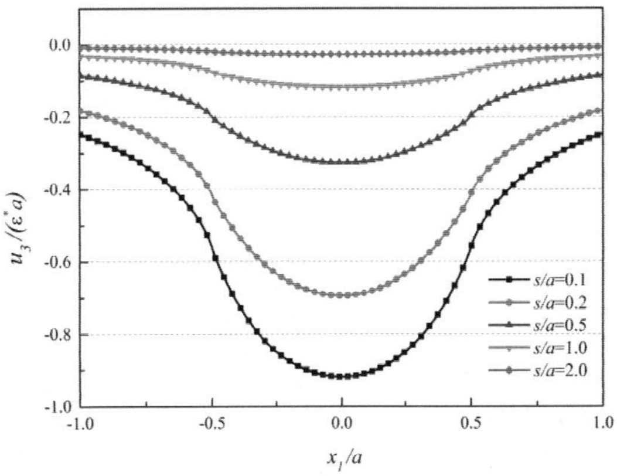

Figure 4. Variation of dimensionless surface displacement along the x_1 axis for different depth.

In the computation, the domain is discretized into 251×251 grids for the baseline case. Next, the computational domain is magnified in order to investigate the effects of the size of the computational domain. The variation of relative dimensionless surface displacement along the x_1 axis with different value of magnification $NFCTX$ is shown in figure 7. It should be noted that the resulting displacement show noticeable discrepancy from the exact solution without magnifying the size of the computational domain, but it tends to agree with the exact solution with magnification $NFCTX = 3$. Further increase of the magnification factor $NFCTX$ would distinctly the occupied more computational time but shows no distinct influence (figure 7). Accordingly, there is no necessary to further enlarge the magnification factor. Therefore, in order to improve the accuracy of calculation, the size of the computational domain must be extended to some degree.

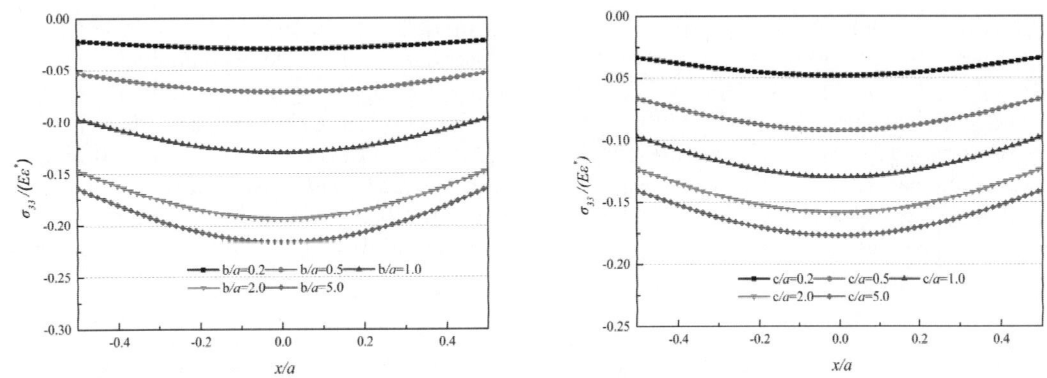

(a) (b)

Figure 5. Comparison of dimensionless surface redundant traction along the x_1 axis for different shapes: (a) b / a change; (b) c / a change.

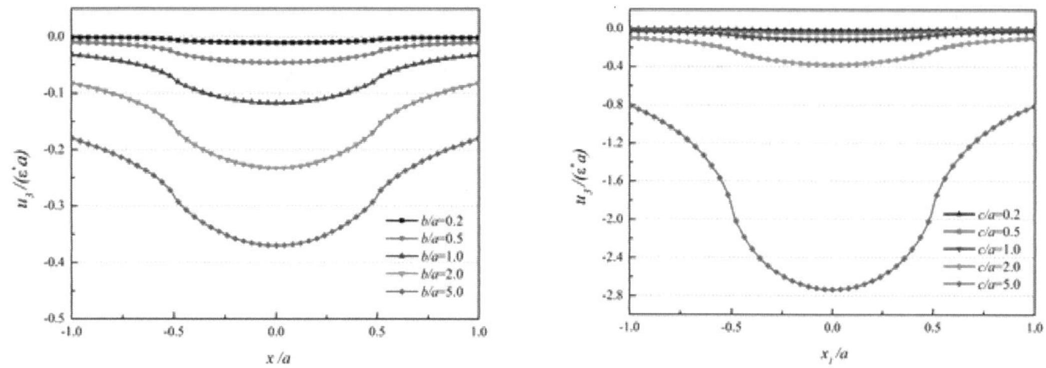

(a) (b)

Figure 6. Comparison of dimensionless surface displacement along the x_1 axis for different shapes: (a) b / a change; (b) c / a change.

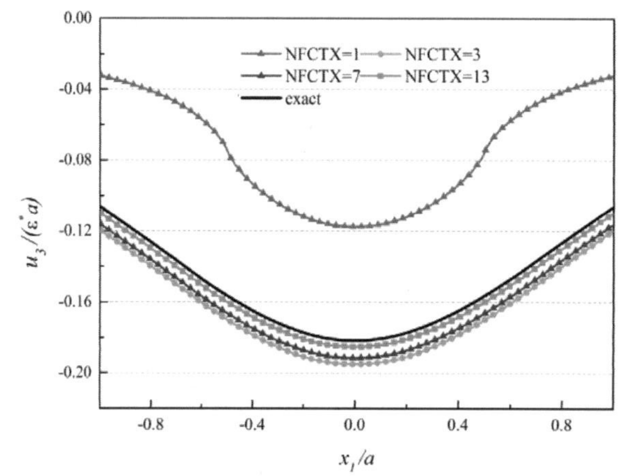

Figure 7. Variation of dimensionless surface displacement along the x_1 axis for different meshes.

4. Conclusions

The elastic field of an inclusion in a half space with eigenstrain is often complex and intricate because of the mathematical difficulties arisen from the effect of the free surface. Notwithstanding, the current work demonstrates that the stress field of an arbitrarily shaped inclusion in a half space can be solved by the method of images. The Voigt notation and the Eshelby tensor were utilized for simplifying the numerical computation and computer programming. A number of interesting parametric studies are provided in this paper in order to explore their effects on the surface displacement. It is shown that the surface displacement of the inclusion is much closer to the exact solution by magnifying the size of the computational domain. Numerical computations for half-space problems always require a large computational domain to guarantee its accuracy.

References

[1] Eshelby J D, P. Roy. Soc. Lond. Ser. A, **241** (1226), 376-396(1957)
[2] Jin X, Lyu D, Zhang X, Zhou Q, Wang Q, Keer L M, ASME. J. Appl. Mech, **83**(12), 121010-121010-12(2016)
[3] Chiu Y P, ASME. J. Appl. Mech, **44**(4), 587-590 (1977)
[4] Chiu Y P, ASME. J. Appl. Mech, **45**(2), 302-306 (1978)
[5] Zhou K, Chen W, Keer L M, Wang Q, Comput Methods Appl Math, **198**(9-12), 885-892(2009)

[6] Liu S, Jin X, Wang Z, Keer L M, Wang Q, Int. J. Plasticity, **35**(Supplement C), 135-154 (2012)
[7] Zhou Q, Jin X, Wang Z, Yang Y, Wang J, Keer L M, Wang Q, Tribol. Int, **93**, 124-136(2016)
[8] Mura T, Micromechanics of defects in solids. 2nd ed. (Springer, Dordrecht, The Netherlands, 1987)
[9] Jin X, Niu F, Zhang X, Zhou Q, Lyu D, Keer L M, Hu Y, Tribol. Int, **103**, 331-342(2016)

Acknowledgments
This work is supported by Fundamental Research Funds for the Central Universities (2018CDYJSY0055 and Nos.106112017CDJQJ328839). The authors are grateful to the Graduate Research and Innovation Foundation of Chongqing, China (Grant Nos. CYB18020 and CYB17025). X.J. would also like to acknowledge the support from the National Natural Science Foundation of China (Grant Nos. 51475057 and 51875059), and the State Key Laboratory of Mechanical Transmissions through funding (SKLMT-ZZKT-2017M15).

IOP Conf. Series: Materials Science and Engineering **576** (2019) 012038 doi:10.1088/1757-899X/576/1/012038

Non-destructive testing Techniques based on Failure Analysis of Steam Turbine Blade

Li Tian, Yan Hai, Zeng Qingyue and Yin Qin

Suzhou Nuclear Power Research Institute, 518120, Shen Zhen, China

Corresponding Author's E-mail: 565314177@qq.com

Abstract. Turbine blades are core components in turbine. It is of great significance to diagnose the blade failures at early stage and regularly inspect them at run time. Based on the analysis of blade failures, a comprehensive application summarize of different non-destructive testing techniques in blade inspection was made in this paper. The non-destructive testing techniques include visual testing, liquid penetrant testing, magnetic particle testing, radiological testing, conventional ultrasonic testing, ultrasonic phased array testing, eddy current testing and metal magnetic memory testing. The applicability and limitation of different methods were demonstrated through comparison and analysis.

1. Introduction

Turbine blades are core components in turbine, which play important roles in energy transformation. Its operation status directly decides the efficiency and safety of turbine and even the entire power plant. The operating conditions of the turbine are very complex. The turbine bears centrifugal force, steam power, steam exciting force, corrosion and vibration, and high speed erosion in wet steam area [1]. It is more likely to cause blade damage or even breakage. Once the breakage happens, the damaged blade may damage other blades. Even it will break the dynamic balance of turbine rotor, which may lead to unstable shaft of steam turbine and cause vibration of turbine rotor. In severe case, emergency shutdown may happen. Therefore, it is very important to study the non-destructive testing techniques based on the failure analysis of steam turbine blades.

2. Failure Analysis of Steam Turbine Blade

2.1. Failure modes of Steam Turbine Blade

In these exceptional cases of steam turbine, failure modes caused by blade damage make a great percentage. The main reason for blade failure is fatigue failure caused by microcosmic stress concentration [2]. Its basic characteristics appear as follows. Under the action of alternating stress less than the limit of static stress strength, many types of inner microdefects, such as dislocation, slippage and phase transformation induced by temperature or stress, were germinated and turned into macroscopic cracking. With crack extension, fracture may occur. The fatigue crack of blade usually generate from thin-wall. It is a dangerous part of microcosmic stress concentration.

Fatigue fracture can be classified into mechanical fatigue fracture and corrosion fatigue fracture. A correlational study [3] showed that the reason for the cracking and fracture of steam turbine blade is corrosion fatigue fracture. The fatigue crack originals are the pitting holes on the blade surface.

Cracks generate under the influence of corrosive environment and exciting forces. It is mainly shown as intergranular stress corrosion and intergranular fracture at fatigue source and fatigue propagation with corrosion characteristics at crack growth regions. Therefore, many factors will influence the generation of the blade fatigue fracture, including the loading, structural strength, temperature and corrosivity of the blade.

2.2. Stress Analysis of Steam Turbine Blade

A steam turbine blade consists of the blade root, blade top and working region. From the technological structure of view, the blades can be divided into several types. They are constant chord blade with filling-up piece, straight blade with massive blade root, variable chord blade and variable chord twisted blade. By Analyzing the stresses of different types of blades (with or without shroud and lacing wire), it is observed that blades with different structures have different stress condition, and stress conditions of different parts on the same blade are also different. We takes variable chord blade as an example, the stress condition of steam turbine blade is calculated [4].

A stress analysis of the blades with a shroud or a lacing wire was shown in table 1. It is s a reference for blade force comparison with a shroud or a lacing wire. Stress analyses of the blades without a shroud or a lacing wire was shown in table 2, while the stress analysis of the blades with a shroud was shown in table 3.

Table 1. Stress analyses of the blade without a shroud or a lacing wire

Part	Force types	Force value*	Vibration
Root	Centrifugal tensile stress; Centrifugal bending stress; Gas flow bending stress; Dynamic stress;	Most serious	More serious vibration
Top	Dynamic stress	Slightest	Mainly A0 vibration
Working region	Centrifugal tensile stress; Centrifugal bending stress; Gas flow bending stress; Dynamic stress;	Common	
*note: As a reference for blade force comparison with a shroud or a lacing wire.			

Table 2. Stress analyses of the blade with a shroud

Part	Force types	Force value	Vibration
Root	Centrifugal tensile stress; Centrifugal bending stress; Gas flow bending stress; Dynamic stress;	Tensile stress increases Dynamic stress decreases	Vibration is lighter Mainly A0 vibration
Top	Tensile stress; Pressure stress; Torque;	Dynamic stress of some blades increase considerably	
Working region	Centrifugal tensile stress; Centrifugal bending stress; Gas flow bending stress; Dynamic stress; Torque;	Gas flow bending stress increases Dynamic stress decreases	

Table 3. Stress analyses of the blade with a lacing wire

Part	Force types	Force value	Vibration
Root	Centrifugal tensile stress; Centrifugal bending stress; Gas flow bending stress; Dynamic stress;	Integral stress decreases	Vibration is lighter Mainly A0 vibration
Top	Bending stress	The minimum stress	
Working region	Centrifugal tensile stress; Centrifugal bending stress; Gas flow bending stress; Dynamic stress; Torque;	Gas flow bending stress and dynamic stress decrease	
Lacing hole	Tensile stress; Pressure stress; Torque;	Mainly dynamic stress, easily causing stress concentration	

The study [3] indicated that under the operation situation of the steam turbine, the rotating blades bear the maximum static stress and alternating stress. In addition, the blades face forced vibration under the excitation force, which will increase the risk of blade fracture. According to the analysis of the blade actual operation situation, its fracture mode is not simply belonging to fatigue fracture, the environmental medium and the surface salt also exacerbate the crack producing and spreading.

3. Non-destructive testing methods of steam turbine blades

There are various reasons for the failure of steam turbine blades, such as the damage caused by the unqualified specific characteristics. During the actual operation process, it is common that the nonconforming frequency of the blade could generate resonance and result in resonance damages. Defective design, poor materials and immaturity processing technic will also cause the blades failure. In addition, operational issues of the steam turbine are the essential reason for the blade failure, such as the operation deviation from the rated power, overload operation, low temperature, poor steam quality, too high or too low vacuum degree, water-entry impact and frequent starts and stops. It can be seen from the above reasons the steam turbine blades may suffer from damages at each stage of the design, manufacturing and operation. Therefore, it is very important to periodically inspection the blades during the refueling outages, find the failure blades with defects or deformity timely and repair or replace them, to avoid a serious accident and heavy losses of economy.

3.1. Surface Examination

Visual testing is used to test the physical dimension, structural integrity and shape defect of the blade. It is also capable of finding out the defects and other details on the blade surface. There are many advantages of visual examination. The principle is sample. It is easy to operate and should not be affected by the material, the structure, the shape nor the size. The inspection result is simple, rapid, reliable and sensitive with repeatability. However, due to limited to human vision and the equipment resolution, the slightest defects on the blade surface may be ignored during the inspection process.

Liquid permeation method is used to test the surface-breaking defects on the blade surface. The permeated solution has strong permeability. Because of the wetting action and capillary action, when brushed on the tested area, the permeated solution will infiltrate into the surface-breaking defects on the blade surface. Then the permeated solution is wiped away and the developer with strong sorption ability is uniformly sprayed. Under the capillary action, the permeated solution remaining in the defect is adsorbed to the blade surface and shows the location and shape of the defects. The inspection principle of liquid permeation method is easy. The inspection sensitive is comparatively high. However, the inspection range of the liquid permeation method is definitely limited. The technique can only test surface-breaking defects and have a high demand on the surface roughness and clearness.

Magnetic particle testing is another surface examination method. The blade inspection process of magnetic particle testing technology is as follows. When the ferromagnetic blade is magnetized, due to discontinuity, local distortion of magnetic field lines on the surface and near surface occurs, lead to the generation of leakage flux. By adsorbing the magnetic particles on the blade surface, visible magnetic particle indication occurs and shows the position, shape and size of the defects. A related study inspected the steam turbine rotor blades by using the method of fluorescent magnetic particle testing with proper inspection techniques [1]. The factors of affecting the sensitivity are researched, including the number of turns of the magnetized coils, magnetic suspension concentration and magnetizing time. Results indicate that the fine defects at the blade tip, the blade root and the blade surface can be detected reliably by using the method of fluorescent magnetic particle testing. It has high testing sensitivity and reliability, which is suitable for defect inspection of irregular parts of blade.

3.2. Radiological testing

X-ray radiographic testing is one of the important methods for turbine blade defect inspection at early stage. When x-rays penetrate through the examined objects, the absorbing ability of the defective part is different from that of non-defective parts. In general, the intensity of the X-ray transmitting through the defective part is higher than that of the non-defective parts. The differences in film blackness can show the defects in the examined components. The results can be directly observed on the films and the films are suitable for long-term preservation. Because the turbine blade is thin and has variable thickness with technology of high quality requirements, the flaw detection sensitivity and film definition are important factors during the X-ray radiographic testing. The early study on the turbine blade inspection by using the X-ray radiographic testing method was around the flaw detection sensitivity [5-6].By choosing the proper hardness of X-rays, focal length, trans-illumination mode, film type and using the intensifying screen and double film technique, the defect in the turbine blades are successfully inspected with quite high sensitivity.

3.3. Ultrasonic testing

Ultrasonic testing is a common nondestructive testing method to inspect the volume defects. Ultrasonic waves can travel through the material. Ultrasonic testing has advantages, such as short time, flexible operation and low cost. Meanwhile, it has very high requirement for the shape, the surface condition of the blades and the experience of technical personnel. Therefore, ultrasonic testing is feasible for blade surface and blade root section, but not appropriate in effectually inspecting the blade roots. At present, ultrasonic phased array testing is used for the full coverage inspection of the turbine blade roots.

The common types of blade root include the T type, the mushroom shaped type, the fork shaped type and the fir type. According to the stress analysis, during the running process the stress mainly concentrated in the joint part of the blade root and the rotor wheel groove. The stress generally reaches the maximum on the surface of the first or second root, easy to cause the root cracks. Additionally, the roots have complex structure and small spacing. It is hard to inspect the roots by using the traditional ultrasonic testing technique. Therefore, it is widely to study the inspection techniques of different types of roots based on the ultrasonic phased array testing technique. Zhang Yicheng et al inspected the fir type rotor blade roots by using the PAUT technique [7]. Researchers first determine the key ultrasound scan area according to the stress analysis. By the means of simulation, two types of ultrasonic phased array testing probes, transverse wave and longitudinal wave probes, are optimized and designed. The turbine blade with passing hollow is the simulated test block. Several scanning patterns are set up. The inspection results indicate that the designed ultrasonic phased array testing technology is highly accurate and effective. It can be used to effectively inspect the turbine blade root. Zhang Wuneng et al designed a special phased array ultrasonic transducer [8].By using a matched wedge block, the mushroom shaped roots were well scanned. With the accurate beam index, the inspection system enhances the inspection efficiency and can achieve the crack defects in a quicker, more intuitive and more effective way. Huang Qiaosheng et al inspected the mushroom shaped blade roots by using the method of ultrasonic phased array testing technique [9], aimed at studying the judgment method of defect signals. Wang Jun studied the quantitative method of PAUT results of the mushroom shaped blade roots [10], giving that -6dB quantitative method is most suitable for measuring the defect length of the mushroom shaped blade roots. He Cheng et al inspected the T type rotor blade roots by using the PAUT technique [11]. In consideration of that the T type blade roots are buried in the impeller rim, it is of great significance to study the online inspection of T type roots. To realize a full coverage PAUT of cracked areas of T type roots, the researchers developed special small-size linear array phased array probe. By using its CAD advanced artifact diagram import module, the T type roots were loaded on to S-scanning mode images. The inspection results are comprehensive, clear and easy to judge.

3.4. Eddy current testing

Eddy current testing technique is an efficient surface and near-surface defect inspection method. When the eddy current probe moves on the blade surface, if there are defects on the blade surface or its conductivity, permeability and shape change, the coil impedance of the eddy current probe will change correspondingly. The features of eddy current signals can show the defects on the blade surface.

At present eddy current testing technique is a positive try, which will become an additional inspection of the turbine blade.Shao Zebo et al inspected the blade simulated crack with the methods of eddy current testing technique [12]. By quantitatively studying the corresponding relation between the defect sizes and response signals, the experimental parameters used for practical application are confirmed. However, the inspection sensitivity of conventional point-type probe is related to the defect direction. Therefore, for the special blade parts, such as the blade groove, some defects may be skipped over when the defect direction is consistent with the eddy current direction. Based on the structural features of the blade grooves, flexibility and copying array probes are made [13] and used to inspect the blade body and blade grooves of low-pressure rotor last stage blade. It has features of easy operation, no-pollution and high sensitivity and reliability.

3.5. Magnetic memory testing

Based on the leakage magnetic field detection of the components, metal magnetic memory testing technique is used to determine the stress concentration and damage degree of components. It can detect locations where macroscopic defects are not present. Early diagnosis is helpful to evaluate the safety of equipment accurately. Magnetic memory testing is the only nondestructive detection method that can accurately identify the stress and deformation concentration areas and other dangerous areas. In this, it is often used to diagnose microscopic defects and early failure and damage, preventing sudden fatigue damage.

For turbine blades, the main reason of blade fracture is fatigue failure caused by microstress concentration. The conventional methods are mainly used in detecting the macroscopic cracks of turbine blades. Therefore, magnetic memory technique is an effective method to find stress concentration zone, which can provide early diagnosis for blade damages. Dai Guang et al inspected the various stages of rotor blades with a stress concentration magnetic detector [2]. Based on the leakage magnetic field and scattering magnetic field gradient of the detected component, the position and severity of blade stress concentration are analyzed. Results show that the most dangerous stress concentration areas of blade is the blade root, the joint between the blade and the girdle and the vicinity of the rib hole. This provides a reference for the early diagnosis and maintenance strategy of turbine blades. Another inspection found cracks in the stress concentration line of the blade [14], it is verified that the metal magnetic memory detection method is very reliable for the initial diagnosis of stress concentration in turbine blades. Researches on the application of metal magnetic memory testing technique include metal equipment strength and reliability diagnosis, quality inspection and sorting of mechanical parts and pressure vessel strength and life diagnosis. Development of intelligent metal magnetic memory diagnostic equipment and signal processing software will also contribute to the improvement of metal magnetic memory testing technique.

3.6. Comparison and analysis of NDT methods

Comparison and analysis of different NDT methods are shown in table 4.

Table 4. Stress analyses of the blade with a lacing wire

NDT Methods	Defect Type	Examination area	Sensitivity	Efficiency and Reliability
VT	Surface Defect	Blade Body	Lower	Higher
PT	Surface-Breaking Defect	Blade Body, Lacing Hole	Lower	Lower
MT	Surface Slight Defect	Blade Edge	Higher	Lower
RT	Volume Defect	Blade Weld	Higher	Low Efficiency & High Reliability
ET	Near Surface Defect	Blade Edge, Lacing Hole	Higher	Higher
UT	Volume Defects	Blade Body	Lower	Lower
PAUT		Blade Root	Higher	Lower
MMM	Micro And Macro Defects	Blade Edge	Very High	Higher

4. Summary

Nondestructive testing technologies play important role in turbine blade inspection. Different NDT methods have their own advantages and limitations during the blade detection process. The main characteristics are summarized as follows.

(1) Traditional nondestructive testing methods play an important role in the inspection of surface and near-surface defects.

(2) The application of phased-array ultrasonic technology solves the problem of high leakage rate and low inspection efficiency in conventional ultrasonic inspection of blades.

(3) The metal magnetic memory technology can be used to diagnose the micro-defects, the early failure and damage. It has a broad prospect in early fault diagnosis of the blades.

(4) The development direction of nondestructive testing technology for blade inspection is that: the intellectualization of the inspection equipment, high efficiency of the inspection process, the digitalization of test results and realizing the early fault diagnosis and prediction.

5. References

[1] Fu Qianfa, Li Peng, Li Qiuda, Ge Liang, Zhang Ming, *Nondestructive Testing*, **37** (8) (2015) 80-82.

[2] Dai Gaung, Ding Kebei, Wang Wenjiang, Test and Application of the Magnetic Memory of the Vanes of a Turbine, *Chemical Machinery*, **30** (5) (2003) 276-278.

[3] Hu Shubing, Fu Qin, Chen Yanyu, The Fracture Analysis of Turbine Blade, *Material Protection*, **47** (5) (2015) 172-176.

[4] Wang Wenjiang,*The Search of the Magnetism Memory Technology and Its Application Method*, Da Qing: Northeast Petroleum University, (2003) 41-44.

[5] Li Fangying, Duan Qiande, Mo Xueyu, Radiographic Inspection of Welds in Steam Turbine Blades, *Nondestructive Testing*, **2** (1982) 20-24.

[6] Jiang Chuanhai, Zhu Dixin, X-ray Inspection of the Lashing Joint Weld on Large Steam Turbine Blades, *Power Engineer*, **5** (1987) 49-52+64.
[7] Zhang Yicheng, Qiu Jinjie, Cai Jiapan, Nie Yong, Lu Wei, Phased Array Inspection for Nuclear Power Fir Style Turbine Blade Root, *Nondestructive Testing*, **36** (9) (2014) 38-41.
[8] Zhang Wuneng, Wang Yi, Jin Feng, Li Shitao, Yang Xu, Liu Wensheng, Phased Array Inspection for Bacterial Type Turbine Blade Root, *Nondestructive Testing*, **37** (3) (2013) 44-45.
[9] Huang Qiaosheng, Peng Bicao, Shen Dingjie, Yang Xiangwei, Research for Ultrasonic Phased Array test of Steam Turbine's Mushroom Shaped Blade, *Thermal turbine*, **42** (2) (2013)138-140.
[10] Wang Jun, Imperfection Sizing Method of Turbine Straddle Tree Blade Roots Using Phased Array Ultrasonic Technology, *Nondestructive Testing*, **39** (1) (2017) 19-23.
[11] He Cheng, Li Wensheng, Zhao Jiangping, Yin Lu, Ji Xuanrong, Chen Xiurong, Ultrasonic Phased Array Testing of T-type Turbine Blade Root, *Nondestructive Testing*, **39** (5) (2017) 33-36+48.
[12] Shao Zebo, Song Shubo, Eddy Current Inspection of Turbine Blades, *Nondestructive Testing*, **24** (10) (2002) 444 - 445+453.
[13] Guo Derui, Application of Arrays Eddy Current for turbine Blades and Blade Groove, *China Plant Engineer*, **6** (2018) 92-95.
[14] Chi Yongbin, Liu Yuzhe, Hu Xianlong, Ma Delong, Metal Magnetic Memory Testing of Turbine Blades, *Nondestructive Testing*, **24** (10) (2002) 440-442.

IOP Conf. Series: Materials Science and Engineering **576** (2019) 012039 doi:10.1088/1757-899X/576/1/012039

Research on identification method of multiple cracks in a shaft by vibration characteristics

Wenke Dai, Wenguang Liu* and Ahmed Tousif Alif

School of Aeronautical Manufacturing Engineering, Nanchang Hangkong University, 330063 Nanchang, China

Corresponding Author's E-mail: liuwg14@nchu.edu.cn

Abstract. As the main transmission device in mechanical equipment, the shaft plays an important role in the function realization and stable operation of the machine, but it is particularly prone to failure due to the long-term alternating or random axial load, bending moment and torsion, so it is of great significance to find the crack on the shaft in its early state, to reduce economic losses and prevent catastrophic accidents. The vibration characteristics of the circular cross-section shaft with multiple cracks are studied, and a damage detection method is proposed. Based on the basic beam theory and the local compliance theory, the stress intensity factors(SIFs) for the cracked shaft subjected to axial force, shear force and bending moment are derived, and the evaluation of corresponding compliance coefficient is given. The differential equation of motion is discussed. Results indicate that the derived stress intensity factor is in good agreement with the experimental results. As the shaft has cracks, the mode shape will show discontinuous at the crack location. This method can identify the crack position, depth and the number of cracks.

1. Introduction

Shaft is one of the common components which is widely used in a dynamic machine. It plays an important role in the realization of function and stable operation of dynamic machine. Therefore, the reliability of shaft is very significant to the machine. Due to the environment of vibration and repeated force, cracks are often initiated in a shaft, making the machine work unstable and reducing the strength of the shaft. Especially when the shaft works in a resonant environment, the crack will be propagated rapidly which may lead some accidents. To guarante the safety reliability of the shaft, it is of great significance to find the crack position and crack size in the shaft in its early state. Numerous studies have been done about the identification method of crack. Zhong et al. proposed a novel non-projection vision-based system, the purpose of this system is to realize simultaneous measurement of the radial and axial displacements with high accuracy and good reliability by using a tailored artificial constant density sinusoidal fringe pattern. Results indicated that this method was an effective and accurate technique for real-time vibration monitoring of rotating shaft[1]. Zhang et al. studied damage detection method of a beam based on the coherence function of random vibration. Results indicated that this method can be used to detect single-crack and multi-crack of the cantilever beam accurately and reliably[2]· Gounaris et al. simulated the cracks using local flexibility theory, and established a finite element model of the cracked beam, and analyzed the dynamic response of a cantilever beam under harmonic excitation. Results indicated that the modal frequencies and modal shape are greatly affected

IOP Conf. Series: Materials Science and Engineering **576** (2019) 012039 doi:10.1088/1757-899X/576/1/012039

by the presence of medium cracks[3]. An sandwich panel with some cracks was used as the object by Zhang et al. to study three identification methods of cracks. Results shown that the crack location can be accurately identified according to the diffrence of vibration shape[4]. A simple and effective identification algorithm of crack damage before a break occurs was employed by Cruz-Vega et al. to identify several levels of crack of a rotor bar under different load conditions[5]. A vector due to measured values of vibration response of the rotor was used to establish the system matrix by Ratan et al., results indicated that this method can even detect and locate cracks smaller than 4% of the diameter of the shaft[6]. Hwang detected the location and the size of crack damage by minimizing the differences between trials and analysis of FRFs[7]. Yang et al. studied the effects of open crack on the vibration characteristics of elastic beams[8]. Steady-state response study displayed that the combined frequency and longitudinal response of the rotational and torsional excitations in the lateral response are reversed. The frequency of excitation can be used to detect the oblique cracks in shaft of the rotor system[9]. Bovsunovsky et al. studied a simple procedure to evalute the effectiveness of vibration diagnostics of damage. The procedure is based on the determination of the change of shaft's compliance caused by a crack via the use of linear fracture mechanics[10].

Although many works about crack detection method for shaft have been done by the researchers in the past few decades. However, most of the works only consider single crack in the model, the influence of the neutral axis on the stress intensity factor are seldom considered. In this article, the model of SIFs is derived based on the elastic beam theory, and the local compliance model is also established. An identification method of multiple cracks in shaft is discussed in the end.

2. Calculation of SIFs

The cracked shaft model is shown in Figure 1. The radius of the shaft is assumed to be 'R' and the crack depth are 'a1'and 'a2'. The shaft is subjected to the shear forces 'S1' and 'S2' bending moments 'M1' and 'M2', and axial tensile forces 'N1'and 'N2'.

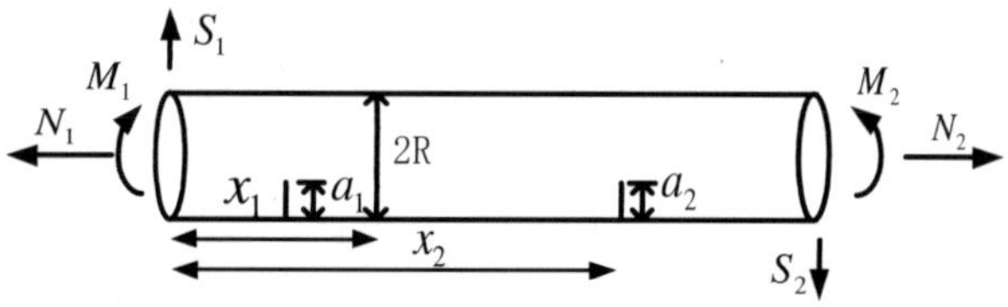

Figure 1. Cracked shaft model

When the crack grows, strain energy concentration occurs in the vicinity of the crack tip, the energy release due to crack propagation will cause structural compliance changes. It is assumed that the structural crack region is in the elastic phase. Based on the linear elastic fracture mechanics theory, the strain energy release rate is defined as the strain energy dissipated during increase the crack surface area.

$$G = \frac{\partial U}{\partial A} \tag{1}$$

The strain energy release rate due to crack growth can also be expressed as

$$G = \frac{1}{E'}(K_{\mathrm{I}}^2 + K_{\mathrm{II}}^2 + (1+\nu)\ K_{\mathrm{III}}^2) \tag{2}$$

where E' is elastic modulus, E' = E/(1-v2) for plane strain, E'=E for plane stress. According to the beam theory, a shaft subjected to a bending moment M, the strain energy due to Δb is

$$\Delta V_e = \frac{M^2}{2}\frac{1}{EI}\Delta b \tag{3}$$

where ΔU denotes the strain energy required for crack expansion. When Δb→0

$$\frac{\partial U}{\partial b} = -\frac{M^2}{2E}\left(\frac{1}{I^c} - \frac{1}{I}\right) \tag{4}$$

Assuming that the bending stiffness of the fracture section is 'EIc'and the bending stiffness of the uncracked section is 'EI'. Therefore the strain energy due to crack growth is expressed as

$$\Delta U = \frac{M^2}{2}\left(\frac{1}{EI} - \frac{1}{EI^c}\right)\Delta b \tag{5}$$

The strain energy take the following form when it is subjected to a axial tension or shear

$$\frac{\partial U}{\partial b} = -\lambda\frac{S^2}{2u}\left(\frac{1}{A^c} - \frac{1}{A}\right) \tag{6}$$

where A is the uncracked cross-section area of the shaft, 'Ac'is the area of the cracked cross section, and u is the shear modulus. According to the method proposed by Kienzler and Herrmann[11]. The energy release rate $\partial U/\partial a$ due to crack growth is related to that one due to crack widening $\partial U/\partial b$ can be expressed as

$$\frac{\partial U}{\partial a} = 2\beta\frac{\partial U}{\partial b} \tag{7}$$

Therefore

$$\frac{\partial U}{\partial a} = -4R\frac{K^2}{E'}\sqrt{\eta - \eta^2} \tag{8}$$

where $\eta = a/2R$ is crack ratio, Subsituting (4),(5),(6) into (9), we obtain

$$K_{\mathrm{I}_N} = \frac{N}{2}\sqrt{\frac{\beta'}{R\sqrt{\eta-\eta^2}}\left(\frac{1}{A^c} - \frac{1}{A}\right)} \tag{9}$$

$$K_{\mathrm{I}_M} = \frac{M}{2}\sqrt{\frac{\beta'}{R\sqrt{\eta-\eta^2}}\left(\frac{1}{I^c} - \frac{1}{I}\right)} \tag{10}$$

$$K_{\mathrm{II}_S} = \frac{S}{2}\sqrt{\frac{2\lambda\beta'(1+v)}{R\sqrt{\eta-\eta^2}}\left(\frac{1}{A^c} - \frac{1}{A}\right)} \tag{11}$$

where $\beta' = \beta E'/E$ and β is the slope of the stress diffusion lines. We first consider the shaft subjected to a tension. The position of the neutral axis changes because of the presence of cracks. The offset distance is

$$e = 16R^3(\eta-\eta^2)^{\frac{3}{2}}/3A^c \tag{12}$$

Then, a combined tension and bending moment occurs

$$K_{\mathrm{I}_N} = \frac{N}{2}\sqrt{\frac{\beta'}{R\sqrt{\eta-\eta^2}}\left(\frac{1}{A_c} - \frac{1}{A}\right)} + \frac{eN}{2}\sqrt{\frac{\beta'}{R\sqrt{\eta-\eta^2}}\left(\frac{1}{I^c} - \frac{1}{I}\right)} \tag{13}$$

therefore

$$K_{\mathrm{I}_N} = \frac{N}{R^{3/2}} F_N(\eta) \tag{14}$$

where

$$F_N(\eta) = \frac{1}{2}\sqrt{\frac{1}{\Delta}(\frac{1}{\Pi_1}-\frac{1}{\pi})} + \frac{8\Delta^3}{3\Pi_1}\sqrt{\frac{1}{\Delta}(\frac{1}{4}\Pi_2+(1-2\eta)^3\Delta-\frac{256\Delta^6}{9\Pi_1})^{-1}-\frac{4}{\pi}} \tag{15}$$

with

$$\Pi_2 = \pi - \cos^{-1}(1-2\eta) - 2(1-2\eta)\sqrt{\eta-\eta^2}$$

$$\Pi_2 = \pi - \cos^{-1}(1-2\eta) - 2(1-2\eta)\sqrt{\eta-\eta^2} \tag{16}$$

$$\Delta = \sqrt{\eta-\eta^2}$$

Similarly, when the shaft is subjected to a bending moments, we obtain

$$K_{\mathrm{I}_M} = \frac{M}{2}\sqrt{\frac{\beta'}{R\sqrt{\eta-\eta^2}}(\frac{1}{I^c}-\frac{1}{I})} = \frac{M}{R^{5/2}} F_M(\eta) \tag{17}$$

where

$$F_M(\eta) = \frac{1}{2}\sqrt{\frac{1}{\Delta}((\frac{1}{4}\Pi_2+(1-2\eta)^3\Delta-\frac{256\Delta^6}{9\Pi_1})^{-1}-\frac{4}{\pi})} \tag{18}$$

When the shaft is subjected to a shear, its stress intensity factor is

$$K_{\mathrm{II}_S} = \frac{S}{2}\sqrt{\frac{2\lambda\beta'(1+v)}{R\sqrt{\eta-\eta^2}}(\frac{1}{A^c}-\frac{1}{A})} = \frac{S}{R^{3/2}} F_S(\eta) \tag{19}$$

where

$$F_S(\eta) = 0.8832\sqrt{\frac{1}{\Delta}(\frac{1}{\Pi_1}-\frac{1}{\pi})} \tag{20}$$

3. Local compliance of the cracked shaft

According to the Castigliano's theorem, if strain energy is expressed as a function related to the force, and then the strain energy can be divided into the displacement ui in the direction of the force with respect to the forces as follows

$$u_i = \frac{\partial U}{\partial Q_i} \tag{21}$$

where Qi is the force direction of the displacement ui. The compliance coefficients of the shaft subjected to the axial tensile, bending moment and the shear force is given as

$$c_{ij} = \frac{\partial u_i}{\partial Q_j} = \frac{\partial^2}{\partial Q_i Q_j}\int_A \frac{1}{E'}[(K_{\mathrm{I}_N}+K_{\mathrm{I}M})^2 + K_{\mathrm{II}_S}^2]dA \tag{22}$$

Substituting equations (9), (10) and (11) into equation (23), the conpliance coefficents can be computed as

$$c_{11} = \frac{2(1-v^2)}{ER^3}\int_0^{a}\int_{-b}^{+b} F_N^2(\mu/D)dxd\mu \tag{23}$$

$$c_{22} = \frac{2(1-v^2)}{ER^3}\int_0^{a}\int_{-b}^{b} F_S^2(\mu/D)dxd\mu \tag{24}$$

$$c_{33} = \frac{2(1-v^2)}{ER^3}\int_0^{a}\int_{-b}^{b} F_M^2(\mu/D)dxd\mu \tag{25}$$

$$c_{13} = \frac{2(1-v^2)}{ER^4}\int_0^{a}\int_{-b}^{b} F_{MN}(\mu/D)dxd\mu \tag{26}$$

where

$$b = \sqrt{R^2 - y^2} = \sqrt{2Rz - \mu^2} \tag{27}$$

Therefore, the compliance matrix of nodal point is determined as

$$\mathrm{c} = \begin{bmatrix} c_{11} & 0 & c_{13} \\ 0 & c_{22} & 0 \\ c_{13} & 0 & c_{33} \end{bmatrix} \tag{28}$$

therefore

$$K_f = c^{-1} \tag{29}$$

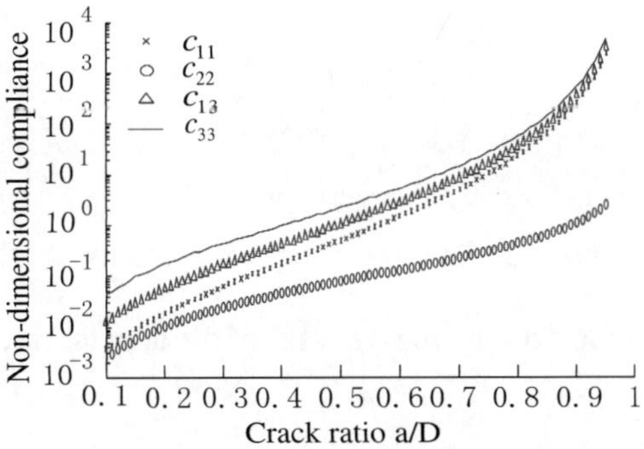

Figure 2. Compliance coefficients

Figure 2 displays the effects of the cracks on compliance coefficients. Results show that the compliance coefficient increases with the increasing of the crack depth. And the local flexibility coefficient is the largest when the shaft is subjected to the bending moment, and that is minimum one when subjected to shear to the same crack form of the shaft. When an uncracked shaft element subjected to a bending moment M, a shear force S and a axial tensile N at the same time, the strain energy is

$$E = \frac{1}{2EI}(M^2 l - MSl^2 + \frac{S^2 l^3}{3}) + \frac{N^2 l}{2EA} + \frac{S^2 l}{\mu\Lambda} \tag{30}$$

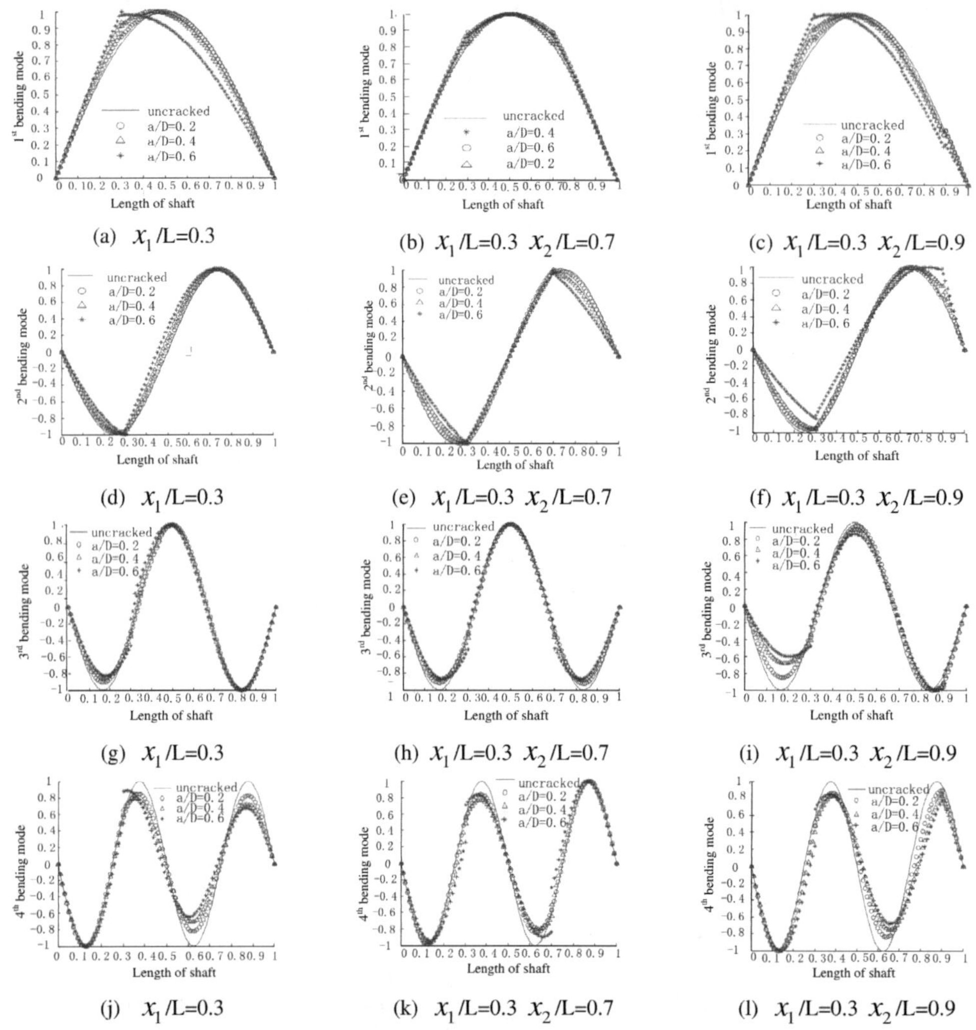

(a) x_1/L=0.3 (b) x_1/L=0.3 x_2/L=0.7 (c) x_1/L=0.3 x_2/L=0.9

(d) x_1/L=0.3 (e) x_1/L=0.3 x_2/L=0.7 (f) x_1/L=0.3 x_2/L=0.9

(g) x_1/L=0.3 (h) x_1/L=0.3 x_2/L=0.7 (i) x_1/L=0.3 x_2/L=0.9

(j) x_1/L=0.3 (k) x_1/L=0.3 x_2/L=0.7 (l) x_1/L=0.3 x_2/L=0.9

Figure 3. First four bending modes for shafts

According to the equation (30), the stiffness matrix of the uncracked element can be expressed as

$$K=\begin{bmatrix} \frac{AE}{l} & 0 & 0 & -\frac{AE}{l} & 0 & 0 \\ 0 & \frac{12EI}{(1+\Gamma)l^3} & \frac{-6EI}{(1+\Gamma)l^2} & 0 & -\frac{12EI}{(1+\Gamma)l^3} & \frac{6EI}{(1+\Gamma)l^2} \\ 0 & -\frac{6EI}{(1+\Gamma)l^2} & \frac{(4+\Gamma)EI}{(1+\Gamma)l} & 0 & \frac{6EI}{(1+\Gamma)l^2} & \frac{(2-\Gamma)EI}{(1+\Gamma)l} \\ -\frac{AE}{l} & 0 & 0 & -\frac{AE}{l} & 0 & 0 \\ 0 & -\frac{12EI}{(1+\Gamma)l^3} & \frac{6EI}{(1+\Gamma)l^2} & 0 & \frac{12EI}{(1+\Gamma)l^3} & \frac{6EI}{(1+\Gamma)l^2} \\ 0 & \frac{-6EI}{(1+\Gamma)l^2} & \frac{(2-\Gamma)EI}{(1+\Gamma)l} & 0 & \frac{6EI}{(1+\Gamma)l^2} & \frac{(4+\Gamma)EI}{(1+\Gamma)l} \end{bmatrix} \tag{31}$$

Assuming that the cracked section is modeled as two planes joined together by a line-spring with zero length. Therefore, the dynamic equation of the cracked shaft can be expressed as

$$M\ddot{x}+(K+K_f)x=0 \tag{32}$$

IOP Conf. Series: Materials Science and Engineering **576** (2019) 012039 doi:10.1088/1757-899X/576/1/012039

where ***M*** is the mass matrix of the system, ***K*** is the stiffness matrix.The problem of determining the natural vibration frequencies and the associated mode shapes of a system is transformed into solving the eigenproblem of the homogeneous linear system, such as

$$(K-\omega^2 M)\ A=0 \tag{33}$$

4. Results and discussion

The geometry and material properties of the cracked shaft is shown in Table 1.

Table 1. Basic Parameters of shaft

Basic Parameters	value
Young's modulus E	210Gpa
Poisson's ratio v	0.33
Mass density ρ	7900kg/m^3
Diameter d	0.1m
Length L	1m

Assuming that the boundary conditions of the shaft are simply supported on both end. Crack position is set as x1/L=0.3. Shafts having two cracks, the crack positions are set as x_1/L=0.3, x_2/L=0.7 or x_2/L=0.9. And the crack depth is set as η=0.2, 0.4 and 0.6.

Figure 3 shows the first four bending modes of the shaft with single crack and double cracks respectively. When a crack is in the shaft, the vibration mode changes, a discontinuity occurs at the crack position. Comparing these curves in Figure 3, it can be found that the mode shape appears discontinuous at the 0.3. With the increasing of modal frequencies, discontinuous interval increases too. When the crack position is at 0.3 and 0.7, the vibration mode appears discontinuous at these two positions. When the position of the first crack remains unchanged, and the location of the second crack is changed to the position at 0.9. Discontinuous appear at the position of 0.3 and 0.9, it shows that this character not only can be used to identify the location of cracks, but also can be used to identify the number of cracks. It is noted that with the increaseing of the crack depth, discontinuity also become larger. This is helpful to find the crack and crack depth. However, it is also found that this phenomenon is not obvious when the cracks are located close to the nodal position.

5. Conclusion

With the help of theoretical calculations, an identification method of cracks in a shaft is discussed. The main conclusions are as follows

- Based on the fracture mechanics and basic beam theory, the stress intensity factors of a shaft subjected to axial tension, bending moment and shear force are derived. The local compliance model of the transverse crack shaft is calculated by using the Castigliano's theorem and the expressions of local compliance of the cracked shaft are given.
- The vibration modal shapes of the cracked shaft are calculated by the differential equation of motion. The variation of the vibration mode of a shaft having different crack locations and crack numbers are discussed. The model established in this paper is very convenient for the damage identification of the shaft.

6. References

[1] Zhong J F, Zhong S F, Zhang Q K 2019 *Mech. Syst. Signal. Pr.* **115** 1
[2] Zhang Y F, Wang S S, Gan S L 2016 *J. Vib. Shock.*, **35** 11

[3] Gounaris G, Dimarogonas A 1988 *Comput. Struct.* **28** 3
[4] Zhang Y J, Yan Y H, Li Y Q 2012 *J. Northeast. Univ.* **33** 3
[5] Cruz-Vega I, Rangel-Magdaleno J, Ramirez-Cortes J and Peregrina-Barrtto H 2017 *J. Mech. Sci. Technol.* **31** 6 :
[6] Ratan S, Baruh H and Rodriguez J 1996 *J. Sound. Vib.* **194** 1
[7] Hwang Y H and Kim 2004 *J. Sound, Vib.* **270** 1
[8] Yang X F, Swamidas A S J, Seshadri R 2001 *J. Sound. Vib.* **244** 2
[9] Lin Y L and Chu F 2010 *Mech. Syst. Signal. Pr.* **24** 2
[10] Bovsunovsky A P 2018 *Electr. Pow. Syst. Res.* **154** 1
[11] Kienzler R and Herrmann G 1986 *Acta. Mecb.* **62**

Acknowledgment
This research work is supported by the National Science Foundation of China (51565039) and the Science and Technology Project of Jiangxi Provincial Education Department (GJJ170584).

IOP Conf. Series: Materials Science and Engineering **576** (2019) 012040 doi:10.1088/1757-899X/576/1/012040

Research on Key Patent Licensing Strategy of 3D Printing Technology

Qiang Wang[1,2], Haiying Liu[1,2], Ping Wu[1,2] , Yu Cao[1,2] and Bo Xu[1,2]

[1] Ordos Institute of Technology, 017000 Erdos, China
[2] Erdos College of Inner Mongolia University, Erdos 017000, China

Corresponding Author's E-mail: lhy_liuhaiying@163.com

Abstract. Patent licensing is one of the most important means to gain profits from the commercial application of patents. Based on the analysis of the market sales price, volume and profit of key patented 3D printing technology products, the profit of patentee under the mode of exclusive licensing, sole licensing and general licensing is studied in order to provide reference for formulating the best licensing strategy for key patents of 3D printing technology and speed up the application of patents .

1. Introduction

On May 29, 2016, the Central Committee of the Communist Party of China and the State Council issued the Outline of National Innovation-Driven Development Strategy, proposing to improve the ability of intellectual property creation, utilization, protection and management. On January 18, 2018, according to the data released by the State Intellectual Property Office, in 2017, the volume of patent applications for inventions in China was 1382,000, of which 63.3% were from enterprises. Patent is one of the important forms of scientific research and innovation. The main purpose of patent application by enterprises is to realize their economic value. It is an important way to realize the economic value of patent application by promoting the transformation of patent application using effective methods.

Patent licensing is one of the important means to obtain profits from the commercial application of patents. Scholars have made many contributions to patent licensing and have made some achievements. Ozel et al. [1] studied the impact of different licensing modes on social welfare with a model of a large scientific and technological institution and four manufacturers, and then made specific suggestions on patent licensing. Thomas [2] studies compulsory licensing strategies for pharmaceutical patents in developing countries and proposes alternative strategies. Simon [3] conducted a questionnaire survey on companies in the pharmaceutical and electronic industries in Japan, studied the patterns of patent licensing and external licensing, and found that reciprocal technology sharing agreements are evolving into more formal relationships, and enterprises are promoting the improvement of their technology brokerage capabilities. Rumyana [4] studied the nature and significance of invention patents for enterprise strategy, and proposed that enterprises classify patents and licenses according to various standards. Pia et al. [5] studied the relationship between innovation model and patent strategy and licensing practices of eight ICT companies in 2004, and found that companies with different attitudes towards innovation are also pursuing funds and benefites from innovation. The common point is that intellectual property rights are useful only when they are in line with business strategies. Shen Chengran et al. [6] studied the two strategies adopted by the original manufacturer under patent protection in the patent perfection market when facing the competition of re-manufacturer, i.e. unauthorized re-manufacturing and licensed re-manufacturing. Zhang [7] Through the investigation of 350 patent licensing transactions in Chinese universities, found that the duration of patent examination is positively correlated with the delay of patent licensing. Liu et al. [8] From the perspective of 3D printing patent information between China and the United States, this paper makes a comparative study of 3D printing patents between China and the United States, and puts forward some feasible suggestions for the development of China's 3D printing industry. Ke et al. [9] discussed the price contract of technology license in view of the bargaining power of technology licensee. Tian et al. [10] constructed a two-cycle model to study and compare the decision-making of supply chain members under two different patent licensing strategies of fixed fee and unit fee. Based on the analysis of the characteristics and influencing factors of patent centralized licensing of technology standard alliance, Based on the complete monopoly model and Cournot oligopoly competition model, the product market model is constructed to analyze the patentee's income, and then the optimal licensing strategy is studied.

2. Market game model

2.1. Market demand function model

Assuming that there are n independent product manufacturers in the key product market, the counter-demand function of the product market is as follows:

$$P = a - b\sum_{i=1}^{n} q_i \tag{1}$$

In the formula P is the selling price; q_i is the sales volume of product manufacturer i, $i = 1,2,\ldots,n$; a, b is the coefficient of the counter-demand function of the product market, a>0, b>0; c is the marginal production cost of the product manufacturer, c>0.

2.2. Complete monopoly model

In the market model of complete monopoly of key products, only one product manufacturer has obtained patent license for key products. For consumers, it belongs to the typical complete monopoly market. Therefore, this paper uses the model of complete monopoly market to simulate the product market, in which only one product manufacturer is licensed. The profit function of a single product manufacturer in the product market is as follows:

$$\pi^{dz} = \left[a - bq^{dz} - c\right] q^{dz} \tag{2}$$

According to the first-order condition of profit maximization, it can be concluded that the selling price of a single product manufacturer is:

$$p^{dz} = \frac{a+c}{2} \tag{3}$$

The sales volume of a single product manufacturer is:

$$q^{dz} = \frac{a-c}{2b} \tag{4}$$

The profits of a single product manufacturer is:

$$\pi^{dz} = \frac{(a-c)^2}{4b} \tag{5}$$

2.3. Cournot oligopoly competition model

In the Cournot oligopoly competition model of key products, there are two or more product manufacturers who have obtained product patent licenses, there are many product manufacturers, which will break monopoly, and there is competition in the market. At this time, Cournot oligopoly competition model can be used to simulate the product market. Assuming that n product manufacturers operate independently of each other, each product manufacturer determines its own sales volume in order to maximize its own profits, depending on the sales volume of each other. At this time, the profit function of the first product manufacturer is:

$$\pi_i^c = q_i(a - b\sum_{i=1}^{n} q_i^c) - q_i^c c \tag{6}$$

In formula π_i is the profit of enterprise i. At this time, according to the first-order condition of profit maximization, it can be concluded that the market equilibrium price in the product market is:

$$p^c = \frac{a+nc}{n+1} \tag{7}$$

The sales volume of product manufacturer i is:

$$q_i^c = \frac{a-c}{(n+1)b} \tag{8}$$

The profit of product manufacturer i is:

$$\pi_i^c = \frac{(a-c)^2}{(n+1)^2 b} \tag{9}$$

Then, when $n \geq 2$, Sales price of product manufacturer:

$$p^c = \frac{a+nc}{n+1} < p^{dz} = \frac{a+c}{2} \tag{10}$$

Total sales volume:

$$q^c = \sum_{i=1}^{n} q_i^c = \frac{n}{n+1}\frac{a-c}{b} > q^{dz} = \frac{1}{2}\frac{a-c}{b} \tag{11}$$

Total profit:

$$\pi^c = \sum_{i=1}^{n} \pi_i^c = \frac{n}{(n+1)^2}\frac{(a-c)^2}{b} < \pi^{dz} = \frac{1}{4}\frac{(a-c)^2}{b} \tag{12}$$

Thus, under the model of complete monopoly, the sales price of products in the product market is increased, the total sales volume is reduced, and the total profit of the product industry is increased, which will increase the profits of product patent owners.

3. Patentee profit model

3.1. Profit model of patentee with production capacity

3.1.1. Profit model of exclusive licensing. Under the exclusive licensing mode, when a patent is granted to a single product manufacturer, who will be unique user of the patent. No one else (including the patentee himself) can use the patent. At this time, the exclusive licensing profit model can be expressed as:

$$\omega_y^{el} = \delta_y^{el}\pi^{dz} = \delta_y^{el}\frac{(a-c)^2}{4b} \tag{13}$$

Where δ_y^{el} is profit sharing ratio of patentee from licensed single product manufacturer under the exclusive licensing model, $0 \le \delta_y^{el} \le 100\%$.

At this time, the key product can obtain monopoly profits. Since the patentee has production capacity, the best strategy is to set the profit ratio as $\delta_y^{el} = 100\%$, or if it produces by itself, it will only license the internal production department of the company, rather than the external producers.

3.1.2. Profit model of sole licensing. Under the sole licensing mode, only the patentee and the licensee can use the patent. The patentee's profits are divided into two parts, the profits from manufacturing products for himself, and the income from licensing a single product manufacturer. At this time, there are two product manufacturers, and the Sole license profit can be expressed as:

$$\begin{aligned}\omega_y^{ep} &= \frac{1}{2}\pi^c + \delta_y^{ep}\frac{1}{2}\pi^c \\ &= \frac{1}{2}\left(1+\delta_y^{ep}\right)\frac{n}{(n+1)^2}\frac{(a-c)^2}{b} \quad (n=2) \\ &= \frac{1+\delta_y^{ep}}{9}\frac{(a-c)^2}{b}\end{aligned} \tag{14}$$

Where δ_y^{ep} is the proportion of the profits obtained by the patentee from the licensed product manufacturer under the sole license mode, $0 \le \delta_y^{ep} \le 100\%$.

3.1.3. Profit model of general licensing. Under the general licensing mode, the patentee can license the patent to multiple product manufacturers at the same time. Both the patentee and the licensed product manufacturers can use the patent. The patentee's profits are divided into two parts, one is the profits from making products for himself, the other is the income from licensing multiple product manufacturers. At this time, the general licensing profit model can be expressed as:

$$\begin{aligned}\omega_y^{gp} &= \frac{1}{n}\pi^c + \delta_y^{gp}\frac{n-1}{n}\pi^c \\ &= \left(\frac{1}{n} + \delta_y^{gp}\frac{n-1}{n}\right)\frac{n}{(n+1)^2}\frac{(a-c)^2}{b} \\ &= \frac{1+\delta_y^{gp}(n-1)}{(n+1)^2}\frac{(a-c)^2}{b}\end{aligned} \tag{15}$$

Where δ_y^{gp} is the proportion of the profits obtained by the patentee from the licensed product manufacturer under the general licensing mode. $0 \le \delta_y^{gp} \le 100\%$.

3.2. Profit model of patentee without production capacity

3.2.1. Profit model of exclusive licensing. Under the exclusive licensing mode, when a patent is granted to a single product manufacturer, only the manufacturer of the product can use the patent. No one else (including the patentee himself) can use the patent. At this time, the exclusive licensing profit model can be expressed as:

$$\omega^{el} = \delta^{el}\pi^{dz} = \frac{\delta^{el}}{4}\frac{(a-c)^2}{b} \tag{16}$$

Where δ^{el} is the proportion of the profits obtained by the patentee from the licensed manufacturer of a single product under the exclusive licensing mode, $0 \le \delta^{el} \le 100\%$.

3.2.2. Profit model of sole licensing. Under the sole licensing model, only patentee and licensees can use the patent. Since the patentee does not have production capacity itself, only one product manufacturer in the market produces the barrier-type product, and the sole license profit can be expressed as:

$$\omega^{ep} = \delta^{ep}\pi^{dz} = \frac{\delta^{ep}}{4}\frac{(a-c)^2}{b} \tag{17}$$

Where δ_y^{ep} is the proportion of the profits obtained by the patentee from the licensed product manufacturer under the sole license mode, $0 \le \delta_y^{ep} \le 100\%$.

3.2.3. Profit model of general licensing. Under the general licensing model, the patentee can license the patent to multiple product manufacturers at the same time, and the patentee and the manufacturer of the licensed product can use the patent. The profit of the patentee is the revenue share obtained from the license of multiple product manufacturers. At this point, the general license profit model can be expressed as:

$$\omega^{gp} = \delta^{gp}\pi^{c} = \delta^{gp}\frac{n}{(n+1)^2}\frac{(a-c)^2}{b} \tag{18}$$

Where δ_y^{gp} is the proportion of the profits obtained by the patentee from the licensed product manufacturer under the general licensing mode. $0 \le \delta_y^{gp} \le 100\%$.

4. Numerical simulation

4.1. Problem description

Taking the 3D printer market as an example, it is assumed that there is one 3D printing patent R&D enterprise and eight 3D printer manufacturers in a certain regional market. The sales volume of each 3D printer manufacturer is q_i; the margin of the 3D printer manufacturer is the production cost c=3000 yuan/set, and the inverse demand function of the 3D printer market is $P = 15000 - 2\sum_{i=1}^{n} q_i (yuan/set)$.

4.2. Profit Model of Patentee with Production Capacity

When the patentee has its own production capacity, the internal 3D printer manufacturer is licensed under the exclusive licensing mode with a profit sharing ratio of 100%; in the sole licensing mode, there are internal 3D printer manufacturers and licensed 3D printer manufacturers. A total of 2, the external profit share ratio is set to 15%; in the ordinary licensing mode, there are 9 internal 3D printer manufacturers and 8 licensed 3D printer manufacturers, the external profit share ratio is set to 15%.
According to the above model, the relevant market data of the capacity-type patentee in the three modes is shown in Table 1.

Table 1. Market data of patentee with capacity

mode	Price	Volume	Sales profit	Patentee profit
Exclusive	9000	3000	1800	1800
Sole	7000	4000	1600	920
General	4200	5400	648	158.4

4.3. Profit Model of Patentee without Production Capacity

When the patentee does not have its own capacity, the single external 3D printer manufacturer is licensed under the exclusive licensing model with a profit share ratio of 15%; in the sole license mode, there is only one licensed 3D printer manufacturer. The profit share ratio is set to 15%; under the normal license mode, there are 8 licensed 3D printer manufacturers with a profit share ratio of 15%. According to the above model, the relevant market data of the three types of non-capacity patent holders are shown in Table 2.

Table 2. Market data of patentee without capacity

mode	Price	Volume	Sales profit	Patentee profit
Exclusive	9000	3000	1800	270
Sole	9000	3000	1800	270
General	4334	5333	711.4	182.3

It can be draw from Table 2 that under the same profit sharing ratio, the exclusive license and the sole license are same, so the exclusive license is the optimal strategy under this condition, because the patent holder still enjoys the right. In addition, in the general license, when the volume of licensed manufacturers increased to 9, the patentee's income decreased by 239,000 yuan. Obviously, with the increase of the license manufacturer, the patentee's income is decreasing; when the profit share ratio increased by 30%, and the patentee's income increased by 311,000 yuan. Obviously, as the profit share ratio increases, the patentee's income increases.

5. Conclusion

In the context of technological innovation, the formulation of patent licensing strategies has an important impact on patent rights, consumers and manufacturers.

(1)For a patent holder with production capacity, because the patent of a key product is monopolistic, the maximum monopoly profit can be obtained by a single license to the manufacturer of the internal product;

(2)For those who do not have a capacity-type patentee, they can obtain a maximum monopoly profit share by granting a single license to the external product manufacturer, and retain the rights to the patent;

(3)In the Cournot oligopoly market of product manufacturers, for the patentee, as the license manufacturer increases, the patentee's income is decreasing; as the profit share ratio increases, the patentee's income is increasing.

Further research can be conducted in the following directions:

(1)Building models that take more factors into account, such as complementary technologies alternative technologies, that affect patent value;

(2)Conducting empirical studies, to verify the effectiveness of the model through case studies.

6. References

[1] Ozel S O, Penin J. Exclusive or open? An economic analysis of university intellectual property patenting and licensing strategies. *Journal of Innovation Economics,* (3)(2016)

[2] Hemphill T A. Pharmaceutical patent expropriation and technology strategy: strategic options to compulsory licensing . *Technology Analysis and Strategic Management,*,32(1)(2010)

[3] Collinson S, Kato H, Yoshihara H. Technology strategy revealed: Patterns and influences of patent-licensing behaviour in Japanese firms. *International Journal of Technology Management,* 30(3)(2005)

[4] Georgieva R. Patent and Licensing Strategies in Management of Industrial Companies . *Rumiana Georgieva,* 6(2011)

[5] Laukkanen P H, Soininen A. Different paths of appropriation-patent strategies and licensing practices for closed and open innovation. *International Journal of Intellectual Property Management,*4(3)(2011)

[6] Sheng C R,Liu M W, Xiong Z K. Competitive Strategy for the Manufacturer and Re-manufacturer Under Patent Licensing . *Operations Research and Management Science,* 23(2) (2016)

[7] Zhang R J. Research on Patent License Delay of Chinese Universities. *Science and Technology Management Research,*36(19) (2016)

[8] Lix X, Yu X,Zhang B. China-US 3D Printing Patent Comparison Study and Implications for Chinese Industry Development . *Journal of Intelligence,* 34(5) (2015)

[9] Ke Z Y,Wen Q T. The Bargaining Model of Technology Licensing. *Mathematics in Practice and Theory,* 49(9) (2017)

[10] Tian H, Meng Z J. Re-manufacturing Licensing Decision Research Based on Differential Pricing Strategies and Recycling Competition. *Journal of Industrial Technological & Economics,* 36(10) 8(2017)

Acknowledgments

This work was supported by National Natural Science Foundation of China (Grant nos. 71563031) ; Science and Technology Plan of Ordos City (Grant nos. YY201610045); Project of Ordos Institute of Technology (Grant nos.KYZD2017003) and Higher Education Science and Technology Research Project of Inner Mongolia (Grant nos.NJSY19262).

IOP Conf. Series: Materials Science and Engineering **576** (2019) 012041 doi:10.1088/1757-899X/576/1/012041

A novel way of avoiding stylus drop down from model while scanning

Wu Lei

School of Information Science and Technology. North China Univ. of Tech., 100144. Beijing. China

Corresponding author: stone.wu@ncut.edu.cn

Abstract. Based on analysing the basic feature of scanning process, this paper proposes a new architecture to regulate autonomous scanning speed with the nominal vector D_N. In addition, explains and demonstrates how to introduce the nominal vector D_N into the tracer scanning for successive and self-adjusting scanning speed and autonomy contact level between stylus of trace and model according to the curved surface. The technical activities and experimental results shown that the parameter of nominal vector D_N play a proper role in the case of retaining tracer contact on curved surface especially when stylus scanning from horizontal plane to the vertical plane.

1. Introduction

Stylus is employed in the shape of unknown curved surface and hence the stylus may be experiencing the surface for the first time. The stylus may not have been in a similar situation before. In addition, in actual applications, the stylus normally suffers from drop-down and perturbations inflicted upon its control system, making it extremely important to propose a new architecture which would be able to cope with such problems. The nominal deflection of stylus is known as a proper solution due to its remarkable capabilities. A novel approach based on nominal deflection of stylus has been presented in this study for the scanning of such stylus in their concerted motion. Simulation results demonstrate the efficacy of the proposed method.

Typically, the tracer provides center-of-tool information that directly translates into tool-path data. However, surface data of the model can be acquired using the appropriate stylus and stylus-deflection values. In addition to the *XYZ* data for each point, the system calculates the appropriate surface normal. Each point contains *XYZ* coordinates and *IJK* vector data at that point. Calculating the position of center of stylus, the numerical control (hereinafter referred to as *NC*) can generate normal vector and record space position values of the models. Both vector and position values can be used to generate *NC* codes for machine tool to directly machining parts.

As we known, at one hand, scanning speeds are absolutely limited for scanning with high speed at some smooth geometric models; at the other hand, sometimes scanning speeds cannot be autonomous regulated when scanning from the horizontal plane to the vertical plane for avoiding stylus drop-down from the model. Therefore, the functions of the tracer must be studied both autonomously and dynamically to adjustment the scanning speed for optimize the scanning process and avoiding stylus drop-down from the model.

2. Literature review

Measurement methods of curved surface can be divided into contact and non-contact methods. In actual contact measurement, analog stylus and trigger probe are often used. Non-contact measurement mainly adopts laser triangulation method. In production practice, contact analog measurement is widely used in engineering due to its advantages of high measurement accuracy, mature technology and low cost.

Laser is typical one of the non-contact measurement equipment. Laser scanning technology also has some disadvantages, for example, in order to prevent stray reflection, the surface of the workpiece needs to be sprayed a layer of white powder before measurement. If the inclination of the measured surface exceeds 45°C, the radiation blind area (known as the shadow effect) will appear. Thus, in terms of blind area, the laser scanning is incapable of accurately measuring the workpiece surface. Although the shadow effect can be reduced or eliminated through multi-axis control, the cost will be increased but the accuracy is still low. The scanning process effected by a various of factors and no more reported about this method to be widely applied in real production.

There are two distinct types of contact measurement: tracer scanning and touch-probe digitizing. Tracer scanning uses a tracer stylus corresponding in size and shape to the anticipated tool that will be used to mill the workpiece. The electronics sensor of stylus senses the deflection conducted the stylus continuously scans the surface of the model and collecting data of the surface. Typically, the data is center-of-stylus information that is accepted by a CNC milling machine as an NC program or cutter path.

Touch-probe digitizing uses a Renishaw probe typical of CMMs and provides actual surface data. The surface points are used to automatically construct a math model of the surface, commonly referred to as wireframe. Wireframe data usually is used by designers and tooling manufacturers in CAD/CAM environment. Probe belongs to digital measure equipment, which can only know his probe at reset and not at reset, but cannot generate normal vector on probe. Copy or digitizing measure equipment belong to analog measure equipment. During scanning, the stylus of copy or digitizing constant touch the model and record the model position and the position of the center-of-stylus. The normal vector can be calculated by the data of the center-of-stylus. The efficiency of copy or digitizing is higher than probe, but the accuracy is lower than probe.

As far as contact measurement equipment of analog is concerned, there are a various of contact measurement equipment of analog and are widely applied in the copy milling and digitizing machine tools. The typical analog measurement equipment from abroad can be listed as follow. The Gettys with the swing movable structure is from the United States. Both the Accura-trace 6 of Makino and the Tv-330b of Fanuc are from Japan, the P5/K5/K2 of Fidia from Italy, three types of them are all with the parallel movable structure. Both the setups of copy milling control system and tracer from Fidia dominant the fields for a long time. In China, in the early 1980s, a team led by Mr. Lu jiechi from Dalian university of technology first carried out the research and development of the copy instrument and offered a prototype DIGIT-02C of NC copy system and corresponding tracer to be mounted on milling machine tool of type X4450 of their own, and milling machine tool of type XKF718 from Changzheng machine tool company. Institute of Optics & Electronics of Chengdu from Chinese Academy of Sciences in the 1990s also studied and provided a prototype FCG-1 of tracer to be mounted on milling machine tool of type XKF718 from Changzheng machine tool company. But from now, no literatures and reports show that the above mentioned two tracers to be widely mounted on copy milling machine tools[1,2].

3. The principle of scanning

3.1. Normal vector of stylus

Let us consider the situation when stylus with radius R scanning on a semicircle surface of an object. Figure 1 shows the initial point P_1 of the stylus just contact the model. In the process of tracking and scanning, the stylus contacts and presses against the surface of the model. Under the action of normal reaction force, a space pressure deviation is generated at the center C of the stylus. Assuming that there is no friction between the stylus and the model surface, the line $\overline{CO}$ is just the normal direction D_N of the model at the contact point P_1 between the stylus and the model surface. The V_N is the normal speed at the contact point of P_1, to be applied to keep the stylus in contact with the model. The V_T is the tangential speed to be employed to keep relative motion of the stylus and scanning the model. The V is the resultant speed of the V_N and the V_T, as well as the scanning speed of the stylus, and also the feed speed of copy milling machining. The speed vector V of the stylus motion is derived from the displacement detected by the tracing stylus.

The DR represents the amount of actual deflection along the direction of the normal vector, which is confined within a specified range.

The DN expresses the amount of expectation deflection (also known as nominal vector), which is determined by the characteristics and criterion of the tracer. If the value of DN is large, the pressure of the stylus against the model will be greater, and the possibility of damage the model will also be increase. If the value of DN is small and the scanning speed of the stylus is fast, the stylus cannot keep always contact with the model and the possibility of drop-down from the model is increase greatly. Both the DN of nominal deflection and the DR of actual deflection can construct the following equation:

$$e_n = D_R - D_N. \tag{1}$$

e_n: an important reference value to regulate the moving of machine tool around the value of D_N along the tangential direction with the speed of V_N to shift stylus up or down.

When en > 0 (overpressure of the stylus), the stylus shifts a tiny displacement up along the direction of normal vector of the model (leaving the model); the stylus tends to decrease the value of en, which conduct the stylus decrease the level of pressing the model and the stylus should be given properly additional force with the normal direction of leaving to the object.

When en < 0 (underpressure of the stylus), the stylus moves along the direction of tangent of the model and shifts a tiny displacement down along the negative direction of normal vector (approach to the model) at the same time. The stylus tends to increase the value of en, which conduct the stylus increase the level of contact the model and the stylus should be given properly additional force with the normal direction of pointing to object.

3.2. The motion of the stylus without friction

Tracking and scanning on the model surface, the stylus detects and feedbacks information about the position and situation of the model surface to the *NC*. *NC* by the information drive the tracer movement and make the stylus always maintain the specified pressure and deviation and along the surface of the model for scanning. Generally, the speed of the stylus is proportional to the amount of the deviation perceived at that sampling point of P_1. Supposed the tangential speed is V_T, its components in the directions of X and Y is V_{TX} and V_{TY}. The e_n can also be decomposed along the X and Y into e_{nx} and e_{ny}. Refer to figure 1, the follow formals can be deduced respectively.

$$V_{TX} = \pm V_T * D_{Ry} / D_R \tag{2}$$

$$V_{TY} = \mp V_T * D_{Rx} / D_R \tag{3}$$

When the model is on the right side of the stylus, the upper symbol is taken; when the model is on the left side of the stylus, the lower symbol is taken. In the coordinate system, both V_{TX} and V_{TY} with signed symbol, and both D_{Rx} and D_{Ry} also with signed symbol, that means, the above formulas can be applied to any quadrants.

3.3. The motion of the stylus with friction

In fact, at one hand, due to the friction between the stylus and the model surface is forever exist, the normal line between the line of CP_1 and the vector line of OP_1 at the contact point of P_1 is no longer coincident. On the other hand, due to the existence of e_n and the value of e_n constantly altering and accumulating, in order to improve the accuracy of scanning and reduce the possibility of stylus drop-down from the model, e_n needs to be automatically regulated in timely. The curve of $\widehat{P_1P_2}$ is anticipated route, refer to figure 1. If the *NC* fails to automatic drive the stylus moving on the model within frequent of f_T, the stylus tendency shifts a tiny target displacement of the $\overline{P_1P_3}$ not the arc length of the $\widehat{P_1P_2}$. If the scanning speed is still calculated by the equations (2) and (3), the value of the e_n will inevitable decrease to less than zero, so that the stylus probability drop-down from the model. Therefore, an additional force must be employed in the direction of $\overline{CP_2}$ to make the resultant speed *V* of the stylus parallel to the tangent line of the model surface[3-9].

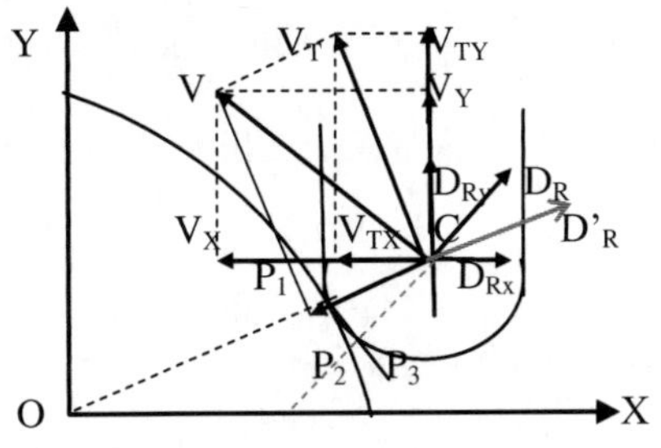

Figure 1. The motion of the stylus

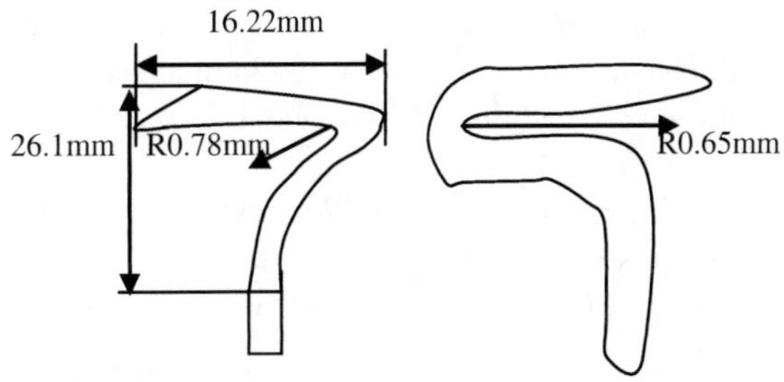

Figure 2. A steel noodle

The direction and size of V_N should be guaranteed that the value of D_R is rough equal to D_N ($e_n = D_R - D_N$). On the other hand, the value of the parameter V_N is regulated by e_n in order to making the value of D_R close to the value of D_N. The correction formal by the way of proportion adjustment to regulate the V_N is shown as below:

$$V_N = K * D_R \tag{4}$$

K is the proportional coefficient, the components of V_N in the coordinate axis of *X* and *Y* are respectively V_{NX} and V_{NY}, the following equations can be get respectively by equations (4).

$$V_{NX} = V_N * D_{Rx} / D_R \tag{5}$$

$$V_{NY} = V_N * D_{Ry} / D_R \tag{6}$$

By the equations (2) and (5), and (3) and (6), the following equations can be obtained respectively.

$$V_X = V_{TX} + V_{NX} = \pm V_T * D_{Ry} / D_R + V_N * D_{Rx} / D_R \tag{7}$$

$$V_Y = V_{TY} + V_{NY} = \mp V_T * D_{Rx} / D_R + V_N * D_{Ry} / D_R \tag{8}$$

4. Technical activity

4.1. A process before scanning to make deflection to zero and get gains of x, y and z axis

When the stylus does not touch model, the value of the deflection should be roughly zero. Being the varied of weight and shape of the tracer, even if the tracer dose not contact the model, the value of the deflection will not always be zero. In order to get accurate scanning data, based on the principle of establishing a nominal deflection of the stylus with respect to its cantered zero-position, some compensation should be added to the tracer. One of the compensation activity is achieved through execute special command by software way. If the deflection is too great to be compensated by software way, hardware way should be employed by adjusting the potentiometer for *x*, *y* and *z* respectively in the electric circuit.

4.2. The normal technical activity

In many cases, there are many different size of stylus used. Copy milling requires a certain relationship between the diameter of the tool and the size of the stylus of the tracer. In the actual processing of a batch of parts, because there is wear, a small amount of compensation need to make for tool, otherwise the parts milled will not be qualified.

When a new tool replaced, there exists the shap and the dimensions between the original the stylus does not exactly conform to the new tool, for this reason, a inconvenient method is to replace the stylus. This is inconvenient, however, that means a lot of extra styluses are needed.

In addition, there always exists tool wear when milling a large parts, and also exists the phenomenal that the accuracy of the initial processing parts is not the same accuracy of the subsequent processing parts with the same tool.That means tool compensation need to be made.

Accordingly, the stylus and tool should be given some restriction in the actual milling in order to get accurate scanning data:

① The shape of the stylus and tool must be similar.

② The dimension and size of the stylus and tool have to be the same. The tolerance is allowed that the dimension of the tool is 10% smaller than the stylus as defined by the following:

$$R_{DEF} = R_{stylus} - R_{tool}. \tag{9}$$

The above formula shows: If the radius of stylus is shorter than the tool, the tool will cut too much raw material, which could result in interference with the walls of the part, especially when milling near internal edges, and the problem of over deflection will appear.

In generally, copy milling is taken as roughing and not for the final finishing. In order to get accurate milling finishing, digitizing method (also refer to as differ copying) to be used. Normal vector and x, y and z position of the model are recorded by digitizing on the disc of the computer. While milling, the NC takes the normal vector as proportion to be added as 3D dimension tool compensation, which consist of the milling program to be executed by NC. The compensation can be drawn by the following expression:

$$R_{com} = R_{stylus} - R_{tool} \tag{10}$$

Comparing formula (9) with formula (10), we can know that the digitizing way can get more accurate milling by adding compensation. In formula (9): R_{DEF} represents the allowance for copy milling. In formula (10), R_{com} represents the value of compensation to be added in milling. Though formula (9) is similar to formula (10), but the idea is total different[10-14].

4.3. Experimental activity

In the following example, a steel noodle product is shown at figure2. The radius of the smallest R_{tool} = 0.5mm and R_{DEF} = 0.75. Under a given tolerance, the small products need to be digitized before milling. The shape of steel noodle is a more comprehensive product. It is clear that this small product can't be used with copy milling method.

Before scanning, multiple-trace boundaries limits have to be established for automatic scanning of the part. When the stylus recognizes a trace boundary, the mill machine tool will automatically stop, index over, and continue digitizing in the opposite direction for lace mode, or retract to a plateau and continue scanning in the same direction for box mode. When the system completes scanning a defined surface patch, the tracer will retract to a safe position and automatically continue digitizing another area of the part. Trace boundaries can be viewed or modified. Another scanning feature is anticipation: a command that adjusts scanning speed when the stylus senses a sudden change in the surface.

The step of production is as following:

①. Execution the zero process of stylus before scanning.

②. Digitizing a similar prototype of steel noodle by scanning with tracer to recording NC codes.

③. Machining steel noodle with the above codes.

5. Conclusions

As nominal vector D_N to be introduced in tracer scanning on curved surface, the tracer can autonomous regulate the scanning speed of stylus of tracer and hold contact level between stylus and the object. The formula deducing and experimental results shown that the parameter of nominal vector D_N play a proper role in the case of retaining tracer contact on curved surface.

① Using DN can better realize to self-adjust scanning speed when stylus encounter special curved surface, and improve the accuracy and efficiency of scanning.

② The stratagem of machining with special curved surface is: first collect workpiece data by digitizing, then carry out milling, so as to obtain a higher accuracy.

6. References

[1] Hongwei Liu, Hua Xiang, Jihong Chen，Rui Yang, Measurement and compensation of machine tool geometry error based on Abbe principle, *The International Journal of Advanced Manufacturing Technology* **98** (2018)2769-2774

[2] Yank Jinn, Fend Qibo, Research progress on volumetric geometric error measurement of numerical control (NC) machine tools, *Chinese Journal of Scientific Instrument*, Vol. 38 No. 8 Aug. (2017) 1901-1911

[3] Wang Pingjiang et al., The development of special CNC system for profiling lathe based on automatic measurement and process, *Chinese Modern Manufacturing Engineering*, **9** (2015) 74-80

[4] WANG Dong at al., Manual Feed Reciprocating Tracing Technique of CNC Tracing System, *Journal of Zhengzhou University of Technology*, Vol .20 Vo.1 Mar.(1999)94-96

[5] Keiichi Shirase et al., Digital copy milling- autonomous milling process control without an NC program, *Robotics and Computer-Integrated Manufacturing* **21** (2005) 312-317

[6] Keiichi Shirase et al., Prototyping of Autonomous CNC Machine Tool Based on Digital Copy Milling Concept, *The 41st CIRP Conference on Manufacturing Systems* (2008) 391-394

[7] Keiichi Shirase et al., Machining Strategy to Adapt Cutting Conditions under Digital Copy Milling Concept, *Journal of Advanced Mechnical Design Systems and Manufacturing*, **5**(2010) 924-935

[8] Toshihiko HIROOKA et al., Surface Roughness Control Based on Digital Copy Milling Concept to Achieve Autonomous Milling Operation, *3rd CIRP Conference on Process Machine Interactions (3rd PMI)*，Procedia CIRP 4 (2012) 35-40

[9] Zhao Haitao et al., Simulation of thermal behavior of a CNC machine tool spindle, International *Journal of Machine Tools&Manufacture* **47** (2007) 1003-1010

[10] M Ristic et al., Contact probe radius compensation using computer aided design models, *Proc Instn Mech Engrs* Vol 215 Part B(2001) 819-834

[11] Xie Chuanbo, Accuracy Analysing for FCG-1 Tracer, *Chinese Journal of Optic-Electronic Engineering*, Vol .17 No.6 Dec. (1990) 58-62

[12] Liu Zhiqi et al., Applied Method for Improving the Stability and Precision of NC-Copying Process, *Chinese Journal of Zhengzhou University of Technology*, Vol.19 No.4 Dec.(1998)5-8

[13] Liu Pingyi et al., Theoretical analysis and experiment of all terrain adaptive locomotion vehicle, *Transactions of the Chinese Society of Agricultural Engineering*, Vol.28 Supp.2 Oct. (2012) 134-139

[14] Song Zhen et al., Research on the Algorithm of CNC Profiling Processing, *J. Huazhong Univ. of Sci.&Tech.* Vol. 26 No. 8 Aug. (1998)13-15

Acknowledgements

The authors wound like to acknowledge the financial supported by "Beijing Key Laboratory on Integration and Analysis of Large-scale Stream Data, Beijing, China 100144" and "Social Science Foundation of Beijing, (Project No.: 18JYB015)".

IOP Conf. Series: Materials Science and Engineering **576** (2019) 012042 doi:10.1088/1757-899X/576/1/012042

Prediction of critical heat flux on surface under boiling in case of IVR-ERVC by using CFD simulation

Rasool Urfa[1], Daogang Lu[1], Noor Farooq Asran[2], Ansari Munib[1] and Rasool Aazim[1]

Beijing Key Laboratory of Passive Safety Technology for Nuclear Energy,
[1]North China Electric Power University, Beijing 102206, China,
[1]Pakistan Institute of Engineering and Applied Sciences.

E-mail: urfarasool@ncepu.edu.cn

Abstract. The efficacy of in-vessel through external reactor vessel cooling (IVR-ERVC) is strongly correlated with the critical heat flux (CHF) on the lower head of reactor pressure vessel (RPV). The high limit of CHF prevents the RPV from failure by extracting much heat out of the vessel through the circulation of working fluid in external flow channel. In present study, the deionized water is allowed to flow through curved flow channel having radius of curvature of 368mm and hydraulic diameter of 40mm with square cross-section. The heat flux of $1500kWm^{-2}$ is applied perpendicular to the convex wall of the channel. The CHF on the wall of flow channel has been predicted by incorporating CFD based two-phase (liquid-vapor) boiling model. The investigations on spatial variation of pressure, temperature, and velocity and heat flux acquired as a result of numerical simulations have been discussed in detail. The CHF and average HTC is predicted as $1.798e+04kWm^{-2}$ and $38.981kWm^{-2}K^{-1}$ respectively.

1. Introduction

The researchers have proposed after Three Mile Island accidents, that a pressure vessel can enhance ample heat transfer when submerged in the coolant to retrofit in-vessel retention of melting core. IVR-ERVC is considered as a requisite strategy for banishing utmost heat out of the RPV by introducing external flow channel outside of the RPV. The coolant while flowing through the ERVC extracts heat out of the pressure vessel and keeps its temperature low[1-3]. The CHF is the maximum limit which is bearable by the vessel. As far as the CHF on the outer surface of the vessel under boiling remains higher than the wall heat flux from the melting core, the temperature of outer vessel could be retained near saturation temperature of water in order to maintain the integrity of RPV. It is therefore, necessary to predict the CHF limit on the outer surface of vessel in order to do necessary measures for its enhancement and for the enhancement of boiling heat transfer.

Many experimental and theoretical studies on the enhancement of boiling heat transfer have been conducted in literature [4-8]. Presently, the broadly used CFD approach in engineering technologies for the simulation of two-phase flows with profound amount of void fraction was the Eulerian approach[9]. In this model, the conservation of mass, momentum and energy of both vapor and liquid phase had been considered separately and weighted by the volume fraction which depicted the assembled mean probability of the inhabitation for each phase at a particular point with respect to time and space. The sink and the source are the interchange terms between phases in the conservation

IOP Conf. Series: Materials Science and Engineering **576** (2019) 012042 doi:10.1088/1757-899X/576/1/012042

equations. These referents comprised of empirical or analytical correlations, depicting the interfacial forces, mass flux and the heat flux as a function of mean flow aspects. The mass and momentum transfer between two phases were analyzed by the bubble dynamics up to the moderate level of void fraction. The characteristics bubble diameter was used in order to consider deformation. The turbulence intensity in the flow channel and near the walls was strongly affected by the bubble dynamics at high Reynolds number. In case of diabatic boiling flows, when the heat was transferred from heated wall to the bulk fluid at high rates due to the generation and departure of vapor bubbles, the source term illustrating the mechanics of those phenomena at heated wall cannot be neglected.

In this paper, the two-phase boiling flow in case of IVR-ERVC has been studied by CFD simulation. Moreover, the elaborated discussion referring to the spatial distribution of heat flux, velocity, temperature and pressure has also been proposed, along with the prediction on CHF of IVR-ERVC by incorporating CFD vapor-liquid boiling model.

2. Mathematical modelling for CFD

There are several different mechanisms in which heat is transferred from hot body to the surrounding fluid i.e. in boiling. At the part of the heating surface where no bubble inhabits, the heat flows to the subcooled liquid directly like a single-phase flow. The part of the heating surface where the generation of vapors occurs at the nucleation sites extracts heat out of the surface. With the escaping of bubbles, the liquid mixing mechanism exists. The cold liquid from the bulk comes into contact with the heating surface as a result of recirculation around the detaching bubble, this phenomenon conducts additional cooling. In terms of two sets of conservation equations, governing the balance of mass, momentum, and energy of each phase the two-phase subcooled boiling model has been established. In governing equations, the interaction terms couple the interface transfer of mass, momentum, and energy. The water-liquid is taken as continuous phase whereas water-vapor is illustrated as a dispersed or secondary phase.

Conservation of mass,

$$\nabla.(\rho_m u_m) = 0 \tag{1}$$

Mass average velocity (u_m),

$$u_m = \sum_{k=1}^{2} \frac{\alpha_k \rho_k u_k}{\rho_m} \tag{2}$$

Mixture density (ρ_m),

$$\rho_m = \sum_{k=1}^{2} \alpha_k \rho_k \tag{3}$$

Conservation of momentum(p)[10],

$$\nabla.(\rho_m u_m u_m) = -\nabla p + \nabla[\mu_m(\nabla u_m + \nabla u_m^T)] + \rho_m g + \nabla(\sum_{k=1}^{2} \alpha_k \rho_k u_{dr,k} u_{dr,k}) \tag{4}$$

The viscosity of mixture (μ_m),

$$\mu_m = \sum_{k=1}^{2} \alpha_k \mu_k \tag{5}$$

The drift velocity ($u_{dr,k}$) for k-phase,

$$u_{dr,k} = u_k - u_m \tag{6}$$

The slip velocity (u_{vl}),

$$u_{vl} = u_v - u_l \tag{7}$$

The drift velocity is correlated with the slip velocity as follow,

$$u_{dr,k} = u_{vl} - \sum_{k=1}^{2} \frac{\alpha_k \rho_k}{\rho_m} u_{lk} \tag{8}$$

The relative velocity consisting of diffusion term as a result of dispersion by the turbulent [11]:

$$u_{vl} = \frac{(\rho_v - \rho_m) d_v^2 a}{18 \mu_{leff} f_{drag}} - \frac{v_{t,m}}{\alpha_v \sigma_D} \nabla \alpha_l \tag{9}$$

Where $v_{t,m}$ is the turbulent viscosity of mixthe ture, σ_D , is the Prandtl dispersion coefficient. The drag force f_{drag} , is considered as[12]:

$$f_{drag} = \begin{cases} 1 + 0.15 Re_v^{0.687} & Re_v \leq 1000 \\ 0.0183 Re_v & Re_v > 1000 \end{cases} \tag{10}$$

The Reynolds number,

$$Re_v = \frac{u_m d_v}{v_m} \tag{11}$$

The acceleration (a),

$$a = g - (u_m \nabla) u_m \tag{12}$$

The energy equation for the mixture is as follows;

$$\nabla . \sum_{k=1}^{2} (\alpha_k u_k \rho_k h_k) = \nabla . \left(k_{eff} \nabla T \right) + Q_w \tag{13}$$

$$k_{eff} = \left[\sum_{k=1}^{2} \alpha_k (k_k + k_t) \right] \tag{14}$$

The effective conductivity (k_{eff}) and the turbulent thermal conductivity (k_t) are related as;

$$k_t = \frac{C_P \mu_t}{Pr_t} \tag{15}$$

The heat source i.e. the wall heat flux $Q_w$which is equivalent to the latent heat required for phase change. C_P is specific heat capacity.By incorporating continuity equation for the phases, the void fraction for dispersed phase i.e. vapors is computed as;

$$\nabla . (\alpha_v \rho_v v_m) = -\nabla . \left(\alpha_v \rho_v v_{dr,v} \right) + \sum_{k=1}^{2} (m_{lv} - m_{vl}) \tag{16}$$

'm', is the mass transfer through evaporation and condensation. The turbulence model (k-ε) has been used in this investigation due to complex modeling in the multiphase simulation. It is mainly effectual for turbulent core flows. The equations for turbulent kinetic energy and dissipation rate are represented as;

$$\nabla . (\rho_m u_m k) = \nabla . \left(\frac{\mu_{t,m} \nabla k}{\sigma_k} \right) + G_{k,m} - \rho_m \epsilon \tag{17}$$

$$\nabla . (\rho_m u_m \epsilon) = \nabla . \left(\frac{\mu_{t,m} \nabla \epsilon}{\sigma_\epsilon} \right) - C_2 \rho_m \frac{\epsilon}{k} (C_{1\epsilon} G_{k,m} - C_{2\epsilon} \rho_m \epsilon) \tag{18}$$

The turbulent viscosity and turbulent kinetic viscosity are taken as[13];

$$\mu_{t,m} = C_\mu \rho_m \frac{k^2}{\epsilon} \tag{19}$$

$$G_{k,m} = \mu_{t,m} (\nabla u_m + \nabla u_m^T) : \nabla u_m \tag{20}$$

σ_k, σ_ϵ, $C_{1\epsilon}$, $C_{2\epsilon}$, C_μ equal to 1, 1.3, 1.44, 1.92, 0.09 respectively are the empirical constants[14].The effective viscosity is given as;

$$\mu_{eff} = \mu_{leff} + \mu_{t,m} \tag{21}$$

The wall heat flux (Q_w) is constituted into three parts:

$$Q_w = Q_c + Q_e + Q_q \tag{22}$$

The first part is convective heat flux (Q_c):

$$Q_c = A_1 St \rho_l C_{pl} u_l (T_w - T_l);\ A_1 = 1 - A_2 \tag{23}$$

Where $A_1$and A_2 are the fraction of the wall surface under the influence of liquid and vapor formed on the wall surface respectively. T_l , and u_l are the liquid temperature and velocity at the cell closed to the wall. St is the Stanton number.

$$S_t = \frac{Nu}{Re_l Pr_l} \tag{24}$$

The active nucleation density is correlated as;

$$n = [210(T_w - T_{sat})]^{1.805} \tag{25}$$

The second part is evaporation heat flux (Q_e) for the generation of vapor bubbles:

$$Q_e = nf\left(\frac{\pi d_{BW}^3}{6}\right)\rho_v h_{vl} \tag{26}$$

The bubble frequency (f) is driven from the correlation:

$$f = \sqrt{\frac{4g(\rho_{leff} - \rho_v)}{3 d_{BW} \rho_{leff}}} \tag{27}$$

As depicted that the frequency is dependent on bubble diameter and phase density. The average of multiple sizes of bubbles is usually considered as a single size bubble in literature. The bubble departure diameter is computed as:

$$d_{BW} = \min\left[1.4, 0.6. \exp\left(-\frac{\Delta T_{sub}}{45}\right)\right] \tag{28}$$

The third part is quenching heat flux (Q_q) representing heat flux from heating wall to the bulk liquid by departing or collapsing bubbles during the bubble eruption cycle. The Q_q is calculated from the temperature difference of wall and the bulk liquid multiplied by the heat transfer coefficient. Whereas, K=2[15]

$$Q_q = \left(\frac{2}{\pi}\sqrt{k_{leff}\rho_{leff}C_{pleff}f}\right)A_2(T_w - T_l) \tag{29}$$

$$A_2 = nK\left(\frac{\pi d_{BW}^2}{4}\right) \tag{30}$$

By accounting the mass transfer from the liquid to the vapor phase on the heating wall is:

$$m_{wv} = \frac{Q_e}{h_{vl} + C_{pleff}(T_{sat} - T_l)} \tag{31}$$

The mass transfer rate from liquid to the vapor (evaporation) and from vapor to the liquid (condensation) inside the fluid is evaluated by using the following equations:

$$m_{lv} = \begin{cases} \frac{r_{lv}\alpha_l \rho_{leff}(T_l - T_{sat})}{T_{sat}} & ; T_l \geq T_{sat} \\ 0 & ; T_l < T_{sat} \end{cases} \tag{32}$$

$$m_{vl} = \begin{cases} \frac{r_{vl}\alpha_v \rho_{veff}(T_{sat} - T_l)}{T_{sat}} & ; T_v \geq T_{sat} \\ 0 & ; T_v < T_{sat} \end{cases} \tag{33}$$

Phase volume fraction is represented as α, density is ρ whereas r_{lv} and $r_{vl}$are mass transfer time parameters with unit ms^{-1}[16]. The aforementioned boiling model is coupled with the CFD simulation where the wall boundary condition is the wall heat flux taken as $1500 kWm^{-2}$ and the reservoir temperature is considered as 20°C and 80°C. By considering the velocity, void fraction and heating wall temperature distribution under sub cooling effect, the heat transfer from the heating wall to the flowing fluid is examined, CHF and the HTC are predicted.

3. Numerical simulation

For the numerical simulation of CFD, the discretization of the governing equations has been done by considering second order upwind scheme. The SIMPLE (semi-implicit method for pressure linked equations) algorithm is adopted for the coupling of pressure and velocity. The k-$\mathcal{E}$ model is used for modeling turbulence in flow boiling. The illustration of geometric parameters of curved flow channel with downward facing heating wall and the boundary conditions are summarized in Table-1.

Table 1. Specifications of curved channel and boundary conditions

Outlet gauge pressure, P_2	21.16	[kPa]
Inlet velocity, v_l	0.347	[ms^{-1}]
Inlet temperature, T_i	293	[K]
Heat flux, Q	1500	[kWm^{-2}]
Volumetric flow rate,	6.626	[m^3s^{-1}]
Hydraulic diameter, d_h	0.040	[m]
Bend angle of channel, θ	90°	[degrees]
Radius of curvature	0.368	[m]

The fluid is allowed to flow from the inlet at the bottom of the channel to the outlet on the top of the channel. The inlet boundary condition is set with a fluid velocity of 0.347ms^{-1}, and temperature of 293K whereas the outlet boundary condition is set at a pressure of 21.61kPa. The supplied heat flux to the convex wall of the channel is 1500kWm^{-2}. The concave wall of the channel is kept as adiabatic due to insulation effect.

4. Results and discussions

4.1. Variation of pressure field

Figure.1 depicts non uniform pressure distribution of water coolant through curved flow channel from inlet to the outlet under constant heating through its convex wall. The static pressure observed to be decreased from 7.75E5Pa to 1.36E4Pa while flowing upward along flow field. The pressure gradient through curved channel is plotted in Figure.2. It can be seen in plot that from inlet to the outlet of the channel the pressure initially dropped because of the utilization of energy for the generation of secondary phase (vapors) but after reaching a certain point i.e. nearby 0.13m, the pressure gradient suddenly raised due to increasing interactions between two phases with the continuous formation of vapor phase under boiling.

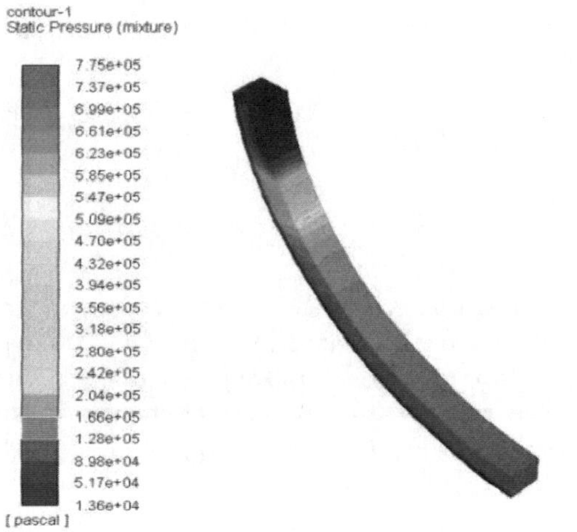

Figure 1. Pressure distribution

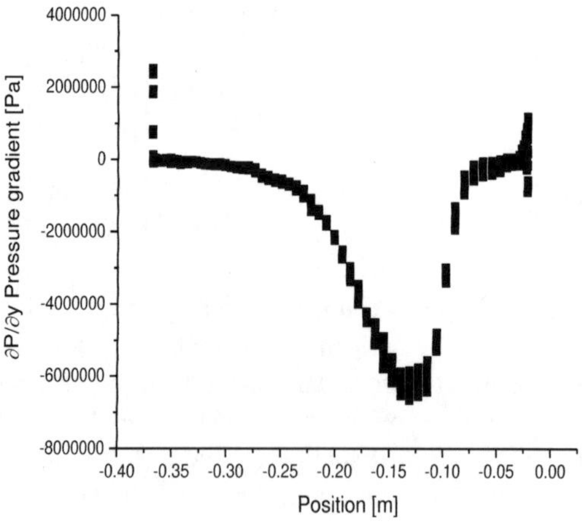

Figure 2. Pressure gradient

4.2. Variation of temperature field

Figure. 3 shows distribution of temperature on heated wall of the channel. It varied from 293K to 20200K all the way from inlet to the outlet. The generation and accumulation of vapors on heated surface at consistent heat flux supplied perpendicular to the convex wall formed vapor blanket near outlet of the channel which consequently increased wall temperature due to low thermal conductivity of vapor phase. The distribution of void fraction on heated wall is plotted against position in Figure.4. The void fraction can be seen concentrated from 0.20m to 0.10m due to the occurance of large number of vapors which could not extract heat out of the heated wall and acquired temperature upto 20200K.

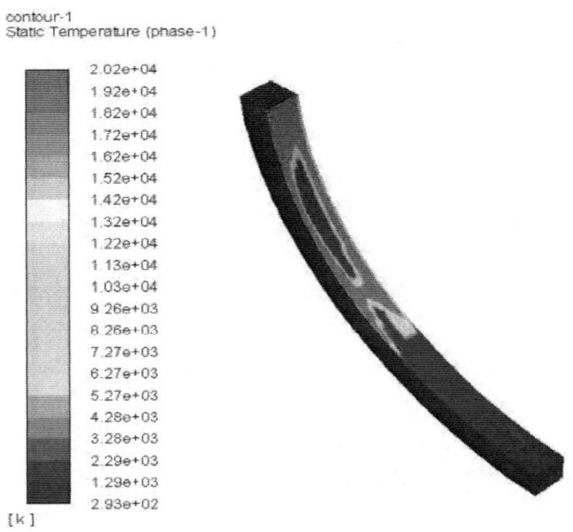

Figure 3. Temperature distribution

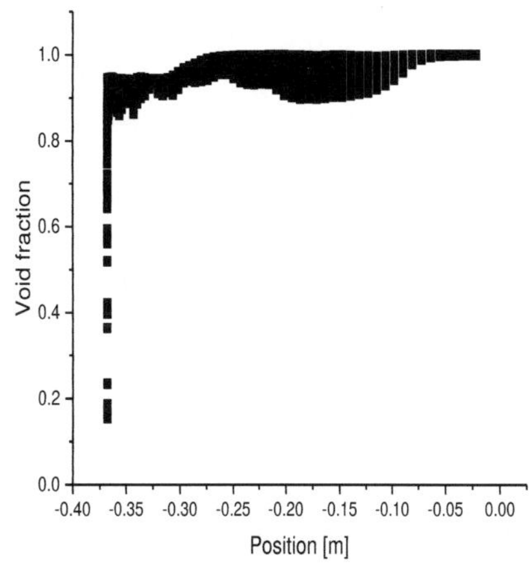

Figure 4. Void fraction distribution on heated wall

4.3. Variation of velocity field

The variation of fluid velocity is illustrated in Figure. 5. It can be seen in the figure that the flow rate in the premises of heated wall of the channel is considerably higher than the premises of adiabatic wall due high energy and upward slip velocity of vapors as compared to liquid phase. The buoyancy effect kept vapors to stay near upper heated wall in a wide range.

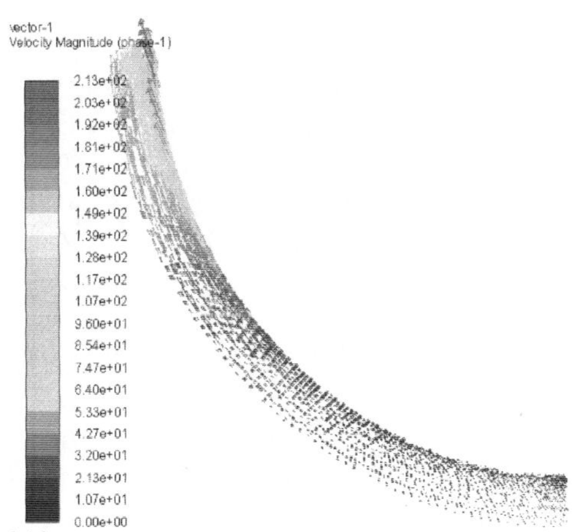

Figure 5. Velocity distribution

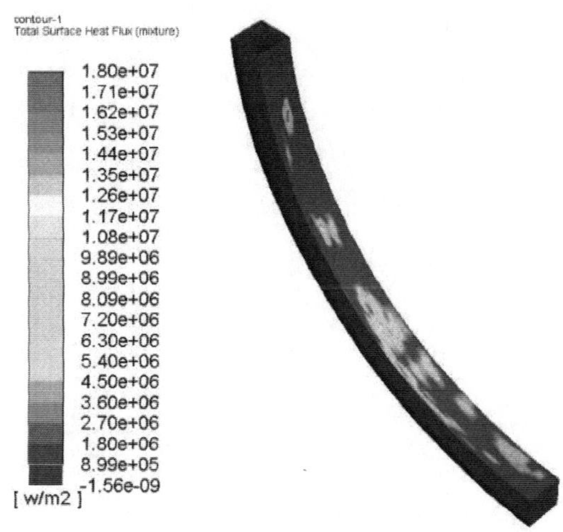

Figure 6. Heat flux distribution

4.4. The prediction of CHF
The distribution of heat flux on heated wall of the curved channel is illustrated in Figure.6. The continuous supply of heat flux increased formation of vapors, after a certain limit the accumulation of vapors created vapor blanket and entrapped heat within the channel by blocking liquid pathways and rewetting. As a result, dry spots started to originate on heated surface which then extended into dry patches thus, wall temperature increased abruptly. The increased rate of vapor formation induced thermal resistance and wall stresses which brought boiling to the transition regime followed by the turbulent regime and gradually to the utmost limit of bearable heat flux i.e. critical heat flux. The critical heat flux (CHF) on heated surface in present case is calculated as 1.798e+04kWm^{-2} near outlet of the channel whereas, the average heat transfer coefficient (HTC) is calculated as 38.981kWm^{-2}K^{-1}.

5. Conclusions
A two-phase CFD boiling model has been used to numerically investigate the flow boiling characteristics of water coolant flowing through the external curved flow channel by considering temperature dependent quantities and in the turbulent flow regime. It has been concluded that the pressure gradient initially decreases from inlet to the outlet of the channel due to the utilization of energy in the formation of secondary phase then after reaching a certain limit the pressure gradient started to increase with increasing interaction of vapor and liquid phase. The flow velocity appeared greater near the heating wall than that of adiabatic wall due to the effect of boiling. Under the influence of buoyancy, the vapors accumulated on the upper heating wall of the channel, incorporated thermal resistance augmented the wall temperature because of low thermal conductivity of vapor phase and eventually the boiling crises occurred which can be seen initiated near the outlet of the channel. The CHF and average HTC is predicted as 1.798e+04kWm^{-2} and 38.981kWm^{-2}K^{-1} respectively.

6. Abbreviations
The following abbreviations are used in this manuscript:

ρ	*Density [kgm^{-3}]*	C_p	*Specific heat [Jkg^{-1}K^{-1}]*
u	*Velocity [ms^{-1}]*	m	*Inter-phase mass transfer*
α	*Volume fraction of liquid or vapor phase*	$k-\epsilon$	*Turbulence dissipation rate - turbulence kinetic energy [Jkg^{-1}]*
μ	*Kinematic viscosity [m^{2}s^{-1}]*	Nu	*Nusselt number*
v	*Dynamic viscosity [Pas]*	Pr	*Prandtl number*
u_{dr}	*Drift velocity [ms^{-1}]*	St	*Stanton number*
u_{vl}	*Slip velocity [ms^{-1}]*	Re	*Reynolds number*
σ	*Surface tension [Nm^{-1}]*	n	*Nucleation site density [m^{-2}]*
f_{drag}	*Drag force [N]*	r	*Mass transfer time[s^{-1}]*
d	*Diameter[m]*	Q_w	*Wall heat flux [Wm^{-2}]*
a	*Acceleration [ms^{-2}]*	Q_e	*Heat flux due to evaporative [Wm^{-2}]*
g	*Gravity [ms^{-2}]*	Q_c	*Heat flux due to convection [Wm^{-2}]*
h	*Enthalpy [Jkg^{-1}]*	Q_q	*Heat flux due to quenching [Wm^{-2}]*
k	*Thermal conductivity [Wm^{-1}K^{-1}]*	A_s	*Surface area [m^2]*
α	*Volume fraction of vapor or liquid*	A_1	*Area fraction of wall under the influence of liquid [m^2]*
T	*Temperature [K]*	A_2	*Area fraction of wall under the influence of vapor [m^2]*
ΔT	*Sub cooled [K]*	f	*Bubble departure frequency[s^{-1}]*
Q	*Heat flux [Wm^{-2}]*	d_{BW}	*Bubble departure diameter[m]*

$\dot{Q}$	*Heat transfer rate [kW]*	h_{vl}	*Latent heat of vaporization [Jkg^{-1}]*
G	*Mass flux [$kgm^{-2}s^{-1}$]*	V	*Voltage [V]*
P_1	*Inlet pressure of channel [Pa]*	I	*Current [A]*
P_2	*Outlet pressure of the channel [Pa]*	θ	*Bend angle of the channel [degree]*

Subscripts

v	*Vapor*	t	*Turbulent*
l	*Liquid*	b	*Bulk*
m	*Mixture*	dr	*Drift*
k	*Summation index*	e	*Evaporation*
sat	*Saturated*	c	*Convection*
eff	*Effective*	q	*Quenching*

7. References

[1] Rempe, J.L., et al., *In-vessel retention-recent efforts and future needs*. 2004: Idaho National Laboratory.

[2] Rempe, J.L., et al., *Insights from investigations of in-vessel retention for high powered reactors*. 2005: Idaho National Laboratory.

[3] Rempe, J.L., et al., *In-vessel retention strategy for high power reactors.* Final Report, Idaho National Engineering and Environmental Laboratory, Report No. INEEL/EXT-04-02561, 2003.

[4] Henry, R.E. and H.K. Fauske, *External cooling of a reactor vessel under severe accident conditions.* Nuclear Engineering and Design, 1993. **139**(1): p. 31-43.

[5] Chu, T.Y., et al., *Ex-vessel boiling experiments: laboratory-and reactor-scale testing of the flooded cavity concept for in-vessel core retention Part I: Observation of quenching of downward-facing surfaces.* Nuclear Engineering and Design, 1997. **169**(1-3): p. 77-88.

[6] Chu, T.Y., et al., *Ex-vessel boiling experiments: laboratory-and reactor-scale testing of the flooded cavity concept for in-vessel core retention Part II: Reactor-scale boiling experiments of the flooded cavity concept for in-vessel core retention.* Nuclear Engineering and Design, 1997. **169**(1-3): p. 89-99.

[7] Theofanous, T.G. and S. Syri, *The coolability limits of a reactor pressure vessel lower head.* Nuclear Engineering and Design, 1997. **169**(1-3): p. 59-76.

[8] Theofanous, T.G., et al., *Quantification of Limits to Coolability in ULPU-2000 Configuration IV.* CRSS-02.05, 2002. **3**.

[9] Drew, D.A. and S.L. Passman, *Theory of multicomponent fluids*. Vol. 135. 2006: Springer Science & Business Media.

[10] Fluent, A., *Ansys fluent theory guide.* ANSYS Inc., USA. **15317**: p. 724-746.

[11] Manninen, M., V. Taivassalo, and S. Kallio, *On the mixture model for multiphase flow*. 1996, Technical Research Centre of Finland Finland.

[12] Schiller, L., *A drag coefficient correlation.* Zeit. Ver. Deutsch. Ing., 1933. **77**: p. 318-320.

[13] Lucas, D., E. Krepper, and H.M. Prasser, *Use of models for lift, wall and turbulent dispersion forces acting on bubbles for poly-disperse flows.* Chemical Engineering Science, 2007. **62**(15): p. 4146-4157.

[14] Launder, B.E. and D.B. Spalding, *The numerical computation of turbulent flows*, in *Numerical Prediction of Flow, Heat Transfer, Turbulence and Combustion*. 1983, Elsevier. p. 96-116.

[15] Kenning, D.B.R., *Fully-developed nucleate boiling: overlap of areas of influence and interference between bubble sites.* International Journal of Heat and Mass Transfer, 1981. **24**(6): p. 1025-1032.

[16] Abedini, E., et al., *Numerical investigation of subcooled flow boiling of a nanofluid.* International Journal of Thermal Sciences. **64**: p. 232-239.

IOP Conf. Series: Materials Science and Engineering **576** (2019) 012043 doi:10.1088/1757-899X/576/1/012043

Determination of the Thermal Stress Produced by a Thermo-poro-inclusion of Spheroid-shaped Near a Free Surface

X Zhang[1], X Jin[1,2], P Li[1], D Lyu[1], and R Zhang[1]

[1]State Key Laboratory of Mechanical Transmissions, Chongqing University, Chongqing 400030, China

[2]College of Aerospace Engineering, Chongqing University, Chongqing, 400030, China

Corresponding Author's Email: jinxq@cqu.edu.cn

Abstract. Eshelby's inclusion with thermo-porous eigenstrain is a classical micromechanics problem, and may find wide applications in various engineering fields including mechanical transmission system, composite materials and geophysical structures. In this paper, the elastic field produced by thermo-porous spheroidal inclusion placed vertically near a free boundary surface is investigated. The current work complements our previous publication on the horizontally aligned spheroidal inclusion [Computers & Geosciences. 2019;122:15-24], and hence concludes that a complete elasticity solution with respect to the displacement, strain and stress can be obtained in closed-form. The elastic field of spheroidal inclusion due to thermal expansion is provided, and comparative study of inclusion due to fluid withdrawal leading to a decrease in pore pressure is discussed. The present half-space inclusion model can be used to determine the elastic fields of porous materials caused by produced by pore pressure fluctuation and temperature alteration.

1. Introduction

The Eshelby inclusion problem [1, 2] has broad applications in modern science and technology, as exemplified by the materials science and geophysical engineering [3]. However, a majority of the studies concern the inclusion in a full-space, whereas those on half-space problems are less reported in existing literatures due to mathematical complexity. Mindlin and Cheng [4] solved the spherical inclusion with uniform dilatational thermal expansion by utilizing the Galerkin vector stress function. Chiu [5] formulated the closed-form solution of uniformly distributed cuboidal inclusion by employing the method of images. Seo and Mura [6] studied the elastic stress field produced by a uniform thermal ellipsoidal inclusion in a semi-infinite space. The exterior solution of the inclusion problem is usually more intricate. Healy [7] and Manoylov et al. [8] respectively employed the spheroid-shaped inclusion model to solve the strain and stress fields of a penny-shaped crack, and the elastic properties of the porous material. Jin et al. [9] derived the displacement, deformation gradient, strain and stress Eshelby tensors and presented in a simpler and geometrically meaningful form. Lyu et al. [10] proposed analytical solutions for the complete elastic fields with respect to the displacements, strains, and stresses, produced by an ellipsoidal thermal inclusion in a half-space.

Due to the involved derivatives of elliptic integrals of the first and second kinds as well as tremendous mathematical difficulties arisen from the effect of the free boundary surface, the existing analytical solutions to half-space inclusion problems are usually lengthier and hence difficult for engineering applications. In a companion paper [11], the closed-form solution for horizontally aligned spheroidal inclusion subjected to the thermo-porous eigenstrain near free surface is obtained. Following this line of work, we present a complete elasticity solution corresponding to the displacement, strain and stress for the vertical alignment case. The current solution complements our previous publication on the horizontally placed spheroidal inclusion with thermo-porous eigenstrain [11] and concludes that a complete elasticity solution can be analytically presented in closed-form. In addition, several illustrating examples are provided to validate the present solution.

2. Formulation

2.1. Thermo-porous eigenstrain

Eigenstrain is the stress-free transform strain which encompasses a wide range of nonelastic strains including plastic strain, misfit strain and lattice mismatch, etc. As typical applications in geophysical engineering, the thermoelastic and poroelastic eigenstrains are employed to characterize the temperature change and pore pressure fluctuation, respectively. For an isotropic thermo-poroelastic medium, the eigenstrain, ε_{ij}^*, of poroelastic and thermoelastic couplings may be written as [11]

$$\varepsilon_{ij}^* = \frac{1-2\nu}{2\mu(1+\nu)}\delta_{ij}\alpha_B \Delta P + \delta_{ij}\lambda_T \Delta T \tag{1}$$

where the change of pore pressure and temperature are denoted by ΔP and ΔT, and μ, ν are the shear modulus and Poisson's ratio. Kronecker's delta, δ_{ij}, is defined as 1 if $i=j$ and 0 if $i \neq j$. Moreover, Boit pore-pressure coefficient is denoted by α_B, and coefficient of thermal expansion, λ_T, is used to describe the tendency of matter to change in shape, area, and volume in response to a change in temperature.

2.2. Thermo-porous ellipsoidal inclusion in a half-space

Consider a semi-infinite matrix (Figure. 1) containing a thermo-porous ellipsoidal inclusion, Ω, which is bounded by

$$\frac{x_1^2}{a_1^2} + \frac{x_2^2}{a_2^2} + \frac{(x_3-c)^2}{a_3^2} \leq 1 \qquad (c \geq x_3) \tag{2}$$

where $a_I\ (I=1,2,3)$ is the semi-axis of the ellipsoidal inclusion, *c* represents the depth of the inclusion. It is noted that the boundary surface $(x_3 = 0)$ is free from any external force. The uniform thermo-porous eigenstrain (Eq. (1)), is prescribed in Ω and vanished in the remainder of the matrix. (i.e. $\varepsilon_{11}^* = \varepsilon_{22}^* = \varepsilon_{33}^* = \varepsilon^*$ and $\varepsilon_{12}^* = \varepsilon_{13}^* = \varepsilon_{23}^* = 0$).

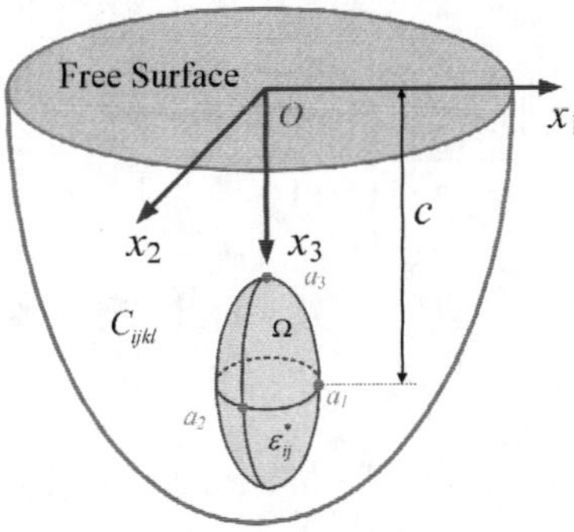

Figure 1. An ellipsoidal inclusion in a half-space.

The exterior displacement solution of the thermo-porous ellipsoidal inclusion in a semi-infinite space may be analytically presented as

$$u_i(\mathbf{x})=\frac{(1+\nu)\varepsilon^*}{1-\nu}\begin{Bmatrix} x_i\mathcal{J}_I(\lambda)-2x_3\left[\rho_{123}^* n_3^* n_i^* - \delta_{3i}\mathcal{J}_I(\lambda^*)\right] \\ +(3-4\nu)\left[x_i\mathcal{J}_I(\lambda^*)-2\delta_{3i}x_I\mathcal{J}_I(\lambda^*)\right]\end{Bmatrix} \tag{3}$$

for i=1, 2

$$u_i(\mathbf{x})=\frac{(1+\nu)\varepsilon^*}{1-\nu}\begin{Bmatrix} (x_i-c)\mathcal{J}_I(\lambda)-2x_3\left[\rho_{123}^* n_3^* n_i^* - \delta_{3i}\mathcal{J}_I(\lambda^*)\right] \\ +(3-4\nu)\begin{bmatrix}(x_i+c)\mathcal{J}_I(\lambda^*) \\ -2\delta_{3i}(x_I+c)\mathcal{J}_I(\lambda^*)\end{bmatrix}\end{Bmatrix} \tag{4}$$

for i=3, and λ, λ^* are the maximum positive root of cubic equations with respect to the imaginary and imaginary confocal ellipsoids. The function $\mathcal{J}_I$, ρ_{123}^* and unit normal vectors n_i, n_i^* are omitted here due to length, and interested readers are referred to [10, 11].

The exterior strain solution may be expressed as

$$\varepsilon_{ij}(\mathbf{x})=\frac{(1+\nu)\varepsilon^*}{1-\nu}\begin{Bmatrix} \underline{\underline{\delta_{ij}\mathcal{J}_I(\lambda)-\rho_{123}n_in_j}}-\delta_{3I}\delta_{ij}(4-8\nu)\mathcal{J}_3(\lambda^*) \\ +(3-4\nu)\left[\delta_{ij}\mathcal{J}_I(\lambda^*)-\rho_{123}^* n_i^* n_j^*\right] \\ -4\nu\rho_{123}^*\left(\delta_{3i}n_I^* n_j^*+\delta_{3j}n_i^* n_J^*\right) \\ +2\rho_{123}^*\left[M_{ij}(\lambda^*)+N_{ij}(\lambda^*)\right]\end{Bmatrix} \tag{5}$$

for i=1, 2, 3, and terms M_{ij}, N_{ij} are

$$M_{ij}(\lambda^*)=\frac{x_3n_3^*n_i^*n_j^*}{\sqrt{H(\lambda^*)}}\begin{pmatrix} \dfrac{1-4n_1^*n_1^*}{a_1^2+\lambda^*}+\dfrac{1-4n_2^*n_2^*}{a_2^2+\lambda^*}+\dfrac{1-4n_3^*n_3^*}{a_3^2+\lambda^*} \\ +\dfrac{2}{a_I^2+\lambda^*}+\dfrac{2}{a_J^2+\lambda^*}+\dfrac{2}{a_3^2+\lambda^*}\end{pmatrix} \tag{6}$$

$$N_{ij}(\lambda^*)=\frac{\left(\delta_{3i}n_j^*+\delta_{3j}n_i^*\right)c}{\sqrt{H(\lambda^*)}\left(a_3^2+\lambda^*\right)}-\frac{\delta_{ij}x_3n_3^*}{\sqrt{H(\lambda^*)}\left(a_I^2+\lambda^*\right)} \tag{7}$$

According to Hooke's law, the exterior stress solution may also be determined in an explicit form

$$\sigma_{ij}^{out}(\mathbf{x})=\frac{2\mu(1+\nu)\varepsilon^*}{1-\nu}\begin{Bmatrix} \underline{\underline{\delta_{ij}\mathcal{J}_I(\lambda)-\rho_{123}n_in_j}} \\ \underline{-(4-8\nu)\delta_{3I}\delta_{ij}\mathcal{J}_I}(\lambda^*) \\ +(3-4\nu)\left[\delta_{ij}\mathcal{J}_I(\lambda^*)-\rho_{123}^* n_i^* n_j^*\right] \\ -4\nu\left(\delta_{3I}+\delta_{3J}\right)\rho_{123}^* n_i^* n_j^* \\ +4\nu\delta_{ij}\left[\rho_{123}^* n_3^* n_3^*-\mathcal{J}_3(\lambda^*)\right] \\ +2\rho_{123}^*\left[M_{ij}(\lambda^*)+N_{ij}(\lambda^*)\right]\end{Bmatrix} \tag{8}$$

2.3. In the case of spheroidal inclusion

By setting $a_1=a_2\neq a_3$ and $\alpha=a_3/a_1$, λ for the spheroidal inclusion is thus derived explicitly as

$$\lambda = \frac{1}{2}\left\{ \begin{array}{l} x_1^2 + x_2^2 + \left(x_3 - c\right)^2 - a_1^2 - a_2^2 + \\ \sqrt{\left[x_1^2 + x_2^2 + \left(x_3 - c\right)^2 + a_1^2 - a_2^2\right]^2 + 4\left(a_2^2 - a_1^2\right)x_1^2} \end{array} \right\} \tag{9}$$

and the corresponding $\mathcal{I}_3$ function is

$$\mathcal{I}_3(\lambda) = \frac{\alpha}{\left(1-\alpha^2\right)^{\frac{3}{2}}}\left(\tan\varpi - \varpi\right) \tag{10}$$

for oblate spheroid $(\alpha < 1)$, and

$$\mathcal{I}_3(\lambda) = \frac{\alpha}{\left(\alpha^2 - 1\right)^{\frac{3}{2}}}\left[-\ln\left(\tan\frac{\omega}{2}\right) - \cos\omega\right] \tag{11}$$

for prolate spheroid $(\alpha > 1)$. Moreover, quantities in Eqs. (10) and (11) are

$$\varpi = \arccos\sqrt{\frac{a_3^2 + \lambda}{a_1^2 + \lambda}}, \quad \omega = \arcsin\sqrt{\frac{a_1^2 + \lambda}{a_3^2 + \lambda}} \tag{12}$$

The elastic field caused by a thermo-porous spheroidal ellipsoidal inclusion near free surface may be analytically obtained. The displacement solution for the exterior field is

$$u_1(\mathbf{x}) = \frac{(1+\nu)\varepsilon^*}{1-\nu}\left\{ \begin{array}{l} \frac{x_1}{2}\left[\rho_{13} - \mathcal{I}_3(\lambda)\right] - 2x_3\rho_{13}^* n_1^* n_3^* \\ +\left(\frac{3}{2} - 2\nu\right)x_1\left[\rho_{13}^* - \mathcal{I}_3(\lambda^*)\right] \end{array} \right\} \tag{13}$$

$$u_2(\mathbf{x}) = \frac{(1+\nu)\varepsilon^*}{1-\nu}\left\{ \begin{array}{l} \frac{x_2}{2}\left[\rho_{13} - \mathcal{I}_3(\lambda)\right] - 2x_3\rho_{13}^* n_2^* n_3^* \\ +\left(\frac{3}{2} - 2\nu\right)x_2\left[\rho_{13}^* - \mathcal{I}_3(\lambda^*)\right] \end{array} \right\} \tag{14}$$

$$u_3(\mathbf{x}) = \frac{(1+\nu)\varepsilon^*}{1-\nu}\left\{ \begin{array}{l} (x_3 - c)\mathcal{I}_3(\lambda) - 2x_3\rho_{13}^* n_3^* n_3^* \\ +\left[2x_3 + (4\nu - 3)(x_3 + c)\right]\mathcal{I}_3(\lambda^*) \end{array} \right\} \tag{15}$$

The corresponding exterior strain and stress solutions can be derived in closed-form through Eqs. (5)and (8). The detail formulation is omitted here due to length.

3. Results and discussions

The stress field of a thermal spheroidal inclusion in a half-space was reported by Seo and Mura [6]. In this section, the elastic fields corresponding to the displacement, strain and stress for the thermo-porous inclusion near free surface is carried out with $E = 210\text{GPa}$ and $\nu = 0.3$. The stress distributions for the spheroidal inclusion with thermal eigenstrain may serve as a benchmark example to validate the present solution.

For the spheroidal inclusion ($3a_1 = 3a_2 = a_3$) centered at ($0,\ 0,\ a_3$) subjected to thermal eigenstrain, the displacements along the x_1 -direction are plotted in Figure. 2. The displacement components are normalized by $u_0 = \lambda_T \Delta T a_1$. It can be seen from Figure. 2 that the displacements are continuous across the interface between the inclusion and surrounding matrix. The displacement u_1 represents tension both inside and outside the inclusion, while u_3 shows compression.

The strain distributions along the x_1-direction are illustrated in Figure. 3. The strain components are normalized by $\varepsilon_0 = \lambda_T \Delta T$, and the strains in the interior field are no longer constant due to the existence of the free surface, as opposed to the case of full-space [9].

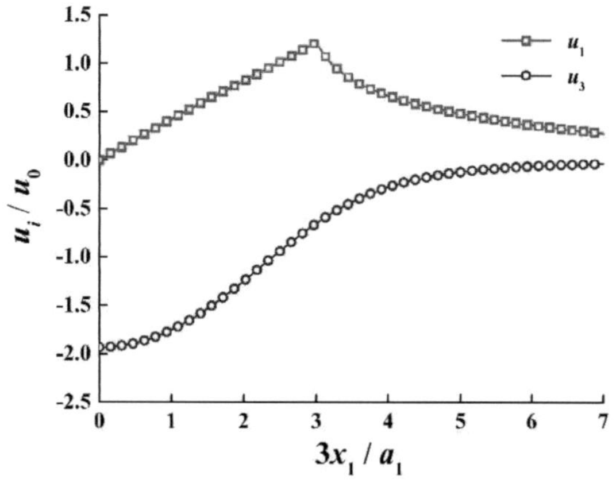

Figure 2. Displacements of a thermal spheroidal inclusion

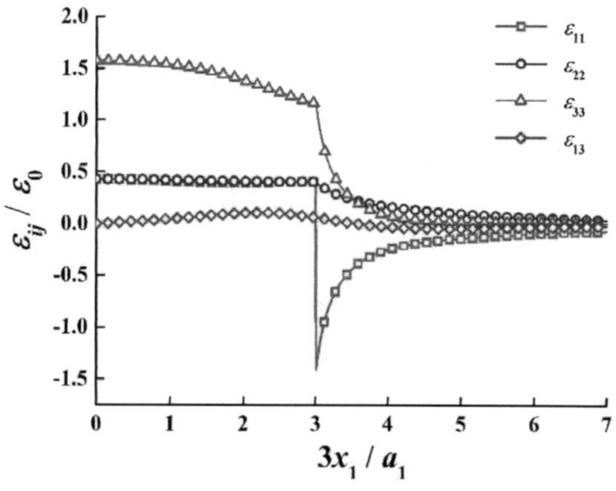

Figure 3. Strains of a thermal spheroidal inclusion

The stresses are normalized by $\sigma_0 = E\lambda_T \Delta T / (1-\nu)$, and the variations along the x_1-direction are shown in Figure. 4. The normal stress components σ_{22} and σ_{33} suffer discontinuities across the interface, and become almost zero for $x_1 / a_3 > 7$.

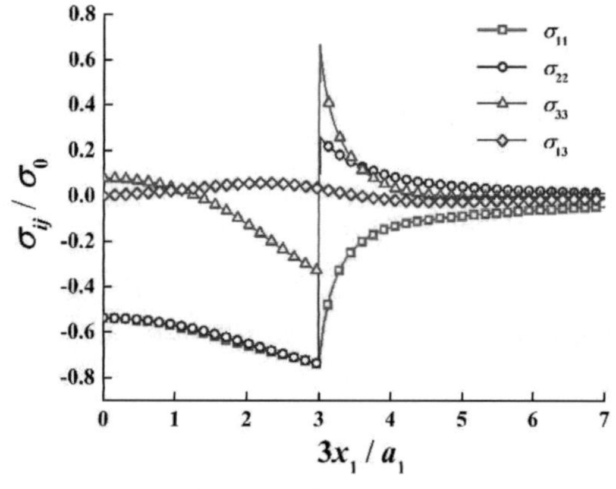

Figure 4. Stresses of a thermal spheroidal inclusion

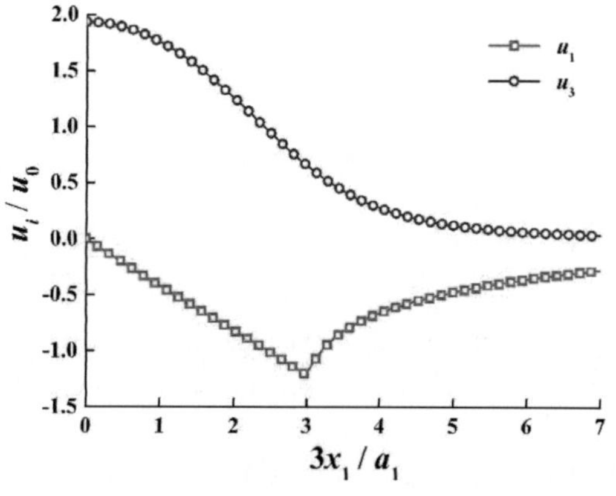

Figure 5. Displacements of a spheroidal inclusion subjected to the porous eigenstrain in a half-space

The results of the elastic field for the spheroidal inclusion subjected to the porous eigenstrain are studied, where the size of porous inclusion and depth location are identical to the case of thermal inclusion. In this work, all the displacement, strain and stress components are normalized respectively by $u_0 = -(1-2\nu)\alpha_B \Delta P / Ea_1$, $\varepsilon_0 = -(1-2\nu)\alpha_B \Delta P / E$ and $\sigma_0 = -\alpha_B \Delta P$. The corresponding elasticity solution with respect to the displacements, strains and stresses along the x_1 -direction are illustrated in Figures. 5-7 in sequence. It is noted that the prescribed thermal eigenstrain is caused by temperature change resulting in a localized heating, while the porous eigenstrain is produced by fluid migration leading to a decrease in pore pressure.

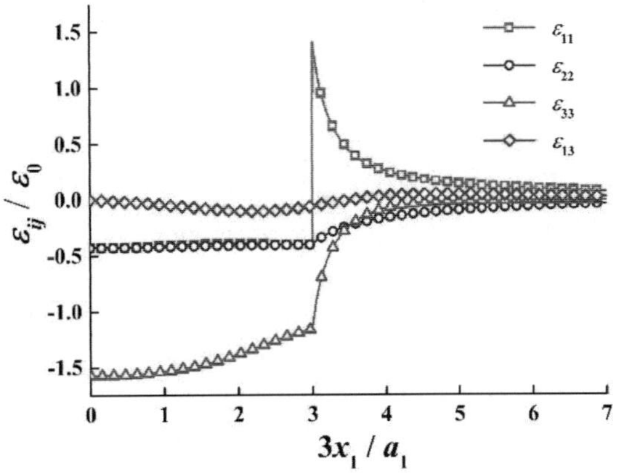

Figure 6. Strains of a spheroidal inclusion with porous eigenstrain near the surface

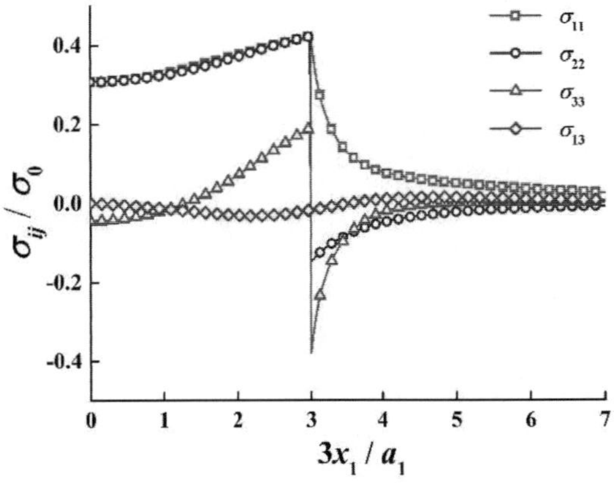

Figure 7. Stresses of a porous spheroidal inclusion in a semi-infinitely extended matrix

4. Concluding remarks

The current work investigates the solution of the complete elastic field for a spheroidal inclusion with thermo-porous eigenstrain vertically placed in a half-space. This complements our previous publication [11] on the horizontally aligned thermo-porous spheroidal inclusion in a half-space. The elastic fields for both the prolate and oblate spheroidal inclusions are derived based on the degenerate case of ellipsoidal inclusion near free surface. The present formulation of the displacements, strains and stresses are presented in closed-form, for convenience of engineering applications.

Jump conditions are discussed to interpret that the discontinuities are only depending on the terms without superscript "*". The effects of geometry for both oblate and prolate spheroidal inclusions are examined in response to the alteration of eigenstrains which include the thermal eigenstrain due to localized heating and porous eigenstran caused by fluid withdrawal leading to a decrease in pore pressure. It is demonstrated that the proposed inclusion model can be employed to predict the elastic fields of porous materials produced by thermal and porous eigenstains near a free surface.

5. References

[1] J. Eshelby, P Roy Soc A-Math Phy, **241**(1226), 376-396 (1957)
[2] J. Eshelby, P Roy Soc A-Math Phy, **252**(1271), 561-569 (1959)
[3] K. Zhou, H. Hoh, X. Wang, L. Keer, J. Pang, B. Song, Q. Wang, Mech Mater, **60**, 144-158 (2013)
[4] R. Mindlin, D. Cheng, J Appl Phys, **21**(9), 931-933 (1950)
[5] Y. Chiu, J Appl Mech-T ASME, **45**(2), 302-306 (1978)
[6] K. Seo, T. Mura, J Appl Mech-T ASME, **46**(3), 568-572 (1979)
[7] D. Healy, Comput Geosci, **35**(10), 2170-2173 (2009)
[8] A. Manoylov, F. Borodich, H. Evans, P Roy Soc A-Math Phy, **469**(2154), 20120689 (2013)
[9] X. Jin, D. Lyu, X. Zhang, Q. Zhou, Q. Wang, L. Keer, J Appl Mech-T ASME, **83**(12), 121010-12 (2016)
[10] D. Lyu, X. Zhang, P. Li, D. Luo, Y. Hu, X. Jin, L. Zhang, L. Keer, J Appl Mech-T ASME, **85**(5), 051005-8 (2018)
[11] X. Zhang, D. Lyu, P. Li, X. Jin, P. Liaw, L. Keer, Comput Geosci, **122**, 15-24 (2019)

IOP Conf. Series: Materials Science and Engineering **576** (2019) 012043 doi:10.1088/1757-899X/576/1/012043

Acknowledgments
This work is supported by Fundamental Research Funds for the Central Universities (2018CDYJSY0055 and 106112017CDJQJ328839), and the Graduate Research and Innovation Foundation of Chongqing, China (Nos. CYB17025 and CYB18020). X.J. would like to acknowledge the support from the National Natural Science Foundation of China (Nos. 51475057 and 51875059), and the State Key Laboratory of Mechanical Transmissions (SKLMT-ZZKT-2017M15).

IOP Conf. Series: Materials Science and Engineering **576** (2019) 012044 doi:10.1088/1757-899X/576/1/012044

Numerical study of natural convection in a rectangular cavity with variation of cavity aspect ratios and cavity inclination angles

U Prasopchingchana

Department of Mechanical Engineering, Burapha University, Chonburi, Thailand, 20131

Corresponding author: uthai@eng.buu.ac.th

Abstract. Effects of cavity aspect ratios and cavity inclination angles to natural convection in a rectangular cavity are numerically investigated. Investigation is performed at the Rayleigh number (*Ra*) equal to 10^4, the cavity aspect ratios from 1 to 50 and the cavity inclination angles from 0 to 180°. Consequently, Heat transfer enhancement or decreasing due to the effects is exposed. In addition, streamline contours in the rectangular cavity are illustrated. Multi-cellular flow figuring on the appropriate conditions is exhibited. A new correlation of the average Nusselt number, the cavity aspect ratio and the cavity inclination angle is formulated at *Ra* equal to 10^4.

1. Introduction

In many engineering applications relating to thermal energy, forced convection plays as an important role in operating conditions. But in critical conditions, natural convection will be more important for safety [1]. For example, when water circulating pumps of nuclear reactors do not work or ventilation fans of computers fail. In addition, operating or ventilation with natural convection is cost free. In recent decades, energy conservation has been emphasized. And many scientists have been investigating this phenomenon, because natural convection is a phenomenon that can affect both energy saving and energy consumption.

Natural convection in the gap of double glazed units is a case where the phenomenon can enhance or decrease heat gain from solar energy. To understand the causes of this phenomenon, the gap of double glazed units is assumed to be a rectangular cavity filled with air, and the simulation of air flow in the cavity is performed. At high cavity aspect ratios with appropriate Rayleigh numbers, the phenomenon of multi-cellular [2-7] air flow obviously appears. Moreover, many authors [2-26] performed the numerical simulations and experiments of natural convection in a cavity with the various ranges of cavity aspect ratios and cavity inclination angles.

Most [18, 20-23] of the authors mentioned in the foregoing paragraphs performed the numerical simulations of natural convection with the Boussinesq approximation. Although the density in the body force terms is a weak function of the pressure, the pressure still affects to the air density at the high difference of the pressure in the tall cavity.

The objective of this article is to numerically study on the effects of cavity aspect ratios and cavity inclination angles to the results of heat transfer through a rectangular cavity filled with real air, a

IOP Conf. Series: Materials Science and Engineering **576** (2019) 012044 doi:10.1088/1757-899X/576/1/012044

function of the temperature and pressure, which the ranges of the study cover greater than the former literature.

2. Problem details

Schematic details of the problem shown in Figure 1. are a two dimensional rectangular cavity filled with air, and the height and width of the cavity are h and b, respectively. The cavity aspect ratio is AR ($AR = h/b$). The temperatures of the left and right walls of the cavity are T_H and T_C, respectively, and the bottom and top walls of the cavity are adiabatic. Then, the velocities at the boundaries of the cavity are the no-slip condition. The Greek letter θ symbolizes the inclination angle of the cavity. Thermal radiation is excluded to the problem.

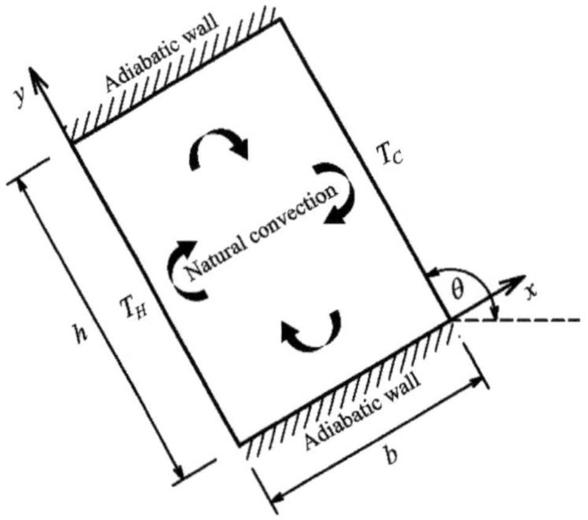

Figure 1. Schematic geometry of problem

3. Mathematical models

Natural convection in the cavity occurs by circulation of air. Since air flow in the cavity is quite slow, it is reasonable to assume that the flow is incompressible except the densities in the body force terms which induce the buoyancy force. The governing equations defining air flow in the cavity are the energy equation with the viscous dissipation term, the continuity equation and the momentum equations in the x and y directions. The governing equations can be expressed as:

$$\rho c_p\left(\frac{\partial T}{\partial t}+u\frac{\partial T}{\partial x}+v\frac{\partial T}{\partial y}\right)=k\left(\frac{\partial^2 T}{\partial x^2}+\frac{\partial^2 T}{\partial y^2}\right)+\mu\left(2\left(\left(\frac{\partial u}{\partial x}\right)^2+\left(\frac{\partial v}{\partial y}\right)^2\right)+\left(\frac{\partial u}{\partial y}+\frac{\partial v}{\partial x}\right)^2\right) \tag{1}$$

$$\frac{\partial u}{\partial x}+\frac{\partial v}{\partial y}=0 \tag{2}$$

$$\rho\left(\frac{\partial u}{\partial t}+u\frac{\partial u}{\partial x}+v\frac{\partial u}{\partial y}\right)=-\frac{\partial p}{\partial x}+\mu\left(\frac{\partial^2 u}{\partial x^2}+\frac{\partial^2 u}{\partial y^2}\right)+\rho_{i,j}g\cos\theta \tag{3}$$

$$\rho\left(\frac{\partial v}{\partial t}+u\frac{\partial v}{\partial x}+v\frac{\partial v}{\partial y}\right)=-\frac{\partial p}{\partial y}+\mu\left(\frac{\partial^2 v}{\partial x^2}+\frac{\partial^2 v}{\partial y^2}\right)-\rho_{i,j}g\sin\theta \tag{4}$$

where T, u, v and p are the temperature, velocity components in the horizontal and vertical directions and pressure, respectively. Then, ρ, c_p, k, μ and g are the density, specific heat at the constant pressure, thermal conductivity, viscosity and gravitational acceleration, respectively. And $\rho_{i,j}$ is the compressible density.

The air properties except the densities in the body force terms in the momentum equations are imposed from the average temperature between the hot and cold walls of the cavity.

The difference of the vertical wall temperatures causes the difference of the air densities in the body force terms in the momentum equations, and then the buoyancy force results in air circulation in the cavity. Heat transfer from the hot wall to the cold wall occurs due to air circulation called natural convection. The equation of the air densities in the body force terms is a function of the temperature and pressure of air which is obtained from polynomial curve fitting. The data of the air densities, temperatures, and pressures for curve fitting are obtained from the National Institute of Standards and Technology (NIST) Standard Reference Database 23, Version 9.0. The air density equation can be written as $\rho_{i,j} = \sum_{mp=1}^{mpden}\left(\left(\sum_{mt=1}^{mtden}\left(a_{mt,mp} T_{i,j}^{mt-1}\right)\right) p_{i,j}^{mp-1}\right)$ where $a_{mt,mp}$ are the coefficients of the equation, and *mtden* and *mpden* are equal to 6. The determination coefficient (R^2) of the equation is 0.999.

The final solution of the problem is the Nusselt number, a dimensionless parameter showing the performance of heat transfer through the cavity. The average Nusselt number of the cavity is calculated from [25] $\overline{Nu} = \frac{1}{h}\int_0^h \frac{\left(q''|_{x=0} + q''|_{x=b}\right)}{2q''_{ref}} dy$. And heat flux on the differentially heated walls of the cavity is obtained from $q'' = -k\frac{dT}{dx}$. Then, reference heat flux which air in the cavity is assumed to be as a solid matter is received from $q''_{ref} = k\frac{\left(T_H - T_C\right)}{b}$.

4. Numerical implementation

To achieve the purpose of the study, an in-house code was developed by using the FORTRAN computing language. The finite volume method with the LIP scheme [25] were adopted to discretize the values and the derivative values of the fluid variables on the cell faces and the cell centers in the governing equations. The SIMPLE algorithm was used to couple the continuity equation and the momentum equations to determine the pressure field of air flow. The transient condition approached to steady condition was employed to determine the solutions of the problem. The dimensionless time step of the code was 0.001. The convergent criterion for the code was that the maximum residual value of the dimensionless fluid variables at the present iteration and the previous iteration are less than or equal to 10^{-4} in ten time continuously. But there were some unstable cases [23], so that for these cases, the code was run until the dimensionless time was equal to 10,000. The Cartesian coordinate with the non-uniform grids were used. The fine grids were applied to the nearby area of the cavity walls, and the coarse grids were applied to the core area of the cavity. In addition, the grid independence test was simultaneously performed to ensure that the solutions did not have any variation due to the grid sizes.

5. Code validation

To ensure that the code gives correct solutions, code validation was carried out by comparison the solutions of the code to the benchmark and published numerical solutions [6, 8, 11-13, 15, 17, 21, 25, 26] and the experimental solutions [9] in the cases of natural convection in a square cavity and a tall cavity with vertical differentially heated walls. The results of the comparison have validated the code.

Table 1. displays comparison of the solutions computed from the code with the benchmark and published numerical solutions of natural convection in a square cavity at the Rayleigh number equal to 104. The maximum differences of the solutions computed from the code with the benchmark and published numerical solutions for the average Nusselt number $\left(\overline{Nu}\right)$, the maximum horizontal velocity component $\left(u_{max}^{**}\right)$ occurring on the vertical cavity centerline $\left(0.5, y^{**}\right)$ and the maximum vertical velocity component $\left(v_{max}^{**}\right)$ occurring on the horizontal cavity centerline $\left(x^{**},0.5\right)$ are 3.59%, 2.15% and 2.28%, respectively.

Table 2. shows comparison of the solution computed from the code with and the experimental solution and the published numerical solutions of natural convection in a rectangular cavity at the Rayleigh number and the aspect ratio equal to 1.1 × 10^4 and 16, respectively. The maximum differences of the average Nusselt number obtained from the code with the average Nusselt numbers obtained from the experimental and numerical implementation are 6.45% and 0.37%, respectively.

Figure 2. shows temperature and streamline contours of natural convection in a square cavity at the Rayleigh number equal to 10^4. Overview of Figure 2. is similar the temperature and streamline contours shown in [10].

Figure 3. displays temperature and streamline contours of natural convection in a rectangular cavity at the Rayleigh number and the aspect ratio equal to 1.1 × 10^4 and 16, respectively. The contour patterns of Figure 3. look like the patterns of the temperature and streamline contours shown in [6].

Table 1. Comparison of the solutions computed from the code with the benchmark and published numerical solutions of natural convection in a square cavity at $Ra = 10^4$.

Author	$\overline{Nu}$	$u_{max}^{**}\left(0.5,y^{**}\right)$	$v_{max}^{**}\left(x^{**},0.5\right)$
Present work	2.2488	15.8429 (0.5, 0.8093)	19.1875 (0.1254, 0.5)
[10]	2.243	16.178 (0.5, 0.823)	19.617 (0.119, 0.5)
[12]	2.2415	16.1838 (0.5, 0.8232)	19.6165 (0.1191, 0.5)
[11]	2.201	16.18 (0.5, 0.832)	19.44 (0.113, 0.5)
[13]	2.245	-	-
[15]	2.286	16.163 (0.5, 0.828)	19.569 (0.125, 0.5)
[17]	2.168	-	-
[25]	2.263	-	-
[26]	2.2448	16.1799	19.6254

Table 2. Comparison of the average Nusselt number obtained from the code with the average Nusselt numbers obtained from the experimental and numerical implementation of natural convection in a rectangular cavity at $Ra = 1.1 \times 10^4$ and $AR = 16$.

Author	$\overline{Nu}$	Implementation
Present work	1.5294	Numerical
[9]	1.4307	Experimental
[6]	1.525	Numerical
[21]	1.5351	Numerical

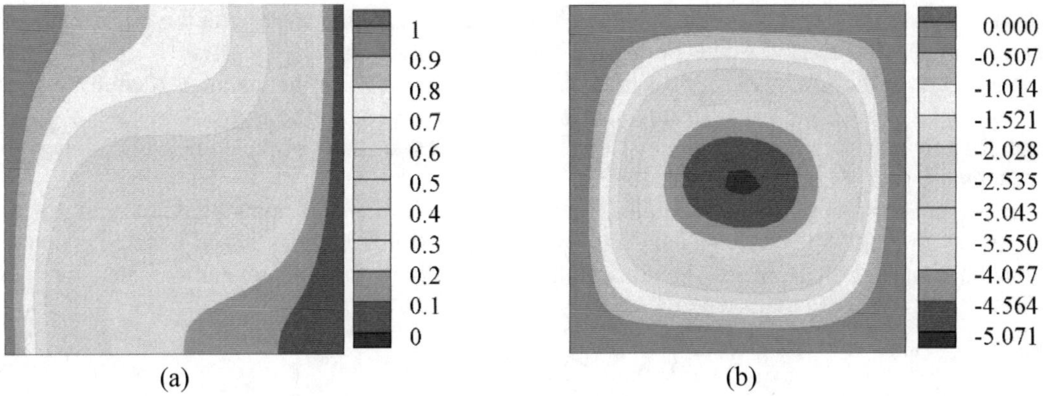

Figure 2. Temperature and streamline contours of natural convection in a square cavity at $Ra = 10^4$ (a) temperature contour and (b) streamline contour

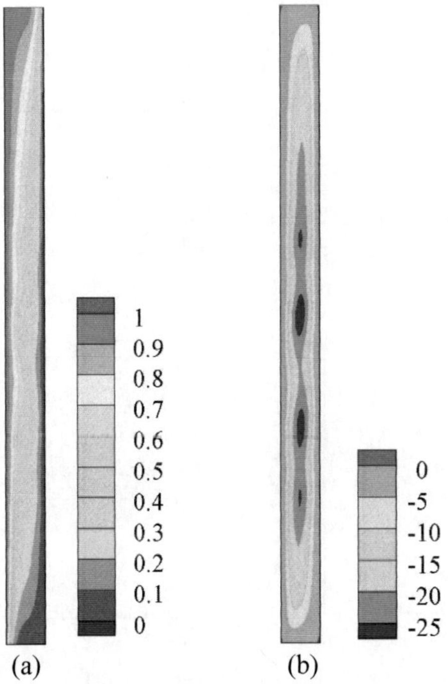

Figure 3. Temperature and streamline contours of natural convection in a rectangular cavity at $Ra = 1.1 \times 10^4$ and $AR = 16$ (a) temperature contour and (b) streamline contour

6. Results and discussion

Figure 4. and 5. display the variation of the local Nusselt numbers $\left(Nu = \frac{q''_{y^{**}}}{q''_{ref}} \right)$ on the hot and cold walls with the dimensionless height $\left(y^{**} = \frac{y}{b} \right)$ of the cavity at the different cavity aspect ratios and cavity inclination angles, and the Rayleigh number is equal to 10^4. The wavy graph lines shown in Figure 4.(b), 4.(c), 5.(b) and 5.(c) are appeared because of the multi-cellular effect. The wavy graph lines are more apparent, when the cavity aspect ratios and cavity inclination angles are higher. While the highest local Nusselt numbers occur at the bottom corner on the hot wall and top corner on the cold wall of the cavity, the lowest local Nusselt numbers occur at the top corner on the hot wall and bottom corner on the cold wall of the cavity

Figure 6. shows the variation of the average Nusselt numbers of the cavity with the different cavity inclination angles and cavity aspect ratios and the Rayleigh number is equal to 10^4. The average Nusselt numbers at the low cavity aspect ratios have higher values than the average Nusselt numbers at the high cavity aspect ratios, when the cavity inclination angles are between 0° to 150°. And the average Nusselt numbers at the different cavity aspect ratios are almost the same value, when the cavity inclination angles are greater than 150° to 180°.

Figure 7. and 8. show streamline contours in the cavity at the different cavity inclination angles and cavity aspect ratios, and the Rayleigh number is equal to 10^4. Multi-cellular obviously figure, when the cavity inclination angles become 60° to 180° and the cavity aspect ratios are greater than 30.

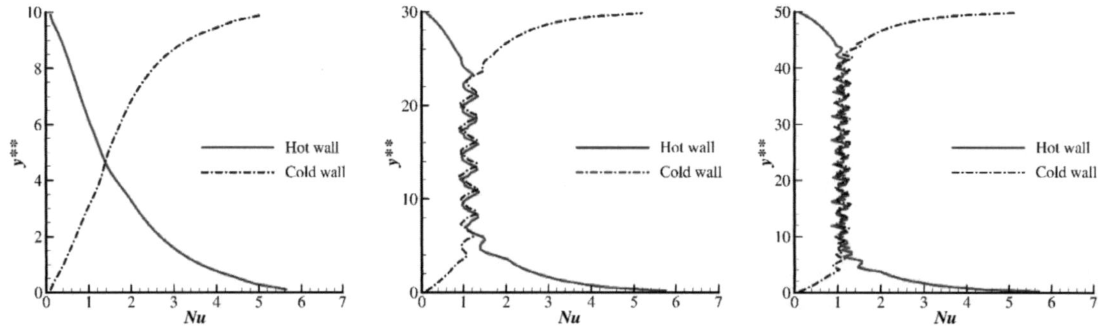

(a) (b) (c)

Figure 4. Variation of the local Nusselt numbers along the differential heated walls at $Ra = 10^4$, $\theta = 90°$ and the different cavity spect ratios (a) $AR = 10$, (b) $AR = 30$ and (c) $AR = 50$

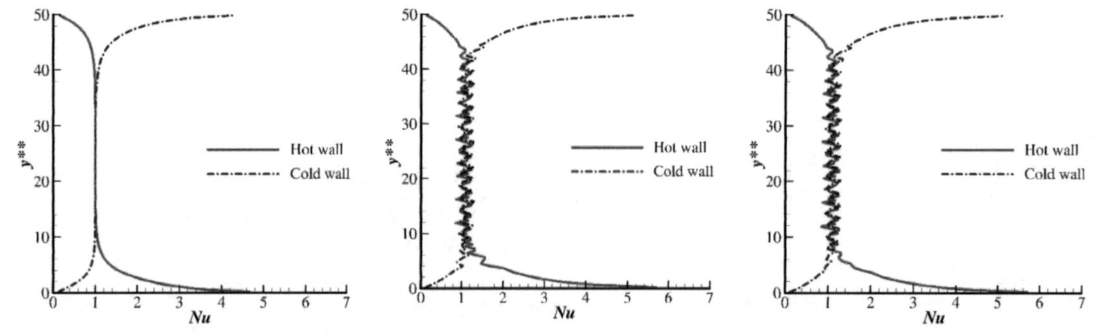

(a) (b) (c)

Figure 5. Variation of the local Nusselt numbers along the differential heated walls at $Ra = 10^4$, $AR = 50$ and the different cavity inclination angles (a) $\theta = 45°$, (b) $\theta = 90°$ and (c) $\theta = 135°$

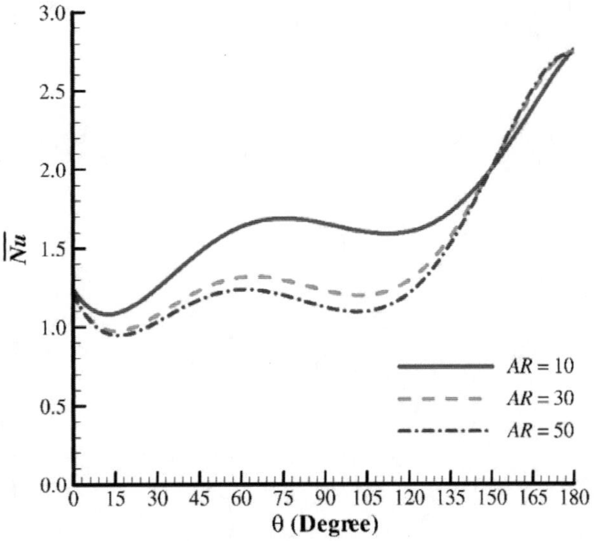

Figure 6. Variation of the average Nusselt numbers with the different cavity inclination angles and cavity aspect ratios at $Ra = 10^4$

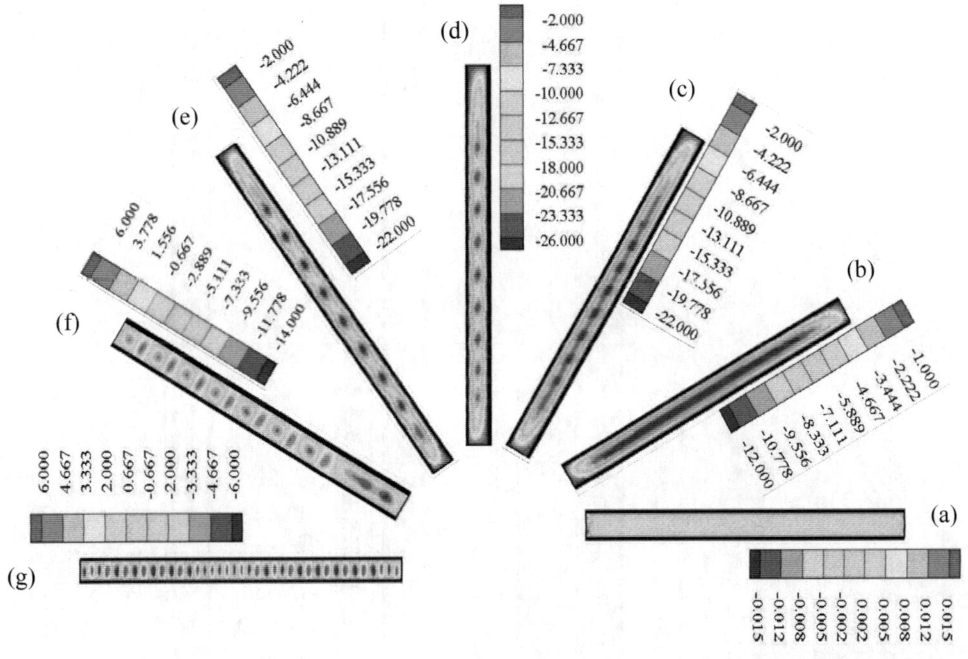

Figure 7. Streamline contours in the cavity at $Ra = 10^4$, $AR = 30$ and the different cavity inclination angles (not to scale) (a) $\theta = 0°$, (b) $\theta = 30°$, (c) $\theta = 60°$, (d) $\theta = 90°$, (e) $\theta = 120°$, (f) $\theta = 150°$, and (g). $\theta = 180°$

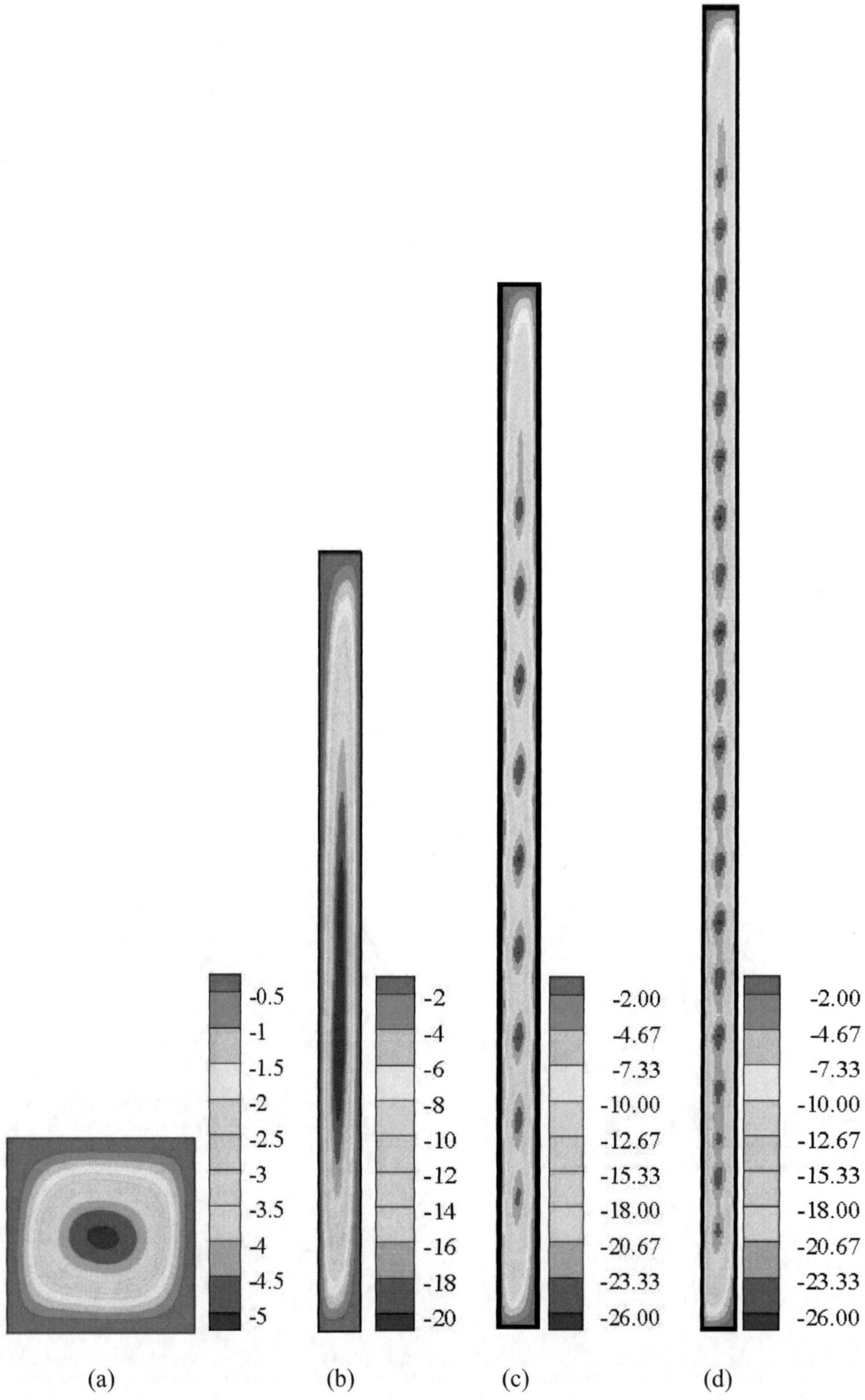

Figure 8. Streamline contours in the cavity at $Ra = 10^4$, $\theta = 90^{\circ}$ and the different cavity aspect ratios (not to scale) (a) $AR = 1$, (b) $AR = 10$, (c) $AR = 30$ and (d) $AR = 50$

7. Conclusion

From the simulation, a correlation of the average Nusselt number, the cavity aspect ratio and the cavity inclination angle is formulated in a power law form. The correlation can be written as $\overline{Nu} = 1.73367\,AR^{-0.07768}\left(1.0725+\cos\theta\right)^{-0.22963}$ at $1 \le AR \le 50$, $0^o \le \theta \le 180^o$ and $Ra = 10^4$. The coefficient of determination (R^2) of the correlation is 0.77. The 72 sets of data are employed for forming the correlation.

8. Acknowledgment

The author wishes to acknowledge Thai National e-Science Infrastructure Consortium for providing computing resources that have contributed to the research results reported within this paper.

9. References

[1] Baïri A, Zarco-Pernia E and Maria J-M G d 2014 *Appl. Therm. Eng.* **63** 304
[2] Lee Y and Korpela S A 1983 *J. Fluid Mech.* **126** 91
[3] Wakitani S 1998 *Phys. Fluids* **10** 1924
[4] Lartigue B, Lorente S and Bourret B 2000 *Int. J. Heat Mass Tran.* **43** 3157
[5] Kim S and Kim M C 2002 *KSME Int. J.* **16** 94
[6] Zhu Z J and Yang H X 2003 *Heat Mass Tran.* **39** 579
[7] Ganguli A A, Pandit A B and Joshi J B 2009 *Chem. Eng. Res. Des.* **87** 711
[8] Davis G D V 1968 *Int. J. Numer. Meth. Fluids* **11** 1675
[9] ElSherbiny S M, Raithby G D and Hollands K G T 1982 *J. Heat Transfer* **104** 96
[10] Davis G D V 1983 *Int. J. Numer. Meth. Fluids* **3** 249
[11] Markatos N C and Pericleous K A 1984 *Int. J. Heat Mass Tran.* **27** 755
[12] Saitoh T and Hirose K 1989 *Comput. Mech.* **4** 417
[13] Barakos G, Mitsoulis E and Assimacopoulos D 1994 *Int. J. Numer. Meth. Fluids* **18** 695
[14] Ismail K A R and Henriquez J R 2005 *Int. J. Heat Mass Tran.* **48** 461
[15] Dixit H N and Babu V 2006 *Int. J. Heat Mass Tran.* **49** 727
[16] Cheikh N B, Beya B B and Lili T 2007 *C. R. Mecanique* **335** 113
[17] Baïri A 2008 *Energ. Convers. Manag.* **49** 771
[18] Nicolas A, Baez E and Bermudez B 2011 *Phys. Lett. A* **375** 2683
[19] Cooper D, Craft T J, Esteifi K and Iacovides H 2012 *Int. J. Heat Mass Tran.* **55** 6321
[20] Saeedi M and Oskouie S N 2012 *World Appl. Sci. J.* **17** 882
[21] Báez E and Nicolás A 2013 *Phys. Lett. A* **377** 2270
[22] Cheong H T, Siri Z and Sivasankaran S 2013 *Int. J. Heat Mass Tran.* **45** 75
[23] Omranian A, Craft T J and Iacovides H 2014 *Int. J. Heat Mass Tran.* **77** 1
[24] Alloui Z and Vasseur P 2015 *Int. J. Heat Mass Tran.* **84** 912
[25] Prasopchingchana U and Manewattana T 2015 *Eng. J.* **19** 133
[26] Zhao B and Tian Z 2016 *Int. J. Heat Mass Tran.* **98** 313

On the Inclusion Model of Localized Heating in an Elastic Plane

P Li[1], X Jin[1,2], X Zhang[1], D Luo[1],and Y An[1]

[1]State Key Laboratory of Mechanical Transmissions, Chongqing University, Chongqing 400030, China
[2]College of Aerospace Engineering, Chongqing University, Chongqing, 400030, China

Corresponding Author's E-mail: jinxq@cqu.edu.cn

Abstract. Failure localization in a variety of mechanical structures may be ascribed to elevated temperature, which may be effectively analyzed by employing the inclusion model. This work presents an explicit solution to the plane thermal inclusion problem, based on the customized Green's function. A contour integral representation is further developed so as to provide an effective and straightforward approach for treating an arbitrarily shaped inclusion. Several benchmark examples are examined to validate the present solution.

1. Introduction

A common challenge in engineering design and mechanical manufacturing is to understand the failure mechanisms due to elevated temperature in a localized concentration. In micromechanics, such problem may be effectively analyzed by employing the inclusion model [1, 2]. The inclusion here refers to a subdomain with identical material as the surrounding matrix, and the thermal expansion strain can be regarded as eigenstrain [2] which is a generic name for nonelastic strains such as plastic strains, misfit strains and phase transformation. Other typical eigenstrains in engineering material may exist in the forms of precipitates, twinnings and martensitic transformations [3-5], to name a few. The mechanical behavior of materials may be seriously affected especially when these inclusions undergo a localized elevated temperature.

In literature, most of the previous thermal inclusion studies are mainly focused on the strain/stress solutions. Rodin et al. [6] derived an algorithmic closed-form solution for the polygonal and polyhedral inclusions. Downes and Faux [7] obtained a unified Green's function for interior and exterior stress fields and then proposed a simple and straightforward method [8] to evaluate the stress and strain distributions. Nozaki and Taya [9, 10] investigated the elastic fields for a polygonal or polyhedral inclusion. Jin et al. [11] developed Green's functions for a point eigenstrain in an infinite plane and then demonstrated their applications in evaluating the interior and exterior stress fields. Trotta et al. [4, 5] developed the analytical formulae of the Eshelby tensor for an arbitrary polygonal inclusion. Xie et al. [12] reported the study of the inclusion experiment, which however is beyond the scope of this research.

In this work, an explicit solution to the displacement field caused by a point thermal eigenstrain excitation is derived in closed-form, and then utilized to evaluate the displacement field in an infinite plane. Based on the method of Green's function, the interior and exterior displacement is represented in a unified area integral. With the assistance of contour integral, analytical solutions are derived for arbitrary polygonal inclusion subjected to uniform thermal eigenstrains. Benchmark examples are provided to validate the effectiveness of the present method of solution. The present method provides an effective method to evaluate the displacement field produced by randomly dispersed inclusions in an infinite plane (Figure. 1).

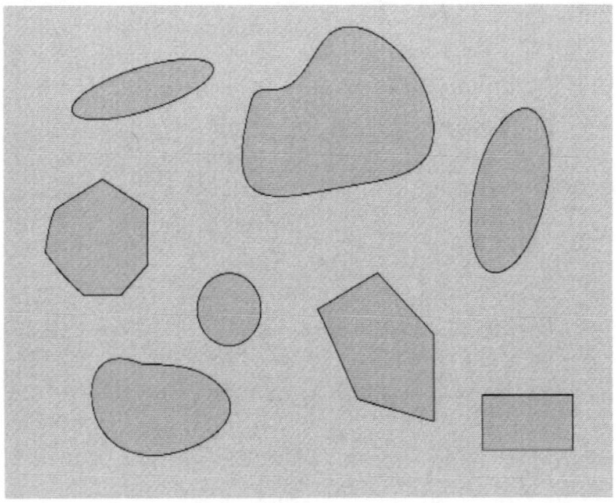

Figure 1. Schematic illustration of the randomly dispersed inclusions in an infinite plane.

2. Formulation

2.1. Unit point eigenstrain solution

Consider a point eigenstrain excitation located at the source point (ξ,η) in an infinitely extended isotropic medium (Figure. 2). Jin et al. [11] has obtained Green's functions for the stresses at a field point (x, y) in an infinite plane. The methodology may also be applied to solve the displacement in this study.

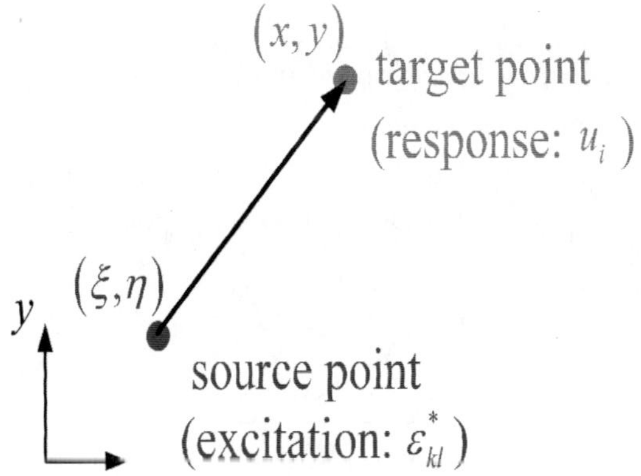

Figure 2. Excitation-response of a point eigenstrain.

Firstly, the analytical solution to the displacement field associated with a uniform rectangular eigenstrain ε^* may be obtained by superposing the contributions from the primitive functions.

$$u(x, y;\xi,\eta) = \varepsilon^* \sum_{\alpha=1}^{2}\sum_{\beta=1}^{2}(-1)^{\alpha+\beta} w(x_\alpha, y_\beta) \tag{1}$$

where $w\left(x_\alpha, y_\beta\right)$ is the primitive function and may be determined from [13]. The excitation at a source point may be obtained using a limit process [11] when the dimensions of side lengths $\Delta x \times \Delta y$ approach to zero.

$$\varepsilon^* = \frac{1}{\Delta x \Delta y} \tag{2}$$

It is worthwhile to note that the eigenstrain density at Δx=0, $\Delta y = 0$ may be regarded as the limit of distributed eigenstrains [11, 14]. In other words, the limit in Eq. (2) may be expressed as the delta function [14]. The point eigenstrain solution may be consequently determined as

$$G\left(x, y; \xi, \eta\right) = \lim_{\Delta x, \Delta y \to 0} \frac{1}{\Delta x \Delta y} \sum_{\alpha=1}^{2} \sum_{\beta=1}^{2} \left(-1\right)^{\alpha+\beta} w\left(x_\alpha, y_\beta\right) \tag{3}$$

In view of the definition of the partial derivatives, Eq. (3) may be reduced to

$$G\left(x, y; \xi, \eta\right) = w_{,xy}\left(x - \xi, y - \eta\right) \tag{4}$$

The corresponding displacement caused by the aforementioned point eigenstrain excitation may be expressed in a tensorial form as:

$$u_i(x, y) = G_{ikl}(x, y; \xi, \eta) \varepsilon_{kl}^*(\xi, \eta) \tag{5}$$

where $G_{ikl}(x, y; \xi, \eta)$ is termed as Green's function, and is only associated with the relative position between the field point and the location of the source excitation,

$$G_{ij}(x, y; \xi, \eta) = G_{ij}(x - \xi; y - \eta) \tag{6}$$

For an elastic isotropic material, the eigenstrain prescribed in the source point (ξ, η) may be represented as [15]

$$\varepsilon_{ij}^* = \lambda_T \left(T - T_0\right) \delta_{ij} \tag{7}$$

where λ_T is the coefficient of linear thermal expansion; T and T_0 are the temperatures of the current and the initial state; δ_{ij} is the Kronecker delta.

When the eigenstrain is uniformly distributed in an inclusion, i.e. $\varepsilon_{11}^* = \varepsilon_{22}^* = \varepsilon^*$, $\varepsilon_{12}^* = 0$, the displacement components for the present thermal case are formulated as:

$$\begin{cases} u_1 = \dfrac{2\varepsilon^*}{\pi\left(1+\kappa\right)} \dfrac{X}{X^2 + Y^2} \\ u_2 = \dfrac{2\varepsilon^*}{\pi\left(1+\kappa\right)} \dfrac{Y}{X^2 + Y^2} \end{cases} \tag{8}$$

where κ is Kolosov's constant,

$$\begin{cases} \kappa = 3 - 4\nu, & \text{in plane strain} \\ \kappa = \dfrac{3-\nu}{1+\nu}, & \text{in plane stress} \end{cases} \tag{9}$$

with ν being Poisson's ratio, and X, Y depending on the relative position between the response and excitation points, i.e.

$$X = x - \xi;\ Y = y - \eta \tag{10}$$

2.2. Contour integral representation

In view of the method of Green's function, the displacement for an arbitrarily shaped inclusion Ω subjected to any distributed eigenstrain may be evaluated as

$$\begin{bmatrix} u_1 \\ u_2 \end{bmatrix} = \iint_\Omega \begin{bmatrix} G_{111} & G_{122} & G_{112} \\ G_{211} & G_{222} & G_{212} \end{bmatrix} \begin{bmatrix} \varepsilon_{11}^* \\ \varepsilon_{22}^* \\ \varepsilon_{12}^* \end{bmatrix} d\xi d\eta \tag{11}$$

It should be emphasized that the interior and exterior displacement fields can be determined simultaneously by Eq. (11). Moreover, since the eigenstrain is uniformly distributed in the present thermal case and $\varepsilon_{11}^*=\varepsilon_{22}^*=\varepsilon^*$, $\varepsilon_{12}^*=0$, the term ε^* may be moved outside the integral sign, leading to the area integral of the point Green's function. The resulting expression for the current case may be written as

$$u_i = \varepsilon^* \iint_\Omega G_i d\xi d\eta \tag{12}$$

where G_i is the point eigenstrain solution for the thermal case, and their details are referred to Eq. (8). By applying the Green's theorem over Eq. (12), the area integral may be converted to a contour integral along the boundary $\partial\Omega$ of the inclusion domain.

$$\iint_\Omega G_i d\xi d\eta = \oint_{\partial\Omega} K_i d\xi \qquad (i=1,2) \tag{13}$$

where $\partial\Omega$ is a closed curve and oriented in the positive direction with the kernel function K_i being derived as

$$\begin{cases} K_1 = -4\arctan\dfrac{Y}{X} \\ K_2 = -2\ln\left(X^2+Y^2\right) \end{cases} \tag{14}$$

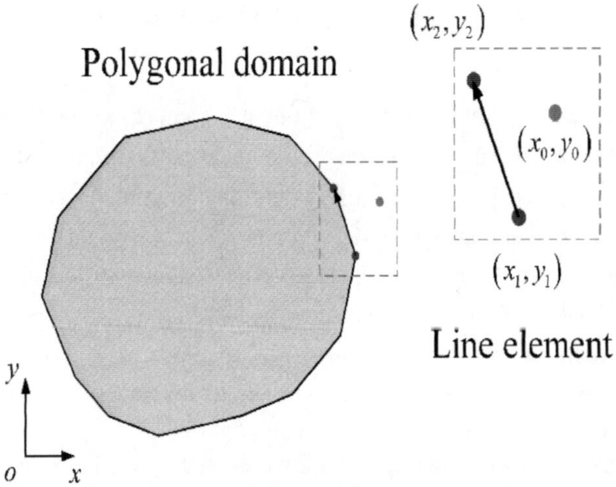

Figure 3. Schematic illustration of a polygonal domain and a line element.

A typical line element having endpoints (x_1, y_1) and (x_2, y_2) constituting a part of the closed boundary curve is shown in Figure. 3. The field point is represented as (x_0, y_0) and the excitation is distributed in the polygonal domain. The elementary solution may be solved in closed-form, following the strategy proposed by Downes and Faux [7]. Therefore, for a polygonal inclusion of arbitrary shape, the resultant displacement may be obtained by summing the contributions from each line element.

3. Results and discussions

Based on the contour integral solution, a computer code is programed using the FORTRAN language with double precision calculation. The rectangular inclusion associated with the thermal expansion is examined to validate the present solution. The rectangle having dimensions $\Delta x \times \Delta y$ and centered at the origin of the Cartesian coordinate systems is demonstrated in Figure. 4.

In the benchmark study, the thermal eigenstrain is chosen as $\varepsilon_{11}^{*}=\varepsilon_{22}^{*}=\varepsilon_{0}$, $\varepsilon_{12}^{*}=0$ and the side lengths are set to be $\Delta x = \Delta y$. Moreover, $\nu = 0.3$ and a vertical target line is placed at $x = \Delta x/4$. The exact displacement may be obtained from [13] by considering the degenerate 3D case of a cuboidal inclusion. Comparisons of the normalized displacement components along the vertical line are illustrated in Figure. 5, and the solution of the present method perfectly agrees with the exact solution.

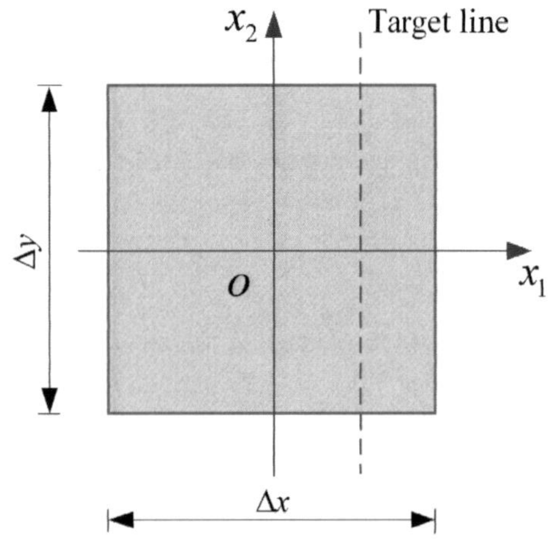

Figure 4. Schematic illustration of the rectangular inclusion and the locations of the target line.

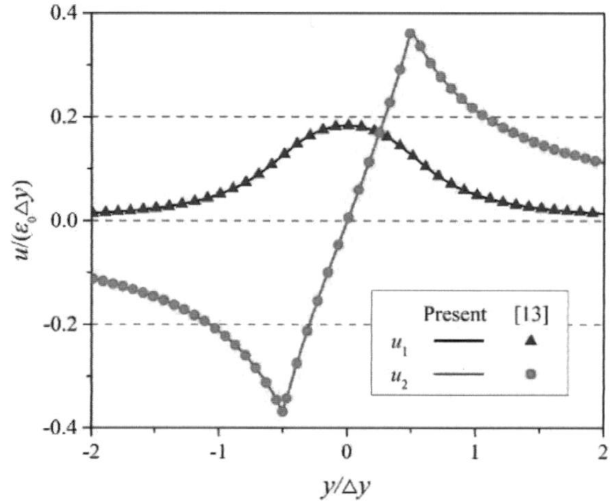

Figure 5. Verification of the present analytical solutions along the target line with the results in reference [13] for a rectangular thermal inclusion.

As shown in Figure. 5, the displacement component u_1 increase moderately from a minimum value at $(\Delta x/4, -2\Delta y)$ outside the cuboid and reaches its peak at the center of the inclusion, and then decreases as $y/\Delta y$ becomes larger. It can be found that the component u_1 is symmetric with respect to the y-axis. For the displacement component u_2, it undergoes a fast decline and arrives at the turning point when the field point varies along the target line with $y/\Delta y$ ranging from $-2\Delta y$ to $\Delta y/2$. It can be readily observed that the component u_2 is linearly distributed within the cuboidal inclusion. In contrast with u_1, the displacement component u_2 is antisymmetric with respect to the y-axis. Furthermore, it may be verified from the calculation that the displacement components u_1 and u_2 are continuous across the interface between the inclusion and the matrix.

It should be noted that the current work provides a closed-form analytical solution for arbitrary polygonal inclusions. Since the geometrical boundary surrounded by closed curves can be approximated by polygonal shape, the present solution is also applicable to derive a highly accurate numerical solution for thermal inclusions of arbitrary shape. In order to demonstrate the effectiveness of the present method for the inclusion bounded by arbitrary closed curves, numerical benchmark for a circular dilatational expansion is performed (Figure. 6).

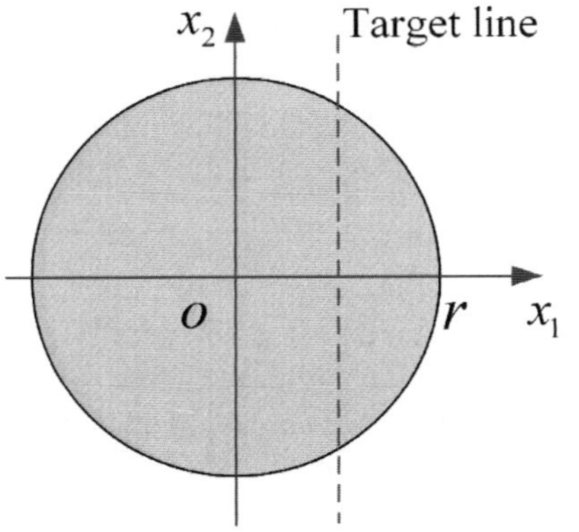

Figure 6. Schematic illustration of the circular inclusion and the target line.

In this work, the circular inclusion is centered at $(0,0)$ with radius r and the target line is $x = 2/r$. The circular inclusion domain is approximated by a regular polygon of 1000 sides. The problem has a closed form solution [16], which is compared with the present numerical results in Figure. 7. It is demonstrated that the present solution shows satisfactory agreement with the exact solution.

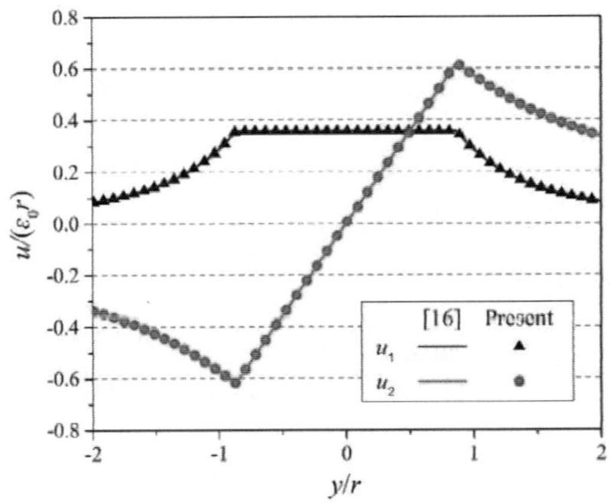

Figure 7. Verification of the normalized displacements along the target line for a circular thermal inclusion.

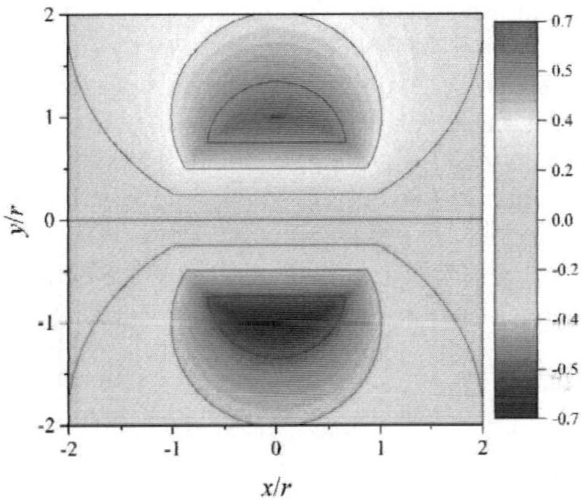

Figure 8. Contour plot of the normalized displacement component for a regular polygonal thermal inclusion with 1000 sides.

As demonstrated in Figure. 7, the pure dilatational strains in this work cause positive displacement component u_1 along the target line. When the dimensionless distance y/r varies along the vertical line from $-2\Delta y$ to the origin of the Cartesian coordinates, the displacement component u_1 increases gradually and tends toward a constant value inside the inclusion domain. Also, with increasing values of y/r, the component u_2 drops rapidly from the ending point $(\Delta x/4, -2\Delta y)$ to the interface, and then exhibits a linear distribution inside the inclusion domain. The corresponding contour plot of the normalized displacement component $u_2/(\varepsilon_0 r)$ for a regular polygonal thermal inclusion with 1000 sides is demonstrated in Figure. 8.

It is worthwhile to note that the total strain is compatible and may be determined from the aforementioned displacement components. Since the above results inside the thermal domain are either constant or linear, the induced strains are distributed uniformly in the inclusion. Once the strains are obtained, the stresses can be achieved directly by taking advantage of Hook's law. Therefore, the

IOP Conf. Series: Materials Science and Engineering **576** (2019) 012045 doi:10.1088/1757-899X/576/1/012045

stress field is also uniform inside the circular domain. This interesting phenomenon was firstly found in Eshelby's pioneering work [1], where the induced strains inside the ellipsoidal inclusion are uniform when a uniform eigenstrain is distributed in an ellipsoidal region contained in an infinitely extended isotropic medium. The results in this study immediately verify this remarkable property.

4. Conclusions

The quantitative determination of the displacement may be of significance in many engineering designs, and the problem may be of practical interest in the presence of the localized heating. However, the analytical solution for the displacements produced by an arbitrarily shaped inclusion has not been fully examined previously. Based on the inclusion model, this work presents an effective, elegant and versatile approach to solve the thermal displacement field in an infinite plane. The point eigenstrain excitation is analyzed and then employed as Green's function for evaluating the response field. The area integral for the present study may be converted to a contour integral by virtue of the Green's theorem. For any polygonal inclusion subjected to uniform thermal eigenstrain, the resultant solution may be derived in closed-form via the method of contour integral. The examples of the rectangular and circular thermal inclusions are examined to illustrate the effectiveness of the present solution.

5. References

[1] Eshelby J D, P. Roy. Soc. Lond. Ser. A, **241**(1226), 376-396 (1957)
[2] Mura T, *Micromechanics of defects in solids. 2nd ed.* (Springer, Dordrecht, The Netherlands, 1987)
[3] Nozaki H, Horibe T, Taya M, JSME Int. J. Ser. A, **44**(4), 472-482 (2001)
[4] Trotta S, Marmo F, Rosati L, *Compos. Part B- Eng.*, **106**, 48-58 (2016)
[5] Trotta S, Marmo F, Rosati L, *Compos. Part B- Eng.*, **115**, 170-181 (2017)
[6] Rodin G J, J. Mech. Phys. Solids, **44**(12), 1977-1995 (1996)
[7] Downes J, D. Faux A, J. appl. Phys., **77**(6): 2444-2447 (1995)
[8] Downes J R, Faux D A, E. P. O'Reilly, J. appl. Phys., **81**(10), 6700-6702 (1997)
[9] Nozaki H, Taya M, J. appl. Phys., **64**(3), 495-502 (1997)
[10] Nozaki H, Taya M, J. appl. Phys., **68**(3), 441-452 (2001)
[11] Jin X, Keer L M, Wang Q, Int. J. Solids Struct., **46**(21), 3788-3798 (2009)
[12] Xie L, Zhou Q, Jin X, Wang Z, Jiang C, Lu W, Wang J, Wang Q, Int. J. Fatigue, **66**(Supplement C), 127-137 (2014)
[13] Liu S, Jin X, Wang Z, Keer L M, Wang Q, Int. J. Plasticity, **35**, 135-154 (2012)
[14] Stakgold I, *Green's Functions and Boundary Value Problems. 3rd ed.* (Springer, Wiley, New York, 2011)
[15] Segall P, Fitzgerald S D, Tectonophysics, **289**(1), 117-128 (1998)
[16] Jin X, Zhang X, Li P, Xu Z, Hu Y, Keer L M, J. Appl. Mech.- T. ASME, **84**(7), 074501-074501-6 (2017)

Acknowledgments

This work is supported by Fundamental Research Funds for the Central Universities (2018CDYJSY0055). P.L. and X.Z. are grateful to the Graduate Research and Innovation Foundation of Chongqing, China (Grant Nos. CYB18020 and CYB17025). X.J. would acknowledge the support from Fundamental Research Funds for the Central Universities (No. 106112017CDJQJ328839), the National Natural Science Foundation of China (Grant Nos. 51475057 and 51875059), and the State Key Laboratory of Mechanical Transmissions through funding (SKLMT-ZZKT-2017M15).

IOP Conf. Series: Materials Science and Engineering **576** (2019) 012046 doi:10.1088/1757-899X/576/1/012046

Enhancement of critical heat flux under subcooled boiling of deionized water in a curved flow channel

Rasool Urfa[1], Daogang Lu[1], Noor Farooq Asran[2], Ansari Munib[1] and Rasool Aazim[1]

Beijing Key Laboratory of Passive Safety Technology for Nuclear Energy,
[1]North China Electric Power University, Beijing 102206, China,
[2]Pakistan Institute of Engineering and Applied Sciences.

Email: urfarasool@ncepu.edu.cn

Abstract-The numerical calculations are performed to predict critical heat flux (CHF) during subcooled boiling of deionized water through 90° curved flow channel with square cross-section. Two types of flow regimes have been observed at low and high flow velocities. In case of low flow velocities, the continuous accumulation of vapors formed a vapor blanket on heating surface which consequently reduced heat dissipation out of the surface and triggered CHF due to drying out of liquid between vapor blanket and heating wall. On the other hand, coolant circulation at higher velocities hindered the formation of vapor blanket and allowed subcooled liquid for rewetting. The consistent rewetting led the formation of small and separate vapor blankets. The enhancement in CHF has been observed with increased degree of subcooling and by increasing flow velocity in present study.

1. Introduction

The understanding of subcooled boiling heat transfer with respect to the flow rate is crucial to examine the phenomena of CHF in a channel. The convection boiling in the channel under subcooled diabatic condition has been used in search of effective cooling in worldwide applications especially in high power electronic devices and in nuclear reactors. Various researchers have proposed correlations for the determination of CHF on channel's wall by varying flowrates. Authors observed that the flow velocity plays a significant role in affecting the start of boiling superheat and the nucleate boiling heat transfer till CHF. Therefore, it is necessary to investigate the accurate measurement of subcooled boiling heat transfer in view of fluid velocity to clarify CHF mechanism[1, 2]. According to Kutateladze[3] and Zuber[4] the occurrence of CHF is the result of hydrodynamic instability.The Katto et al.[5]investigated that the boiling in a small heated flow channel is somehow same as that of the external flow. They proposed a CHF correlation for external flow by modifying Kutateladze equation in eq. (1) as,

$$\frac{q_m}{\rho_v h_{vl} u} = \left[\frac{\rho_v}{\rho_l}, \frac{\rho_l u^2 d}{\sigma}, \frac{g(\rho_l - \rho_v)d}{\rho_l u^2}, \frac{\mu_v}{\mu_l}, \frac{\rho_l u d}{\mu_l}\right] \quad (1)$$

The d and U are the characteristic length and velocity of flow boiling condition. It had been reported in literature that the physical peculiar of CHF at low flow velocities is closer to the pool boiling [6-12]. By ignoring the viscosity and gravity for high velocity case, the aforementioned equation had been suggested for low flow rates and high rates in eq. (2) and eq. (3) respectively.

$$\frac{q_m}{\rho_v h_{lv}}\left[\frac{\sigma g(\rho_l-\rho_v)}{\rho_v}\right]^{-\frac{1}{4}} = f(\frac{\rho_v}{\rho_l}, d', u') \quad (2)$$

$$\frac{q_m}{G h_{vl}} = f\left[\frac{\rho_v}{\rho_l}, \frac{G^2 d}{\rho_l \sigma}\right] \quad (3)$$

Whereas, $d' = d\left[\frac{g(\rho_l-\rho_v)}{\sigma}\right]^{-\frac{1}{2}}; u' = u\left[\frac{\sigma g(\rho_l-\rho_v)}{\rho_v^2}\right]^{-\frac{1}{4}}$

The study of CHF for saturated flow boiling conducted by Katto and Kurata[13] with water and R-113 as working fluid and by considering fluid velocity from 1.25-10ms^{-1} and from 0.5-12.5ms^{-1} as presented in eq. (4) and eq. (5) respectively,

$$\frac{q_m}{G h_{vl}} = 0.1869\left(\frac{\rho_v}{\rho_l}\right)^{0.559}\left(\frac{\sigma \rho_l}{G^2 L}\right)^{0.264} \quad (4)$$

$$\frac{q_m}{G h_{vl}} = 0.66\left(\frac{\rho_v}{\rho_l}\right)^{0.604}\left(\frac{\sigma \rho_l}{G^2 L}\right)^{0.415} \quad (5)$$

For pool boiling and low fluid velocity it has been reported that the CHF is directly proportional to the degree of subcooling[14, 15] whereas for flow boiling with higher velocity the nonlinear relationship with CHF has been observed[6, 16]

The objective of present study is to measure subcooled pool boiling heat transfer by varying fluid velocity and to predict CHF in the curved flow channel numerically by using CFD two-phase (liquid-vapor) boiling model. The constant heat flux is supplied perpendicular to the convex wall of the channel while deionized water as a coolant is allowed to flow through the channel. The specifications of flow channel and boundary conditions are enlisted in Table 1.

Table 1. Specifications of curved channel and boundary conditions

Inlet gauge pressure, P_1	24.50	[kPa]
Outlet gauge pressure, P_2	21.16	[kPa]
Inlet velocity, v_l	0.001,0.005,0.010	[ms^{-1}]
Inlet temperature, T_i	293, 353	[K]
Heat flux, Q	1500	[kWm^{-2}]
Hydraulic diameter, d_h	0.040	[m]
Bend angle of channel, θ	90°	[degrees]
Radius of curvature	0.368	[m]

2. Mathematical modeling for numerical solution

There are several different mechanisms in which heat is transferred from hot body to the surrounding fluid i.e. in boiling. At the part of the heating surface where no bubble inhabits, the heat flows to the subcooled liquid directly like a single-phase flow. The part of the heating surface where the generation of vapors occurs at the nucleation sites extracts heat out of the surface. With the escaping of bubbles, the liquid mixing mechanism exists. The cold liquid from the bulk comes into contact with the heating surface as a result of recirculation around the detaching bubble, this phenomenon conducts additional cooling. In terms of two sets of conservation equations, governing the balance of mass, momentum, and energy of each phase the two-phase subcooled boiling model has been established. In governing equations, the interaction terms couple the interface transfer of mass, momentum, and energy. The water-liquid is taken as continuous phase whereas water-vapor is illustrated as a dispersed or secondary phase. The k-Ɛ turbulence model[17] in a square cross section with hydraulic diameter of 0.040m and radius of curvature of 0.368m has been numerically solved for boiling water with heat flux of 1500kWm^{-2} by incorporating CFD code.

Conservation of mass,

$$\nabla.(\rho_m u_m) = 0 \quad (6)$$

Mass average velocity (u_m),

$$u_m = \sum_{k=1}^{2} \frac{\alpha_k \rho_k u_k}{\rho_m} \quad (7)$$

Mixture density (ρ_m),

$$\rho_m = \sum_{k=1}^{2} \alpha_k \rho_k \quad (8)$$

Conservation of momentum[18],

$$\nabla.(\rho_m u_m u_m) = -\nabla p + \nabla[\mu_m(\nabla u_m + \nabla u_m^T)] + \rho_m g + \nabla(\sum_{k=1}^{2} \alpha_k \rho_k u_{dr,k} u_{dr,k}) \quad (9)$$

The viscosity of mixture (μ_m),

$$\mu_m = \sum_{k=1}^{2} \alpha_k \mu_k \quad (10)$$

The drift velocity ($u_{dr,k}$) for k-phase,

$$u_{dr,k} = u_k - u_m \quad (11)$$

The slip velocity (u_{vl}),

$$u_{vl} = u_v - u_l \quad (12)$$

The drift velocity is correlated with the slip velocity as follow,

$$u_{dr,k} = u_{vl} - \sum_{k=1}^{2} \frac{\alpha_k \rho_k}{\rho_m} u_{lk} \quad (13)$$

The relative velocity consisting of diffusion term as a result of dispersion by the turbulent [19]:

$$u_{vl} = \frac{(\rho_v - \rho_m) d_v^2 a}{18 \mu_{leff} f_{drag}} - \frac{v_{t,m}}{\alpha_v \sigma_D} \nabla \alpha_l \quad (14)$$

Where $v_{t,m}$ is the turbulent viscosity of mixthe ture, σ_D , is the Prandtl dispersion coefficient. The drag force f_{drag} , is considered as[20]:

$$f_{drag} = \begin{cases} 1 + 0.15 Re_v^{0.687} & Re_v \leq 1000 \\ 0.0183 Re_v & Re_v > 1000 \end{cases} \quad (15)$$

The Reynolds number,

$$Re_v = \frac{u_m d_v}{v_m} \quad (16)$$

The acceleration (a),

$$a = g - (u_m \nabla) u_m \quad (17)$$

The energy equation for the mixture is as follows;

$$\nabla.\sum_{k=1}^{2}(\alpha_k u_k \rho_k h_k) = \nabla.(k_{eff} \nabla T) + Q_w \quad (18)$$

$$k_{eff} = [\sum_{k=1}^{2} \alpha_k (k_k + k_t)] \quad (19)$$

The effective conductivity (k_{eff}) and the turbulent thermal conductivity (k_t) are related as;

$$k_t = \frac{C_P \mu_t}{Pr_t} \quad (20)$$

The heat source i.e. the wall heat flux $Q_w$which is equivalent to the latent heat required for phase change. C_P is specific heat capacity.By incorporating continuity equation for the phases, the void fraction for dispersed phase i.e. vapors is computed as;

$$\nabla.(\alpha_v \rho_v v_m) = -\nabla.(\alpha_v \rho_v v_{dr,v}) + \sum_{k=1}^{2}(m_{lv} - m_{vl}) \quad (21)$$

'm', is the mass transfer through evaporation and condensation. The turbulence model (k-ε) has been used in this investigation due to complex modeling in the multiphase simulation. It is mainly effectual for turbulent core flows. The equations for turbulent kinetic energy and dissipation rate are represented as;

$$\nabla.(\rho_m u_m k) = \nabla.\left(\frac{\mu_{t,m} \nabla k}{\sigma_k}\right) + G_{k,m} - \rho_m \epsilon \quad (22)$$

$$\nabla.(\rho_m u_m \epsilon) = \nabla.\left(\frac{\mu_{t,m} \nabla \epsilon}{\sigma_\epsilon}\right) - C_2 \rho_m \frac{\epsilon}{k}(C_{1\epsilon} G_{k,m} - C_{2\epsilon} \rho_m \epsilon) \quad (23)$$

The turbulent viscosity and turbulent kinetic viscosity are taken as[21];

$$\mu_{t,m} = C_\mu \rho_m \frac{k^2}{\epsilon} \quad (24)$$

$$G_{k,m} = \mu_{t,m}(\nabla u_m + \nabla u_m^T) : \nabla u_m \quad (25)$$

$\sigma_k, \sigma_\epsilon, C_{1\epsilon}, C_{2\epsilon}, C_\mu$ equal to 1, 1.3, 1.44, 1.92, 0.09 respectively are the empirical constants[22].The effective viscosity is given as;

$$\mu_{eff} = \mu_{leff} + \mu_{t,m} \tag{26}$$

The wall heat flux (Q_w) is constituted into three parts:

$$Q_w = Q_c + Q_e + Q_q \tag{27}$$

The first part is convective heat flux (Q_c):

$$Q_c = A_1 St \rho_l C_{pl} u_l (T_w - T_l);\ A_1 = 1 - A_2 \tag{28}$$

Where $A_1$and A_2 are the fraction of the wall surface under the influence of liquid and vapor formed on the wall surface respectively. T_l , and u_l are the liquid temperature and velocity at the cell closed to the wall. St is the Stanton number.

$$S_t = \frac{Nu}{Re_l Pr_l} \tag{29}$$

The active nucleation density is correlated as;

$$n = [210(T_w - T_{sat})]^{1.805} \tag{30}$$

The second part is evaporation heat flux (Q_e) for the generation of vapor bubbles:

$$Q_e = nf\left(\frac{\pi d_{BW}^3}{6}\right)\rho_v h_{vl} \tag{31}$$

The bubble frequency (f) is driven from the correlation:

$$f = \sqrt{\frac{4g(\rho_{leff} - \rho_v)}{3 d_{BW} \rho_{leff}}} \tag{32}$$

As depicted that the frequency is dependent on bubble diameter and phase density. The average of multiple sizes of bubbles is usually considered as a single size bubble in literature. The bubble departure diameter is computed as:

$$d_{BW} = \min\left[1.4, 0.6.\exp\left(-\frac{\Delta T_{sub}}{45}\right)\right] \tag{33}$$

The third part is quenching heat flux (Q_q) representing heat flux from heating wall to the bulk liquid by departing or collapsing bubbles during the bubble eruption cycle. The Q_q is calculated from the temperature difference of wall and the bulk liquid multiplied by the heat transfer coefficient. Whereas, K=2[23]

$$Q_q = \left(\frac{2}{\pi}\sqrt{k_{leff}\rho_{leff}C_{pleff}f}\right)A_2(T_w - T_l) \tag{34}$$

$$A_2 = nK\left(\frac{\pi d_{BW}^2}{4}\right) \tag{35}$$

By accounting the mass transfer from the liquid to the vapor phase on the heating wall is:

$$m_{wv} = \frac{Q_e}{h_{vl} + C_{pleff}(T_{sat} - T_l)} \tag{36}$$

The mass transfer rate from liquid to the vapor (evaporation) and from vapor to the liquid (condensation) inside the fluid is evaluated by using the following equations:

$$m_{lv} = \begin{cases} \frac{r_{lv}\alpha_l \rho_{leff}(T_l - T_{sat})}{T_{sat}}; & T_l \geq T_{sat} \\ 0 & ; T_l < T_{sat} \end{cases} \tag{37}$$

$$m_{vl} = \begin{cases} \frac{r_{vl}\alpha_v \rho_{veff}(T_{sat} - T_l)}{T_{sat}}; & T_v \geq T_{sat} \\ 0 & ; T_v < T_{sat} \end{cases} \tag{38}$$

Phase volume fraction is represented as α, density is ρ whereas r_{lv} and $r_{vl}$are mass transfer time parameters with unit ms^{-1}[24]. The aforementioned boiling model is coupled with the CFD simulation where the wall boundary condition is the wall heat flux taken as $1500kWm^{-2}$ and the reservoir temperature is considered as 20˚C and 80˚C. By considering the velocity, void fraction and heating wall temperature distribution under sub cooling effect, the heat transfer from the heating wall to the flowing fluid is examined and the HTC is predicted.

3. Results and discussion

The numerical calculations have been performed by taking deionized water as a coolant through one sided heated curved channel at gauge pressure of 24.50kPa covering flow velocity of $0.001ms^{-1}$, $0.005ms^{-1}$, and $0.010ms^{-1}$ along with subcooling range of 20-80C (293-353K). The boiling process prior and after CHF under the influence of varying flow conditions has been discussed. The CHF mechanism is associated with inlet velocity and subcooling. The effect of inlet subcooling and flow velocity on CHF is illustrated in Figure 1. The substantial enhancement in CHF can be seen in Figure 1 with increasing flow velocity from 0.001 to $0.010ms^{-1}$and with increased degree of inlet subcooling from 293K to 353K. The subcooling caused generation of small vapor bubbles which rarely spreaded to the neighboring nucleation sites as compared to the larger bubbles. The CHF trend in view of inlet velocity with subcooling temperature of 353K (i.e. when inlet temperature was 20C) is found higher than the subcooling temperature of 293K (i.e. when inlet temperature was 80C). High flow velocity imposed shear and drag forces on independent vapor growing on heated wall, as a result rapid departure of bubbles out of the wall occurred with high frequency followed nucleation. In case of high velocity, the entrained bubbles used to be smaller than that of fluid flowing with low velocity. The breakdown of vapor blanket into smaller portions at high velocity is due to Helmholtz instability which occurs at the interface of liquid core-vapor blanket parallel to the heating wall due to large difference in velocity between liquid core and vapor blanket.

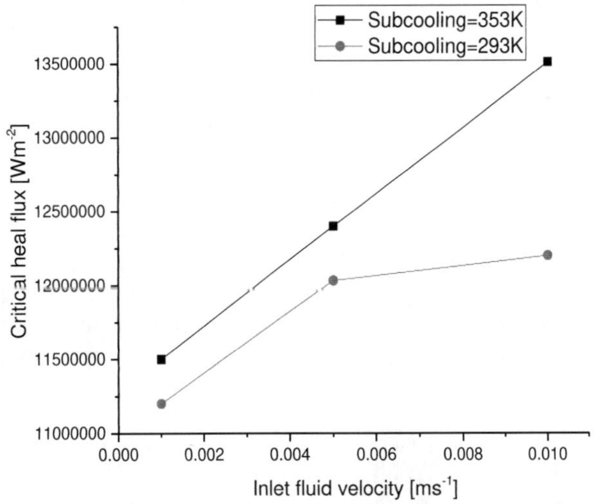

Figure 1. The effect of in-let fluid velocity and sub-cooling on CHF

Table 2. Outcomes

Sub-cooling [K]	Inlet velocity[ms^{-1}]	CHF [Wm^{-2}]
353	0.001	1.150e7
353	0.005	1.240e7
353	0.010	1.351e7
293	0.001	1.122e7
293	0.005	1.203e7
293	0.010	1.220e7

The predicted values of CHF limit at three different velocities and two subcooling temperatures are tabulated in Table 2. The flow velocity and subcooling have strong influence on boiling. Figure 2. and

IOP Conf. Series: Materials Science and Engineering **576** (2019) 012046 doi:10.1088/1757-899X/576/1/012046

Figure 3. depict velocity distribution and temperature distribution at maximum CHF limit of 1.351e7Wm^{-2} with inlet velocity of 0.010ms^{-1} and subcooling temperature of 353K. The bubble boundary thickness reduced with increasing velocity. The low velocity of flowing fluid leads dry out following development of continuous vapor blanket whereas CHF in the range of high velocity triggered by drying out of vapor blankets which screened smaller area of the heated surface. The intense vapor generation and two-phase turbulence raised temperature of heating wall.

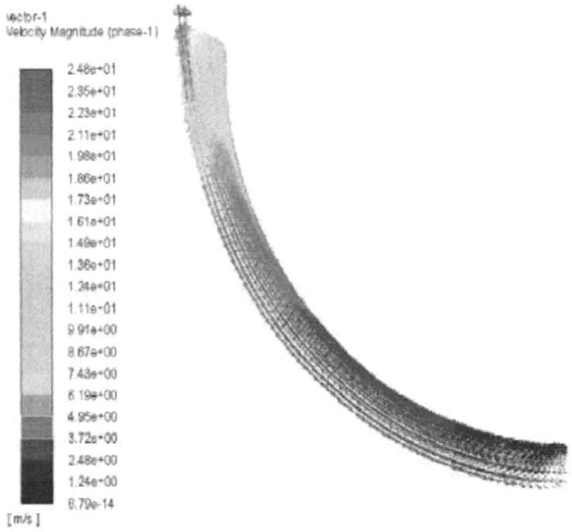

Figure 2. Velocity distribution at v_{in}=0.010ms-1; ΔT_{sat}=353K

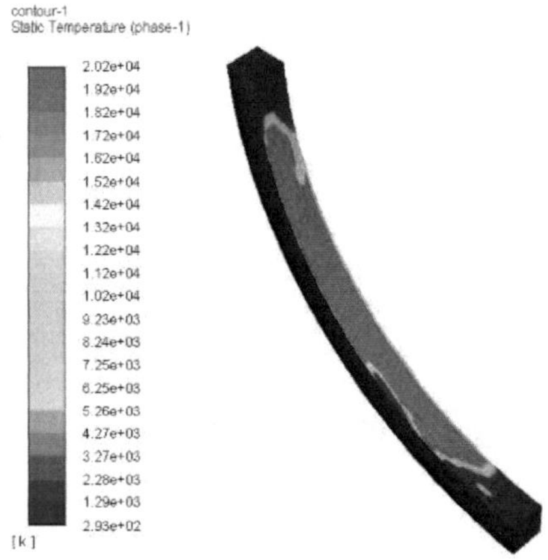

Figure 3. Temperature distribution v_{in}=0.010ms^{-1}; ΔT_{sat}=353K

Figure 4. depicts distribution of heat flux where maximum critical heat flux is achieved near the upstream due to the accumulation of vapor bubbles on upper wall i.e. on the heating wall under the influence of buoyancy. CHF is characterized by the formation of vapor blanket on heating surface which appeared thick near upstream. The large bundles of vapors seized heat dissipation due to its low thermal conductivity by disconnecting heating surface from rewetting. The occurrence of thick vapor blanket induced thermal resistance which consequently became a cause of boiling crisis.

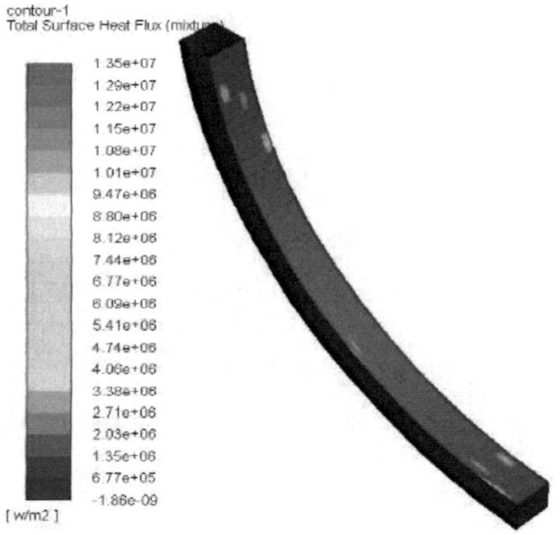

Figure 4. Heat flux distribution v_{in}=0.010ms^{-1}; ΔT_{sat}=353K

4. Conclusions

The CFD based two-phase (liquid-vapor) numerical simulations have been performed to determine the effects of flow velocity and subcooling on CHF from flowing fluid (DI water) through the 90° curved flow channel with square cross section having 40mm of hydraulic diameter and 368mm of radius of curvature while a constant heat flux is applied perpendicular to the convex wall of the channel. It has been concluded that the higher limit of CHF is achieved with higher flow velocities. The augmenting velocity enlarged the contact of liquid to the heating surface by reducing the thickness of vapor layer. The enhancement of CHF is reported by varying inlet fluid velocity from 0.001-0.010ms^{-1}. Moreover, the CHF is observed to increase with the increased degree of subcooling from 293-352K. The maximum value of CHF limit is achieved at 0.010ms^{-1} with inlet subcooling temperature of 353K.

5. Abbreviations

The following abbreviations are used in this manuscript:

ρ	*Density [kgm^{-3}]*	Q_e	*Heat flux due to evaporative [Wm^{-2}]*
u	*Velocity [ms^{-1}]*	Q_c	*Heat flux due to convection [Wm^{-2}]*
α	*Volume fraction of liquid or vapor phase*	Q_q	*Heat flux due to quenching [Wm^{-2}]*
μ	*Kinematic viscosity [m^{2}s^{-1}]*	A_s	*Surface area [m^2]*
v	*Dynamic viscosity [Pas]*	A_1	*Area fraction of wall under the influence of liquid [m^2]*
u_{dr}	*Drift velocity [ms^{-1}]*	A_2	*Area fraction of wall under the influence of vapor [m^2]*
u_{vl}	*Slip velocity [ms^{-1}]*	f	*Bubble departure frequency[s^{-1}]*
σ	*Surface tension [Nm^{-1}]*	d_{BW}	*Bubble departure diameter[m]*
f_{drag}	*Drag force [N]*	h_{vl}	*Latent heat of vaporization [Jkg^{-1}]*
d	*Diameter[m]*	G	*Mass flux [kgm^{-2}s^{-1}]*
a	*Acceleration [ms^{-2}]*	P_1	*Inlet pressure of channel [Pa]*
g	*Gravity [ms^{-2}]*	P_2	*Outlet pressure of the channel [Pa]*
h	*Enthalpy [Jkg^{-1}]*	V	*Voltage [V]*
k	*Thermal conductivity [Wm^{-1}K^{-1}]*	I	*Current [A]*
α	*Volume fraction of vapor or liquid*	θ	*Bend angle of the channel [degree]*
T	*Temperature [K]*	***Subscripts***	
ΔT	*Sub cooled [K]*	v	*Vapor*
Q	*Heat flux [Wm^{-2}]*	l	*Liquid*
$\dot{Q}$	*Heat transfer rate [kW]*	m	*Mixture*
C_p	*Specific heat [Jkg^{-1}K^{-1}]*	k	*Summation index*
m	*Inter-phase mass transfer*	sat	*Saturated*
$k-\epsilon$	*Turbulence dissipation rate - turbulence kinetic energy [Jkg^{-1}]*	eff	*Effective*
Nu	*Nusselt number*	t	*Turbulent*
Pr	*Prandtl number*	b	*Bulk*
St	*Stanton number*	dr	*Drift*
Re	*Reynolds number*	e	*Evaporation*
n	*Nucleation site density [m^{-2}]*	c	*Convection*
r	*Mass transfer time[s^{-1}]*	q	*Quenching*
Q_w	*Wall heat flux [Wm^{-2}]*		

IOP Conf. Series: Materials Science and Engineering **576** (2019) 012046 doi:10.1088/1757-899X/576/1/012046

6. References:

[1] Hata, K. and S. Masuzaki, *Subcooled boiling heat transfer in a short vertical SUS304-tube at liquid Reynolds number range 5.19□— 104 to 7.43□— 105.* Nuclear Engineering and Design, 2009. **239**(12): p. 2885-2907.

[2] Hata, K. and S. Masuzaki. *Subcooled Boiling Heat Transfer in a Short Vertical SUS304-Tube at High Liquid Reynolds Number.* in *17th International Conference on Nuclear Engineering.* 2009. American Society of Mechanical Engineers.

[3] Kutateladze, S.S., *Heat transfer in condensation and boiling.* AEC-tr-3770, 1959.

[4] Zuber, N., *Hydrodynamic aspects of boiling heat transfer (thesis).* 1959, Ramo-Wooldridge Corp., Los Angeles, CA (United States); Univ. of California, Los Angeles, CA (United States).

[5] Katto, Y., *Critical heat flux*, in *Advances in heat transfer.* 1985, Elsevier. p. 1-64.

[6] Vliet, G.C. and G. Leppert, *Critical heat flux for nearly saturated water flowing normal to a cylinder.* Journal of Heat Transfer, 1964. **86**(1): p. 59-66.

[7] Cochran, T.H. and C.R. Andracchio, *Forced-convection peak heat flux on cylindrical heaters in water and refrigerant 113.* 1974.

[8] Lienhard, J.H. and R. Eichhorn, *Peak boiling heat flux on cylinders in a cross flow.* International Journal of Heat and Mass Transfer, 1976. **19**(10): p. 1135-1142.

[9] Lienhard, J.H. and M.M. Hasan, *On predicting boiling burnout with the mechanical energy stability criterion.* Journal of Heat Transfer, 1979. **101**(2): p. 276-279.

[10] Yilmaz, S. and J.W. Westwater, *Effect of velocity on heat transfer to boiling Freon-113.* Journal of Heat Transfer, 1980. **102**(1): p. 26-31.

[11] Katto, Y., et al., *Critical heat flux on a uniformly heated cylinder in a cross flow of saturated liquid over a very wide range of vapor-to-liquid density ratio.* International Journal of Heat and Mass Transfer, 1987. **30**(9): p. 1971-1977.

[12] Katto, Y. and Y. Haramura, *Critical heat flux on a uniformly heated horizontal cylinder in an upward cross flow of saturated liquid.* International Journal of Heat and Mass Transfer, 1983. **26**(8): p. 1199-1205.

[13] Katto, Y. and C. Kurata, *Critical heat flux of saturated convective boiling on uniformly heated plates in a parallel flow.* International Journal of Multiphase Flow, 1980. **6**(6): p. 575-582.

[14] Meyer, G., E.S. Gaddis, and A. Vogelpohl. *Critical heat flux on a cylinder of large diameter in a cross flow.* in *Proc. 8th Int. Heat Transfer Conf.* 1986.

[15] Kutateladze, S.S., *Fundamentals of heat transfer.* 1964.

[16] Vliet, G.C. and G. Leppert, *Critical heat flux for subcooled water flowing normal to a cylinder.* Journal of Heat Transfer, 1964. **86**(1): p. 68-74.

[17] Jones, W.P. and B.E. Launder, *The prediction of laminarization with a two-equation model of turbulence.* International Journal of Heat and Mass Transfer, 1972. **15**(2): p. 301-314.

[18] Fluent, A., *Ansys fluent theory guide.* ANSYS Inc., USA. **15317**: p. 724-746.

[19] Manninen, M., V. Taivassalo, and S. Kallio, *On the mixture model for multiphase flow.* 1996, Technical Research Centre of Finland Finland.

[20] Schiller, L., *A drag coefficient correlation.* Zeit. Ver. Deutsch. Ing., 1933. **77**: p. 318-320.

[21] Lucas, D., E. Krepper, and H.M. Prasser, *Use of models for lift, wall and turbulent dispersion forces acting on bubbles for poly-disperse flows.* Chemical Engineering Science, 2007. **62**(15): p. 4146-4157.

[22] Launder, B.E. and D.B. Spalding, *The numerical computation of turbulent flows*, in *Numerical Prediction of Flow, Heat Transfer, Turbulence and Combustion.* 1983, Elsevier. p. 96-116.

[23] Kenning, D.B.R., *Fully-developed nucleate boiling: overlap of areas of influence and interference between bubble sites.* International Journal of Heat and Mass Transfer, 1981. **24**(6): p. 1025-1032.

[24] Abedini, E., et al., *Numerical investigation of subcooled flow boiling of a nanofluid.* International Journal of Thermal Sciences. **64**: p. 232-239.